CRITICAL STABILITY CONSTANTS

Volume 3: Other Organic Ligands

CRITICAL STABILITY CONSTANTS

CRITICAL STABILITY CONSTANTS

Volume 3: Other Organic Ligands

**by Arthur E. Martell
and Robert M. Smith**

Department of Chemistry
College of Science
Texas A & M University
College Station, Texas

PLENUM PRESS · NEW YORK AND LONDON

Library of Congress Cataloging in Publication Data

Martell, Arthur Earl, 1916-
  Critical stability constants.

  On vol. 2 and 4 Smith's name appears first on t.p.
  Includes bibliographical references.
  CONTENTS: v. 1. Amino acids.—v. 2. Amines.—v. 3. Other organic
ligands—v. 4. Inorganic complexes.
  1. Chemical equilibrium—Tables, etc. 2. Complex compounds—Tables,
etc. I. Smith, Robert Martin, 1927-   joint author. II. Title. [DNLM:
1. Amino acids. 2. Chemistry, Physical. QD503 M376c]
QD503.M37   541'.392   74-10610
ISBN 0-306-35213-3 (v. 3)

© 1977 Plenum Press, New York
A Division of Plenum Publishing Corporation
227 West 17th Street, New York, N.Y. 10011

Printed in the United States of America

PREFACE

Over the past fifteen years the Commission on Equilibrium Data of the Analytical Division of the International Union of Pure and Applied Chemistry has been sponsoring a noncritical compilation of metal complex formation constants and related equilibrium constants. This work was extensive in scope and resulted in the publication of two large volumes of *Stability Constants* by the Chemical Society (London). The first volume, edited by L. G. Sillen (for inorganic ligands) and by A. E. Martell (for organic ligands), was published in 1964 and covered the literature through 1962. The second volume, subtitled Supplement No. 1, edited by L. G. Sillen and E. Hogfeldt (for inorganic ligands) and by A. E. Martell and R. M. Smith (for organic ligands), was published in 1971 and covered the literature up to 1969. These two large compilations attempted to cover all papers in the field related to metal complex equilibria (heats, entropies, and free energies). Since it was the policy of the Commission during that period to avoid decisions concerning the quality and reliability of the published work, the compilation would frequently contain from ten to twenty values for a single equilibrium constant. In many cases the values would differ by one or even two orders of magnitude, thus frustrating readers who wanted to use the data without doing the extensive literature study necessary to determine the correct value of the constant in question.

Because of difficulties of this nature, and because of the general lack of usefulness of a noncritical compilation for teaching purposes and for scientists who are not sufficiently expert in the field of equilibrium to carry out their own evaluation, we have decided to concentrate our efforts in this area toward the development of a critical and unique compilation of metal complex equilibrium constants. Although it would seem that decisions between available sets of data must sometimes be arbitrary and therefore possibly unfair, we have found the application of reasonable guidelines leads directly to the elimination of a considerable fraction of the published data of doubtful value. Additional criteria and procedures that were worked out to handle the remaining literature are described in the *Introduction* of this book. Many of these methods are quite similar to those used in other compilations of critical data.

In cases where a considerable amount of material has accumulated, it is felt that most of our critical constants will stand the test of time. Many of the data listed, however, are based on only one or a very few literature references and are subject to change when better data come along. It should be fully understood that this compilation is a continually changing and growing body of data, and will be revised from time to time as new results of these systems appear in the literature.

The scope of these tables includes the heats, entropies, and free energies of all reactions involving organic and inorganic ligands. The magnitude of the work is such that far more than a thousand book pages will be required. In order that the material be available in convenient form, the amino acid complexes are presented in Volume 1 and amine complexes (which do not contain carboxylic acid functions) are included in Volume 2. The remaining organic complexes are the subject of Volume 3. Volume 4 comprises the inorganic complexes.

Texas A&M University
College Station, Texas

Arthur E. Martell
Robert M. Smith

CONTENTS

INTRODUCTION

Purpose

This compilation of metal complex equilibrium (formation) constants and the corresponding enthalpy and entropy values represent the authors' selection of the most reliable values among those available in the literature. In many cases wide variations in published constants for the same metal complex equilibrium indicate the presence of one or more errors in ligand purity, in the experimental measurements, or in calculations. Usually, the nature of these errors is not readily apparent in the publication, and the reader is frequently faced with uncertainties concerning the correct values. In the course of developing noncritical compilations of stability constants, the authors have long felt that these wide variations in published work constitute a serious impediment to the use of equilibrium data. Thus these critical tables were developed in order to satisfy what is believed to be an important need in the field of coordination chemistry.

Scope

These tables include all organic and inorganic ligands for which reliable values have been reported in the literature. The present volume is restricted to organic ligands other than amino acids and amines.

Values determined in nonaqueous solutions as well as values involving two or more different ligands (i.e., "mixed ligand" complexes) have not been included in this compilation but may be included in a subsequent volume. Mixed ligand complexes containing hydrogen or hydroxide ions are included since these ions are derived from the solvent and are therefore potentially always available. In general, data were compiled for only those systems that involve metal ion equilibria. Data on potentially important ligands for which only acid—base equilibria are presently available are given in a separate table.

Selection Criteria

When several workers are in close agreement on a particular value, the average of their results has been selected for that value. Values showing considerable scatter have been eliminated. In cases where the agreement is poor and few results are available for comparison, more subtle methods were needed to select the best value. This selection was often guided by a comparison with values obtained for other metal ions with the same ligand and with values obtained for the same metal ion with similar ligands.

While established trends among similar metal ions and among similar ligands were valuable in deciding between widely varying values, such guidelines were used cautiously, so as not to overlook occasionally unexpected real examples of specificity or anomalous behavior.

When there was poor agreement between published values and comparison with other metal ions and ligands did not suggest the best value, the results of more experienced research groups who had supplied reliable values for other ligands were selected. When such assurances were lacking, it was sometimes possible to give preference to values reported by an investigator whc had published other demonstrably reliable values obtained by the same experimental method.

In some cases the constants reported by several workers for a given group of metal ions would have similar relative values, but would differ considerably in the absolute magnitudes of the constants. Then a set of values from one worker near the median of all values reported were selected as the best constants. By this method it is believed that internal consistency was preserved to a greater extent than would be obtained by averaging reported values for each individual metal ion. When an important constant was missing from the selected set of values, but was available in another set of values not selected for this compilation, the missing constant was obtained by adjusting the nonselected values by a common factor, which was set so as to give the best agreement between the two groups of data.

Values reported by only one investigator are included in these tables unless there was some reason to doubt their validity. It is recognized that some of these values may be in error, and that such errors will probably not be detected until the work is repeated by other investigators, or until more data become available for analogous ligands or other closely related metal ions. Some values involving unusual metal ions have been omitted because of serious questions about the form of their complexes.

Papers deficient in specifying essential reaction conditions (e.g., temperature, ionic strength, nature of supporting electrolyte) were not employed in this compilation. Also used as a basis for disqualification of published data is lack of information on the purity of the ligand. Frequent deficiencies are lack of calibration of potentiometric apparatus, and failure to define the equilibrium quotients reported in the paper. Papers in which both temperature and ionic strength are not controlled have been omitted from the bibliography.

A bibliography for each ligand is included so that the reader may determine the completeness of the literature search employed in the determination of critical values. The reader may also employ these references to make his own evaluation if he has any questions or reservations concerning this compilation.

Arrangement

The arrangement of the tables is based on the placement of similar ligands together. Ligands containing carboxylic acid functional groups are placed together except for aminocarboxylic acids, which are in Volume 1, and phenolic carboxylic acids, which are listed with the phenols. Within each group of tables, ligands with a smaller number of coordinating groups are placed before those with a larger number of coordinating groups. Next there is a table of protonation constants for ligands for which no stability constants or only questionable metal stability constants are reported. Finally, there is a list of other ligands considered but not included in the tables for various reasons.

Metal Ions

The metal ions within each table are arranged in the following order: hydrogen, alkali metals, alkaline earth metals, lanthanides (including Sc and Y), actinides, transition metals, and posttransition metals. Within each group the arrangement is by increasing oxidation state of the metal, and within each oxidation state the arrangement follows the periodic table from top to bottom and from left to right. An exception is that Cu^+, Ag^+, Pd^{2+}, and Pt^{2+} are included with the posttransition metals.

Equilibrium

An abbreviated equilibrium quotient expression in the order products/reactants is included for each constant, and periods are used to separate distinct entities. Charges have been omitted as these can be determined from the charge of the metal ion and the abbreviated ligand formulas (such as HL) given after the name. Water has not been included in the equilibrium expressions since all of the values cited are for aqueous solutions. For example, $ML_2/M \cdot L^2$ for Cu^{2+} and acetic acid would represent the equilibrium: $Cu^{2+} + 2CH_3CO_2^- \rightleftharpoons (CH_3CO_2)_2Cu$. The symbol M represents the metal ion given in the first column and may include more than one atom as in the case of Hg_2^{2+}. The symbol H_{-1} (H_{-2}, etc.) is used for the ionization from the ligand of a proton that would not be ionized in the absence of the metal ion.

Equilibria involving protons are written as stability constants (protonation constants) rather than as ionization constants to be consistent with the metal complex formation constants. Consequently the $\triangle H$ and $\triangle S$ values have signs opposite to those describing ionization constants.

Solids and gases are identified by (s) and (g) respectively and are included for identification purposes even though they are not involved in the equilibrium quotient.

Log K Values

The log K values are the logarithms of the equilibrium quotients given in the second column at the specified conditions of temperature and ionic strength. The selected values are those considered to be the most reliable of the ones available. In some cases the value is the median of several values and in other cases it is the average of two or more values. The range of other values considered reliable is indicated by + or − quantities describing the algebraic difference between the other values and the selected value. The symbol ±0.00 indicates that there are one or more values which agree exactly with the stated value to the number of significant figures given. Values considered to be of questionable validity are enclosed in parentheses. Such values are included when the evidence available is not strong enough to exclude them on the basis of the above criteria. Values concerning which there is considerable doubt have been omitted.

The log K values are given for the more commonly reported ionic strengths. The ionic strengths most used are 0.1, 0.5, 1.0, 2.0, 3.0, and 0. Zero ionic strength is perhaps more important from a theoretical point of view, but several assumptions are involved in extrapolating or calculating from the measured values. The Davies equation is often used to calculate constants to zero from low-ionic-strength measurements. It was established from results obtained with monovalent and divalent ions and its extension to trivalent ions is extremely questionable.

The temperature of 25°C was given preference in the tables because of its widespread use in equilibrium measurements and reporting other physical properties. When available, enthalphy changes ($\triangle H$) were used to calculate log K at 25°C when only measurements at other temperatures were available.

Other temperatures frequently employed are 20°C, 30°C, and 37°C. These are not included in the tables when there is a lack of column space and $\triangle H$ is available, since they may be calculated using the $\triangle H$ value. Values at other temperatures, especially those at 20°C and 30°C, were converted to 25°C to facilitate quantitative comparisons with the 25°C values listed.

Equilibria involving protons have been expressed as concentration constants in order to be more consistent with the metal ion stability constants which involve only concentration terms. Concentration constants may be determined by calibrating the electrodes with solutions of known hydrogen ion concentrations or by conversion of pH values using the appropriate hydrogen ion activity coefficient. When standard buffers are used, mixed constants (also known as Bronsted or practical constants) are obtained which include both activity and concentration terms. Literature values expressed as mixed constants have been converted to concentration constants by using the hydrogen ion activity coefficients determined in KCl solution before inclusion in the tables. In some cases, papers were omitted because no indication was given as to the use of concentration or mixed constants. Some papers were retained despite this lack of information when it could be ascertained which constant was used by comparing to known values or by personal communication with the authors. For those desiring to convert the listed protonation constants to mixed constants, the following values should be added to the listed values at the appropriate ionic strength (the tabulation applies only to single proton association constants):

Ionic strength	Increase in log K
0.05	0.09
0.10	0.11
0.15	0.12
0.2	0.13
0.5	0.15
1.0	0.14
2.0	0.11
3.0	0.07

The values in the tables have not been corrected for complexation with medium ions for the most part. There are insufficient data to make corrections for most of the ligands, and in order to make values between ligands more comparable, the correction has not been made in the few cases where it could be made. In general the listed formation constants at constant ionic strength include competition by ions from KNO_3 and $NaClO_4$ and are somewhat smaller than they would be if measured in solutions of tetraalkylammonium salts.

Limited comparisons were made between values at different ionic strengths using observed trends. With simple carboxylic acids, the stability constants usually decrease with increasing ionic strength and generally reach a minimum at an ionic strength of about 0.5. Then the constants were observed to increase through ionic strength of at least 3.0.

Protonation constants with a log value of 12 or larger at an ionic strength of 0.10 have been discarded, and the metal constants using these protonation constants have been rearranged to remove their dependence upon them. These extremely high values must have been measured at a higher ionic strength, and their use at 0.10 ionic strength would introduce considerable inaccuracy. Values measured at relatively high ionic strength and then corrected to zero ionic strength have been retained.

Equilibria involving B(III), As(III), Ge(IV), and Te(VI) complexes with polyhydroxy ligands have been written showing the loss of a proton on complex formation. Thus the equilibrium quotient $M(OH)_2(H_{-2}L) \cdot H/M(OH)_3 \cdot L$ is employed for the reaction of B(III) with glycerol as a representation for the reaction. These equilibria are often reported in the literature with the ionization constant of the metal species $(H_2MO_3 \cdot H/M(OH)_3)$ divided into this quotient, thus eliminating the proton from the complex formation reaction $(M(OH)_2(H_{-2}L)/H_2MO_3 \cdot L)$.

Enthalpy Values

The enthalpy of complexation values ($\triangle H$) listed in the tables have the units kcal/mole because of the widespread use of these units by workers in the field. These may be converted to SI units of KJ/mole by multiplying the listed values by 4.184.

Calorimetrically determined values and temperature-variation-determined values from cells without liquid junction were considered of equal validity for the tables. Other temperature-variation-determined values were rounded off to the nearest kcal/mole and were enclosed in parentheses because of their reduced accuracy. Other values considered to be reliable but differing from the listed value were indicated by + or − quantities describing the algebraic difference between the other value and the selected value.

The magnitude of $\triangle H$ may vary with temperature and ionic strength, but usually this is less than the variation between different workers and little attempt has been made to show $\triangle H$ variation with changing conditions except for certain carefully measured equilibria such as the protonation of hydroxide ion and of ammonia. These $\triangle H$ values may be used for estimating log K values at temperatures other than those listed, using the relationship

$$\frac{\triangle H}{2.303RT^3} = \frac{d \log K}{dT}$$

or, at 25°C,

$$\log K_2 = \log K_1 + \triangle H(T_2 - T_1)(0.00246).$$

This assumes that $\triangle C_p$ = 0, which is not necessarily the case. The greater the temperature range employed, the greater the uncertainty of the calculated values.

Entropy Values

The entropy of complexation values ($\triangle S$) listed in the tables have the units cal/mole/degree and have been calculated from the listed log K and $\triangle H$ values, using the expression

$$\triangle G = \triangle H - T\triangle S$$

or, at 25°C,

$$\triangle S = 3.36 \, (1.363 \log K + \triangle H).$$

These entropy values have been rounded off to the nearest cal/mole/degree, except in cases where $\triangle H$ values were quite accurate.

Bibliography

The references considered in preparing each table are given at the end of the table. The more reliable references are listed after the ions for which values are reported. In some tables groups of similar metal ions have been grouped together for the bibliography. The term "Other references" is used for those reporting questionable values, or values at conditions considerably different from those used in the tables, or values for metal ions not included in the tables because of questionable knowledge about the forms of their complexes. These additional references are cited to inform the reader of the extent of the literature search made in arriving at the selected values.

The bibliographical symbols used represent the year of the reference and the first letter of the surnames of the first two listed authors. In cases of duplication, letters a, b, c, etc., or the first letter of the third author's name are employed. The complete reference is given in the bibliography at the end of each volume.

In a work of this magnitude, there will certainly be errors and a few pertinent publications will have been overlooked by the compilers. We should like to request those who believe they have detected errors in the selection process, know of publications that were omitted, or have any suggestions for improvement of the tables, write to:

A. E. Martell, Head
Department of Chemistry
Texas A&M University
College Station, Texas 77843, U.S.A.

It is the intention of the authors to publish more complete and accurate revisions of these tables as demanded by the continually growing body of equilibrium data in the literature.

$$HCO_2H$$

CH_2O_2		Methanoic acid	(formic acid)			HL
Metal ion	Equilibrium	Log K 25°, 0.1	Log K 25°, 1.0	Log K 25°, 0	ΔH 25°, 0	ΔS 25°, 0
H^+	HL/H.L	3.55 −0.01	3.53 ±0.00	3.745 ±0.007	0.04 ±0.04	17.3
		3.49[b]	3.90[e]	4.39[g]		
Mg^{2+}	ML/M.L	0.34[s]		1.43		
Ca^{2+}	ML/M.L	0.27[s]		1.43	(1)[t]	(10)
Sr^{2+}	ML/M.L			1.39	(1)[t]	(10)
Ba^{2+}	ML/M.L			1.38		
Y^{3+}	ML/M.L	1.12				
	$ML_2/M.L^2$	1.95				
La^{3+}	ML/M.L	1.10				
	$ML_2/M.L^2$	2.09				
Ce^{3+}	ML/M.L	1.12				
	$ML_2/M.L^2$	1.85				
Pr^{3+}	ML/M.L	1.14				
	$ML_2/M.L^2$	1.53				
Nd^{3+}	ML/M.L	1.15				
	$ML_2/M.L^2$	1.54				
Sm^{3+}	ML/M.L	1.23				
	$ML_2/M.L^2$	1.65				
Eu^{3+}	ML/M.L	1.40				
	$ML_2/M.L^2$	1.94				
Gd^{3+}	ML/M.L	1.34				
	$ML_2/M.L^2$	1.87				
Tb^{3+}	ML/M.L	1.24				
	$ML_2/M.L^2$	1.74				
Dy^{3+}	ML/M.L	1.21				
	$ML_2/M.L^2$	1.52				
Ho^{3+}	ML/M.L	1.14				
	$ML_2/M.L^2$	1.96				
Er^{3+}	ML/M.L	1.11				
	$ML_2/M.L^2$	1.81				
Tm^{3+}	ML/M.L	1.08				
	$ML_2/M.L^2$	1.60				
Yb^{3+}	ML/M.L	1.06				
	$ML_2/M.L^2$	1.26				
Lu^{3+}	ML/M.L	0.99				
	$ML_2/M.L^2$	2.10				
VO^{2+}	ML/M.L		1.98[u]			
	$ML_2/M.L^2$		2.77[u]			

[b] 25°, 0.5;   [e] 25°, 3.0;   [g] 25°, 5.0;   [s] 30°, 0.4;   [t] 25-35°, 0;   [u] 18°, 1.0

Formic acid (continued)

Metal ion	Equilibrium	Log K 25°, 0.1	Log K 25°, 1.0	Log K 25°, 0	ΔH 25°, 0	ΔS 25°, 0
Th^{4+}	$ML/M.L$		3.09^j			
	$ML_2/M.L^2$		5.15^j			
	$ML_3/M.L^3$		6.73^j			
VO_2^+	$ML/M.L$		1.36^u			
UO_2^{2+}	$ML/M.L$		1.86^j ±0.03			
	$ML_2/M.L^2$		3.05^j ±0.08			
	$ML_3/M.L^3$		3.52^j			
Cr^{2+}	$ML/M.L$			1.07		
Co^{2+}	$ML/M.L$	0.68^s	0.73^d			
	$ML_2/M.L^2$		1.18^d			
Ni^{2+}	$ML/M.L$	0.67^s	$(0.46)^d$	1.04^g		
Cu^{2+}	$ML/M.L$	1.38^s	1.40^d	2.00 ±0.02		
	$ML_2/M.L^2$		1.53^e	1.83^g		
			2.30^d			
	$ML_3/M.L^3$		2.42^e	2.60^g		
			2.2^d			
	$ML_4/M.L^4$		2.68^e	3.0^g		
			1.9^d	3.3^g		
Fe^{3+}	$ML/M.L$		3.1^j			
	$M_3(OH)_2L_6/M^3.(OH)^2.L^6$	20.0^h	19.9^j			
CH_3Hg^+	$ML/M.L$	2.67^b				
Zn^{2+}	$ML/M.L$	0.73^s	0.70^d			
	$ML_2/M.L^2$		1.08^d			
	$ML_3/M.L^3$		1.20^d			
Cd^{2+}	$ML/M.L$	1.15^s	1.04^d			
	$ML_2/M.L^2$		1.23^d			
	$ML_3/M.L^3$		1.75^d			
Hg^{2+}	$ML/M.L$	(5.43)				
Pb^{2+}	$ML/M.L$	1.65^s	1.23^d			
	$ML_2/M.L^2$		2.01^d			
	$ML_3/M.L^3$		1.8^d			
Al^{3+}	$ML/M.L$		1.36			
In^{3+}	$ML/M.L$		2.74^k			
	$ML_2/M.L^2$		4.72^k			
	$ML_3/M.L^3$		5.70^k			
	$ML_4/M.L^4$		6.70^k			

---

^{b} 25°. 0.4; ^{d} 25°, 2.0; ^{e} 25°. 3.0; ^{g} 25°, 5.0; ^{h} 20°, 0.1; ^{j} 20°, 1.0; ^{k} 20°, 2.0; ^{s} 30°, 0.4; ^{u} 18°, 1.0

A. MONO-CARBOXYLIC ACIDS

Formic acid (continued)

Bibliography:

H^+ 31LAa,34HE,40SD,51B,59MR,60FS,63BM, 66PR,69HSB,69RS,70CSS,70IV,71VSa, 72PTa,73IV

Mg^{2+}-Ba^{2+} 56N,70BT

Y^{3+}-Lu^{3+} 66KP

VO^{2+} 73IV

Th^{4+} 72PTa

VO_2^+ 70IV

UO_2^{2+} 62HO,67MN,72MPZ

Cr^{2+} 58YF

Co^{2+}-Cu^{2+},Zn^{2+},Cd^{2+},Pb^{2+} 51LW,57BDM,64SM, 68FP,70BT,70FM,70GF,71BA

Fe^{3+} 59P,64T

CH_3Hg^+ 73LR

Hg^{2+} 73LU

Al^{3+} 75KI

In^{3+} 53S

Other references: 52CM,57HB,58SB,58CP,62T, 65DSa,65GJ,66JGa,67CIH,68RSa,71GK,72PZ, 73TD,73TDa,74MM,75AD,75KPV,75W

$$CH_3CO_2H$$

$C_2H_4O_2$ Metal ion	Equilibrium	Ethanoic acid (acetic acid) Log K 25°, 0.1	Log K 25°, 1.0	Log K 25°, 0	ΔH 25°, 0	HL ΔS 25°, 0
H^+	HL/H.L	4.56 ±0.03	4.57 ±0.04	4.757 ±0.002	0.10 ±0.01	22.1
		4.50[b]±0.02	4.80[d]±0.00	5.015[e]±0.005	-0.75[e]-0.03	20.5[e]
Li^+	ML/M.L			0.26		
Na^+	ML/M.L			-0.18		
Be^{2+}	ML/M.L	1.62				
	$ML_2/M.L^2$	2.36				
Mg^{2+}	ML/M.L	0.51[x]		1.27 ±0.02		
Ca^{2+}	ML/M.L	0.53[x]	0.45[k]	1.18 ±0.06	(1)[t]	(9)
Sr^{2+}	ML/M.L	0.43[x]	0.32[k]	1.14 ±0.06	(1)[t]	(9)
Ba^{2+}	ML/M.L	0.39[x]	0.14[k]	1.07 ±0.09		
Y^{3+}	ML/M.L	1.68	1.59[d]		3.3[d]	18[d]
	$ML_2/M.L^2$	3.17	2.73[d]		5.4[d]	31[d]
	$ML_3/M.L^3$		3.5[d]		5.2[d]	33[d]
	$ML_4/M.L^4$		3.3[k]			
La^{3+}	ML/M.L	1.82	1.59[d]	2.55	2.2[d]	15[d]
	$ML_2/M.L^2$	2.82	2.53[d]	4.12	3.8[d]	24[d]
	$ML_3/M.L^3$	3.53	3.04[d]		4.6[d]	29[d]
	$ML_4/M.L^4$		2.9[k]			
Ce^{3+}	ML/M.L	1.91	1.71[d]-0.01		2.1[d]	15[d]
	$ML_2/M.L^2$	3.09	2.74[d]		3.7[d]	25[d]
	$ML_3/M.L^3$	3.68	3.19[d]		5.1[d]	32[d]
	$ML_4/M.L^4$		3.2[d]			

---

[b] 25°, 0.5;   [d] 25°, 2.0;   [e] 25°, 3.0;   [k] 20°, 2.0;   [t] 25-35°, 0;   [x] temperature not stated, 0.2

Acetic acid (continued)

Metal ion	Equilibrium	Log K 25°, 0.1	Log K 25°, 1.0	Log K 25°, 0	ΔH 25°, 0	ΔS 25°, 0
Pr^{3+}	$ML/M.L$	2.01	1.83[d]		1.7[d]	14[d]
	$ML_2/M.L^2$	3.41	2.86[d]		4.2[d]	27[d]
	$ML_3/M.L^3$		3.33[d]		3.6[d]	27[d]
	$ML_4/M.L^4$		3.3[d]			
Nd^{3+}	$ML/M.L$	2.11	1.92[d] -0.04	2.67	1.7[d]	15[d]
	$ML_2/M.L^2$	3.59	3.06[d] ±0.01	4.54	3.5[d]	26[d]
	$ML_3/M.L^3$		3.56[d] ±0.04		4.4[d]	31[d]
	$ML_4/M.L^4$		3.5[k]			
Sm^{3+}	$ML/M.L$	2.17	2.03[d]	2.84	1.5[d]	14[d]
	$ML_2/M.L^2$	3.76	3.30[d]	4.80	2.9[d]	25[d]
	$ML_3/M.L^3$		3.90[d]		3.8[d]	31[d]
	$ML_4/M.L^4$		3.8[k]			
Eu^{3+}	$ML/M.L$	2.13, 1.94[i]	1.90[d]		1.4[d]	13[d]
	$ML_2/M.L^2$	3.64, 3.19[i]				
	$ML_3/M.L^3$	4.24, 3.79[i]				
Gd^{3+}	$ML/M.L$	2.02	1.86[d]		1.9[d]	15[d]
	$ML_2/M.L^2$	3.47	3.16[d]		3.3[d]	25[d]
	$ML_3/M.L^3$	4.26	3.76[d]		3.5[d]	29[d]
	$ML_4/M.L^4$		3.7[k]			
Tb^{3+}	$ML/M.L$	1.91	1.79[k]			
	$ML_2/M.L^2$	3.23	3.16[k]			
	$ML_3/M.L^3$	4.39	3.81[k]			
Dy^{3+}	$ML/M.L$	1.85	1.71[d]		2.9[d]	18[d]
	$ML_2/M.L^2$	3.16	3.03[d]		4.4[d]	29[d]
	$ML_3/M.L^3$	4.30	3.84[d]		4.3[d]	32[d]
	$ML_4/M.L^4$		3.9[k]			
Ho^{3+}	$ML/M.L$	1.81	1.67[d]		3.2[d]	18[d]
	$ML_2/M.L^2$	3.11	2.92[d]		5.0[d]	30[d]
	$ML_3/M.L^3$	4.27	3.81[d]		4.5[d]	33[d]
	$ML_4/M.L^4$		3.6[k]			
Er^{3+}	$ML/M.L$	1.79	1.64[d]		3.3[d]	18[d]
	$ML_2/M.L^2$	3.06	2.90[d]		5.5[d]	30[d]
	$ML_3/M.L^3$	4.20	3.72[d]		5.2[d]	33[d]
	$ML_4/M.L^4$		3.6[k]			
Tm^{3+}	$ML/M.L$	1.83				
	$ML_2/M.L^2$	3.02				
	$ML_3/M.L^3$	4.17				
Yb^{3+}	$ML/M.L$	1.84	1.68[d]	2.56 +0.01	3.5[d]	20[d]
	$ML_2/M.L^2$	3.13	2.91[d]	4.36 ±0.02	6.1[d]	34[d]
	$ML_3/M.L^3$	4.15	3.62[d]		6.6[d]	39[d]
	$ML_4/M.L^4$		3.6[k]			
Lu^{3+}	$ML/M.L$	1.85				
	$ML_2/M.L^2$	3.16				
	$ML_3/M.L^3$	4.02				

[d] 25°, 2.0;   [i] 20°, 0.5;   [k] 20°, 2.0

A. MONO-CARBOXYLIC ACIDS

Acetic acid (continued)

Metal ion	Equilibrium	Log K 25°, 0.1	Log K 25°, 1.0	Log K 25°, 0	ΔH 25°, 0	ΔS 25°, 0
Am^{3+}	$ML/M.L$	1.99^i	1.96^d		4.3^d	23^d
	$ML_2/M.L_2$	3.28^i				
	$ML_3/M.L_3$	3.9^i				
Cm^{3+}	$ML/M.L$	2.06^i	2.03^d		4.3^d	24^d
	$ML_2/M.L_2$	3.09^i				
Bk^{3+}	$ML/M.L$		2.06^d		4.4^d	24^d
Cf^{3+}	$ML/M.L$		2.11^d		3.8^d	22^d
Th^{4+}	$ML/M.L$		3.89 ± 0.03		2.7^c	27^c
	$ML_2/M.L_2$		6.94 ± 0.03		1.1^c	18^c
	$ML_3/M.L_3$		9.01 ± 0.09		3.3^c	20^c
	$ML_4/M.L_4$		10.3		1.2^c	10^c
	$ML_5/M.L_5$		11.0		0.9^c	6^c
UO_2^{2+}	$ML/M.L$	2.61^h	2.44 ± 0.02		$2.7^c\pm0.2$	20^c
	$ML_2/M.L_2$	4.9^h	4.42 ± 0.04		$1.9^c\pm0.4$	27^c
	$ML_3/M.L_3$	6.3^h	6.43 ± 0.09		$-0.6^c\pm0.4$	27^c
NpO_2^{2+}	$ML/M.L$		2.31^j			
	$ML_2/M.L_2$		4.23^j			
	$ML_3/M.L_3$		6.00^j			
PuO_2^{2+}	$ML/M.L$	2.31	2.13			
	$ML_2/M.L_2$	3.80	3.49			
	$ML_3/M.L_3$		5.01			
Cr^{2+}	$ML/M.L$	1.25^z		1.80		
	$ML_2/M.L_2$	2.15^z		2.92		
Mn^{2+}	$ML/M.L$	0.8^r	0.69	1.40		
Fe^{2+}	$ML/M.L$			1.40		
Co^{2+}	$ML/M.L$	1.10^r	0.81	1.46		
		0.71^s				
Ni^{2+}	$ML/M.L$	0.74^b	$0.83 -0.1$	1.43		
Cu^{2+}	$ML/M.L$	1.83 ± 0.06	$1.71 +0.01$	2.22 ± 0.03	1.0^e	12^e
		1.76^s		1.87^e		
	$ML_2/M.L_2$	3.09	2.71	3.63	1.4^e	19^e
				3.12^e		
	$ML_3/M.L_3$		3.1^j	3.58^e	1.5^e	22^e
	$ML_4/M.L_4$		2.9^j	3.3^e		
Cr^{3+}	$ML/M.L$	4.63^z			$(1)^u$	$(20)^z$
	$ML_2/M.L_2$	7.08^z			$(3)^u$	$(20)^z$
	$ML_3/M.L_3$	9.6^z			$(8)^u$	$(10)^z$
Fe^{3+}	$ML/M.L$	3.38^h	3.2^j			
			3.23^e			
	$ML_2/M.L_2$	6.5^h	6.22^e			
	$ML_3/M.L_3$	8.3^h				
VO_2^+	$ML/M.L$		2.28^j			

---

^{b} 25°, 0.5; ^{c} 25°, 1.0; ^{d} 25°, 2.0; ^{e} 25°, 3.0; ^{h} 20°, 0.1; ^{i} 20°, 0.5; ^{j} 20°, 1.0;

^{k} 20°, 2.0; ^{r} 25°, 0.16; ^{s} 30°, 0.4; ^{u} 25-75°, 0.3; ^{z} 25°, 0.3

Acetic acid (continued)

Metal ion	Equilibrium	Log K 25°, 0.1	Log K 25°, 1.0	Log K 25°, 0	ΔH 25°, 0	ΔS 25°, 0
Ag^+	$ML/M \cdot L$		0.37[e]±0.01	0.73	0.9	6
	$ML_2/M \cdot L^2$		0.14[e]±0.04	0.64 ±0.00	0.9	6
	$ML_3/M \cdot L^3$		-0.3[e]±0.2			
	$M_2L/M^2 \cdot L$			(1.14)		
CH_3Hg^+	$ML/M \cdot L$	3.18[v]				
$(CH_3)_3Pb^+$	$ML/M \cdot L$		0.54			
$(C_2H_5)_3Pb^+$	$ML/M \cdot L$		0.44			
$(C_6H_5)_3Pb^+$	$ML/M \cdot L$	5.4[w]				
Tl^+	$ML/M \cdot L$			-0.11		
Zn^{2+}	$ML/M \cdot L$	1.1 ±0.2	0.63	1.57		
		0.91[s]		0.91[e]	2.0[e]	11[e]
	$ML_2/M \cdot L^2$	1.9		1.36[e]	5.3[e]	24[e]
	$ML_3/M \cdot L^3$			1.57[e]	6.4[e]	30[e]
Cd^{2+}	$ML/M \cdot L$	1.56 ±0.06	1.17 +0.09	1.93	1.8[b]	
	$ML_2/M \cdot L^2$	1.19[b]+0.1	1.23[d]	1.32[e]±0.02	1.5[e]	11[e]
		(2.68)	1.82	3.15	3.2[b]	
	$ML_3/M \cdot L^3$	1.90[b]	1.98[d]	2.26 ±0.06	2.3[e]	18[e]
		2.17[b]	2.04			
			2.13[d]			
	$ML_4/M \cdot L^4$			2.42[e]		
				2.0[e]	2.7[e]	20[e]
Hg^{2+}	$ML/M \cdot L$	5.89	5.55[y]			
	$ML_2/M \cdot L^2$		9.30[y]			
	$ML_3/M \cdot L^3$		13.28[y]			
	$ML_4/M \cdot L^4$		17.06[y]			
Sn^{2+}	$ML/M \cdot L$			3.3[e]		
	$ML_2/M \cdot L^2$			6.0[e]		
	$ML_3/M \cdot L^3$			7.3[e]		
Pb^{2+}	$ML/M \cdot L$	2.15 ±0.05	2.1 ±0.1	2.68		
		1.93[s]	2.17[d]±0.02	2.33[e]	-0.1[e]	10[e]
	$ML_2/M \cdot L^2$	3.5	2.98[y]	4.08		
	$ML_3/M \cdot L^3$		3.18[d]-0.3	3.60[e]	-0.2[e]	16[e]
	$ML_4/M \cdot L^4$		3.4[d]±0.1	3.6[e]	-1.2[e]	13[e]
				2.9[e]		
$(CH_3)_2Pb^{2+}$	$ML/M \cdot L$		2.62			
	$ML_2/M \cdot L^2$		3.62			
$(C_2H_5)_2Pb^{2+}$	$ML/M \cdot L$		2.77			
	$ML_2/M \cdot L^2$		(3.28)			
$(C_3H_7)_2Pb^{2+}$	$ML/M \cdot L_2$		2.94			
	$ML_2 \cdot M \cdot L^2$		3.95			

[b] 25°, 0.5;   [d] 25°, 2.0;   [e] 25°. 3.0;   [s] 30°, 0.4;   [v] 25°, 0.4;   [w] 30°, 0.1;   [y] 30°, 1.0

A. MONO-CARBOXYLIC ACIDS

Acetic acid (continued)

Metal ion	Equilibrium	Log K 25°, 0.1	Log K 25°, 1.0	Log K 25°, 0	ΔH 25°, 0	ΔS 25°, 0
$(C_6H_5)_2Pb^{2+}$	$ML/M.L$		3.50			
	$ML_2/M.L^2$		4.90			
Al^{3+}	$ML/M.L$		1.51			
In^{3+}	$ML/M.L$		3.50^k			
	$ML_2/M.L^2$		5.95^k			
	$ML_3/M.L^3$		7.90^k			
	$ML_4/M.L^4$		9.08^k			
Tl^{3+}	$ML/M.L$			6.17^e		
	$ML_2/M.L^2$			11.28^e		
	$ML_3/M.L^3$			15.10^e		
	$ML_4/M.L^4$			18.3^e		
	$MHL_2/M.HL.L$			7.97^e		
	$MOHL/M.OH.L$			18.41^e		
	$M(OH)_2L/M.(OH)^2.L$			30.1^e		
	$MOHL_2/M.OH.L^2$			22.9^e		

---

^{e} 25°, 3.0; ^{k} 20°, 2.0

Bibliography:

H^+ 31LA,31MS,33HE,33SL,34DW,37HH,37KM, 38CK,45P,48F,51A,51B,52S,53KE,59MR, 59P,60YY,61SM,62KP,62W,64N,64T,66A, 67AD,67CIH,67G,68DM,68ES,69CN,70BL, 70BRS,70CS,71P,71PR,71SK,71VSa,72K, 72M,72PTa,72SI,73LR,74G,74KB,74PD

Li^+,Na^+ 64AM

Be^{2+} 73BS

Mg^{2+}-Ba^{2+} 38CK,46J,56N,64AM,70BT,73RS, 73RSM

Y^{3+}-Lu^{3+} 53F,58S,59Sc,60Sb,62G,64AM, 64G,66AM,66KP,70CS,74GE

Am^{3+}-Cf^{3+} 62G,63Ga,70CS

Th^{4+} 72PTa,75PD

UO_2^{2+} 51A,60Sa,67MN,71AKa,74PD

NpO_2^{2+} 70PTM

PuO_2^{2+} 68ESB,68MPC

Cr^{2+},Fe^{2+} 56YF

Mn^{2+},Co^{2+}-Cu^{2+},Zn^{2+},Cd^{2+},Pb^{2+} 38CK,40EB, 45P,46L,48F,51F,51LW,52F,56BH,57BDM, 47LW,60YY,62KP,63G,64AM,64BS,64SM, 66G,67Ga,68FP,68G,68GJ,70BT,70FM, 70GF,71P,73HH,73O

Cr^{3+} 70TG

Fe^{3+} 59P,61SP,69CN

VO^{2+}66Ia

$Ag+$ 42MA,46L,47MP,51DM,64SM

CH_3Hg^+ 73LR

$(CH_3)_3Pb^+$,$(C_2H_5)_3Pb^+$,$(CH_3)_2Pb^{2+}$,$(C_2H_5)_2Pb^{2+}$, $(C_3H_7)_2Pb^{2+}$,$(C_6H_5)_2Pb^{2+}$ 69PM

$(C_6H_5)_3Pb^+$ 65SM

Tl^+ 37RD

Hg^{2+} 64BS,73LU

Sn^{2+} 74G

Al^{3+} 75KI

In^{3+} 53S

Tl^{3+} 65KY

Other references: 05SA,10J,34FR,38Da,46PS, 52CM,52DA,52SL,53AP,53JA,53MA,53Sa, 53SA,55BA,55GL,56L,57CR,58BG,58BS,58CP, 58SBa,58W,59TK,59TKa,60TKa,60TKO,61NPG, 61RM,62BC,62MM,62SN,63DS,63FK,63FR, 63GAa,63LC,63MA,63NS,63SW,63TS,64L, 64Sb,65HS,65MH,65SMa,65TSO,66AMa,66GA, 67A,67MC,68RSa,68WF,69FT,69KP,69MBJ, 69SCF,69VOP,69W,70Ha,70MC,70MW,70PSb, 70SA,70SR,71GK,71KT,71LD,71MC,71NV, 71SS,71TD,71TDa,71TDG,72TA,73Ab,73SNH, 73VP,74KBM,74KT,74La,75CS,75KPV,75LB, 75NM,75VS,75W

$$CH_3CH_2CO_2H$$

$C_3H_6O_2$		Propanoic acid				HL
Metal ion	Equilibrium	Log K 25°, 0.1	Log K 25°, 1.0	Log K 25°, 0	ΔH 25°, 0	ΔS 25°, 0
H^+	HL/H.L	4.67 ±0.02	4.67 −0.01	4.874 +0.001	0.20 ±0.03	23.0
		4.63[b]	4.89[d]	5.16[e]		
Mg^{2+}	ML/M.L	0.54[x]				
Ca^{2+}	ML/M.L	0.50[x]				
Sr^{2+}	ML/M.L	0.43[x]				
Ba^{2+}	ML/M.L	0.34[x]				
Y^{3+}	ML/M.L	1.88[h]	1.61[d]		3.9[d]	20[d]
	$ML_2/M.L^2$	3.06[h]	2.81[d]		5.9[d]	33[d]
La^{3+}	ML/M.L	1.89[h]	1.53[d]		2.5[d]	15[d]
	$ML_2/M.L^2$	3.07[h]	2.42[d]		4.1[d]	25[d]
Ce^{3+}	ML/M.L	2.05[h]	1.67[d]		2.2[d]	15[d]
	$ML_2/M.L^2$	3.28[h]	2.67[d]		3.9[d]	25[d]
Pr^{3+}	ML/M.L	2.12[h]	1.78[d]		1.9[d]	15[d]
	$ML_2/M.L^2$	3.46[h]	2.86[d]		3.6[d]	25[d]
Nd^{3+}	ML/M.L	2.20[h]	1.93[d]		1.8[d]	15[d]
	$ML_2/M.L^2$	3.42[h]	3.08[d]		3.3[d]	15[d]
Sm^{3+}	ML/M.L	2.21[h]	2.02[d]		1.6[d]	15[d]
	$ML_2/M.L^2$	3.70[h]	3.24[d]		2.9[d]	25[d]
Eu^{3+}	ML/M.L	2.23[h]	1.98[d] −0.05		1.8[d]	15[d]
	$ML_2/M.L^2$	3.75[h]	3.28[d] −0.04		2.8[d]	24[d]
Gd^{3+}	ML/M.L	2.09[h]	1.84[d]		2.2[d]	16[d]
	$ML_2/M.L^2$	3.57[h]	3.17[d]		3.1[d]	25[d]
Tb^{3+}	ML/M.L	2.00[h]	1.73[d]		3.0[d]	18[d]
	$ML_2/M.L^2$	3.59[h]	3.10[d]		3.7[d]	27[d]
Dy^{3+}	ML/M.L	1.93[h]	1.63[d]		3.7[d]	20[d]
	$ML_2/M.L^2$	3.45[h]	2.97[d]		4.5[d]	29[d]
Ho^{3+}	ML/M.L	1.96[h]	1.62[d]		3.9[d]	21[d]
	$ML_2/M.L^2$	3.46[h]	2.85[d]		5.5[d]	32[d]
Er^{3+}	ML/M.L	1.94[h]	1.60[d]		4.0[d]	21[d]
	$ML_2/M.L^2$	3.48[h]	2.72[d]		6.3[d]	34[d]
Tm^{3+}	ML/M.L	1.91[h]	1.61[d]		4.0[d]	21[d]
	$ML_2/M.L^2$	3.44[h]	2.66[d]		6.8[d]	35[d]
Yb^{3+}	ML/M.L	1.93[h]	1.63[d]		3.8[d]	20[d]
	$ML_2/M.L^2$	3.38[h]	2.70[d]		6.5[d]	34[d]
Lu^{3+}	ML/M.L	2.00[h]	1.66[d]		3.8[d]	20[d]
	$ML_2/M.L^2$	3.53[h]	2.78[d]		6.3[d]	34[d]
Th^{4+}	ML/M.L		3.94[j]			
	$ML_2/M.L^2$		7.25[j]			
	$ML_3/M.L^3$		9.44[j]			
	$ML_4/M.L^4$		11.2[j]			

[b] 25°, 0.5;   [d] 25°, 2.0;   [e] 25°, 3.0;   [h] 20°, 0.1;   [j] 20°, 1.0;   [x] temperature not stated, 0.2

Propanoic acid (continued)

Metal ion	Equilibrium	Log K 25°, 0.1	Log K 25°, 1.0	Log K 25°, 0	ΔH 25°, 0	ΔS 25°, 0
UO_2^{2+}	$ML/M.L$		2.53^j			
	$ML_2/M.L^2$		4.68^j			
	$ML_3/M.L^3$		6.49^j			
	$ML_4/M.L^4$		8.25^j			
NpO_2^{2+}	$ML/M.L$		2.45^j			
	$ML_2/M.L^2$		4.46^j			
	$ML_3/M.L^3$		5.49^j			
Co^{2+}	$ML/M.L$		0.72			
			0.78^d			
Ni^{2+}	$ML/M.L$		0.78			
			0.86^d			
	$ML_2/M.L^2$		1.26^d			
Cu^{2+}	$ML/M.L$		1.66	2.22		
				1.86^e		
	$ML_2/M.L^2$		2.62	3.00^e		
Cr^{3+}	$ML/M.L$	4.70^s				
	$ML_2/M.L^2$	7.04^s				
	$ML_3/M.L^3$	9.7^s				
Fe^{3+}	$ML/M.L$		3.4^j			
CH_3Hg^+	$ML/M.L$	3.39^b				
Zn^{2+}	$ML/M.L$	1.01^x	0.85			
			$1.04^d \pm 0.04$			
	$ML_2/M.L^2$		1.23^d			
			1.2^d			
Cd^{2+}	$ML/M.L$		1.19			
			$1.27^d \pm 0.04$			
	$ML_2/M.L^2$		1.86			
			$1.9^d \pm 0.1$			
Pb^{2+}	$ML/M.L$		2.08^d	2.64^t		
	$ML_2/M.L^2$		3.34^d	4.15^t		
Al^{3+}	$ML/M.L$		1.69			
In^{3+}	$ML/M.L$		3.57^k			
	$ML_2/M.L^2$		6.36^k			
	$ML_3/M.L^3$		8.15^k			
	$ML_4/M.L^4$		9.08^k			
	$ML_5/M.L^5$		10.0^k			

---

^{b} 25°, 0.4; ^{d} 25°, 2.0; ^{e} 25°, 3.0; ^{h} 20°, 0.1; ^{j} 20°, 1.0; ^{k} 20°, 2.0; ^{s} 25°, 0.3; ^{t} 35°, 0; ^{x} temperature not stated, 0.2

Bibliography:

H^+ 31LA,33HEa,52EL,59MR,59P,64PKP,67CIH, 69GW,69RS,69SMM,73FP,73LR

Mg^{2+}-Ba^{2+} 38CK

Y^{3+}-Lu^{3+} 64PKP,65CG,71ALN

Th^{4+} 72PTa

UO_2^{2+} 67MN

NpO_2^{2+} 69CMT

Co^{2+}-Cu^{2+},Zn^{2+},Cd^{2+},Pb^{2+} 38CK,51LW,55MA, 64SM,68FP,70FM,70GF,73WK

Propanoic acid (continued)

Cr^{3+} 70TG

Fe^{3+} 59P In^{3+} 53S

CH_3Hg^+ 73LR   Other references: 52CM,58CP,58SBb,63MS,65DSa,
                                                           66AA,67A,68RSa,69LW,71LDa,71SS,71TD,
Al^{3+} 75KI   73SNH,74Ja,75TD,75W

$$CH_3CH_2CH_2CO_2H$$

$C_4H_8O_2$			Butanoic acid			HL
Metal ion	Equilibrium	Log K 25°, 0.1	Log K 25°, 1.0	Log K 25°, 0	ΔH 25°, 0	ΔS 25°, 0
H^+	HL/H.L	4.63 ±0.00	4.62 +0.01	4.819 ±0.001	0.71 ±0.02	24.5
		4.58[b]	4.86[d]	5.13[e]		
Mg^{2+}	ML/M.L	0.53[x]				
Ca^{2+}	ML/M.L	0.51[x]				
Sr^{2+}	ML/M.L	0.36[x]				
Ba^{2+}	ML/M.L	0.31[x]				
Th^{4+}	$ML/M.L$		3.90[j]			
	$ML_2/M.L^2$		7.00[j]			
	$ML_3/M.L^3$		9.74[j]			
UO_2^{2+}	$ML/M.L$		2.58[j]			
	$ML_2/M.L^2$		4.71[j]			
	$ML_3/M.L^3$		6.91[j]			
Co^{2+}	$ML/M.L$		0.62[d]±0.04			
	$ML_2/M.L^2$		0.82[d]±0.06			
Ni^{2+}	$ML/M.L$		0.72[d]±0.01			
	$ML_2/M.L^2$		0.85[d]±0.05			
Cu^{2+}	ML/M.L		1.7[d]±0.2	2.14		
				1.82[e]		
	$ML_2/M.L^2$		2.6[d]±0.2	2.98[d]		
Zn^{2+}	ML/M.L	1.00[x]	0.98[d]±0.02			
	$ML_2/M.L^2$		1.65[d]			
Cd^{2+}	$ML/M.L$		1.25[d]±0.05			
	$ML_2/M.L^2$		1.98[d]±0.05			
Pb^{2+}	$ML/M.L$		2.13[d]±0.05			
	$ML_2/M.L^2$		3.74[d]±0.05			
Al^{3+}	ML/M.L		1.58			

[b] 25°, 0.5;   [d] 25°, 2.0;   [e] 25°, 3.0;   [j] 20°, 1.0;   [x] temperature not stated, 0.2

Bibliography:

H^+ 31LAa,34HS,38CK,40SD,48CD,52EL,59MR, $Co^{2+}-Pb^{2+}$ 38CK,51LW,64SM,68FP,70FM,70GF,
       67CIH,69RRS,73FP   73FP,74GMP,75GM

$Mg^{2+}-Ba^{2+}$ 38CK Al^{3+} 75KI

Th^{4+} 72PTa   Other references: 52CM,58CP,69T,71MSb,

UO_2^{2+} 72MPZ   71SS,75GT,75KM

$$\begin{array}{c} CH_3 \\ | \\ CH_3CHCO_2H \end{array}$$

$C_4H_8O_2$ Metal ion	Equilibrium	2-Methylpropanoic acid (isobutyric acid)				HL
		Log K 25°, 0.1	Log K 25°, 1.0	Log K 25°, 0	ΔH 25°, 0	ΔS 25°, 0
H^+	HL/H.L	4.63 ±0.03	4.64 ±0.00	4.849	0.77 ±0.03	24.8
		4.62[b]±0.02	5.15[e]		0.6[a]	
Y^{3+}	ML/M.L	1.60[b]	1.64[d]		5.4[d]	26[d]
	$ML_2/M.L^2$	2.71[b]	2.79[d]		9.6[d]	42[d]
La^{3+}	ML/M.L	1.64[b]	1.57[d]		3.5[d]	19[d]
	$ML_2/M.L^2$	2.16[b]	2.47[d]		6.0[d]	31[d]
Ce^{3+}	ML/M.L	1.79[b]	1.62[d]		3.3[d]	19[d]
	$ML_2/M.L^2$	2.32[b]	2.72[d]		5.9[d]	32[d]
Pr^{3+}	ML/M.L	1.92[b]	1.80[d]		3.0[d]	18[d]
	$ML_2/M.L^2$	3.18[b]	2.91[d]		5.5[d]	32[d]
Nd^{3+}	ML/M.L	1.98[b]	1.91[d]		2.8[d]	18[d]
	$ML_2/M.L^2$	3.10[b]	3.09[d]		5.2[d]	32[d]
Sm^{3+}	ML/M.L	2.05[b]	2.00[d]		2.7[d]	18[d]
	$ML_2/M.L^2$	3.30[b]	3.25[d]		4.7[d]	31[d]
Eu^{3+}	ML/M.L	1.98[b]	1.98[d]		2.9[d]	19[d]
	$ML_2/M.L^2$	3.10[b]	3.29[d]		4.8[d]	31[d]
Gd^{3+}	ML/M.L	1.87[b]	1.86[d]		3.5[d]	20[d]
	$ML_2/M.L^2$	3.28[b]	3.30[d]		5.1[d]	32[d]
Tb^{3+}	ML/M.L	1.82[b]	1.73[d]		4.4[d]	23[d]
	$ML_2/M.L^2$	2.84[b]	3.12[d]		5.8[d]	34[d]
Dy^{3+}	ML/M.L	1.74[b]	1.65[d]		5.0[d]	25[d]
	$ML_2/M.L^2$	2.57[b]	2.97[d]		6.8[d]	37[d]
Ho^{3+}	ML/M.L	1.70[b]	1.63[d]		5.3[d]	25[d]
	$ML_2/M.L^2$	2.92[b]	2.84[d]		7.9[d]	39[d]
Er^{3+}	ML/M.L	1.69[b]	1.61[d]		5.5[d]	26[d]
	$ML_2/M.L^2$	2.59[b]	2.72[d]		8.9[d]	42[d]
Tm^{3+}	ML/M.L	1.69[b]	1.61[d]		5.4[d]	26[d]
	$ML_2/M.L^2$	2.28[b]	2.67[d]		9.5[d]	44[d]
Yb^{3+}	ML/M.L	2.01	1.62[d]		5.4[d]	25[d]
		1.78[b]			5.5[a]	
	$ML_2/M.L^2$	3.61	2.67[d]		9.4[d]	44[d]
		3.10[b]			6.3[a]	
Lu^{3+}	ML/M.L	1.81[b]	1.65[d]		5.4[d]	26[d]
	$ML_2/M.L^2$	2.32[b]	2.77[d]		9.0[d]	43[d]
Th^{4+}	ML/M.L		3.85[j]			
	$ML_2/M.L^2$		7.30[j]			
UO_2^{2+}	ML/M.L		2.48[j]			
	$ML_2/M.L^2$		4.87[j]			
	$ML_3/M.L^3$		7.20[j]			

---

[a] 25°, 0.1; [b] 25°, 0.5; [d] 25°, 2.0; [e] 25°, 3.0; [j] 20°, 1.0

Isobutyric acid (continued)

Metal ion	Equilibrium	Log K 25°, 0.1	Log K 25°, 1.0	Log K 25°, 0	ΔH 25°, 0	ΔS 25°, 0
Cu^{2+}	$ML/M.L$	1.75		2.17		
	$ML_2/M.L^2$	2.70				
Fe^{3+}	$ML/M.L$		3.6[j]			

-----

[j] 20°, 1.0

Bibliography:
H^+ 31LAa,48CD,52EL,59P,64SM,64SP,68CO,
        70CB,70CSS,71BC
$Y^{3+}-Lu^{3+}$ 64SP,65CG,70CB,71BC
Th^{4+} 72PTa
UO_2^{2+} 72MPZ

Cu^{2+} 51LW,70CB
Fe^{3+} 59P

Other references: 52CM,58CP,65DS,68REa,69RRS,
                    75KM

$$CH_3CH_2CH_2CH_2CO_2H$$

| $C_5H_{10}O_2$ | | Pentanoic acid (valeric acid) | | | | HL |
Metal ion	Equilibrium	Log K 25°, 0.1	Log K 25°, 1.0	Log K 25°, 0	ΔH 25°, 0	ΔS 25°, 0
H^+	$HL/H.L$	4.64	4.62	4.843	0.69 ±0.03	24.5
		4.59[b]	5.17[e]			
Cu^{2+}	$ML/M.L$		1.92[e]	2.12		
	$ML_2/M.L^2$		3.0[e]			

-----

[e] 25°, 3.0

Bibliography:
H^+ 31LAa,52EL,64SM,70CSS
Cu^{2+} 51LW,64SM

Other references: 74CJ,75KM

$$\begin{array}{c} CH_3 \\ | \\ CH_3CHCH_2CO_2H \end{array}$$

| $C_5H_{10}O_2$ | | 3-Methylbutanoic acid (isovaleric acid) | | | | HL |
Metal ion	Equilibrium	Log K 25°, 0.1	Log K 25°, 1.0	Log K 25°, 0	ΔH 25°, 0	ΔS 25°, 0
H^+	$HL/H.L$	4.58	4.56	4.781	1.09 ±0.08	25.6
		4.53[b]				
Cu^{2+}	$ML/M.L$			2.08		

-----

[b] 25°, 0.5

A. MONO-CARBOXYLIC ACIDS

Isovaleric acid (continued)

Metal ion	Equilibrium	Log K 25°, 0.1	Log K 25°, 1.0	Log K 25°, 0	ΔH 25°, 0	ΔS 25°, 0
Cd^{2+}	$ML/M.L$		1.34^e			
	$ML_2/M.L_2$		2.30^e			
	$ML_3/M.L_3$		2.5^e			
	$ML_4/M.L_4$		2.0^e			

---

^{e} 25°, 3.0

Bibliography:

H^+ 32L,48CD,52EL,68CD,70CSS Cd^{2+} 43L

Cu^{2+} 51LW   Other references: 74CJ,75KM

$$\begin{array}{c} CH_3 \\ | \\ CH_3CCO_2H \\ | \\ CH_3 \end{array}$$

$C_5H_{10}O_2$		2,2-Dimethylpropanoic acid (pivalic acid)				HL
Metal ion	Equilibrium	Log K 25°, 0.1	Log K 25°, 1.0	Log K 25°, 0	ΔH 25°, 0	ΔS 25°, 0
H^+	$HL/H.L$	4.83 +0.01	4.79	5.032	0.72 ±0.04	25.5
		4.77^b	5.33^e			
Cu^{2+}	$ML/M.L$		1.87^e	2.19		
	$ML_2/M.L_2$		3.7^e			
CH_3Hg^+	$ML/M.L$	3.50^b				
Hg^{2+}	$ML/M.L$	5.92				

---

^{b} 25°, 0.4; ^{e} 25°, 3.0

Bibliography:

H^+ 31LAa,52EL,64SM,70CSS,73LR,73LU Hg^{2+} 73LU

Cu^{2+} 51LW,64SM   Other reference: 52CM

CH_3Hg^+ 73LR

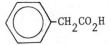

$C_8H_8O_2$		Phenylacetic acid			HL
Metal ion	Equilibrium	Log K 25°, 0.1	Log K 25°, 1.0	Log K 25°, 0	
H^+	$HL/H.L$	4.10 ±0.02	4.09	4.310 ±0.003	
		4.07^b	4.56^e		

---

^{b} 25°, 0.5; ^{e} 25°, 3.0

Phenylacetic acid (continued)

Metal ion	Equilibrium	Log K 25°, 0.1	Log K 25°, 1.0	Log K 25°, 0
Cu^{2+}	ML/M.L		1.61[e]	1.97

[e] 25°, 3.0

Bibliography:

H^+ 31LAb,34DW, 34JV,61KPb,64SM,68RSb, Cu^{2+} 51LW,64SM
71DR

Other references: 68RSa,73KP

$$CH_2=CHCH_2CO_2H$$

$C_4H_6O_2$ But-3-enoic acid (vinylacetic acid)   HL

Metal ion	Equilibrium	Log K 25°, 0.1	Log K 25°, 1.0	Log K 25°, 0
H^+	HL/H.L	4.12[h] 4.06[i]	4.07	3.337 ±0.002
Cu^+	ML/M.L	4.53		

[h] 18°, 0.1;   [i] 18°, 0.5

Bibliography:

H^+ 32La,33IL,37DL Cu^+ 49KAa

$$CH_3CH=CHCO_2H$$

$C_4H_6O_2$ trans-But-2-enoic acid (crotonic acid)   HL

Metal ion	Equilibrium	Log K 25°, 0.1	Log K 25°, 1.0	Log K 25°, 0
H^+	HL/H.L			4.686 ±0.008
Cu^+	ML/M.L	3.20		
Ag^+	ML/M.L		-1.05	

Bibliography:

H^+ 33IL,37GJ Ag^+ 38WL

Cu^+ 49KAa

Other reference: 31Ab

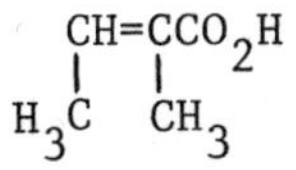

$$CH=CCO_2H$$
$$H_3C \quad CH_3$$

		Log K	Log K	Log K	
		cis-2-Methylbut-2-enoic acid (tiglic acid)			HL
$C_5H_8O_2$					
Metal ion	Equilibrium	Log K 25°, 0.1	Log K 18°, 1.0	Log K 18°, 0	
H^+	HL/H.L	4.77[h] 4.71[i]	4.73	4.96	
Cu^+	ML/M.L	2.32			

[h] 18°, 0.1; [i] 18°, 0.5

Bibliography:

H^+ 32La Cu^+ 49KAa

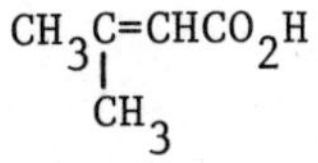

$$CH_3C=CHCO_2H$$
$$CH_3$$

		3-Methylbut-2-enoic acid (β,β-dimethylacrylic acid)		HL
$C_5H_8O_2$				
Metal ion	Equilibrium	Log K 25°, 0.1	Log K 25°, 0	
H^+	HL/H.L		5.119 ±0.001	
Cu^+	ML/M.L	2.04		

Bibliography:

H^+ 33IL,37GJ   Other reference: 32La

Cu^+ 49KAa

$$N\equiv CCH_2CO_2H$$

		Cyanoacetic acid				HL
$C_3H_3O_2N$						
Metal ion	Equilibrium	Log K 25°, 3.0	Log K 25°, 0	ΔH 25°, 0	ΔS 25°, 0	
H^+	HL/H.L	2.63	2.472 ±0.002	0.89	14.3	
Ca^{2+}	ML/M.L		0.55			
Cu^{2+}	ML/M.L	0.87				
	ML_2/M.L_2	1.00				

Bibliography:

H^+ 40SD,56FI,64SM Cu^{2+} 64SM

Ca^{2+} 38Da

$$\langle\!\!\!\!\bigcirc\!\!\!\!\rangle\!-CO_2H$$

$C_7H_6O_2$ — Benzenecarboxylic acid (benzoic acid) — HL

Metal ion	Equilibrium	Log K 25°, 0.1	Log K 25°, 1.0	Log K 25°, 0	ΔH 25°, 0	ΔS 25°, 0
H^+	HL/H.L	4.00 ±0.01 3.96^b±0.03	3.97 -0.01	4.202 ±0.003	-0.11 ±0.01 0.46^u	18.9
Mg^{2+}	ML/M.L		0.1^s			
Ca^{2+}	ML/M.L		0.2^s			
Co^{2+}	ML/M.L		0.55^s			
Ni^{2+}	ML/M.L	0.9	0.55^s			
Cu^{2+}	ML/M.L	1.6	1.51^s			
Ag^+	ML/M.L $ML_2/M.L^2$	0.60^r	0.52 0.54	0.91		
$(C_6H_5)_3Sn^+$	ML/M.L	6.2^t				
$(C_6H_5)_3Pb^+$	ML/M.L	6.2^t				
Zn^{2+}	ML/M.L	0.9	0.74^s			
Cd^{2+}	ML/M.L $ML_2/M.L^2$	1.4	1.15^s 0.99^v 1.76^v			
Pb^{2+}	ML/M.L $ML_2/M.L^2$	2.0	1.99^s 3.30^v			
Al^{3+}	MOHL/M.OH.L	12.09^b				

^{b} 25°, 0.5; ^{r} 25°, 0.2; ^{s} 30°, 0.4; ^{t} 30°, 0.1; ^{u} 40°, 0; ^{v} 30°, 1.0

Bibliography:

H^+ 31LAb,34BK,34DWa,34SM,48CD,52JP,53K, 54DS,57RB,58CP,59FHa,60YY,61KPb,67AD, 67WG,70BT,70LG,70NL,71VSa,72BF,74BS, 74FR,74MKa,74RR,75LS,75O,75TM

Mg^{2+},Ca^{2+} 70BT

Co^{2+}-Cu^{2+},Zn^{2+}-Pb^{2+} 60YY,64JG,65VSa,70BT

$(C_6H_5)_3Sn^+$,$(C_6H_5)_3Pb^+$ 65SM

Ag^+ 49L,67VS

Al^{3+} 70NL

Other references: 51BB,66JGb,68RSb,73SN

$$ClCH_2CO_2H$$

$C_2H_3O_2Cl$ — Chloroacetic acid — HL

Metal ion	Equilibrium	Log K 25°, 0.1	Log K 25°, 1.0	Log K 25°, 0	ΔH 25°, 0	ΔS 25°, 0
H^+	HL/H.L	2.68 ±0.06 2.60^b	2.64 ±0.04 3.02^e	2.865 ±0.004	1.07 ±0.08 0.75^c	16.7
Mg^{2+}	ML/M.L	0.23^r				

^{b} 25°, 0.4; ^{c} 25°, 1.0; ^{e} 25°, 3.0; ^{r} 30°, 0.4

Chloroacetic acid (continued)

Metal ion	Equilibrium	Log K 25°, 0.1	Log K 25°, 1.0	Log K 25°, 0	ΔH 25°, 0	ΔS 25°, 0
Ca^{2+}	$ML/M.L$	0.14^r				
Y^{3+}	$ML/M.L$	1.06				
	$ML_2/M.L^2$	1.93				
La^{3+}	$ML/M.L$	1.12				
	$ML_2/M.L^2$	1.86				
Ce^{3+}	$ML/M.L$	1.18				
	$ML_2/M.L^2$	1.95				
Pr^{3+}	$ML/M.L$	1.26				
	$ML_2/M.L^2$	2.06				
Nd^{3+}	$ML/M.L$	1.31				
	$ML_2/M.L^2$	3.08				
Sm^{3+}	$ML/M.L$	1.32				
	$ML_2/M.L^2$	2.16				
Eu^{3+}	$ML/M.L$	1.38				
	$ML_2/M.L^2$	2.04				
Gd^{3+}	$ML/M.L$	1.26				
	$ML_2/M.L^2$	2.12				
Tb^{3+}	$ML/M.L$	1.24				
	$ML_2/M.L^2$	2.24				
Dy^{3+}	$ML/M.L$	1.21				
	$ML_2/M.L^2$	2.12				
Ho^{3+}	$ML/M.L$	1.15				
	$ML_2/M.L^2$	1.81				
Er^{3+}	$ML/M.L$	1.15				
	$ML_2/M.L^2$	2.06				
Tm^{3+}	$ML/M.L$	1.10				
	$ML_2/M.L^2$	1.91				
Yb^{3+}	$ML/M.L$	1.04				
	$ML_2/M.L^2$	2.15				
Lu^{3+}	$ML/M.L$	1.02				
	$ML_2/M.L^2$	2.29				
Th^{4+}	$ML/M.L$		2.77^j			
	$ML_2/M.L^2$		4.64^j			
	$ML_3/M.L^3$		5.8^j			
	$ML_4/M.L^4$		6.8^j			
UO_2^{2+}	$ML/M.L$		1.43 ±0.01		1.9^c	13^c
	$ML_2/M.L^2$		2.28 ±0.04		1.9^c	17^c
	$ML_3/M.L^3$		2.7 ±0.1		2.0^c	19^c
NpO_2^{2+}	$ML/M.L$		1.33^j			
	$ML_2/M.L^2$		2.11^j			
	$ML_3/M.L^3$		2.8^j			
PuO_2^{2+}	$ML/M.L$		1.16^j			
	$ML_2/M.L^2$		1.61^j			
	$ML_3/M.L^3$		2.0^j			
Co^{2+}	$ML/M.L$	0.23^r	0.2^k ±0.2			

^{c} 25°, 1.0; ^{j} 20°, 1.0; ^{k} 18°, 2.0; ^{r} 30°, 0.4

Chloroacetic acid (continued)

Metal ion	Equilibrium	Log K 25°, 0.1	Log K 25°, 1.0	Log K 25°, 0	ΔH 25°, 0	ΔS 25°, 0
Ni^{2+}	ML/M.L	0.23[r]	0.20[k]			
Cu^{2+}	ML/M.L	1.07[r]	0.91[j]	1.61		
			1.03[e]			
	$ML_2/M.L^2$		1.09[j]			
			1.43[e]			
	$ML_3/M.L^3$		1.45[j]			
Fe^{3+}	ML/M.L		2.1 [j]			
Ag^+	ML/M.L			0.64		
	$ML_2/M.L^2$			0.7 ±0.2		
CH_3Hg^+	ML/M.L	2.19[b]				
Zn^{2+}	ML/M.L	0.56[r]	0.40[k]			
Cd^{2+}	ML/M.L	0.99[r]	0.84[k]			
Hg^{2+}	ML/M.L	4.64				
Pb^{2+}	ML/M.L	1.1 [r]	1.50[k]±0.02			
	$ML_2/M.L^2$		1.8 [k]±0.2			

[b] 25°, 0.4; [e] 25°, 3.0; [j] 20°, 1.0; [k] 28°, 2.0; [r] 30°, 0.4

Bibliography:

H^+ 31LAb,34W,49A,55IP,59P,60JP,62HO,
        64SM,66A,68CO,61NPP,71SK,72PTa,73LR,
        74PD

Mg^{2+},Ca^{2+} 70BT

$Y^{3+}-Lu^{3+}$ 66KP

Th^{4+} 72PTa

UO_2^{2+} 49A,62HO,74PD

NpO_2^{2+} 69CMT

PuO_2^{2+} 69CP

$Co^{2+}-Cu^{2+},Zn^{2+},Cd^{2+},Pb^{2+}$ 48F,51LW,64SM,70BT,
        70FB,70FM,70GF

Fe^{3+} 59P

Ag^+ 49PH,52MT

CH_3Hg^+ 73LR

Hg^{2+} 73LU

Other references: 34FR,50DS,55MAa,63LC,68RSa,
        70BRS,71SS,73La,73SBL,73VB,75C,75KPV

Cl_2CHCO_2H

$C_2H_2O_2Cl_2$ Dichloroacetic acid   HL

Metal ion	Equilibrium	Log K 25°, 0.2	Log K 25°, 0	ΔH 25°, 0	ΔS 25°, 0
H^+	HL/H.L	0.87	1.30	0.2	7
CH_3Hg^+	ML/M.L	1.14[b]			

[b] 25°, 0.4

Bibliography:

H^+ 58J,66A,73LR

CH_3Hg^+ 73LR

Other references: 50DS,69PJa,70BRS,71SS,74Lb

$$Cl_3CCO_2H$$

$C_2HO_2Cl_3$		Trichloroacetic acid	HL
Metal ion	Equilibrium	Log K 25°, 0.1	
H^+	HL/H.L	0.66	
Hg^{2+}	ML/M.L	3.08	

Bibliography: 73LU

Other references: 50DS,70BRS,71BFA,71SS,74Lb,75KPV

$$ClCH_2CH_2CO_2H$$

$C_3H_5O_2Cl$		3-Chloropropanoic acid				HL
Metal ion	Equilibrium	Log K 25°, 0.1	Log K 25°, 1.0	Log K 25°, 0	ΔH 25°, 0	ΔS 25°, 0
H^+	HL/H.L	3.92	3.91 ±0.02	4.11	0.55 ±0.04	20.7
		3.86[b]			0.17[c]	18.5[c]
Th^{4+}	$ML/M.L$		3.50[j]			
	$ML_2/M.L_2$		5.98[j]			
	$ML_3/M.L_3$		8.2 [j]			
UO_2^{2+}	$ML/M.L$		2.07 ±0.01		2.7[c]	19[c]
	$ML_2/M.L_2$		3.58 ±0.00		2.3[c]	15[c]
	$ML_3/M.L_3$		5.1 ±0.1		0.0[c]	7[c]
NpO_2^{2+}	$ML/M.L$		1.88[j]			
	$ML_2/M.L_2$		3.30[j]			
	$ML_3/M.L_3$		(3.6)[j]			
PuO_2^{2+}	$ML/M.L$		1.70[j]			
	$ML_2/M.L_2$		2.95[j]			
	$ML_3/M.L_3$		3.8 [j]			

[b] 25°, 0.5;   [c] 25°, 1.0;   [j] 20°, 1.0

Bibliography:

H^+	33L,48CD,68CO,72PTa,74PD	NpO_2^{2+}	69CMT
Th^{4+}	72PTa	PuO_2^{2+}	70PC
UO_2^{2+}	72MPZ,74PD		

Other references: 66A,67CIH,68Me,74Ja

$$BrCH_2CO_2H$$

$C_2H_3O_2Br$		Bromoacetic acid				HL
Metal ion	Equilibrium	Log K 25°, 0.1	Log K 25°, 1.0	Log K 25°, 0	ΔH 25°, 0	ΔS 25°, 0
H^+	HL/H.L	2.72	2.69	2.902	1.17 ±0.07	17.2
		2.66[b]				

[b] 25°, 0.1

Bromoacetic acid (continued)

Metal ion	Equilibrium	Log K 25°, 0.1	Log K 25°, 1.0	Log K 25°, 0	ΔH 25°, 0	ΔS 25°, 0
Cu^{2+}	ML/M.L			1.59		

Bibliography:

H^+ 33L,55IP,68CO   Other references: 49DW,52CM,66A,71SS

Cu^{2+} 51LW

$$\overset{\overset{\text{Br}}{|}}{CH_3CH_2CHCO_2H}$$

$C_4H_7O_2Br$		2-Bromobutanoic acid	HL

Metal ion	Equilibrium	Log K 25°, 0
H^+	HL/H.L	2.97
Cu^{2+}	ML/M.L	1.46

Bibliography: 51LW

$$ICH_2CO_2H$$

$C_2H_3O_2I$		Iodoacetic acid			HL

Metal ion	Equilibrium	Log K 25°, 0.1	Log K 25°, 1.0	Log K 25°, 0	ΔH 25°, 0	ΔS 25°, 0
H^+	HL/H.L	2.98 2.93[b]	2.95	3.175	1.3 ±0.1	19
Y^{3+}	ML/M.L $ML_2/M.L^2$	1.23 2.20				
La^{3+}	ML/M.L $ML_2/M.L^2$	1.18 2.33				
Ce^{3+}	ML/M.L $ML_2/M.L^2$	1.26 2.32				
Pr^{3+}	ML/M.L $ML_2/M.L^2$	1.34 2.49				
Nd^{3+}	ML/M.L $ML_2/M.L^2$	1.32 2.60				
Sm^{3+}	ML/M.L $ML_2/M.L^2$	1.43 2.57				
Eu^{3+}	ML/M.L $ML_2/M.L^2$	1.48 2.50				
Gd^{3+}	ML/M.L $ML_2/M.L^2$	1.36 2.49				
Tb^{3+}	ML/M.L $ML_2/M.L^2$	1.15 2.56				

---

[b] 25°, 0.5

Iodoacetic acid (continued)

Metal ion	Equilibrium	Log K 25°, 0.1	Log K 25°, 1.0	Log K 25°, 0	ΔH 25°, 0	ΔS 25°, 0
Dy^{3+}	$ML/M.L$	1.11				
	$ML_2/M.L^2$	2.42				
Ho^{3+}	$ML/M.L$	1.23				
	$ML_2/M.L^2$	2.51				
Er^{3+}	$ML/M.L$	1.26				
	$ML_2/M.L^2$	2.29				
Tm^{3+}	$ML/M.L$	1.18				
	$ML_2/M.L^2$	2.39				
Yb^{3+}	$ML/M.L$	1.28				
	$ML_2/M.L^2$	2.29				
Lu^{3+}	$ML/M.L$	1.32				
	$ML_2/M.L^2$	2.06				

Bibliography:

H^+ 33L,55IP,68CO   Other reference: 66A

Y^{3+}-Lu^{3+} 66KP

$$ICH_2CH_2CO_2H$$

$C_3H_5O_2I$		3-Iodopropanoic acid				HL
Metal ion	Equilibrium	Log K 25°, 0.1	Log K 25°, 1.0	Log K 25°, 0	ΔH 25°, 0	ΔS 25°, 0
H^+	HL/H.L	3.90 3.85[b]	3.88	4.10 ±0.02	(1.4)	(23)
Cu^{2+}	ML/M.L			1.91		

[b] 25°, 0.5

Bibliography:

H^+ 33L,66A Cu^{2+} 51LW

$$O_2NCH_2CO_2H$$

$C_2H_3O_4N$		Nitroacetic acid				HL
Metal ion	Equilibrium	Log K 18°, 0.1	Log K 18°, 0.4	Log K 18°, 0	ΔH 18°, 0	ΔS 18°, 0
H^+	HL/H.L	1.46	1.34	1.68	(1)[r]	(11)
Be^{2+}	ML/M.L		0.26			
Mg^{2+}	ML/M.L		-0.19			
Ca^{2+}	ML/M.L		-0.30			

[r] 10-18°, 0

Nitroacetic acid (continued)

Metal ion	Equilibrium	Log K 18°, 0.1	Log K 18°, 0.4	Log K 18°, 0	ΔH 18°, 0	ΔS 18°, 0
Co^{2+}	ML/M.L		0.0			
Ni^{2+}	ML/M.L		0.06			
Cu^{2+}	ML/M.L		0.44			
Zn^{2+}	ML/M.L		0.03			
Cd^{2+}	ML/M.L		0.19			
Pb^{2+}	ML/M.L		0.14			
Al^{3+}	ML/M.L		0.48			

Bibliography:

H^+ 34P,49P $Be^{2+}-Al^{3+}$ 49P

$C_7H_5O_4N$				Nitrobenzoic acid			HL
Isomer	Metal ion	Equilibrium	Log K 30°, 1.0	Log K 25°, 0	ΔH 25°, 0	ΔS 25°, 0
2-Nitro	H^+	HL/H.L		2.179 ±0.006	3.36	21.3
	Ag^+	ML/M.L	0.30			
		$ML_2/M.L^2$	0.15			
3-Nitro	H^+	HL/H.L		3.449 ±0.001	-0.37 ±0.05	14.6
	Ag^+	ML/M.L	0.38			
		$ML_2/M.L^2$	0.18			
4-Nitro	H^+	HL/H.L		3.442 ±0.001	-0.43 ±0.3	14.3
	Ag^+	ML/M.L	0.32			
		$ML_2/M.L^2$	0.48			

Bibliography:

H^+ 37DLa,39EW,51BB,67WG,72BF,74MK Ag^+ 67VS

$$H_2O_3POCH_2CH_2CO_2H$$

| $C_3H_7O_6P$ | | 2-Hydroxypropanoic acid 2-dihydrogen phosphate (phosphoenolpyruvic acid) | H_3L |

Metal ion	Equilibrium	Log K 25°, 0.1
H^+	HL/H.L	(6.35)[a]
	$H_2L/HL.H$	(3.4)[a]
K^+	ML/M.L	1.08[a]
Mg^{2+}	ML/M.L	2.26[a]
Mn^{2+}	ML/M.L	2.75[a]
Co^{2+}	ML/M.L	2.54[a]
Ni^{2+}	ML/M.L	2.34[a]
Zn^{2+}	ML/M.L	2.96[a]
Cd^{2+}	ML/M.L	2.96[a]

[a] $(C_3H_7)_4NI$ used as background electrolyte.

Bibliography: 57WB

$$\overset{\displaystyle OPO_3H_2}{\underset{\displaystyle HOCH_2CHCO_2H}{|}}$$

| $C_3H_7O_7P$ | | D-2,3-Dihydroxypropanoic acid 2-dihydrogen phosphate (D-2-phosphoglyceric acid) | H_3L |

Metal ion	Equilibrium	Log K 25°, 0.1
H^+	HL/H.L	(7.00)[a]
	$H_2L/HL.H$	(3.55)[a]
K^+	ML/M.L	1.18[a]
Mg^{2+}	ML/M.L	2.45[a]
Mn^{2+}	ML/M.L	3.09[a]
Co^{2+}	ML/M.L	2.97[a]
Ni^{2+}	ML/M.L	2.88[a]
Zn^{2+}	ML/M.L	3.40[a]
Cd^{2+}	ML/M.L	3.40[a]

[a] $(C_3H_7)_4NI$ used as background electrolyte.

Bibliography: 57WB

$$\text{(benzene ring with } OPO_3H_2 \text{ and } CO_2H)$$

$C_7H_7O_6P$	2-Hydroxybenzoic acid 2-dihydrogenphosphate (salicyl phosphate)				H_3L
Metal ion	Equilibrium	Log K 25°, 0.1		ΔH 25°, 0.1	ΔS 25°, 0.1
H^+	HL/H.L	6.61		$(-6)^a$	(10)
	H_2L/HL.H	3.69		$(-4)^a$	(3)
VO^{2+}	ML/M.L	5.81		$(-5)^a$	(9)
	ML/MOHL.H	5.7			
	$(MOL)_2/(MOL)^2$	2.3			

[a] 25-35°, 0.1

Bibliography: 66MM   Other reference: 53CG

$$\begin{array}{c} OH \\ | \\ CH_2CO_2H \end{array}$$

$C_2H_4O_3$		Hydroxyacetic acid (glycolic acid)				HL
Metal ion	Equilibrium	Log K 25°, 0.1	Log K 25°, 1.0	Log K 25°, 0	ΔH 25°, 0	ΔS 25°, 0
H^+	HL/H.L	3.63 ±0.03	3.62 ±0.01	3.831 +0.001	-0.16 ±0.05	17.0
		3.57^b±0.01	3.74^d±0.03	3.91^e ±0.02	-0.56^d	15.3^d
Li^+	ML/M.L			-0.11		
Mg^{2+}	ML/M.L	0.92^x		1.33		
Ca^{2+}	ML/M.L	1.11^x		1.62 ±0.03		
Sr^{2+}	ML/M.L	0.80^x				
Ba^{2+}	ML/M.L	0.66^x		1.01 ±0.03		
Y^{3+}	ML/M.L	2.79	2.47^d-0.09		-0.1^d ±0.0	11^d
	ML_2/M.L^2	4.88	4.40^d-0.05		-0.2^d -0.2	19^d
	ML_3/M.L^3	5.8	5.7^d-0.1		-0.9^d -0.7	23^d
	ML_4/M.L^4		6.3^d		-0.9^d	26^d
	ML_5/M.L^5		6.3^k			
La^{3+}	ML/M.L	2.55	2.18^d-0.04		-0.6^d -0.2	8^d
	ML_2/M.L^2	4.24	3.75^d+0.2		-1.1^d -0.1	14^d
	ML_3/M.L^3	5.0	4.79^d		-1.7^d	16^d
	ML_4/M.L^4		5.1^d		-2.2^d	16^d
	ML_5/M.L^5		4.8^k			

[b] 25°, 0.5;   [d] 25°, 2.0;   [e] 25°, 3.0;   [k] 20°, 2.0;   [x] temperature not stated, 0.2

Hydroxyacetic acid (continued)

Metal ion	Equilibrium	Log K 25°, 1.0	Log K 25°, 1.0	Log K 25°, 0	ΔH 25°, 0	ΔS 25°, 0
Ce^{3+}	$ML/M.L$	2.69	2.34^d-0.07		-0.8^d	8^d
	$ML_2/M.L^2$	4.54	4.00^d+0.01		-1.6^d	13^d
	$ML_3/M.L^3$	5.3	5.12^d±0.00		-2.3^d	15^d
	$ML_4/M.L^4$		5.5^d		-3.0^d	16^d
	$ML_5/M.L^5$		5.3^k			
Pr^{3+}	$ML/M.L$	2.78	2.43^k			
	$ML_2/M.L^2$	4.68	4.19^k			
	$ML_3/M.L^3$	5.86	5.4^k			
	$ML_4/M.L^4$		5.9^k			
	$ML_5/M.L^5$		5.7^k			
Nd^{3+}	$ML/M.L$	2.89	2.50^d-0.10		-1.2^d +0.2	7^d
	$ML_2/M.L^2$	4.85	4.31^d-0.05		-2.2^d +0.8	12^d
	$ML_3/M.L^3$	6.11	5.6^d -0.3		-3.5^d	14^d
	$ML_4/M.L^4$		6.0^d		-4.0^d	14^d
	$ML_5/M.L^5$		5.7^k			
Sm^{3+}	$ML/M.L$	2.91	2.55^d-0.09		-1.0^d +0.1	8^d
	$ML_2/M.L^2$	5.01	4.50^d-0.10		-2.4^d +0.7	14^d
	$ML_3/M.L^3$	6.57	5.9^d±0.0		-3.7^d	13^d
	$ML_4/M.L^4$		6.4^d		-4.9^d	14^d
	$ML_5/M.L^5$		6.0^k			
Eu^{3+}	$ML/M.L$	2.93 2.57^i	2.52^d-0.07	3.49	-0.8^d	8^d
	$ML_2/M.L^2$	5.07 4.61^i	4.56^d-0.15		-1.7^d	15^d
	$ML_3/M.L^3$	6.52 5.91^i	5.8^d			
	$ML_4/M.L^4$	6.4^i				
Gd^{3+}	$ML/M.L$	2.79	2.54^j 2.47^d-0.01	2.59^l	-0.6^d ±0.0	9^d
	$ML_2/M.L^2$	4.85	4.48^j 4.41^d-0.08	4.60^l	-1.7^d -0.1	14^d
	$ML_3/M.L^3$	6.00	5.85^j 5.74^d+0.15	6.11^l	-3.5^d	15^d
	$ML_4/M.L^4$		6.4^d	6.8^l	-4.3^d	15^d
	$ML_5/M.L^5$		6.0^k	6.8^l		
	$ML_6/M.L^6$			6.6^l		
Tb^{3+}	$ML/M.L$	2.82	2.40^d		-0.1^d	11^d
	$ML_2/M.L^2$	4.91	4.45^d		-0.9^d	18^d
	$ML_3/M.L^3$	6.02				
Dy^{3+}	$ML/M.L$	2.92	2.52^d-0.04		-0.2^d +0.1	11^d
	$ML_2/M.L^2$	4.97	4.47^d-0.03		-0.6^d -0.9	18^d
	$ML_3/M.L^3$	6.55	5.9^d-0.3		-1.7^d	21^d
	$ML_4/M.L^4$		6.5^d		-2.6^d	21^d
	$ML_5/M.L^5$		6.3^k			

d 25°, 2.0;   i 20°, 0.5;   j 20°, 1.0;   k 20°, 2.0;   l 20°, 3.0

Hydroxyacetic acid (continued)

Metal ion	Equilibrium	Log K 25°, 0.1	Log K 25°, 1.0	Log K 25°, 0	ΔH 25°, 0	ΔS 25°, 0
Ho^{3+}	$ML/M{\cdot}L$	2.99	2.54[d] -0.05		-0.1[d]	11[d]
	$ML_2/M{\cdot}L^2$	5.04	4.47[d] ±0.06		-0.5[d]	19[d]
	$ML_3/M{\cdot}L^3$	6.57	5.9[d] ±0.0		(-2.9)[d]	(16)[d]
	$ML_4/M{\cdot}L^4$		6.4[k]			
	$ML_5/M{\cdot}L^5$		6.5[k]			
Er^{3+}	$ML/M{\cdot}L$	3.00	2.60[d] +0.04		-0.2[d] +0.1	11[d]
	$ML_2/M{\cdot}L^2$	5.19	4.57[d] -0.05		-0.6[d] +0.1	19[d]
	$ML_3/M{\cdot}L^3$	6.84	6.0[d]		-1.3[d]	23[d]
	$ML_4/M{\cdot}L^4$		6.5[d]		-1.4[d]	25[d]
	$ML_5/M{\cdot}L^5$		6.5[k]			
Tm^{3+}	$ML/M{\cdot}L$	3.06	2.62[d]		-0.3[d]	11[d]
	$ML_2/M{\cdot}L^2$	5.33	4.65[d]		-0.9[d]	18[d]
	$ML_3/M{\cdot}L^3$	7.00	6.1[d]		-1.7[d]	22[d]
Yb^{3+}	$ML/M{\cdot}L$	3.13	2.72[d] -0.10		-0.3[d] -0.1	11[d]
	$ML_2/M{\cdot}L^2$	5.37	4.8[d] +0.01		-0.8[d] -0.4	19[d]
	$ML_3/M{\cdot}L^3$	7.11	6.3[d] -0.1		-1.7[d] -1	23[d]
	$ML_4/M{\cdot}L^4$		6.8[d]		-0.6[d]	29[d]
	$ML_5/M{\cdot}L^5$		7.0[k]			
Lu^{3+}	$ML/M{\cdot}L$	3.15	2.67[d]		-0.6[d]	10[d]
	$ML_2/M{\cdot}L^2$	5.48	4.76[d]		-1.2[d]	18[d]
	$ML_3/M{\cdot}L^3$	7.28				
Am^{3+}	$ML/M{\cdot}L$	2.82[i]	2.59[d]		-1.3[d]	8[d]
	$ML_2/M{\cdot}L^2$	4.86[i]	4.37[d]			
	$ML_3/M{\cdot}L^3$	6.3[i]				
Cm^{3+}	$ML/M{\cdot}L$	2.85[i]	2.59[d]		-0.9[d]	9[d]
	$ML_2/M{\cdot}L^2$	4.75[i]	4.54[d]			
Bk^{3+}	$ML/M{\cdot}L$		2.64[d]		-1.2[d]	8[d]
	$ML_2/M{\cdot}L^2$		4.68[d]			
Cf^{3+}	$ML/M{\cdot}L$		2.72[s]			
	$ML_2/M{\cdot}L^2$		4.67[s]			
Th^{4+}	$ML/M{\cdot}L$		3.98[j]			
	$ML_2/M{\cdot}L^2$		7.36[j]			
	$ML_3/M{\cdot}L^3$		9.95[j]			
	$ML_4/M{\cdot}L^4$		11.95[j]			
NpO_2^{+}	$ML/M{\cdot}L$	1.51				
UO_2^{2+}	$ML/M{\cdot}L$		2.40[j] ±0.02			
	$ML_2/M{\cdot}L^2$		3.96[j] -0.01			
	$ML_3/M{\cdot}L^3$		5.19[j] ±0.01			
NpO_2^{2+}	$ML/M{\cdot}L$		2.37[j]			
	$ML_2/M{\cdot}L^2$		3.95[j]			
	$ML_3/M{\cdot}L^3$		5.00[j]			
PuO_2^{2+}	$ML/M{\cdot}L$	2.43	2.16[j]			
	$ML_2/M{\cdot}L^2$	3.79	3.45[j]			
	$ML_3/M{\cdot}L^3$		4.26[j]			
Mn^{2+}	$ML/M{\cdot}L$			1.58		

[d] 25°, 2.0;   [i] 20°, 0.5;   [j] 20°, 1.0;   [k] 20°, 2.0;   [s] 53°, 2.0

Hydroxyacetic acid (continued)

Metal ion	Equilibrium	Log K 25°, 1.0	Log K 25°, 1.0	Log K 25°, 0	ΔH 25°, 0	ΔS 25°, 0
Fe^{2+}	ML/M.L		1.33			
Co^{2+}	$ML/M.L$	1.68	1.48^d	1.97 ±0.01		
	$ML_2/M.L^2$		2.29^d	3.01		
	$ML_3/M.L^3$		2.52^d			
Ni^{2+}	$ML/M.L$		1.69^d	2.26		
	$ML_2/M.L^2$		2.70^d			
	$ML_3/M.L^3$		3.05^d			
Cu^{2+}	ML/M.L		2.31 ±0.05	2.90 ±0.02		
			2.40^d	2.50^e		
	$ML_2/M.L^2$		3.72 ±0.02	4.66		
				4.20^e		
	$ML_3/M.L^3$			4.27^e		
Ag^+	$ML/M.L$			0.30^e		
	$ML_2/M.L^2$			0.36^e		
Fe^{3+}	ML/M.L		2.90			
	$ML/MH_{-1}L.H$		1.31			
	$MH_{-1}L_2/MH_{-1}L.L$		2.41			
	$MH_{-1}L_3/MH_{-1}L_2.L$		1.5			
Zn^{2+}	$ML/M.L$	1.98 ±0.04	1.82^d ±0.10	2.38		
	$ML_2/M.L^2$		2.92^d ±0.02			
	$ML_3/M.L^3$		3.2^d ±0.3			
Cd^{2+}	$ML/M.L$		1.47^d ±0.05	1.87		
	$ML_2/M.L^2$		2.0^d ±0.2			
Pb^{2+}	$ML/M.L$		2.01	2.23^e		
	$ML_2/M.L^2$		2.94	3.24^e		
	$ML_3/M.L^3$			3.2^e		
	$MOHL.OH/M(OH)_3.L$		-0.7			
	$ML_2.OH/M(OH)_3.L^2$		-0.2			
In^{3+}	$ML/M.L$	3.15^u	2.93^k			
	$ML_2/M.L^2$		5.52^k			
	$ML_3/M.L^3$		7.30^k			
	$ML_4/M.L^4$		7.95^k			

^{d} 25°, 2.0; ^{e} 25°, 3.0; ^{k} 20°, 2.0; ^{u} 25°, 0.3

Bibliography:

H^+ 31LAb,36N,38CK,48F,53KEa,54DM,59S, 62CTC,62G,63MPa,64G,64SM,67CIH,68ES, 71BV,72CD,72DC,73FP

Li^+ 54DM

Mg^{2+}-Ba^{2+} 38CK,38Da,52CM,54DM,75DN

Y^{3+}-Lu^{3+} 59Sb,59Sc,60Sb,60SV,61CC,61S, 62BC,62G,63LM,64G,64PK,66CF,72CD

Am^{3+}-Cf^{3+} 62G,63Ga,72CD

Th^{4+} 73MBM

NpO_2^+ 69ES

UO_2^{2+} 53A,74MT

NpO_2^{2+} 72PTM

PuO_2^{2+} 68ES,70PC

Mn^{2+},Co^{2+},Ni^{2+},Zn^{2+} 38CK,54EM,57LW,58SL, 65SMa,70FB,70FM,70Lc

Fe^{2+},Fe^{3+} 73BL

Cu^{2+} 48F,51LW,54EM,63MPa,64SM,72BV

Ag^+ 64SM

Pb^{2+} 65BWa,71BV

Hydroxyacetic acid (continued)

In^{3+} 53S,60WT

Other references: 24FR,52B,55PP,57Lc,
    62CTG,62MM,62SBb,63LC,64L,65BK,65MM,
    66JG,68RSa,69BBa,69W,70KKM,70La,70RB,
    70SG,72NB,73KP,73SM,73SS,74HO,74K,
    74PS,75BP,75PS,75V

$$\begin{array}{c} \text{OH} \\ | \\ CH_3CHCO_2H \end{array}$$

$C_3H_6O_3$		D-2-Hydroxypropanoic acid (lactic acid)				HL
Metal ion	Equilibrium	Log K 25°, 0.1	Log K 25°, 1.0	Log K 25°, 0	ΔH 25°, 0	ΔS 25°, 0
H^+	HL/H.L	3.66 ±0.03	3.64 ±0.01	3.860 ±0.002	0.08 ±0.02	17.9
		3.61^b	3.81^d±0.01			
Li^+	ML/M.L			-0.20		
Mg^{2+}	ML/M.L $ML_2/M.L^2$	0.93^x	0.73 $(1.30)^z$	1.37		
Ca^{2+}	ML/M.L $ML_2/M.L^2$	1.07^x	0.90 $(1.24)^z$	1.45 ±0.03		
Sr^{2+}	ML/M.L $ML_2/M.L^2$	0.70^x	0.53 $(0.69)^z$	0.97 ±0.01		
Ba^{2+}	ML/M.L $ML_2/M.L^2$	0.55^x	0.34 $(0.42)^z$	0.71 ±0.07		
Sc^{3+}	ML/M.L			5.2		
Y^{3+}	ML/M.L $ML_2/M.L^2$ $ML_3/M.L^3$	3.02^h $(5.33)^{h,z}$ $(6.95)^{h,z}$	2.53^d $(4.70)^{d,z}$ $(6.12)^{d,z}$			
La^{3+}	ML/M.L $ML_2/M.L^2$ $ML_3/M.L^3$	2.60^h $(4.34)^{h,z}$ $(5.74)^{h,z}$	2.27^d $(3.95)^{d,z}$ $(5.1)^{d,z}$	3.3	-1.6^d $(-3)^d$ $(-5)^d$	5^d $(9)^d$ $(6)^d$
Ce^{3+}	ML/M.L $ML_2/M.L^2$ $ML_3/M.L^3$	2.76^h $(4.72)^{h,z}$ $(5.95)^{h,z}$	2.33^d $(4.10)^{d,z}$ $(5.21)^{d,z}$		-1.7^d $(-1)^d$ $(-6)^d$	5^d $(15)^d$ $(2)^d$
Pr^{3+}	ML/M.L $ML_2/M.L^2$ $ML_3/M.L^3$	2.85^h $(4.90)^{h,z}$ $(6.10)^{h,z}$				
Nd^{3+}	ML/M.L $ML_2/M.L^2$ $ML_3/M.L^3$	2.87^h $(4.97)^{h,z}$ $(6.37)^{h,z}$	2.47^d $(4.37)^{d,z}$ $(5.60)^{d,z}$		-3.2^d $(-1)^d$	1^d $(15)^d$
Sm^{3+}	ML/M.L $ML_2/M.L^2$ $ML_3/M.L^3$	2.88^h $(5.09)^{h,z}$ $(6.35)^{h,z}$	2.56^d $(4.58)^{d,z}$ $(5.90)^{d,z}$		-2.3^d $(-2)^d$ $(-9)^d$	4^d $(15)^d$ $(-3)^d$

^{b} 25°, 0.5; ^{d} 25°, 2.0; ^{h} 20°, 0.1; ^{x} temperature not stated, 0.2; ^{z} optical isomerism not stated

2-Hydroxypropanoic acid (continued)

Metal ion	Equilibrium	Log K 25°, 0.1	Log K 25°, 1.0	Log K 25°, 0	ΔH 25°, 0	ΔS 25°, 0
Eu^{3+}	$ML/M.L$	2.95[h]	2.53[d] -0.05		-1.9[d] +0.9	5[d]
	$ML_2/M.L^2$	(5.18)[h,z]	(4.60)[d,z] -0.04		(-2)[d] +0.1	(14)[d]
	$ML_3/M.L^3$	(6.43)[h,z]	(5.88)[d,z] -0.1		(-5)[d] ±0.0	(10)[d]
Gd^{3+}	$ML/M.L$	2.89[h]	2.53[d]		-2.0[d]	5[d]
	$ML_2/M.L^2$	(5.04)[h,z]	(4.63)[d,z]		(-2)[d]	(13)[d]
	$ML_3/M.L^3$	(6.24)[h,z]	(5.91)[d,z]		(-5)[d]	(12)[d]
Tb^{3+}	$ML/M.L$	2.90[h]	2.61[d]		-1.6[d]	7[d]
	$ML_2/M.L^2$	(5.20)[h,z]	(4.73)[d,z]			
	$ML_3/M.L^3$	(6.35)[h,z]	(6.01)[d,z]			
Dy^{3+}	$ML/M.L$	3.01[h]	2.71[d]		-2.2[d]	5[d]
	$ML_2/M.L^2$	(5.35)[h,z]	(4.76)[d,z]			
	$ML_3/M.L^3$	(6.67)[h,z]	(6.75)[d,z]			
Ho^{3+}	$ML/M.L$	3.02[h]	2.71[d]			
	$ML_2/M.L^2$	(5.42)[h,z]	(4.97)[d,z]			
	$ML_3/M.L^3$	(6.83)[h,z]	(6.55)[d,z]			
Er^{3+}	$ML/M.L$	3.16[h]	2.77[d]		-1.9[d]	7[d]
	$ML_2/M.L^2$	(5.62)[h,z]	(5.11)[d,z]		(-1)[d]	(20)[d]
	$ML_3/M.L^3$	(7.20)[h,z]	(6.70)[d,z]		(-5)[d]	(15)[d]
Tm^{3+}	$ML/M.L$	3.19[h]				
	$ML_2/M.L^2$	(5.71)[h,z]				
	$ML_3/M.L^3$	(7.43)[h,z]				
Yb^{3+}	$ML/M.L$	3.23[h]	2.85[d]		-2.2[d]	7[d]
	$ML_2/M.L^2$	(5.82)[h,z]	(5.27)[d,z]		(-3)[d]	(12)[d]
	$ML_3/M.L^3$	(7.58)[h,z]	(6.96)[d,z]		(-2)[d]	(16)[d]
Lu^{3+}	$ML/M.L$	3.27[h]		3.85		
	$ML_2/M.L^2$	(5.88)[h,z]				
	$ML_3/M.L^3$	(7.78)[h,z]				
Am^{3+}	$ML/M.L$		2.52[d]			
	$ML_2/M.L^2$		(4.76)[d,z]			
	$ML_3/M.L^3$		(6.0)			
Th^{4+}	$ML/M.L$		4.21[j]	5.5		
	$ML_2/M.L^2$		(7.78)[j,z]			
	$ML_3/M.L^3$		(10.54)[j,z]			
	$ML_4/M.L^4$		(12.90)[j,z]			
NpO_2^+	$ML/M.L$	1.75				
UO_2^{2+}	$ML/M.L$		2.76			
	$ML_2/M.L^2$		(4.43)[z]			
	$ML_3/M.L^3$		(5.77)[z]			
Mn^{2+}	$ML/M.L$	1.19[r]	0.92	1.43		
	$ML_2/M.L^2$		(1.46)[z]			
	$ML_3/M.L^3$		(1.6)[z]			
Co^{2+}	$ML/M.L$		1.38 ±0.03 / 1.39[k]	1.90		
	$ML_2/M.L^2$		2.37 ±0.07 / 2.36[k]			
	$ML_3/M.L^3$		(2.7)[z] ±0.2 / 2.7[k]			

---

d 25°, 2.0;   h 20°, 0.1;   j 20°, 1.0;   k 20°, 2.0;   r 25°, 0.2;   z optical isomerism not stated.

2-Hydroxypropanoic acid (continued)

Metal ion	Equilibrium	Log K 25°, 0.1	Log K 25°, 1.0	Log K 25°, 0	ΔH 25°, 0	ΔS 25°, 0
Ni^{2+}	$ML/M.L$		1.64 ± 0.05	2.22		
	$ML_2/M.L^2$		$(2.76)^z \pm 0.09$			
	$ML_3/M.L^3$		$(3.1)^z \pm 0.1$			
Cu^{2+}	$ML/M.L$	2.55^p	2.45 ± 0.05	3.02		
			$2.43^d \pm 0.01$			
	$ML_2/M.L^2$		$(4.08)^z \pm 0.1$	$(4.84)^z$		
			$(4.08)^{d,z} \pm 0.04$			
	$ML_3/M.L^3$		$(4.7)^z \pm 0.4$			
			$(4.5)^{d,z}$			
VO^{2+}	$ML/M.L$		2.68^j			
	$ML_2/M.L^2$		4.83^j			
Zn^{2+}	$ML/M.L$	1.86^x	1.7 ± 0.1	2.22 ± 0.02		
	$ML_2/M.L^2$		$(2.8)^z \pm 0.1$	$(3.75)^z$		
	$ML_3/M.L^3$		$(3.4)^z \pm 0.3$			
Cd^{2+}	$ML/M.L$		1.30 ± 0.09	1.70		
			$1.40^d \pm 0.08$			
	$ML_2/M.L^2$		$(2.1)^z \pm 0.2$			
			$(2.0)^{d,z}$			
	$ML_3/M.L^3$		$(2.5)^z \pm 0.3$			
			$(2.6)^{d,z} \pm 0.1$			
Pb^{2+}	$ML/M.L$		$1.99 \; -0.01$	2.78		
			2.15^d	2.26^e		
	$ML_2/M.L^2$		$(2.88)^z \pm 0.10$			
			$(3.15)^{d,z}$	$(3.30)^{e,z}$		
	$ML_3/M.L^3$		$(4.3)^{d,z}$	$(3.3)^{e,z}$		
	$MOHL.OH/M(OH)_3.L$		-0.7			

---

[d] 25°, 2.0; [e] 25°, 3.0; [j] 20°, 1.0; [p] 30°, 0.1; [x] temperature not stated, 0.2;
[z] optical isomerism not stated.

Bibliography:

H^+ 31LAb,36NS,37MT,38CK,65JL,66SPa,67TG, UO_2^{2+} 67TG,74MT
          71BV,72DC,73FP,73LSa,73MBM,74FP,74MT

Li$^+$ 54DMa Mn^{2+}-Cu^{2+},Zn^{2+},Cd^{2+},Pb^{2+} 38CK,54DMa,54EM,
                                                      57LW,64CC,64DC,65BWa,65DSa,65SF,67TG,
Mg^{2+}-Ba^{2+} 38CK,38Da,52CM,54DMa,65VT   68FP,71BVa,72LN,72SSF,73WK

Sc^{3+} 73LSa VO^{2+} 65JL

Y^{3+}-Lu^{3+} 60SV,61CC,64DV,64PKK,66CF,66GG Other references: 46J,52B,52SL,53BB,54MP,
          68WZ,71ALN,73LSa   55MAa,57V,58KY,58PM,60RE,62CM,62GLa,
                                                      64RME,65BK,65BW,65DS,67ES,67MN,68RSa,
Am^{3+} 71AL   70AB,70SG,71Ha,71WC,71SP,72ADa,72PSS,
                                                      73SM,73SS,74FP,75BP,75PS,75W
Th^{4+} 73LSa,73MBM

NpO_2^+ 69ES

$$\begin{array}{c} \text{OH} \\ | \\ \text{CH}_3\text{CH}_2\text{CHCO}_2\text{H} \end{array}$$

$C_4H_8O_3$ DL-2-Hydroxybutanoic acid   HL

Metal ion	Equilibrium	Log K 25°, 0.1	Log K 25°, 1.0	Log K 25°, 0
H^+	HL/H.L	3.68 ±0.01	3.80^d	
Y^{3+}	ML/M.L	2.85		
	$ML_2/M.L^2$	$(5.16)^z$		
	$ML_3/M.L^3$	$(6.87)^z$		
La^{3+}	ML/M.L	2.31		
	$ML_2/M.L^2$	$(4.06)^z$		
	$ML_3/M.L^3$	$(5.06)^z$		
Ce^{3+}	ML/M.L	2.44		
	$ML_2/M.L^2$	$(4.27)^z$		
	$ML_3/M.L^3$	$(5.38)^z$		
Pr^{3+}	ML/M.L	2.58		
	$ML_2/M.L^2$	$(4.46)^z$		
	$ML_3/M.L^3$	$(5.80)^z$		
Nd^{3+}	ML/M.L	2.63		
	$ML_2/M.L^2$	$(4.54)^z$		
	$ML_3/M.L^3$	$(6.02)^z$		
Sm^{3+}	ML/M.L	2.74		
	$ML_2/M.L^2$	$(4.82)^z$		
	$ML_3/M.L^3$	$(6.38)^z$		
Eu^{3+}	ML/M.L	2.76		
	$ML_2/M.L^2$	$(4.85)^z$		
	$ML_3/M.L^3$	$(6.42)^z$		
Gd^{3+}	ML/M.L	2.77		
	$ML_2/M.L^2$	$(4.91)^z$		
	$ML_3/M.L^3$	$(6.57)^z$		
Tb^{3+}	ML/M.L	2.85		
	$ML_2/M.L^2$	$(5.08)^z$		
	$ML_3/M.L^3$	$(6.82)^z$		
Dy^{3+}	ML/M.L	2.90		
	$ML_2/M.L^2$	$(5.19)^z$		
	$ML_3/M.L^3$	$(6.98)^z$		
Ho^{3+}	ML/M.L	2.93		
	$ML_2/M.L^2$	$(5.31)^z$		
	$ML_3/M.L^3$	$(7.07)^z$		
Er^{3+}	ML/M.L	2.99		
	$ML_2/M.L^2$	$(5.41)^z$		
	$ML_3/M.L^3$	$(7.30)^z$		
Tm^{3+}	ML/M.L	3.03		
	$ML_2/M.L^2$	$(5.49)^z$		
	$ML_3/M.L^3$	$(7.44)^z$		

---

[z] optical isomerism not stated.

2-Hydroxybutanoic acid (continued)

Metal ion	Equilibrium	Log K 25°, 0.1	Log K 25°, 1.0	Log K 25°, 0
Yb^{3+}	$ML/M.L$	3.10		
	$ML_2/M.L^2$	$(5.47)^z$		
	$ML_3/M.L^3$	$(7.47)^z$		
Lu^{3+}	$ML/M.L$	3.13		
	$ML_2/M.L^2$	$(5.68)^z$		
	$ML_3/M.L^3$	$(7.76)^z$		
NpO_2^+	$ML/M.L$	1.62		
Co^{2+}	$ML/M.L$		1.43^d	
	$ML_2/M.L^2$		$(1.83)^{d,z}$	
Ni^{2+}	$ML/M.L$		1.72^d	
	$ML_2/M.L^2$		$(2.91)^{d,z}$	
Cu^{2+}	$ML/M.L$		$2.66^d \pm 0.03$	
	$ML_2/M.L^2$		$(4.5)^{d,z} \pm 0.2$	
Zn^{2+}	$ML/M.L$		1.72^d	
	$ML_2/M.L^2$		$(3.03)^{d,z}$	
	$ML_3/M.L^3$		$(3.85)^{d,z}$	
	$ML_4/M.L^4$		$(4.24)^{d,z}$	
Cd^{2+}	$ML/M.L$		1.23^d	
	$ML_2/M.L^2$		$(2.15)^{d,z}$	
	$ML_3/M.L^3$		$(2.26)^{d,z}$	
	$ML_4/M.L^4$		$(2.45)^{d,z}$	
Pb^{2+}	$ML/M.L$		2.10^d -0.01	2.04^e
	$ML_2/M.L^2$		$(2.75)^{d,z} \pm 0.0$	$(2.88)^{e,z}$

^{d} 25°, 2.0; ^{e} 25°, 3.0; ^{z} optical isomerism not stated.

Bibliography:

H^+	65FP,69ES,73FPG	Zn^{2+}	73FP
Y^{3+}-Lu^{3+}	65FP	Cd^{2+}	73NP
NpO_2^+	69ES	Pb^{2+}	66WB,73NP,73PG
Co^{2+}-Cu^{2+}	73GP,74GM,75FPG,75GMP	Other references:	62CM,62SBb

$$CH_3CH_2CH_2\overset{\overset{\displaystyle OH}{|}}{C}HCO_2H$$

$C_5H_{10}O_3$		DL-2-Hydroxypentanoic acid		HL

Metal ion	Equilibrium	Log K 25°, 0.1	Log K 25°, 1.0
H^+	$HL/H.L$	(3.75)	(3.59)
Y^{3+}	$ML/M.L$		2.46
La^{3+}	$ML/M.L$		1.98
Pr^{3+}	$ML/M.L$		2.24

2-Hydroxypentanoic acid (continued)

Metal ion	Equilibrium	Log K 25°, 0.1	Log K 25°, 1.0
Nd^{3+}	ML/M.L		2.31
Sm^{3+}	ML/M.L		2.40
Eu^{3+}	ML/M.L		2.43
Gd^{3+}	ML/M.L		2.45
Tb^{3+}	ML/M.L		2.53
Dy^{3+}	ML/M.L		2.63
Ho^{3+}	ML/M.L		2.64
Er^{3+}	ML/M.L		2.68
Yb^{3+}	ML/M.L		2.76
Lu^{3+}	ML/M.L		2.76
NpO_2^{+}	ML/M.L	1.59	

Bibliography:

H^{+} 68GG,69ES Y^{3+}-Lu^{3+} 68GG NpO_2^{+} 69ES

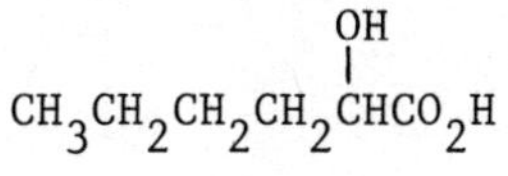

$$\overset{\displaystyle OH}{\underset{\displaystyle |}{CH_3CH_2CH_2CH_2CHCO_2H}}$$

$C_6H_{12}O_3$		DL-2-Hydroxyhexanoic acid	HL

Metal ion	Equilibrium	Log K 25°, 0.1
H^{+}	HL/H.L	3.68
NpO_2^{+}	ML/M.L	1.63

Bibliography: 69ES

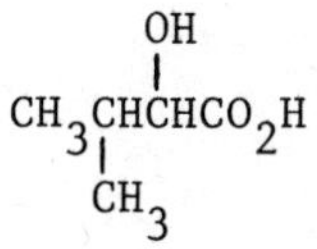

$$\overset{\displaystyle OH}{\underset{\displaystyle \underset{\displaystyle CH_3}{|}}{\overset{\displaystyle |}{CH_3CHCHCO_2H}}}$$

$C_5H_{10}O_3$		DL-2-Hydroxy-3-methylbutanoic acid	HL

Metal ion	Equilibrium	Log K 25°, 0.1	Log K 20°, 0.2
H^{+}	HL/H.L	3.69	

2-Hydroxy-3-methylbutanoic acid (continued)

Metal ion	Equilibrium	Log K 25°, 0.1	Log K 20°, 0.2
Y^{3+}	$ML/M.L$	2.72	2.60
	$ML_2/M.L^2$	$(4.86)^z$	$(4.95)^z$
	$ML_3/M.L^3$	$(6.47)^z$	
La^{3+}	$ML/M.L$	2.16	
	$ML_2/M.L^2$	$(3.73)^z$	
	$ML_3/M.L^3$	$(4.93)^z$	
Ce^{3+}	$ML/M.L$	2.28	2.23
	$ML_2/M.L^2$	$(3.99)^z$	$(3.50)^z$
	$ML_3/M.L^3$	$(5.10)^z$	$(4.8)^z$
Pr^{3+}	$ML/M.L$	2.38	
	$ML_2/M.L^2$	$(4.14)^z$	
	$ML_3/M.L^3$	$(5.39)^z$	
Nd^{3+}	$ML/M.L$	2.43	
	$ML_2/M.L^2$	$(4.22)^z$	
	$ML_3/M.L^3$	$(5.45)^z$	
Sm^{3+}	$ML/M.L$	2.57	
	$ML_2/M.L^2$	$(4.51)^z$	
	$ML_3/M.L^3$	$(5.91)^z$	
Eu^{3+}	$ML/M.L$	2.61	
	$ML_2/M.L^2$	$(4.62)^z$	
	$ML_3/M.L^3$	$(6.22)^z$	
Gd^{3+}	$ML/M.L$	2.62	
	$ML_2/M.L^2$	$(4.66)^z$	
	$ML_3/M.L^3$	$(6.14)^z$	
Tb^{3+}	$ML/M.L$	2.70	
	$ML_2/M.L^2$	$(4.84)^z$	
	$ML_3/M.L^3$	$(6.44)^z$	
Dy^{3+}	$ML/M.L$	2.77	
	$ML_2/M.L^2$	$(4.97)^z$	
	$ML_3/M.L^3$	$(6.67)^z$	
Ho^{3+}	$ML/M.L$	2.80	
	$ML_2/M.L^2$	$(5.04)^z$	
	$ML_3/M.L^3$	$(6.74)^z$	
Er^{3+}	$ML/M.L$	2.85	
	$ML_2/M.L^2$	$(5.14)^z$	
	$ML_3/M.L^3$	$(6.91)^z$	
Tm^{3+}	$ML/M.L$	2.90	
	$ML_2/M.L^2$	$(5.20)^z$	
	$ML_3/M.L^3$	$(7.05)^z$	
Yb^{3+}	$ML/M.L$	2.95	
	$ML_2/M.L^2$	$(5.35)^z$	
	$ML_3/M.L^3$	$(7.24)^z$	
Lu^{3+}	$ML/M.L$	2.99	
	$ML_2/M.L^2$	$(5.41)^z$	
	$ML_3/M.L^3$	$(7.31)^z$	

z optical isomerism not stated.

Bibliography:

H^+ 65PP $Y^{3+}-Lu^{3+}$ 60SV,65PP

$$\begin{array}{c} H_3C \;\; OH \\ \backslash \;\;\; | \\ H_3C-CCHCO_2H \\ / \\ H_3C \end{array}$$

$C_6H_{12}O_3$ DL-2-Hydroxy-3,3-dimethylbutanoic acid   HL

Metal ion	Equilibrium	Log K 25°, 0.1
H^+	HL/H.L	3.87
Y^{3+}	ML/M.L	2.63
	$ML_2/M.L^2$	$(4.68)^z$
	$ML_3/M.L^3$	$(6.13)^z$
Ce^{3+}	ML/M.L	2.14
	$ML_2/M.L^2$	$(3.72)^z$
	$ML_3/M.L^3$	$(4.90)^z$
Pr^{3+}	ML/M.L	2.23
	$ML_2/M.L^2$	$(3.85)^z$
	$ML_3/M.L^3$	$(4.7)^z$
Nd^{3+}	ML/M.L	2.28
	$ML_2/M.L^2$	$(3.90)^z$
	$ML_3/M.L^3$	$(5.28)^z$
Sm^{3+}	ML/M.L	2.45
	$ML_2/M.L^2$	$(4.34)^z$
	$ML_3/M.L^3$	$(5.52)^z$
Eu^{3+}	ML/M.L	2.52
	$ML_2/M.L^2$	$(4.41)^z$
	$ML_3/M.L^3$	$(5.90)^z$
Gd^{3+}	ML/M.L	2.54
	$ML_2/M.L^2$	$(4.48)^z$
	$ML_3/M.L^3$	$(5.90)^z$
Tb^{3+}	ML/M.L	2.63
	$ML_2/M.L^2$	$(4.70)^z$
	$ML_3/M.L^3$	$(6.12)^z$
Dy^{3+}	ML/M.L	2.70
	$ML_2/M.L^2$	$(4.75)^z$
	$ML_3/M.L^3$	$(6.36)^z$
Ho^{3+}	ML/M.L	2.73
	$ML_2/M.L^2$	$(4.84)^z$
	$ML_3/M.L^3$	$(6.36)^z$
Er^{3+}	ML/M.L	2.75
	$ML_2/M.L^2$	$(4.90)^z$
	$ML_3/M.L^3$	$(6.51)^z$
Tm^{3+}	ML/M.L	2.80
	$ML_2/M.L^2$	$(5.05)^z$
	$ML_3/M.L^3$	$(6.63)^z$
Yb^{3+}	ML/M.L	2.85
	$ML_2/M.L^2$	$(5.13)^z$
	$ML_3/M.L^3$	$(6.73)^z$

[z] optical isomerism not stated.

2-Hydroxy-3,3-dimethylbutanoic acid (continued)

Metal ion	Equilibrium	Log K 25°, 0.1
Lu^{3+}	$ML/M.L$	2.86
	$ML_2/M.L^2$	$(5.16)^z$
	$ML_3/M.L^3$	$(6.77)^z$

---

^{z} optical isomerism not stated.

Bibliography: 65PP

$$\begin{array}{c} OH \\ | \\ CH_3CCO_2H \\ | \\ CH_3 \end{array}$$

$C_4H_8O_3$ 2-Hydroxy-2-methylpropanoic acid   HL

Metal ion	Equilibrium	Log K 25°, 0.1	Log K 25°, 1.0	Log K 25°, 0	ΔH 25°, 0.1	ΔS 25°, 0.1
H^+	$HL/H.L$	3.79 +0.01 3.75^b	3.77 ±0.01	4.03	-0.3	16
Mg^{2+}	$ML/M.L$		0.81			
	$ML_2/M.L^2$		1.47			
Ca^{2+}	$ML/M.L$		0.92			
	$ML_2/M.L^2$		1.42			
Sr^{2+}	$ML/M.L$		0.55			
	$ML_2/M.L^2$		0.73			
Ba^{2+}	$ML/M.L$		0.36			
	$ML_2/M.L^2$		0.51			
Y^{3+}	$ML/M.L$	3.20^h	2.86^d		-1.2^d	9^d
	$ML_2/M.L^2$	5.79^h	5.44^d		$(-4.1)^d$	$(11)^d$
	$ML_3/M.L^3$	7.51^h	7.20^d		-5	18^d
La^{3+}	$ML/M.L$	2.62^h	2.28^d		-2.0^d	4^d
	$ML_2/M.L^2$	4.42^h	3.97^d		-2.4^d	10^d
	$ML_3/M.L^3$	5.53^h	5.17^d		(-6)	(4)^d
Ce^{3+}	$ML/M.L$	2.80^h	2.43^d		-1.5^d	6^d
	$ML_2/M.L^2$	4.74^h	4.32^d		-1.9^d	13^d
	$ML_3/M.L^3$	5.95^h	5.32^d		(-8)	(-2)^d
Pr^{3+}	$ML/M.L$	2.84^h				
	$ML_2/M.L^2$	4.91^h				
	$ML_3/M.L^3$	6.21^h				
Nd^{3+}	$ML/M.L$	2.88^h	2.60^d ±0.02		-1.6^d	7^d
	$ML_2/M.L^2$	5.02^h	4.62^d ±0.05		-3.3^d	10^d
	$ML_3/M.L^3$	6.30^h	5.95^d ±0.1		-4	13^d
Sm^{3+}	$ML/M.L$	2.99^h	2.75^d			
	$ML_2/M.L^2$	5.39^h	4.90^d			
	$ML_3/M.L^3$	6.77^h	6.48^d			

---

^{b} 25°, 0.5; ^{d} 25°, 2.0; ^{h} 20°, 0.1

2-Hydroxy-2-methylpropanoic acid (continued)

Metal ion	Equilibrium	Log K 25°, 0.1	Log K 25°, 1.0	Log K 25°, 0	ΔH 25°, 0.1	ΔS 25°, 0.1
Eu^{3+}	$ML/M.L$	3.09^h	2.70^d		-1.5^d	7^d
	$ML_2/M.L^2$	5.54^h	4.94^d		-3.3^d	11^d
	$ML_3/M.L^3$	7.32^h	6.52^d		-5^d	13^d
Gd^{3+}	$ML/M.L$	3.08^h	2.82^d			
	$ML_2/M.L^2$	5.51^h	5.15^d			
	$ML_3/M.L^3$	7.19^h	6.77^d			
Tb^{3+}	$ML/M.L$	3.11^h	2.82^d		-1.5^d	8^d
	$ML_2/M.L^2$	5.63^h	5.25^d		$(-5)^d$	$(8)^d$
	$ML_3/M.L^3$	7.43^h	7.03^d		-4^d	20^d
Dy^{3+}	$ML/M.L$	3.27^h				
	$ML_2/M.L^2$	5.90^h				
	$ML_3/M.L^3$	7.87^h				
Ho^{3+}	$ML/M.L$	3.31^h	3.06^d			
	$ML_2/M.L^2$	5.98^h	5.65^d			
	$ML_3/M.L^3$	7.96^h	7.68^d			
Er^{3+}	$ML/M.L$	3.35^h	3.07^d		-1.0^d	11^d
	$ML_2/M.L^2$	6.04^h	5.73^d		$(-5)^d$	$(8)^d$
	$ML_3/M.L^3$	8.13^h	7.80^d		-4^d	24^d
Tm^{3+}	$ML/M.L$	3.52^h	3.05^d			
	$ML_2/M.L^2$	6.22^h	5.15^d			
	$ML_3/M.L^3$	8.39^h				
Yb^{3+}	$ML/M.L$	3.64^h	3.15^d		-0.7^d -1.1^d	14^d 10^d
	$ML_2/M.L^2$	6.42^h	6.00^d		-2.3^d $(-6)^d$	22^d $(6)^d$
	$ML_3/M.L^3$	8.69^h	8.12^d		-5^d	19^d
Lu^{3+}	$ML/M.L$	3.67^h				
	$ML_2/M.L^2$	6.47^h				
	$ML_3/M.L^3$	8.82^h				
Th^{4+}	$ML/M.L$		4.43^j			
	$ML_2/M.L^2$		8.15^j			
	$ML_3/M.L^3$		11.06^j			
	$ML_4/M.L^4$		13.60^j			
UO_2^{2+}	$ML/M.L$		3.18^j -0.1			
	$ML_2/M.L^2$		5.13^j -0.2			
	$ML_3/M.L^3$		6.67^j -0.3			
NpO_2^{2+}	$ML/M.L$		3.15^j			
	$ML_2/M.L^2$		5.25^j			
PuO_2^{2+}	$ML/M.L$		3.04^j			
	$ML_2/M.L^2$		5.00^j			
	$ML_3/M.L^3$		6.00^j			
Mn^{2+}	$ML/M.L$		0.90			
	$ML_2/M.L^2$		1.48			
	$ML_3/M.L^3$		1.7			
Co^{2+}	$ML/M.L$		1.45			
	$ML_2/M.L^2$		2.43			
	$ML_3/M.L^3$		2.7			

^{d} 25°, 2.0; ^{h} 20°, 0.1; ^{j} 20°, 1.0

2-Hydroxy-2-methylpropanoic acid (continued)

Metal ion	Equilibrium	Log K 25°, 0.1	Log K 25°, 1.0	Log K 25°, 0	ΔH 25°, 0.1	ΔS 25°, 0.1
Ni^{2+}	$ML/M.L$		1.67			
	$ML_2/M.L^2$		2.80			
	$ML_3/M.L^3$		3.2			
Cu^{2+}	$ML/M.L$	2.82	2.74			
	$ML_2/M.L^2$	4.62	4.34		-0.2	13
	$ML_3/M.L^3$		5.7		-1.0	19
Zn^{2+}	$ML/M.L$		1.70			
	$ML_2/M.L^2$		2.99			
	$ML_3/M.L^3$		3.4			
Cd^{2+}	$ML/M.L$		1.24			
	$ML_2/M.L^2$		2.16			
	$ML_3/M.L^3$		2.5			
Pb^{2+}	$ML/M.L$		2.03	2.23[e]		
	$ML_2/M.L^2$		3.20	3.23[e]		
	$ML_3/M.L^3$		3.4	3.29[e]		

[e] 25°, 3.0

Bibliography:

H^+ 33La,64SP,67TG,71BC,73MBM,74MT,74PK

$Mg^{2+}-Ba^{2+}$ 65VT

$Y^{3+}-Lu^{3+}$ 60SV,61CC,64DV,64EV,64PKK,
        64SP,66CF,66LN,67LN,68WZ,71BC

Th^{4+} 73MBM

UO_2^{2+} 67TG,74MT

NpO_2^{2+},PuO_2^{2+} 74MT

$Mn^{2+}-Pb^{2+}$ 66WB,67TG,71BC,74PK

Other references: 560C,62GL,62GLa,63LGM,70AL,
        71A,71GCa

$$CH_3CH_2\overset{\overset{\textstyle OH}{|}}{\underset{\underset{\textstyle CH_3}{|}}{C}}CO_2H$$

$C_5H_{10}O_3$		DL-2-Hydroxy-2-methylbutanoic acid			HL

Metal ion	Equilibrium	Log K 25°, 0.1	Log K 18°, 1.0	Log K 18°, 0
H^+	$HL/H.L$	3.73 3.80[h] 3.74[i]	3.76	3.99
La^{3+}	$ML/M.L$	2.34		
	$ML_2/M.L^2$	(3.92)[z]		
	$ML_3/M.L^3$	(5.2)[z]		
Ce^{3+}	$ML/M.L$	2.51		
	$ML_2/M.L^2$	(4.23)[z]		
	$ML_3/M.L^3$	(5.43)[z]		

[h] 18°, 0.1;   [i] 18°, 0.5;   [z] optical isomerism not stated.

2-Hydroxy-2-methylbutanoic acid (continued)

Metal ion	Equilibrium	Log K 25°, 0.1	Log K 18°, 1.0	Log K 18°, 0
Pr^{3+}	$ML/M.L$	2.54		
	$ML_2/M.L^2$	$(4.31)^z$		
	$ML_3/M.L^3$	$(5.42)^z$		
Nd^{3+}	$ML/M.L$	2.65		
	$ML_2/M.L^2$	$(4.49)^z$		
	$ML_3/M.L^3$	$(5.81)^z$		
Sm^{3+}	$ML/M.L$	2.80		
	$ML_2/M.L^2$	$(4.95)^z$		
	$ML_3/M.L^3$	$(6.46)^z$		
Eu^{3+}	$ML/M.L$	2.90		
	$ML_2/M.L^2$	$(5.20)^z$		
	$ML_3/M.L^3$	$(6.80)^z$		
Gd^{3+}	$ML/M.L$	2.94		
	$ML_2/M.L^2$	$(5.28)^z$		
	$ML_3/M.L^3$	$(6.86)^z$		
Tb^{3+}	$ML/M.L$	3.10		
	$ML_2/M.L^2$	$(5.58)^z$		
	$ML_3/M.L^3$	$(7.31)^z$		
Dy^{3+}	$ML/M.L$	3.16		
	$ML_2/M.L^2$	$(5.70)^z$		
	$ML_3/M.L^3$	$(7.42)^z$		
Ho^{3+}	$ML/M.L$	3.24		
	$ML_2/M.L^2$	$(5.87)^z$		
	$ML_3/M.L^3$	$(7.72)^z$		
Er^{3+}	$ML/M.L$	3.32		
	$ML_2/M.L^2$	$(6.05)^z$		
	$ML_3/M.L^3$	$(8.03)^z$		
Tm^{3+}	$ML/M.L$	3.37		
	$ML_2/M.L^2$	$(6.13)^z$		
	$ML_3/M.L^3$	$(8.05)^z$		
Yb^{3+}	$ML/M.L$	3.43		
	$ML_2/M.L^2$	$(6.26)^z$		
	$ML_3/M.L^3$	$(8.29)^z$		
Lu^{3+}	$ML/M.L$	3.45		
	$ML_2/M.L^2$	$(6.30)^z$		
	$ML_3/M.L^3$	$(8.35)^z$		

---

^{z} optical isomerism not stated.

Bibliography:

H^+ 31LAb,65FP,69PC Y^{3+}-Lu^{3+} 65FP,69PC

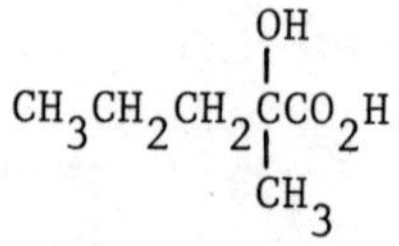

$$\begin{array}{c}\text{OH}\\|\\\text{CH}_3\text{CH}_2\text{CH}_2\text{CCO}_2\text{H}\\|\\\text{CH}_3\end{array}$$

$C_6H_{12}O_3$		DL-2-Hydroxy-2-methylpentanoic acid	HL

Metal ion	Equilibrium	Log K 25°, 1.0
La^{3+}	ML/M.L	2.20
	$ML_2/M.L_2$	$(3.45)^z$
	$ML_3/M.L_3$	$(4.22)^z$
Ce^{3+}	ML/M.L	2.21
	$ML_2/M.L_2$	$(3.87)^z$
	$ML_3/M.L_3$	$(4.90)^z$
	$ML_4/M.L_4$	$(5.71)^z$
Nd^{3+}	ML/M.L	2.38
	$ML_2/M.L_2$	$(4.23)^z$
	$ML_3/M.L_3$	$(5.40)^z$
	$ML_4/M.L_4$	$(6.46)^z$
Sm^{3+}	ML/M.L	2.59
	$ML_2/M.L_2$	$(4.71)^z$
	$ML_3/M.L_3$	$(6.21)^z$
Dy^{3+}	ML/M.L	3.00
	$ML_2/M.L_2$	$(5.47)^z$
	$ML_3/M.L_3$	$(7.33)^z$
	$ML_4/M.L_4$	$(8.45)^z$
Yb^{3+}	ML/M.L	3.29
	$ML_2/M.L_2$	$(5.84)^z$
	$ML_3/M.L_3$	$(8.05)^z$
	$ML_4/M.L_4$	$(9.17)^z$

^{z} optical isomerism not stated.

Bibliography: 64EV

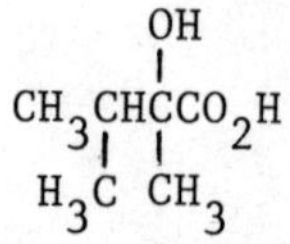

$$\begin{array}{c}\text{OH}\\|\\\text{CH}_3\text{CHCCO}_2\text{H}\\|\ \ |\\\text{H}_3\text{C}\ \ \text{CH}_3\end{array}$$

$C_6H_{12}O_3$		DL-2-Hydroxy-2,3-dimethylbutanoic acid		HL

Metal ion	Equilibrium	Log K 25°, 0.1	Log K 25°, 1.0
H^+	HL/H.L	3.77	3.79
Y^{3+}	ML/M.L	2.94	
	$ML_2/M.L_2$	$(5.26)^z$	
	$ML_3/M.L_3$	$(6.93)^z$	
La^{3+}	ML/M.L	2.07	
	$ML_2/M.L_2$	$(3.60)^z$	
	$ML_3/M.L_3$	$(4.6)^z$	

^{z} optical isomerism not stated.

2-Hydroxy-2,3-dimethylbutanoic acid (continued)

Metal ion	Equilibrium	Log K 25°, 0.1	Log K 25°, 1.0
Ce^{3+}	$ML/M.L$	2.15	
	$ML_2/M.L^2$	$(3.80)^z$	
	$ML_3/M.L^3$	$(4.9)^z$	
Pr^{3+}	$ML/M.L$	2.30	2.25
	$ML_2/M.L^2$	$(4.05)^z$	$(3.87)^z$
	$ML_3/M.L^3$	$(5.2)^z$	$(4.87)^z$
	$ML_4/M.L^4$		$(5.69)^z$
Nd^{3+}	$ML/M.L$	2.38	$(2.57)^z$
	$ML_2/M.L^2$	$(4.17)^z$	$(4.28)^z$
	$ML_3/M.L^3$	$(5.43)^z$	$(5.5)^z$
	$ML_4/M.L^4$		$(6.6)^z$
Sm^{3+}	$ML/M.L$	2.62	
	$ML_2/M.L^2$	$(4.68)^z$	
	$ML_3/M.L^3$	$(6.19)^z$	
Eu^{3+}	$ML/M.L$	2.74	2.68
	$ML_2/M.L^2$	$(4.86)^z$	$(4.65)^z$
	$ML_3/M.L^3$	$(6.45)^z$	$(6.06)^z$
	$ML_4/M.L^4$		$(7.05)^z$
Gd^{3+}	$ML/M.L$	2.78	2.65
	$ML_2/M.L^2$	$(4.98)^z$	$(4.70)^z$
	$ML_3/M.L^3$	$(6.51)^z$	$(6.04)^z$
	$ML_4/M.L^4$		$(6.98)^z$
Tb^{3+}	$ML/M.L$	2.91	
	$ML_2/M.L^2$	$(5.19)^z$	
	$ML_3/M.L^3$	$(6.85)^z$	
Dy^{3+}	$ML/M.L$	2.98	
	$ML_2/M.L^2$	$(5.36)^z$	
	$ML_3/M.L^3$	$(7.06)^z$	
Ho^{3+}	$ML/M.L$	3.03	3.02
	$ML_2/M.L^2$	$(5.42)^z$	$(5.37)^z$
	$ML_3/M.L^3$	$(7.12)^z$	$(6.90)^z$
	$ML_4/M.L^4$		$(8.24)^z$
Er^{3+}	$ML/M.L$	3.07	
	$ML_2/M.L^2$	$(5.54)^z$	
	$ML_3/M.L^3$	$(7.30)^z$	
Tm^{3+}	$ML/M.L$	3.11	
	$ML_2/M.L^2$	$(5.61)^z$	
	$ML_3/M.L^3$	$(7.41)^z$	
Yb^{3+}	$ML/M.L$	3.17	3.12
	$ML_2/M.L^2$	$(5.74)^z$	$(5.56)^z$
	$ML_3/M.L^3$	$(7.56)^z$	$(7.21)^z$
	$ML_4/M.L^4$		$(8.60)^z$
Lu^{3+}	$ML/M.L$	3.19	
	$ML_2/M.L^2$	$(5.79)^z$	
	$ML_3/M.L^3$	$(7.64)^z$	

---

^{z} optical isomerism not stated.

Bibliography: 65PR,65TV

$$CH_3CHCH_2CCO_2H$$

with OH above the fourth carbon, CH_3 below the second carbon and CH_3 below the fourth carbon.

$C_7H_{14}O_3$	DL-2-Hydroxy-2,4-dimethylpentanoic acid	HL

Metal ion	Equilibrium	Log K 25°, 1.0
H^+	HL/H.L	3.78
La^{3+}	$ML/M.L$	2.07
	$ML_2/M.L^2$	$(3.37)^z$
	$ML_3/M.L^3$	$(4.47)^z$
Ce^{3+}	$ML/M.L$	2.23
	$ML_2/M.L^2$	$(3.57)^z$
	$ML_3/M.L^3$	$(4.88)^z$
Pr^{3+}	$ML/M.L$	(2.51)
	$ML_2/M.L^2$	$(4.03)^z$
	$ML_3/M.L^3$	$(5.51)^z$
Nd^{3+}	$ML/M.L$	2.53
	$ML_2/M.L^2$	$(4.42)^z$
	$ML_3/M.L^3$	$(5.74)^z$
Sm^{3+}	$ML/M.L$	2.71
	$ML_2/M.L^2$	$(4.90)^z$
	$ML_3/M.L^3$	$(6.38)^z$
	$ML_4/M.L^4$	$(7.40)^z$
Eu^{3+}	$ML/M.L$	(2.71)
	$ML_2/M.L^2$	$(5.06)^z$
	$ML_3/M.L^3$	$(6.43)^z$
	$ML_4/M.L^4$	$(7.68)^z$
Gd^{3+}	$ML/M.L$	2.77
	$ML_2/M.L^2$	$(5.13)^z$
	$ML_3/M.L^3$	$(6.52)^z$
	$ML_4/M.L^4$	$(7.66)^z$
Dy^{3+}	$ML/M.L$	3.18
	$ML_2/M.L^2$	$(5.64)^z$
	$ML_3/M.L^3$	$(7.39)^z$
	$ML_4/M.L^4$	$(8.57)^z$
Er^{3+}	$ML/M.L$	3.24
	$ML_2/M.L^2$	$(5.87)^z$
	$ML_3/M.L^3$	$(7.76)^z$
	$ML_4/M.L^4$	$(8.85)^z$
Yb^{3+}	$ML/M.L$	3.21
	$ML_2/M.L^2$	$(5.95)^z$
	$ML_3/M.L^3$	$(7.70)^z$
	$ML_4/M.L^4$	$(9.10)^z$

^{z} optical configuration not stated.

Bibliography: 65TV

$$\begin{array}{cc} H_3C & OH \\ \diagdown & | \\ H_3C-CCCO_2H \\ \diagup & | \\ H_3C & CH_3 \end{array}$$

$C_7H_{14}O_3$ | DL-2-Hydroxy-2,3,3-trimethylbutanoic acid | HL

Metal ion	Equilibrium	Log K 25°, 0.1
H^+	HL/H.L	3.94
Y^{3+}	ML/M.L	2.95
	$ML_2/M.L_2$	$(5.21)^z$
	$ML_3/M.L_3$	$(6.73)^z$
La^{3+}	ML/M.L	2.06
	$ML_2/M.L_2$	$(3.54)^z$
	$ML_3/M.L_3$	$(4.3)^z$
Ce^{3+}	ML/M.L	2.08
	$ML_2/M.L_2$	$(3.83)^z$
	$ML_3/M.L_3$	$(4.1)^z$
Pr^{3+}	ML/M.L	2.28
	$ML_2/M.L_2$	$(4.02)^z$
	$ML_3/M.L_3$	$(5.1)^z$
Nd^{3+}	ML/M.L	2.36
	$ML_2/M.L_2$	$(4.12)^z$
	$ML_3/M.L_3$	$(5.4)^z$
Sm^{3+}	ML/M.L	2.66
	$ML_2/M.L_2$	$(4.69)^z$
	$ML_3/M.L_3$	$(6.12)^z$
Eu^{3+}	ML/M.L	2.73
	$ML_2/M.L_2$	$(4.85)^z$
	$ML_3/M.L_3$	$(6.25)^z$
Gd^{3+}	ML/M.L	2.79
	$ML_2/M.L_2$	$(4.99)^z$
	$ML_3/M.L_3$	$(6.35)^z$
Tb^{3+}	ML/M.L	2.94
	$ML_2/M.L_2$	$(5.20)^z$
	$ML_3/M.L_3$	$(6.82)^z$
Dy^{3+}	ML/M.L	2.99
	$ML_2/M.L_2$	$(5.30)^z$
	$ML_3/M.L_3$	$(6.81)^z$
Ho^{3+}	ML/M.L	3.04
	$ML_2/M.L_2$	$(5.39)^z$
	$ML_3/M.L_3$	$(7.02)^z$
Er^{3+}	ML/M.L	3.07
	$ML_2/M.L_2$	$(5.44)^z$
	$ML_3/M.L_3$	$(7.00)^z$
Tm^{3+}	ML/M.L	3.11
	$ML_2/M.L_2$	$(5.52)^z$
	$ML_3/M.L_3$	$(7.18)^z$
Yb^{3+}	ML/M.L	3.18
	$ML_2/M.L_2$	$(5.65)^z$
	$ML_3/M.L_3$	$(7.25)^z$

---

^{z} optical isomerism not stated.

2-Hydroxy-2,3,3-trimethylbutanoic acid (continued)

Metal ion	Equilibrium	Log K 25°, 0.1
Lu^{3+}	$ML/M.L$	3.19
	$ML_2/M.L^2$	$(5.67)^z$
	$ML_3/M.L^3$	$(7.42)^z$

---

z optical isomerism not stated.

Bibliography: 65PP

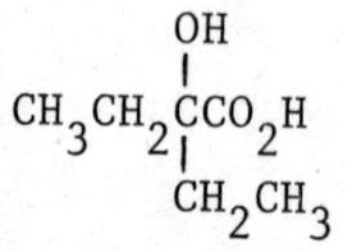

$$\begin{array}{c} OH \\ | \\ CH_3CH_2CCO_2H \\ | \\ CH_2CH_3 \end{array}$$

$C_6H_{12}O_3$ 2-Hydroxy-2-ethylbutanoic acid   HL

Metal ion	Equilibrium	Log K 25°, 0.1	Log K 25°, 1.0	Log K 18°, 0
H^+	$HL/H.L$	3.62	3.63	3.80
Y^{3+}	$ML/M.L$	3.04		
	$ML_2/M.L^2$	5.18		
	$ML_3/M.L^3$	6.36		
La^{3+}	$ML/M.L$	2.01	1.85	
	$ML_2/M.L^2$	3.47	3.11	
	$ML_3/M.L^3$		3.67	
	$ML_4/M.L^4$		4.09	
Ce^{3+}	$ML/M.L$	2.05		
	$ML_2/M.L^2$	3.69		
Pr^{3+}	$ML/M.L$	2.23	(2.31)	
	$ML_2/M.L^2$	3.95	3.80	
	$ML_3/M.L^3$		4.82	
	$ML_4/M.L^4$		5.39	
Nd^{3+}	$ML/M.L$	2.31	2.28	
	$ML_2/M.L^2$	4.15	3.89	
	$ML_3/M.L^3$	4.6	5.10	
	$ML_4/M.L^4$		6.04	
Sm^{3+}	$ML/M.L$	2.69		
	$ML_2/M.L^2$	4.68		
	$ML_3/M.L^3$	5.77		
Eu^{3+}	$ML/M.L$	2.84		
	$ML_2/M.L^2$	4.88		
	$ML_3/M.L^3$	5.97		
Gd^{3+}	$ML/M.L$	2.91	2.71	
	$ML_2/M.L^2$	4.95	4.65	
	$ML_3/M.L^3$	6.09	5.63	
	$ML_4/M.L^4$		6.49	
Tb^{3+}	$ML/M.L$	3.03	3.01	
	$ML_2/M.L^2$	5.17	5.08	
	$ML_3/M.L^3$	6.32	6.45	
	$ML_4/M.L^4$		6.98	

A. MONO-CARBOXYLIC ACIDS   45

2-Hydroxy-2-ethylbutanoic acid (continued)

Metal ion	Equilibrium	Log K 25°, 0.1	Log K 25°, 1.0
Dy^{3+}	$ML/M.L$	3.10	
	$ML_2/M.L^2$	5.27	
	$ML_3/M.L^3$	6.51	
Ho^{3+}	$ML/M.L$	3.13	3.11
	$ML_2/M.L^2$	5.33	5.25
	$ML_3/M.L^3$	6.63	6.71
	$ML_4/M.L^4$		7.4
Er^{3+}	$ML/M.L$	3.16	3.11
	$ML_2/M.L^2$	5.42	5.27
	$ML_3/M.L^3$	6.71	6.60
	$ML_4/M.L^4$		7.55
Tm^{3+}	$ML/M.L$	3.19	
	$ML_2/M.L^2$	5.48	
	$ML_3/M.L^3$	6.82	
Yb^{3+}	$ML/M.L$	3.24	3.10
	$ML_2/M.L^2$	5.55	5.36
	$ML_3/M.L^3$	6.97	6.67
	$ML_4/M.L^4$		7.76
Lu^{3+}	$ML/M.L$	3.25	
	$ML_2/M.L^2$	5.60	
	$ML_3/M.L^3$	7.03	

Bibliography:

H^+ 33La,65FP,65TV $Y^{3+}-Lu^{3+}$ 65FP,65TV

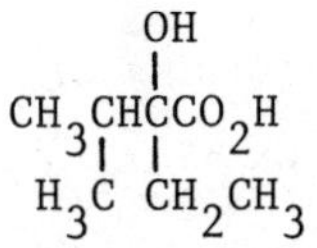

$$\begin{array}{c} OH \\ | \\ CH_3CHCCO_2H \\ | \;\; | \\ H_3C \;\; CH_2CH_3 \end{array}$$

$C_7H_{14}O_3$	DL-2-Hydroxy-2-ethyl-4-methylbutanoic acid		HL

Metal ion	Equilibrium	Log K 25°, 0.1
H^+	$HL/H.L$	3.64
Y^{3+}	$ML/M.L$	2.73
	$ML_2/M.L^2$	$(4.63)^z$
La^{3+}	$ML/M.L$	1.81
	$ML_2/M.L^2$	$(3.08)^z$
Ce^{3+}	$ML/M.L$	1.90
	$ML_2/M.L^2$	$(3.23)^z$
Pr^{3+}	$ML/M.L$	1.97
	$ML_2/M.L^2$	$(3.34)^z$
Nd^{3+}	$ML/M.L$	2.05
	$ML_2/M.L^2$	$(3.69)^z$
Sm^{3+}	$ML/M.L$	2.40
	$ML_2/M.L^2$	$(4.26)^z$

---

z optical isomerism not stated.

2-Hydroxy-2-ethyl-4-methylbutanoic acid (continued)

Metal ion	Equilibrium	Log K 25°, 0.1
Eu^{3+}	$ML/M.L$ $ML_2/M.L^2$	2.57 $(4.40)^z$
Gd^{3+}	$ML/M.L$ $ML_2/M.L^2$	2.63 $(4.49)^z$
Tb^{3+}	$ML/M.L$ $ML_2/M.L^2$	2.75 $(4.68)^z$
Dy^{3+}	$ML/M.L$ $ML_2/M.L^2$	2.80 $(4.76)^z$
Ho^{3+}	$ML/M.L$ $ML_2/M.L^2$	2.81 $(4.81)^z$
Er^{3+}	$ML/M.L$ $ML_2/M.L^2$	2.88 $(4.89)^z$
Tm^{3+}	$ML/M.L$ $ML_2/M.L^2$	2.91 $(4.98)^z$
Yb^{3+}	$ML/M.L$ $ML_2/M.L^2$	2.95 $(5.00)^z$
Lu^{3+}	$ML/M.L$ $ML_2/M.L^2$	2.96 $(4.96)^z$

z

  optical isomerism not stated.

Bibliography: 64PP

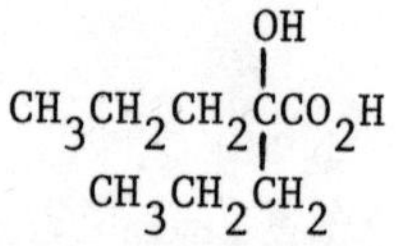

$$\overset{\displaystyle OH}{\underset{\displaystyle CH_3CH_2CH_2}{CH_3CH_2CH_2\overset{|}{\underset{|}{C}}CO_2H}}$$

| $C_8H_{16}O_3$ | 2-Hydroxy-2-propylpentanoic acid | HL |

Metal ion	Equilibrium	Log K 25°, 1.0
H^+	$HL/H.L$	3.80
Pr^{3+}	$ML/M.L$ $ML_2/M.L^2$	2.53 4.0
Nd^{3+}	$ML/M.L$ $ML_2/M.L^2$	2.61 4.4
Sm^{3+}	$ML/M.L$ $ML_2/M.L^2$	2.65 4.82
Eu^{3+}	$ML/M.L$ $ML_2/M.L^2$	2.81 4.99
Gd^{3+}	$ML/M.L$ $ML_2/M.L^2$	2.83 5.03
Dy^{3+}	$ML/M.L$ $ML_2/M.L^2$	3.08 5.57

2-Hydroxy-2-propylpentanoic acid (continued)

Metal ion	Equilibrium	Log K 25°, 1.0
Er^{3+}	$ML/M.L_2$	3.29
	$ML_2/M.L_3^2$	5.53
	$ML_3/M.L^3$	7.7
Yb^{3+}	$ML/M.L_2$	3.36
	$ML_2/M.L_3^2$	5.59
	$ML_3/M.L^3$	7.7

Bibliography: 65TV

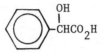

$C_8H_8O_3$		L-Phenylhydroxyacetic acid (mandelic acid)				HL
Metal ion	Equilibrium	Log K 25°, 0.1	Log K 25°, 1.0	Log K 25°, 0	ΔH 25°, 0	ΔS 25°, 0
H^+	HL/H.L	3.19 ±0.00	3.17 ±0.04	3.40 ±0.01	0.05[d]	15.3[d]
		3.12[b]	3.31[d] ±0.01	3.49[e]		
Ca^{2+}	ML/M.L			1.45 +0.01		
Ba^{2+}	ML/M.L			0.77		
Y^{3+}	ML/M.L	2.56				
	$ML_2/M.L^2$	(5.01)[z]				
La^{3+}	ML/M.L	2.28	1.93[d]		-1.4[d]	4[d]
	$ML_2/M.L^2$	(3.81)[z]				
Ce^{3+}	ML/M.L	2.37	2.17[d]		-1.3[d]	6[d]
	$ML_2/M.L^2$	(4.18)[z]				
Pr^{3+}	ML/M.L	2.47	2.30[d]		-1.5[d]	6[d]
	$ML_2/M.L^2$	(4.39)[z]				
Nd^{3+}	ML/M.L	2.52	2.43[d]		-1.4[d]	6[d]
	$ML_2/M.L^2$	(4.51)[z]				
Sm^{3+}	ML/M.L	2.56	2.47[d]		-1.5[d]	6[d]
	$ML_2/M.L^2$	(4.75)[z]				
Eu^{3+}	ML/M.L	2.54	(2.25)[d]		-1.4[d]	6[d]
	$ML_2/M.L^2$	(4.82)[z]				
Gd^{3+}	ML/M.L	2.53	2.42[d]		-1.0[d]	8[d]
	$ML_2/M.L^2$	(4.81)[z]				
Tb^{3+}	ML/M.L	2.59	2.52[d]		-0.9[d]	8[d]
	$ML_2/M.L^2$	(4.90)[z]				
Dy^{3+}	ML/M.L	2.60	2.57[d]		-1.0[d]	8[d]
	$ML_2/M.L^2$	(5.05)[z]				
Ho^{3+}	ML/M.L	2.63	2.54[d]		-1.0[d]	8[d]
	$ML_2/M.L^2$	(5.12)[z]				

[b] 25°, 0.5; [d] 25°, 2.0; [e] 25°, 3.0; [z] optical isomerism not stated.

L-Phenylhydroxyacetic acid (continued)

Metal ion	Equilibrium	Log K 25°, 0.1	Log K 25°, 1.0	Log K 25°, 0	ΔH 25°, 0	ΔS 25°, 0
Er^{3+}	$ML/M.L$	2.66	2.68 [d]		-1.2 [d]	8 [d]
	$ML_2/M.L^2$	(5.24) [z]				
Tm^{3+}	$ML/M.L$	2.65	2.71 [d]		-1.1 [d]	9 [d]
	$ML_2/M.L^2$	(5.43) [z]				
Yb^{3+}	$ML/M.L$	2.84	2.72 [d]		-1.0 [d]	9 [d]
	$ML_2/M.L^2$	(5.44) [z]				
Lu^{3+}	$ML/M.L$	2.82	2.77 [d]		-1.2 [d]	9 [d]
	$ML_2/M.L^2$	(5.55) [z]				
Th^{4+}	$ML/M.L$		3.88 [j]			
	$ML_2/M.L^2$		(6.89) [j,z]			
	$ML_3/M.L^3$		(9.69) [j,z]			
	$ML_4/M.L^4$		(11.98) [j,z]			
UO_2^{2+}	$ML/M.L$		2.57 [j]			
	$ML_2/M.L^2$		(4.10) [j,z]			
	$ML_3/M.L^3$		(5.32) [j,z]			
Co^{2+}	$ML/M.L$		1.22 [k]			
	$ML_2/M.L^2$		1.74 [k]			
	$ML_3/M.L^3$		2.67 [k]			
Ni^{2+}	$ML/M.L$		1.41 [k]			
	$ML_2/M.L^2$		2.26 [k]			
	$ML_3/M.L^3$		2.9 [k]			
Zn^{2+}	$ML/M.L$		1.51 [k]			
	$ML_2/M.L^2$		2.58 [k]			
	$ML_3/M.L^3$		3.36 [k]			
Ge(IV)	$MHL_3/ML_3.H$	(3.72) [z]				
	$MH_2L_3/MHL_3.H$	(2.42) [z]				

---

[d] 25°, 2.0;   [j] 20°, 1.0;   [k] 20°, 2.0;   [z] optical isomerism not stated.

Bibliography:

H^+ 31LAb,38BD,62C,64SM,66SPa,67PN,68FL, 72DC,73MBM,74MT

Ca^{2+},Ba^{2+} 38D,38BD,75DN

Y^{3+}-Lu^{3+} 66SPa,66TV,67PN,68WZ,72DC

Th^{4+} 73MBM

UO_2^{2+} 74MT

Co^{2+}-Zn^{2+} 65LF,68FL

Ge(IV)   62C

Other references: 51BW,57V,61BB,61Sa,62GL, 62GLa,63LG,63LGM,65BK,66KS,66KZ,67Me, 67VA,68PCc,70KAb,70KKM,71PKa,72KA, 73KA,73RM,74CSa,75KAa

$C_9H_{10}O_3$ DL-2-Phenyl-2-hydroxypropanoic acid (atrolactic acid)   HL

Metal ion	Equilibrium	Log K 20°, 0.2	Log K 25°, 1.0	Log K 18°, 0
H^+	HL/H.L			3.53
Pr^{3+}	ML/M.L	2.45	2.40	
	$ML_2/M.L^2$	$(4.20)^z$	$(3.96)^z$	
	$ML_3/M.L^3$	$(5.11)^z$	$(5.32)^z$	
	$ML_4/M.L^4$		$(6.24)^z$	
Nd^{3+}	ML/M.L		2.55	
	$ML_2/M.L^2$		$(4.19)^z$	
	$ML_3/M.L^3$		$(5.61)^z$	
	$ML_4/M.L^4$		$(6.82)^z$	
Sm^{3+}	ML/M.L		2.57	
	$ML_2/M.L^2$		$(4.46)^z$	
	$ML_3/M.L^3$		$(6.00)^z$	
	$ML_4/M.L^4$		$(7.31)^z$	
Eu^{3+}	ML/M.L		2.55	
	$ML_2/M.L^2$		$(4.72)^z$	
Gd^{3+}	ML/M.L		2.54	
	$ML_2/M.L^2$		$(4.61)^z$	
	$ML_3/M.L^3$		$(6.31)^z$	
	$ML_4/M.L^4$		$(7.64)^z$	
Ho^{3+}	ML/M.L		2.97	
	$ML_2/M.L^2$		$(5.35)^z$	
	$ML_3/M.L^3$		$(7.27)^z$	
	$ML_4/M.L^4$		$(9.03)^z$	
Er^{3+}	ML/M.L		3.03	
	$ML_2/M.L^2$		$(5.51)^z$	
	$ML_3/M.L^3$		$(7.52)^z$	
	$ML_4/M.L^4$		$(9.42)^z$	
Yb^{3+}	ML/M.L		3.05	
	$ML_2/M.L^2$		$(5.61)^z$	
	$ML_3/M.L^3$		$(7.68)^z$	
	$ML_4/M.L^4$		$(9.53)^z$	

^{z} optical isomerism not stated.

Bibliography:

H^+ 33La   Other reference: 71GCa
Pr^{3+}-Yb^{3+} 66TV,68WZ

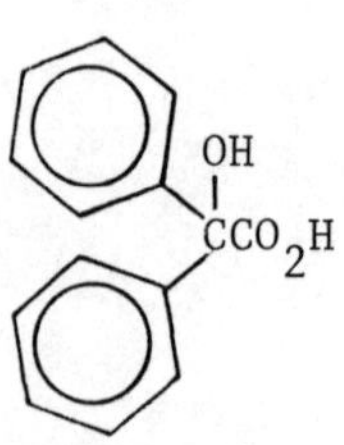

$C_{14}H_{12}O_3$ Diphenylhydroxyacetic acid   HL

Metal ion	Equilibrium	Log K 18°, 0.1	Log K 18°, 1.0	Log K 18°, 0
H^+	HL/H.L	2.87 2.80[i]	2.80	3.05
Pr^{3+}	ML/M.L	2.15[r]		

[i] 18°, 0.5;   [r] 20°, 0.2

Bibliography:

H^+ 31LAb Pr^{3+} 68WZ

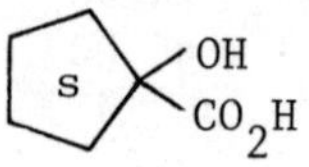

$C_6H_{10}O_3$ 1-Hydroxycyclopentanecarboxylic acid   HL

Metal ion	Equilibrium	Log K 25°, 0.1	Log K 25°, 1.0	Log K 25°, 0
H^+	HL/H.L	3.97 −0.01 3.90[b]	3.92 4.11[d]	4.155
Y^{3+}	ML/M.L	3.00		
	$ML_2/M.L^2$	5.43		
	$ML_3/M.L^3$	7.27		
	$ML_4/M.L^4$	8.96		
La^{3+}	ML/M.L	2.38		
	$ML_2/M.L^2$	4.06		
	$ML_3/M.L^3$	5.01		
Pr^{3+}	ML/M.L	2.60		
	$ML_2/M.L^2$	4.53		
	$ML_3/M.L^3$	5.71		
Nd^{3+}	ML/M.L	2.67		
	$ML_2/M.L^2$	4.63		
	$ML_3/M.L^3$	5.71		
Sm^{3+}	ML/M.L	2.79		
	$ML_2/M.L^2$	4.94		
	$ML_3/M.L^3$	6.44		
	$ML_4/M.L^4$	7.52		
Eu^{3+}	ML/M.L	2.80		
	$ML_2/M.L^2$	5.00		
	$ML_3/M.L^3$	6.57		

[b] 25°, 0.5;   [d] 25°, 2.0

1-Hydroxycyclopentanecarboxylic acid (continued)

Metal ion	Equilibrium	Log K 25°, 0.1	Log K 25°, 1.0	Log K 25°, 0
Gd^{3+}	$ML/M.L$	2.84		
	$ML_2/M.L_2$	5.06		
	$ML_3/M.L_3$	6.71		
	$ML_4/M.L_4$	8.20		
Tb^{3+}	$ML/M.L$	2.90		
	$ML_2/M.L_2$	5.26		
	$ML_3/M.L_3$	6.99		
	$ML_4/M.L_4$	8.22		
Dy^{3+}	$ML/M.L$	2.98		
	$ML_2/M.L_2$	5.45		
	$ML_3/M.L_3$	7.22		
	$ML_4/M.L_4$	8.89		
Ho^{3+}	$ML/M.L$	3.02		
	$ML_2/M.L_2$	5.50		
	$ML_3/M.L_3$	7.37		
	$ML_4/M.L_4$	8.86		
Er^{3+}	$ML/M.L$	3.07		
	$ML_2/M.L_2$	5.65		
	$ML_3/M.L_3$	7.47		
	$ML_4/M.L_4$	9.19		
Tm^{3+}	$ML/M.L$	3.11		
	$ML_2/M.L_2$	5.72		
	$ML_3/M.L_3$	7.71		
	$ML_4/M.L_4$	9.01		
Yb^{3+}	$ML/M.L$	3.18		
	$ML_2/M.L_2$	5.85		
	$ML_3/M.L_3$	7.90		
	$ML_4/M.L_4$	9.22		
Lu^{3+}	$ML/M.L$	3.22		
	$ML_2/M.L_2$	5.94		
	$ML_3/M.L_3$	8.09		
	$ML_4/M.L_4$	9.63		
Co^{2+}	$ML/M.L$	1.57		
	$ML_2/M.L_2$	2.6		
Ni^{2+}	$ML/M.L$	1.82		
	$ML_2/M.L_2$	3.12		
Cu^{2+}	$ML/M.L$	2.80		
	$ML_2/M.L_2$	4.58		
Zn^{2+}	$ML/M.L$	1.89		
	$ML_2/M.L_2$	3.19		
Cd^{2+}	$ML/M.L$	1.45		
	$ML_2/M.L_2$	2.38		

Bibliography:

H^+ 66SPa,68PF

$Co^{2+}-Cd^{2+}$ 67PR

$Y^{3+}-Lu^{3+}$ 66PRb

$C_7H_{12}O_3$		1-Hydroxycyclohexanecarboxylic acid		HL
Metal ion	Equilibrium	Log K 25°, 0.1	Log K 25°, 1.0	Log K 25°, 0
H^+	HL/H.L	3.98 +0.01	3.93	4.159
		3.91^b	4.05^d	
Y^{3+}	$ML/M \cdot L$	2.52		
	$ML_2/M \cdot L^2$	4.65		
	$ML_3/M \cdot L^3$	6.11		
La^{3+}	$ML/M \cdot L$	2.08	1.76	
	$ML_2/M \cdot L^2$	3.70	3.19	
Ce^{3+}	$ML/M \cdot L$	2.20	1.99	
	$ML_2/M \cdot L^2$	3.90	3.48	
Pr^{3+}	$ML/M \cdot L$	2.32	2.05	
	$ML_2/M \cdot L^2$	4.08	3.61	
Nd^{3+}	$ML/M \cdot L$	2.34	2.16	
	$ML_2/M \cdot L^2$	4.26	3.71	
Sm^{3+}	$ML/M \cdot L$	2.45	2.18	
	$ML_2/M \cdot L^2$	4.36	3.91	
	$ML_3/M \cdot L^3$	5.3		
Eu^{3+}	$ML/M \cdot L$	2.48	2.21	
	$ML_2/M \cdot L^2$	4.53	4.07	
	$ML_3/M \cdot L^3$	5.78		
Gd^{3+}	$ML/M \cdot L$	2.45	2.29	
	$ML_2/M \cdot L^2$	4.53	4.17	
	$ML_3/M \cdot L^3$	6.04		
Dy^{3+}	$ML/M \cdot L$	2.53	2.43	
	$ML_2/M \cdot L^2$	4.82	4.33	
	$ML_3/M \cdot L^3$	6.3		
Ho^{3+}	$ML/M \cdot L$	2.67	2.38	
	$ML_2/M \cdot L^2$	4.7	4.67	
	$ML_3/M \cdot L^3$	6.7		
Er^{3+}	$ML/M \cdot L$	2.68	2.48	
	$ML_2/M \cdot L^2$	4.8	4.83	
	$ML_3/M \cdot L^3$	6.8		
Tm^{3+}	$ML/M \cdot L$	2.68		
	$ML_2/M \cdot L^2$	5.0		
	$ML_3/M \cdot L^3$	6.9		
Yb^{3+}	$ML/M \cdot L$	2.76	2.61	
	$ML_2/M \cdot L^2$	5.08	5.06	
	$ML_3/M \cdot L^3$	7.0		
Lu^{3+}	$ML/M \cdot L$	2.78		
	$ML_2/M \cdot L^2$	5.15		
	$ML_3/M \cdot L^3$	7.11		

[b] 25°, 0.5; [d] 25°, 2.0

Bibliography:

H^+ 64PP,68PF $Y^{3+}-Lu^{3+}$ 64PP,67STV

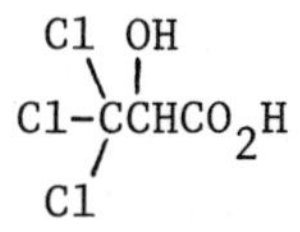

$$\begin{array}{c} Cl \;\; OH \\ | \;\;\; | \\ Cl{-}\overset{|}{\underset{|}{C}}CHCO_2H \\ Cl \end{array}$$

$C_3H_3O_3Cl_3$		DL-3,3,3-Trichloro-2-hydroxypropanoic acid	HL

Metal ion	Equilibrium	Log K 25°, 0
H^+	HL/H.L	2.34
Cu^{2+}	ML/M.L	1.60

Bibliography: 51LW

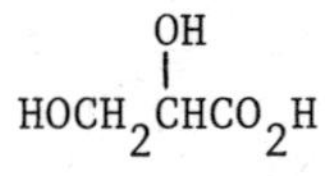

$$\begin{array}{c} OH \\ | \\ HOCH_2CHCO_2H \end{array}$$

$C_3H_6O_4$		DL-2,3-Dihydroxypropanoic acid (glyceric acid)	HL

Metal ion	Equilibrium	Log K 25°(?), 0.2	Log K 25°, 2.0
H^+	HL/H.L	3.52[x]	
Mg^{2+}	ML/M.L	0.86[x]	
Ca^{2+}	ML/M.L	1.18[x]	
Sr^{2+}	ML/M.L	0.89[x]	
Ba^{2+}	ML/M.L	0.80[x]	
Zn^{2+}	ML/M.L	1.80[x]	
Cd^{2+}	ML/M.L		1.60
	$ML_2/M.L^2$		(2.11)
Pb^{2+}	ML/M.L		2.53
	$ML_2/M.L^2$		(3.76)

[x] Temperature not stated.

Bibliography:

H^+-Zn^{2+} 38CK		Cd^{2+},Pb^{2+} 68TF

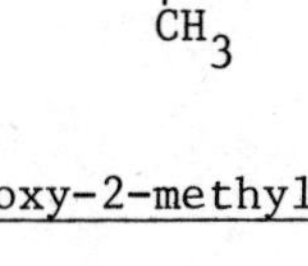

$$\begin{array}{c} OH \\ | \\ HOCH_2CCO_2H \\ | \\ CH_3 \end{array}$$

$C_4H_8O_4$		DL-2,3-Dihydroxy-2-methylpropanoic acid	HL

Metal ion	Equilibrium	Log K 25°, 0.1
H^+	HL/H.L	3.58

DL-2,3-Dihydroxy-2-methylpropanoic acid (continued)

Metal ion	Equilibrium	Log K 25°, 0.1
Y^{3+}	$ML/M.L$	3.05
	$ML_2/M.L^2$	$(5.49)^z$
	$ML_3/M.L^3$	$(7.22)^z$
La^{3+}	$ML/M.L$	2.65
	$ML_2/M.L^2$	$(4.52)^z$
	$ML_3/M.L^3$	$(5.72)^z$
Ce^{3+}	$ML/M.L$	2.83
	$ML_2/M.L^2$	$(4.88)^z$
	$ML_3/M.L^3$	$(6.26)^z$
Pr^{3+}	$ML/M.L$	2.95
	$ML_2/M.L^2$	$(5.14)^z$
	$ML_3/M.L^3$	$(6.57)^z$
Nd^{3+}	$ML/M.L$	2.96
	$ML_2/M.L^2$	$(5.14)^z$
	$ML_3/M.L^3$	$(6.63)^z$
Sm^{3+}	$ML/M.L$	3.05
	$ML_2/M.L^2$	$(5.39)^z$
	$ML_3/M.L^3$	$(7.03)^z$
Eu^{3+}	$ML/M.L$	3.05
	$ML_2/M.L^2$	$(5.45)^z$
	$ML_3/M.L^3$	$(7.11)^z$
Gd^{3+}	$ML/M.L$	3.03
	$ML_2/M.L^2$	$(5.40)^z$
	$ML_3/M.L^3$	$(7.08)^z$
Tb^{3+}	$ML/M.L$	3.04
	$ML_2/M.L^2$	$(5.42)^z$
	$ML_3/M.L^3$	$(7.05)^z$
Dy^{3+}	$ML/M.L$	3.07
	$ML_2/M.L^2$	$(5.47)^z$
	$ML_3/M.L^3$	$(7.14)^z$
Ho^{3+}	$ML/M.L$	3.11
	$ML_2/M.L^2$	$(5.56)^z$
	$ML_3/M.L^3$	$(7.35)^z$
Er^{3+}	$ML/M.L$	3.17
	$ML_2/M.L^2$	$(5.68)^z$
	$ML_3/M.L^3$	$(7.49)^z$
Tm^{3+}	$ML/M.L$	3.22
	$ML_2/M.L^2$	$(5.78)^z$
	$ML_3/M.L^3$	$(7.61)^z$
Yb^{3+}	$ML/M.L$	3.27
	$ML_2/M.L^2$	$(5.85)^z$
	$ML_3/M.L^3$	$(7.67)^z$
Lu^{3+}	$ML/M.L$	3.30
	$ML_2/M.L^2$	$(5.90)^z$
	$ML_3/M.L^3$	$(7.75)^z$

z optical isomerism not stated.

DL-2,3-Dihydroxy-2-methylpropanoic acid (continued)

Metal ion	Equilibrium	Log K 25°, 0.1
Co^{2+}	ML/M.L $ML_2/M.L^2$	1.62 $(2.4)^z$
Ni^{2+}	ML/M.L $ML_2/M.L^2$	1.82 $(2.7)^z$
Cu^{2+}	ML/M.L $ML_2/M.L^2$	2.63 $(4.29)^z$
Zn^{2+}	ML/M.L $ML_2/M.L^2$	1.77 $(2.6)^z$
Cd^{2+}	ML/M.L $ML_2/M.L^2$	1.63 $(2.99)^z$

^{z} optical isomerism not stated.

Bibliography:

$H^+, Co^{2+}-Cd^{2+}$ 74PK $Y^{3+}-Lu^{3+}$ 75PF

HO OH
| |
CH_3CHCCO_2H
|
CH_3

$C_5H_{10}O_4$		DL-2,3-Dihydroxy-2-methylbutanoic acid	HL

Metal ion	Equilibrium	Log K 25°, 0.1	Log K 25°, 0.5
H^+	HL/H.L	3.51	3.47
Y^{3+}	ML/M.L $ML_2/M.L^2$ $ML_3/M.L^3$	2.91 $(5.15)^z$ $(6.58)^z$	
La^{3+}	ML/M.L $ML_2/M.L^2$ $ML_3/M.L^3$	2.57 $(4.35)^z$ $(5.35)^z$	2.42 $(4.08)^z$
Ce^{3+}	ML/M.L $ML_2/M.L^2$ $ML_3/M.L^3$	2.81 $(4.81)^z$ $(5.96)^z$	2.67 $(4.58)^z$ $(5.7)^z$
Pr^{3+}	ML/M.L $ML_2/M.L^2$ $ML_3/M.L^3$	2.96 $(5.11)^z$ $(6.43)^z$	2.79 $(4.77)^z$
Nd^{3+}	ML/M.L $ML_2/M.L^2$ $ML_3/M.L^3$	3.03 $(5.25)^z$ $(6.63)^z$	2.92 $(5.00)^z$ $(6.11)^z$
Sm^{3+}	ML/M.L $ML_2/M.L^2$ $ML_3/M.L^3$	3.14 $(5.46)^z$ $(6.91)^z$	3.03 $(5.27)^z$ $(6.36)^z$
Eu^{3+}	ML/M.L $ML_2/M.L^2$ $ML_3/M.L^3$	3.13 $(5.61)^z$ $(7.21)^z$	3.03 $(5.27)^z$ $(6.42)^z$

^{z} optical isomerism not stated.

DL-2,3-Dihydroxy-2-methylbutanoic acid (continued)

Metal ion	Equilibrium	Log K 25°, 0.1	Log K 25°, 0.5
Gd^{3+}	$ML/M.L$	3.08	2.89
	$ML_2/M.L^2$	$(5.44)^z$	$(5.19)^z$
	$ML_3/M.L^3$	$(7.01)^z$	$(6.20)^z$
Tb^{3+}	$ML/M.L$	3.04	2.85
	$ML_2/M.L^2$	$(5.39)^z$	$(5.17)^z$
	$ML_3/M.L^3$	$(6.95)^z$	$(6.08)^z$
Dy^{3+}	$ML/M.L$	3.00	2.85
	$ML_2/M.L^2$	$(5.36)^z$	$(5.21)^z$
	$ML_3/M.L^3$	$(6.54)^z$	
Ho^{3+}	$ML/M.L$	3.03	2.81
	$ML_2/M.L^2$	$(5.38)^z$	$(5.15)^z$
	$ML_3/M.L^3$	$(6.72)^z$	$(6.20)^z$
Er^{3+}	$ML/M.L$	3.11	2.86
	$ML_2/M.L^2$	$(5.49)^z$	$(5.17)^z$
	$ML_3/M.L^3$	$(7.08)^z$	$(6.28)^z$
Tm^{3+}	$ML/M.L$	3.15	
	$ML_2/M.L^2$	$(5.62)^z$	
	$ML_3/M.L^3$	$(7.34)^z$	
Yb^{3+}	$ML/M.L$	3.21	
	$ML_2/M.L^2$	$(5.76)^z$	
	$ML_3/M.L^3$	$(7.48)^z$	
Lu^{3+}	$ML/M.L$	3.25	
	$ML_2/M.L^2$	$(5.83)^z$	
	$ML_3/M.L^3$	$(7.61)^z$	
Co^{2+}	$ML/M.L$	1.59	
	$ML_2/M.L^2$	$(2.82)^z$	
Ni^{2+}	$ML/M.L$	1.81	
	$ML_2/M.L^2$	$(3.04)^z$	
Cu^{2+}	$ML/M.L$	2.62	
	$ML_2/M.L^2$	$(4.29)^z$	
Zn^{2+}	$ML/M.L$	1.76	
	$ML_2/M.L^2$	$(3.10)^z$	
Cd^{2+}	$ML/M.L$	1.62	
	$ML_2/M.L^2$	$(3.10)^z$	

^{z} optical isomerism not stated

Bibliography:

H^+ 64PP,74PK Co^{2+}-Cd^{2+} 74PK

La^{3+}-Er^{3+} 64PP,75PF

$$\begin{array}{c} \text{OH} \\ | \\ \text{HOCH}_2\text{CCO}_2\text{H} \\ | \\ \text{CH}_2\text{OH} \end{array}$$

$C_4H_8O_5$	2,3-Dihydroxy-2-(hydroxymethyl)propanoic acid		HL

Metal ion	Equilibrium	Log K 25°, 0.5
H^+	HL/H.L	3.29
Y^{3+}	$ML/M.L$	2.65
	$ML_2/M.L^2$	4.67
	$ML_3/M.L^3$	5.26
La^{3+}	$ML/M.L$	2.40
	$ML_2/M.L^2$	3.88
	$ML_3/M.L^3$	4.90
Ce^{3+}	$ML/M.L$	2.61
	$ML_2/M.L^2$	4.45
	$ML_3/M.L^3$	5.98
Pr^{3+}	$ML/M.L$	2.75
	$ML_2/M.L^2$	4.69
	$ML_3/M.L^3$	6.15
Nd^{3+}	$ML/M.L$	2.81
	$ML_2/M.L^2$	4.62
	$ML_3/M.L^3$	6.36
Sm^{3+}	$ML/M.L$	2.86
	$ML_2/M.L^2$	5.07
	$ML_3/M.L^3$	6.51
Eu^{3+}	$ML/M.L$	2.80
	$ML_2/M.L^2$	5.00
	$ML_3/M.L^3$	6.45
Gd^{3+}	$ML/M.L$	2.69
	$ML_2/M.L^2$	4.99
	$ML_3/M.L^3$	6.41
Tb^{3+}	$ML/M.L$	2.71
	$ML_2/M.L^2$	4.88
	$ML_3/M.L^3$	6.58
Dy^{3+}	$ML/M.L$	2.66
	$ML_2/M.L^2$	4.87
	$ML_3/M.L^3$	6.37
Ho^{3+}	$ML/M.L$	2.71
	$ML_2/M.L^2$	4.89
	$ML_3/M.L^3$	6.22
Er^{3+}	$ML/M.L$	2.79
	$ML_2/M.L^2$	4.83
	$ML_3/M.L^3$	6.57
Tm^{3+}	$ML/M.L$	2.85
	$ML_2/M.L^2$	4.97
	$ML_3/M.L^3$	6.51
Yb^{3+}	$ML/M.L$	2.90
	$ML_2/M.L^2$	5.07
	$ML_3/M.L^3$	6.50

2,3-Dihydroxy-2-(hydroxymethyl)propanoic acid (continued)

Metal ion	Equilibrium	Log K 25°, 0.5
Lu^{3+}	ML/M.L	2.94
	ML$_2$/M.L^2	5.19
	ML$_3$/M.L^3	6.90

Bibliography: 64SP

$C_7H_{12}O_6$ 1,3,4,5-Tetrahydroxycyclohexanecarboxylic acid (quinic acid)*   HL

Metal ion	Equilibrium	Log K 25°, 0.1	Log K 25°, 1.0	Log K 25°, 0
H$^+$	HL/H.L	3.36	3.31	3.58
Y^{3+}	ML/M.L	2.67		
	ML$_2$/M.L^2	5.07		
	ML$_3$/M.L^3	6.50		
La^{3+}	ML/M.L	2.35	2.08	
	ML$_2$/M.L^2	4.26	3.70	
	ML$_3$/M.L^3	5.57	4.82	
	ML$_4$/M.L^4		5.62	
Ce^{3+}	ML/M.L	2.52		
	ML$_2$/M.L^2	4.53		
	ML$_3$/M.L^3	5.99		
Pr^{3+}	ML/M.L	2.55	2.20	
	ML$_2$/M.L^2	4.75	4.16	
	ML$_3$/M.L^3	6.11	5.32	
	ML$_4$/M.L^4		6.18	
Nd^{3+}	ML/M.L	2.60		
	ML$_2$/M.L^2	4.85		
	ML$_3$/M.L^3	6.16		
Sm^{3+}	ML/M.L	2.70	2.38	
	ML$_2$/M.L^2	4.98	4.44	
	ML$_3$/M.L^3	6.53	5.88	
	ML$_4$/M.L^4		6.73	
Eu^{3+}	ML/M.L	2.67	2.38	
	ML$_2$/M.L^2	4.96	4.49	
	ML$_3$/M.L^3	6.55	5.97	
	ML$_4$/M.L^4		6.83	
Gd^{3+}	ML/M.L	2.67	2.35	
	ML$_2$/M.L^2	4.93	4.41	
	ML$_3$/M.L^3	6.53	5.87	
	ML$_4$/M.L^4		6.72	
Tb^{3+}	ML/M.L	2.71	2.39	
	ML$_2$/M.L^2	5.06	4.53	
	ML$_3$/M.L^3	6.53	6.01	
	ML$_4$/M.L^4		6.99	

---

* cis-trans and optical isomerism of ring substituents ignored in this case because of remoteness of coordinating groups.

Quinic acid (continued)

Metal ion	Equilibrium	Log K 25°, 0.1	Log K 25°, 1.0	Log K 25°, 0
Dy^{3+}	$ML/M.L$	2.74	2.47	
	$ML_2/M.L^2$	5.19	4.66	
	$ML_3/M.L^3$	6.73	6.14	
	$ML_4/M.L^4$		7.20	
Ho^{3+}	$ML/M.L$	2.80	2.49	
	$ML_2/M.L^2$	5.23	4.75	
	$ML_3/M.L^3$	6.90	6.26	
	$ML_4/M.L^4$		7.37	
Er^{3+}	$ML/M.L$	2.80	2.56	
	$ML_2/M.L^2$	5.31	4.86	
	$ML_3/M.L^3$	6.94	6.43	
	$ML_4/M.L^4$		7.57	
Tm^{3+}	$ML/M.L$	2.89		
	$ML_2/M.L^2$	5.41		
	$ML_3/M.L^3$	7.20		
Yb^{3+}	$ML/M.L$	2.95	2.66	
	$ML_2/M.L^2$	5.54	5.12	
	$ML_3/M.L^3$	7.43	6.84	
	$ML_4/M.L^4$		8.10	
Lu^{3+}	$ML/M.L$	2.95		
	$ML_2/M.L^2$	5.63		
	$ML_3/M.L^3$	7.55		
Cu^{2+}	$ML/M.L$	2.44[r]		2.66

[r] 25°, 0.05

Bibliography:

H^+ 59T,66SPa Cu^{2+} 59T

Y^{3+}-Lu^{3+} 66SPa,67OT

$$\begin{array}{c} OH \\ | \\ HOCH_2CHCHCHCHCO_2H \\ |\ |\ \ \ | \\ HO\ OH\ \ OH \end{array}$$

$C_6H_{12}O_7$	D-2,3,4,5,6-Pentahydroxyhexanoic acid (D-gluconic acid)			HL

Metal ion	Equilibrium	Log K 25°, 0.1	Log K 25°, 1.0	Log K 25°, 0
H^+	$HL/H.L$	3.56 [x]		
Mg^{2+}	$ML/M.L$	0.70 [x]		
Ca^{2+}	$ML/M.L$	1.21[x] +0.01		
Sr^{2+}	$ML/M.L$	1.00[x] +0.01		
Ba^{2+}	$ML/M.L$	0.95 [x]		
Y^{3+}	$ML/M.L$	2.40 [r]		
	$ML_2/M.L^2$	(4.52)[r,z]		

[r] 25°, 0.2;   [x] temperature not stated, 0.2;   [z] optical isomerism not stated.

D-Gluconic acid (continued)

Metal ion	Equilibrium	Log K 25°, 0.1	Log K 25°, 1.0	Log K 25°, 0
La^{3+}	$ML/M.L$	2.32 [r]		
	$ML_2/M.L^2$	(4.25) [r,z]		
Pr^{3+}	$ML/M.L$	2.60 [r]		
	$ML_2/M.L^2$	(4.55) [r,z]		
Nd^{3+}	$ML/M.L$	2.66 [r] -0.01		
	$ML_2/M.L^2$	(4.70) [r,z]		
Sm^{3+}	$ML/M.L$	2.76 [r]		
	$ML_2/M.L^2$	(4.88) [r,z]		
Eu^{3+}	$ML/M.L$	2.74 [r]		
	$ML_2/M.L^2$	(4.97) [r,z]		
Gd^{3+}	$ML/M.L$	2.66 [r]		
	$ML_2/M.L^2$	(4.76) [r,z]		
Tb^{3+}	$ML/M.L$	2.47 [r]		
	$ML_2/M.L^2$	(4.67) [r,z]		
Dy^{3+}	$ML/M.L$	2.40 [r]		
	$ML_2/M.L^2$	(4.57) [r,z]		
Ho^{3+}	$ML/M.L$	2.42 [r]		
	$ML_2/M.L^2$	(4.51) [r,z]		
Er^{3+}	$ML/M.L$	2.50 [r]		
	$ML_2/M.L^2$	(4.53) [r,z]		
Yb^{3+}	$ML/M.L$	2.80 [r]		
	$ML_2/M.L^2$	(4.68) [r,z]		
Lu^{3+}	$ML/M.L$	2.85 [r]		
	$ML_2/M.L^2$	(4.78) [r,z]		
Fe^{2+}	$ML/M.L$		1.0	
Ni^{2+}	$ML/M.L$	1.82		
	$M_2H_{-4}L/M^2.(OH)^4.L$	29.4		
Fe^{3+}	$MH_{-2}L.H^2/M.L$		-5.5	
	$ML/MH_{-2}L.H^2$		2.3	
	$MH_{-2}L/MH_{-3}L.H$		4.0	
	$MH_{-3}L/MOH(H_{-3}L).H$		13.3	
Zn^{2+}	$ML/M.L$	1.70 [x]		
Cd^{2+}	$ML_2/M.L^2$		2.10	
Pb^{2+}	$ML/M.L$	2.6		
	$MH_{-2}L/M.(OH)^2.L$	15.71 [s]	16.17	16.39 [e]

---

[e] $25°$, 3.0;   [r] $25°$, 0.1;   [s] $25°$, 0.15;   [x] temperature not stated, 0.2;   [z] optical isomerism not stated.

Bibliography:

H^+, Zn^{2+}	38CK	Cd^{2+}	57PS
Mg^{2+}-Ba^{2+}	38CK,52SL	Pb^{2+}	56PJ
Y^{3+}-Lu^{3+}	63K,67TK		
Fe^{2+}, Fe^{3+}	55PS		
Ni^{2+}	65JP		

Other references: 51H,55PJ,55PPa,58KY, 62K,62SA,62SKa,63RC,63ZG,64BSb,67MM, 67RK,68LO,71MMb,71ZG,72FP,72Ka,73KM

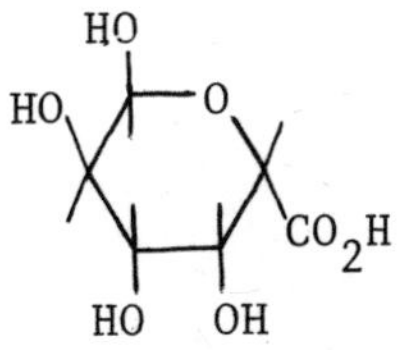

$$HO-\text{(structure)}-CO_2H$$

$C_6H_{10}O_7$ D-Galacturonic acid H_2L

Metal ion	Equilibrium	Log K 37°, 0.15
H^+	HL/H.L	11.42
	$H_2L/HL.H$	3.23
Fe^{2+}	$ML/M.L_2$	9.7
	$ML_2/M.L^2$	18.3

Bibliography: 74CC

$$HOCH_2CH_2CO_2H$$

$C_3H_6O_3$ 3-Hydroxypropanoic acid   HL

Metal ion	Equilibrium	Log K 30°, 0.1	Log K 25°, 1.0	ΔH 25°, 2.0	ΔS 25°, 2.0
H^+	HL/H.L	4.33 ±0.08	4.32[d] 4.56[d]		
Y^{3+}	ML/M.L		1.43[d]	4.6	21
La^{3+}	ML/M.L		1.56[d]	2.6	16
Ce^{3+}	ML/M.L		1.57[d]	2.5	15
Pr^{3+}	ML/M.L		1.62[d]	1.9	14
Nd^{3+}	ML/M.L		1.66[d]	1.5	13
Sm^{3+}	ML/M.L		1.75[d]	2.5	16
Eu^{3+}	ML/M.L		1.64[d]		
Gd^{3+}	ML/M.L		1.61[d]	3.8	20
Tb^{3+}	ML/M.L		1.54[d]	4.4	21
Dy^{3+}	ML/M.L		1.45[d]	4.2	21
Ho^{3+}	ML/M.L		1.48[d]	4.7	22
Er^{3+}	ML/M.L		1.32[d]	4.9	23
Tm^{3+}	ML/M.L		1.45[d]	5.2	24
Yb^{3+}	ML/M.L		1.51[d]	4.5	22
Lu^{3+}	ML/M.L		1.40[d]	4.5	22
Co^{2+}	ML/M.L		0.49[d]		
Ni^{2+}	ML/M.L		0.78[d]		
	$ML_2/M.L^2$		1.32[d]		

[d] 25°, 2.0

3-Hydroxypropanoic acid (continued)

Metal ion	Equilibrium	Log K 30°, 0.1	Log K 25°, 1.0	ΔH 25°, 2.0	ΔS 25°, 2.0
Cu^{2+}	$ML/M.L$	2.05	1.76^d		
	$ML_2/M.L^2$		3.21^d		
Zn^{2+}	$ML/M.L$		0.86^d		
	$ML_2/M.L^2$		1.11^d		
Cd^{2+}	$ML/M.L$		1.15^d		
	$ML_2/M.L^2$		2.20^d		
Pb^{2+}	$ML/M.L$		1.95 2.13^d		
	$ML_2/M.L^2$		2.94 3.11^d		
	$ML_3/M.L^3$		3.56^d		

Bibliography:

H^+ 62CM,62CTC,71BV,72SM,73FP Cd^{2+} 73NP

$Y^{3+}-Lu^{3+}$ 69JC Pb^{2+} 71BVb,73NP

$Co^{2+}-Cu^{2+}$ 65DSa,72SSF   Other references: 62CTG,68RSa,73WK

Zn^{2+} 73FP

$$OH$$
$$|$$
$$CH_3CHCH_2CO_2H$$

$C_4H_8O_3$ DL-3-Hydroxybutanoic acid   HL

Metal ion	Equilibrium	Log K 25°, 0.2	Log K 20°, 1.0	Log K 25°, 0
H^+	$HL/H.L$	4.28 ±0.02	4.35 ±0.00 4.53^d	
Mg^{2+}	$ML/M.L$	0.60^x		
Ca^{2+}	$ML/M.L$	0.60^x		0.82
Sr^{2+}	$ML/M.L$	0.47^x		
Ba^{2+}	$ML/M.L$	0.43^x		
Th^{4+}	$ML/M.L$		3.87	
	$ML_2/M.L^2$		$(6.85)^z$	
	$ML_3/M.L^3$		$(9.01)^z$	
UO_2^{2+}	$ML/M.L$		2.38	
	$ML_2/M.L^2$		$(4.35)^z$	
	$ML_3/M.L^3$		$(6.25)^z$	
Co^{2+}	$ML/M.L$		0.75^d	
	$ML_2/M.L^2$		$(1.15)^{d,z}$	
Ni^{2+}	$ML/M.L$		1.00^d	
	$ML_2/M.L^2$		$(1.36)^{d,z}$	
Cu^{2+}	$ML/M.L$		1.88^d ±0.05	
	$ML_2/M.L^2$		$(2.9)^{d,z}$ ±0.2	

^{d} 25°, 2.0; ^{x} temperature not stated, 0.2; ^{z} optical isomerism not stated.

DL-3-Hydroxybutanoic acid (continued)

Metal ion	Equilibrium	Log K 25°, 0.2	Log K 20°, 1.0	Log K 25°, 0
Zn^{2+}	$ML/M.L$	1.06^x	0.99^d	
	$ML_2/M.L^2$		$(1.71)^{d,z}$	
Cd^{2+}	$ML/M.L$		1.11^d	
	$ML_2/M.L^2$		$(2.20)^{d,z}$	
Pb^{2+}	$ML/M.L$		2.18^d	
	$ML_2/M.L^2$		$(3.0)^{d,z}$	
	$ML_3/M.L^3$		$(3.7)^{d,z}$	

^{d} 25°, 2.0; ^{x} temperature not stated, 0.2; ^{z} optical isomerism not stated.

Bibliography:

H^+ 38CK,62CM,73FP,73MBM,74MT

Mg^{2+}-Ba^{2+} 38CK,38Da

Th^{4+} 73MBM

UO_2^{2+} 74MT

Co^{2+}-Cu^{2+} 74GMP,75FPG,75GM

Zn^{2+} 38CK,73FP

Cd^{2+},Pb^{2+} 73NP

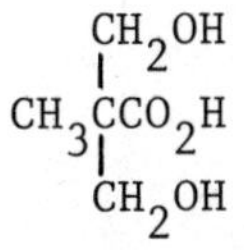

$$CH_3\overset{\displaystyle CH_2OH}{\underset{\displaystyle CH_2OH}{|\atop C|CO_2H}}$$

$C_5H_{10}O_4$		2,2-Bis(hydroxymethyl)propanoic acid		HL

Metal ion	Equilibrium	Log K 25°, 0.1	Log K 25°, 0
H^+	$HL/H.L$	4.39	4.61
Y^{3+}	$ML/M.L$	2.16	
	$ML_2/M.L^2$	3.72	
La^{3+}	$ML/M.L$	2.06	
	$ML_2/M.L^2$	3.54	
	$ML_3/M.L^3$	4.74	
Pr^{3+}	$ML/M.L$	2.30	
	$ML_2/M.L^2$	3.89	
	$ML_3/M.L^3$	5.25	
Nd^{3+}	$ML/M.L$	2.37	
	$ML_2/M.L^2$	3.96	
	$ML_3/M.L^3$	5.28	
Sm^{3+}	$ML/M.L$	2.48	
	$ML_2/M.L^2$	4.15	
	$ML_3/M.L^3$	5.45	
Eu^{3+}	$ML/M.L$	2.46	
	$ML_2/M.L^2$	4.14	
	$ML_3/M.L^3$	5.25	
Gd^{3+}	$ML/M.L$	2.37	
	$ML_2/M.L^2$	4.00	
	$ML_3/M.L^3$	5.23	

2,2-Bis(hydroxymethyl)propanoic acid (continued)

Metal ion	Equilibrium	Log K 25°, 0.1	Log K 25°, 0
Tb^{3+}	$ML/M.L$	2.30	
	$ML_2/M.L^2$	3.90	
Dy^{3+}	$ML/M.L$	2.27	
	$ML_2/M.L^2$	3.91	
Ho^{3+}	$ML/M.L$	2.28	
	$ML_2/M.L^2$	3.92	
Er^{3+}	$ML/M.L$	2.27	
	$ML_2/M.L^2$	3.87	
Tm^{3+}	$ML/M.L$	2.31	
	$ML_2/M.L^2$	3.93	
Yb^{3+}	$ML/M.L$	2.30	
	$ML_2/M.L^2$	3.91	
Lu^{3+}	$ML/M.L$	2.34	
	$ML_2/M.L^2$	3.97	

Bibliography: 70RV

$$HOCH_2CH_2CH_2CO_2H$$

$C_4H_8O_3$		4-Hydroxybutanoic acid		HL

Metal ion	Equilibrium	Log K 20°, 1.0	Log K 25°, 2.0
H^+	$HL/H.L$	4.57 +0.01	4.85
Th^{4+}	$ML/M.L$	3.80	
	$ML_2/M.L^2$	6.65	
UO_2^{2+}	$ML/M.L$	2.34	
	$ML_2/M.L^2$	4.49	
	$ML_3/M.L^3$	6.28	
Co^{2+}	$ML/M.L$		0.48
	$ML_2/M.L^2$		0.56
Ni^{2+}	$ML/M.L$		0.52
	$ML_2/M.L^2$		1.04
Cu^{2+}	$ML/M.L$		1.76 ±0.04
	$ML_2/M.L^2$		2.4 ±0.2
Zn^{2+}	$ML/M.L$		0.96
	$ML_2/M.L^2$		1.56
Cd^{2+}	$ML/M.L$		1.45
	$ML_2/M.L^2$		2.20
Pb^{2+}	$ML/M.L$		2.28
	$ML_2/M.L^2$		3.15
	$ML_3/M.L^3$		3.64

Bibliography:

H^+	73FP,73MBM,74MT	$Co^{2+}-Cu^{2+}$	74GMP,75FPG,75GM
Th^{4+}	73MBM	Zn^{2+} 73FP	
UO_2^{2+}	74MT	Cd^{2+},Pb^{2+} 73NP	

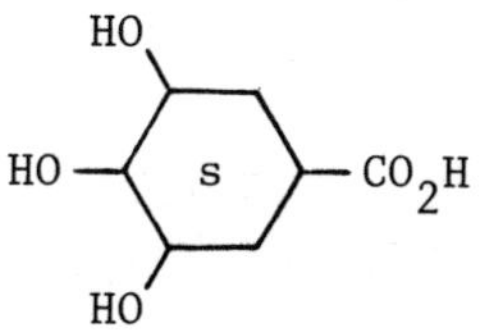

HO

HO ―――― s ――― CO_2H

HO

$C_7H_{12}O_5$	3,4,5-Trihydroxycyclohexanecarboxylic acid (dihydroshikimic acid)*		HL

Metal ion	Equilibrium	Log K 25°, 0.05	Log K 25°, 0
H^+	HL/H.L	4.29	4.39
Cu^{2+}	ML/M.L	1.65	2.08

* isomerism not stated

Bibliography: 59T

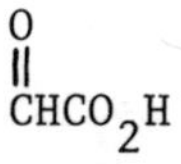

$$\overset{O}{\underset{||}{C}}HCO_2H$$

$C_2H_2O_3$		Oxoacetic acid (glyoxylic acid)				HL
Metal ion	Equilibrium	Log K 20°, 0.1	Log K 25°, 1.0	Log K 25°, 0	ΔH 25°, 0	ΔS 25°, 0
H^+	HL/H.L	3.18 2.83[b]	2.91	3.46	-0.53 -0.64[s]	14
Y^{3+}	ML/M.L	2.56				
	$ML_2/M.L^2$	4.41				
	$ML_3/M.L^3$	5.9				
La^{3+}	ML/M.L	2.36				
	$ML_2/M.L^2$	3.96				
	$ML_3/M.L^3$	4.8				
Ce^{3+}	ML/M.L	2.39				
	$ML_2/M.L^2$	4.17				
	$ML_3/M.L^3$	5.1				
Pr^{3+}	ML/M.L	2.44				
	$ML_2/M.L^2$	4.34				
	$ML_3/M.L^3$	5.4				
Nd^{3+}	ML/M.L	2.48				
	$ML_2/M.L^2$	4.48				
	$ML_3/M.L^3$	5.8				
Sm^{3+}	ML/M.L	2.55				
	$ML_2/M.L^2$	4.59				
	$ML_3/M.L^3$	6.1				
Eu^{3+}	ML/M.L	2.50				
	$ML_2/M.L^2$	4.58				
	$ML_3/M.L^3$	6.1				
Gd^{3+}	ML/M.L	2.49				
	$ML_2/M.L^2$	4.53				
	$ML_3/M.L^3$	6.0				

[b] 25°, 0.5;   [s] 25°, 0.05

Oxoacetic acid (continued)

Metal ion	Equilibrium	Log K 20°, 0.1	Log K 25°, 1.0	Log K 25°, 0	ΔH 25°, 0	ΔS 25°, 0
Tb^{3+}	$ML/M.L$	2.52				
	$ML_2/M.L^2$	4.42				
	$ML_3/M.L^3$	6.1				
Dy^{3+}	$ML/M.L$	2.56				
	$ML_2/M.L^2$	4.46				
	$ML_3/M.L^3$	6.2				
Ho^{3+}	$ML/M.L$	2.58				
	$ML_2/M.L^2$	4.48				
	$ML_3/M.L^3$	6.2				
Er^{3+}	$ML/M.L$	2.60				
	$ML_2/M.L^2$	4.60				
	$ML_3/M.L^3$	6.2				
Tm^{3+}	$ML/M.L$	2.61				
	$ML_2/M.L^2$	4.61				
	$ML_3/M.L^3$	6.3				
Yb^{3+}	$ML/M.L$	2.65				
	$ML_2/M.L^2$	4.73				
	$ML_3/M.L^3$	6.4				
Lu^{3+}	$ML/M.L$	2.68				
	$ML_2/M.L^2$	4.83				
	$ML_3/M.L^3$	6.5				
Ni^{2+}	$ML/M.L$	0.94[b]	1.06			
	$ML_2/M.L^2$		1.32			
Zn^{2+}	$ML/M.L$	0.64[b]	1.06			
	$ML_2/M.L^2$		1.68			

[b] 25°, 0.5

Bibliography:

H^+	64PS,66LHa,67OW,71HL	Ni^{2+},Zn^{2+}	66LHa,71HL
$Y^{3+}-Lu^{3+}$	64PS	Other reference:	75SD

$$\overset{O}{\overset{\|}{CH_3CCO_2H}}$$

$C_3H_4O_3$		2-Oxopropanoic acid (pyruvic acid)				HL

Metal ion	Equilibrium	Log K 25°, 0.1	Log K 25°, 2.0	Log K 25°, 0	ΔH 25°, 0	ΔS 25°, 0
H^+	$HL/H.L$	2.26 2.20[b]	2.45	2.55 ±0.06	-2.90 -3.01[s]	1.9
Ca^{2+}	$ML/M.L$	0.8[r]		1.08		
Sr^{2+}	$ML/M.L$	0.5[r]				
Eu^{3+}	$ML/M.L$		1.97		-1.2[d]	5[d]
	$ML_2/M.L^2$		3.32		-3.6[d]	3[d]
	$ML_3/M.L^3$		3.79			

[b] 25°, 0.5; [d] 25°, 2.0; [r] 25°, 0.16; [s] 25°, 0.05

Pyruvic acid (continued)

Metal ion	Equilibrium	Log K 25°, 0.1	Log K 25°, 2.0	Log K 25°, 0	ΔH 25°, 0	ΔS 25°, 0
Mn^{2+}	ML/M.L	1.26[t]				
Ni^{2+}	ML/M.L	1.12[b]				
	$ML_2/M.L^2$	0.46[b]				
Cu^{2+}	ML/M.L			2.2		
	$ML_2/M.L^2$			4.9		
Zn^{2+}	ML/M.L	1.26[b]				
	$ML_2/M.L^2$	1.98[b]				
Pb^{2+}	ML/M.L		2.04[e]			
	$ML_2/M.L^2$		3.40[e]			

[b] 25°, 0.5; [e] 25°, 3.0; [t] 25°, 0.65

Bibliography:

H^+ 52P,64LS,66LH,67OW,71ALN Cu^{2+} 58GH

Ca^{2+},Sr^{2+} 38Da,52SL Pb^{2+} 69LW

Eu^{3+} 71ALN   Other references: 33CC,33La,50SR,69PG,69RRV

Mn^{2+},Ni^{2+},Zn^{2+} 64LS,66LH

$$\overset{\overset{O}{\parallel}}{CH_3C}OCH_2CO_2H$$

$C_4H_6O_4$		Acetoxyacetic acid	HL

Metal ion	Equilibrium	Log K 30°, 0.4
H^+	HL/H.L	2.81
Mg^{2+}	ML/M.L	0.31
Ca^{2+}	ML/M.L	0.32
Co^{2+}	ML/M.L	0.39
Ni^{2+}	ML/M.L	0.63
Cu^{2+}	ML/M.L	1.22
Zn^{2+}	ML/M.L	0.67
Cd^{2+}	ML/M.L	1.1
Pb^{2+}	ML/M.L	1.17

Bibliography: 70BT

$$\underset{\text{CH}_3\text{CNHCH}_2\text{CO}_2\text{H}}{\overset{\displaystyle O}{\overset{\|}{}}}$$

$C_4H_7O_3N$			N-Acetylglycine				HL
Metal ion	Equilibrium	Log K 25°, 0.1	Log K 25°, 1.0	Log K 25°, 0	ΔH 25°, 0	ΔS 25°, 0	
H^+	HL/H.L	3.47[b] 3.44[b]	3.46	3.670	0.18 ±0.03	17.4	
Mg^{2+}	ML/M.L	0.32[s]					
Ca^{2+}	ML/M.L	0.29[s]					
La^{3+}	ML/M.L	1.61					
Nd^{3+}	ML/M.L	1.86					
Sm^{3+}	ML/M.L	1.88					
Eu^{3+}	ML/M.L		1.95		(4)[t]	(22)[c]	
Tb^{3+}	ML/M.L		1.65		(4)[t]	(21)[c]	
Er^{3+}	ML/M.L	1.51					
Co^{2+}	ML/M.L	0.54[s]					
Ni^{2+}	ML/M.L	0.58[s]					
Cu^{2+}	ML/M.L	1.30[s]					
Zn^{2+}	ML/M.L	0.86[b]					
Cd^{2+}	ML/M.L	1.23[b]					
CH_3Hg^+	ML/M.L	2.68[u]					
Pb^{2+}	ML/M.L $ML_2/M.L^2$	1.38[b] 2.58[b]					

[b] 25°, 0.6;   [c] 25°, 1.0;   [s] 30°, 0.4;   [t] 5-35°, 1.0;   [u] 25°, 0.38

Bibliography:

H^+ 37N,56KK,67Aa,70BT,71JB,71RC,73LR Zn^{2+},Cd^{2+},Pb^{2+} 70BT,72R

$Mg^{2+},Ca^{2+},Co^{2+}-Cu^{2+}$ 70BT CH_3Hg^+ 73LR

$La^{3+}-Er^{3+}$ 71RC   Other references: 60KF,70RS

$$\underset{\text{CH}_3\text{CNHCH}_2\text{CNHCH}_2\text{CO}_2\text{H}}{\overset{\displaystyle O \qquad\;\; O}{\overset{\|\qquad\;\;\; \|}{}}}$$

$C_6H_{10}O_4N_2$		N-Acetylglycylglycine	HL
Metal ion	Equilibrium	Log K 20°, 1.0	
H^+	HL/H.L	3.36	
Cu^{2+}	ML/M.L	1.41	

Bibliography: 60KF

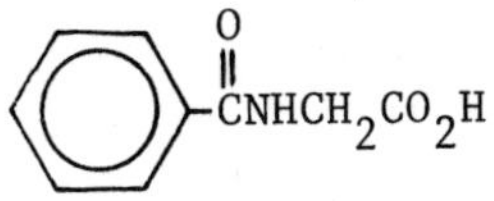

$$\text{(structure: benzene ring)}-\overset{\overset{\text{O}}{\|}}{\text{C}}\text{NHCH}_2\text{CO}_2\text{H}$$

$C_9H_9O_3N$		N-Benzoylglycine (hippuric acid)			HL
Metal ion	Equilibrium	Log K 30°, 0.1	Log K 29°, 1.0	Log K 25°, 0	
H^+	HL/H.L	3.50			
Ca^{2+}	ML/M.L			0.43	
VO^{2+}	$MOHL.H^2/M.HL$	-4.69			
Cd^{2+}	$ML/M.L$		0.95		
	$ML_2/M.L^2$		1.45		
Pb^{2+}	$ML/M.L$		1.04		
	$ML_2/M.L^2$		1.65		

Bibliography:

H^+, VO^{2+} 75STa Cd^{2+}, Pb^{2+} 69GG

Ca^{2+} 50DW   Other reference: 68S

$$\text{H}_3\text{C}-\text{(benzene ring)}-\text{SO}_2\text{NHCHCO}_2\text{H} \quad (\text{with } \overset{\text{CH}_3}{|} \text{ on CH})$$

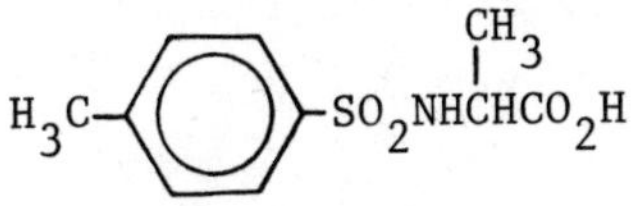

$C_{10}H_{13}O_4NS$		L-2-(4-Methylphenylsulfonylamino)propanoic acid (N-tosylalanine)	H_2L
Metal ion	Equilibrium	Log K 25°, 1.0	
H^+	HL/H.L	12.05	
	$H_2L/HL.H$	3.39	
Cu^{2+}	ML/M.L	8.96	
	ML/MOHL.H	7.05	

Bibliography: 73FB

$$\text{CH}_3\text{OCH}_2\text{CO}_2\text{H}$$

$C_3H_6O_3$		Methoxyacetic acid				HL
Metal ion	Equilibrium	Log K 20°, 0.1	Log K 20°, 1.0	Log K 25°, 0	ΔH 25°, 0	ΔS 25°, 0
H^+	HL/H.L	3.31	3.37	3.570	0.95 ±0.01	19.6
Ca^{2+}	ML/M.L			1.12		
Y^{3+}	$ML/M.L$	2.00				
	$ML_2/M.L^2$	3.11				

Methoxyacetic acid (continued)

Metal ion	Equilibrium	Log K 20°, 0.1	Log K 20°, 1.0	Log K 25°, 0	ΔH 25°, 0	ΔS 25°, 0
La^{3+}	$ML/M.L$	2.03				
	$ML_2/M.L^2$	2.88				
Ce^{3+}	$ML/M.L$	2.06				
	$ML_2/M.L^2$	3.06				
Pr^{3+}	$ML/M.L$	2.07				
	$ML_2/M.L^2$	3.25				
Nd^{3+}	$ML/M.L$	2.11				
	$ML_2/M.L^2$	3.34				
Sm^{3+}	$ML/M.L$	2.13				
	$ML_2/M.L^2$	3.39				
Eu^{3+}	$ML/M.L$	2.12				
	$ML_2/M.L^2$	3.42				
Gd^{3+}	$ML/M.L$	2.06				
	$ML_2/M.L^2$	3.01				
Tb^{3+}	$ML/M.L$	2.05				
	$ML_2/M.L^2$	3.00				
Dy^{3+}	$ML/M.L$	2.05				
	$ML_2/M.L^2$	3.13				
Ho^{3+}	$ML/M.L$	2.07				
	$ML_2/M.L^2$	3.22				
Er^{3+}	$ML/M.L$	2.08				
	$ML_2/M.L^2$	3.23				
Tm^{3+}	$ML/M.L$	2.08				
	$ML_2/M.L^2$	3.23				
Yb^{3+}	$ML/M.L$	2.08				
	$ML_2/M.L^2$	3.36				
Lu^{3+}	$ML/M.L$	2.09				
	$ML_2/M.L^2$	3.37				
Co^{2+}	$ML/M.L$		1.16[d]			
	$ML_2/M.L^2$		1.63[d]			
Ni^{2+}	$ML/M.L$		1.26[d]			
	$ML_2/M.L^2$		2.09[d]			
Cu^{2+}	$ML/M.L$		1.82 1.81[d]	2.01[e]		
	$ML_2/M.L^2$		2.81 3.04[d]	3.34[e]		
	$ML_3/M.L^3$		3.11			
	$ML_4/M.L^4$		2.8			

[d] 25°, 2.0; [e] 25°, 3.0

Bibliography:

H^+ 60K,61Sb,64PK,68CO

Ca^{2+} 38Da

$Y^{3+}-Lu^{3+}$ 64PK

Co^{2+},Ni^{2+} 72TSa

Cu^{2+} 61Sb,64SM,72TSa

Other reference: 74MMa

$$CH_3CH_2OCH_2CO_2H$$

		Ethoxyacetic acid			HL

$C_4H_8O_3$

Metal ion	Equilibrium	Log K 25°, 1.0	Log K 25°, 2.0	Log K 18°, 0
H^+	HL/H.L	3.51		3.65
Co^{2+}	$ML/M.L$		1.06	
	$ML_2/M.L^2$		1.84	
Ni^{2+}	$ML/M.L$	1.02	1.17	
	$ML_2/M.L^2$	1.51	1.91	
	$ML_3/M.L^3$	1.23		
Cu^{2+}	$ML/M.L$	1.79^j	1.74	
	$ML_2/M.L^2$	2.87^j	3.23	
	$ML_3/M.L^3$	3.20^j		
	$ML_4/M.L^4$	2.8^j		
Zn^{2+}	$ML/M.L$		1.13	
	$ML_2/M.L^2$		1.87	
	$ML_3/M.L^3$		1.78	
Cd^{2+}	$ML/M.L$	1.07		
	$ML_2/M.L^2$	1.69		
	$ML_3/M.L^3$	1.54		
	$ML_4/M.L^4$	(1.99)		
Pb^{2+}	$ML/M.L$	1.72		
	$ML_2/M.L^2$	2.65		
	$ML_3/M.L^3$	2.65		

[j] 20°, 1.0

Bibliography:

H^+	33La,61Sb,69S		Zn^{2+}	70Sa
Co^{2+}	72TSb		Cd^{2+}	69S
Ni^{2+}	70S,72TSb		Pb^{2+}	70Sb
Cu^{2+}	61Sb,72TSb		Other references:	74MMa,74MMb

$$\begin{matrix} H_3C \\ \\ H_3C \end{matrix} CHOCH_2CO_2H$$

	2-Propoxyacetic acid	HL

$C_5H_{10}O_3$

Metal ion	Equilibrium	Log K 25°, 1.0
H^+	HL/H.L	3.59
Cu^{2+}	$ML/M.L$	1.78
	$ML_2/M.L^2$	2.89

Bibliography: 72Sd

$C_8H_8O_3$		Phenoxyacetic acid				HL
Metal ion	Equilibrium	Log K 25°, 0.1	Log K 25°, 3.0	Log K 25°, 0	ΔH 25°, 0.1	ΔS 25°, 0.1
H^+	HL/H.L	2.99 ±0.06	3.36	3.15 ±0.02	1.12	18.2
Ni^{2+}	ML/M.L	0.3				
Cu^{2+}	ML/M.L	1.3				
Ag^+	ML/M.L	0.92[r]			-1.1[r]	1[r]
Zn^{2+}	ML/M.L	0.5				
Cd^{2+}	ML/M.L	1.0				
Pb^{2+}	ML/M.L	1.4				

[r] 20°, 0.2, acetate buffer

Bibliography:

H^+ 33La,43HB,62SY,64SM,68PRS,68RSb,69BLP Ag^+ 68PSW,69BLP

Ni^{2+},Cu^{2+},Zn^{2+}-Pb^{2+} 62SY

$C_5H_4O_3$		Oxole-2-carboxylic acid (2-furoic acid)				HL
Metal ion	Equilibrium	Log K 25°, 0.1	Log K 25°, 1.0	Log K 25°, 0	ΔH 25°, 0	ΔS 25°, 0
H^+	HL/H.L	2.97 ±0.02	3.06	3.162 ±0.007	(2)[t]	(20)
Ca^{2+}	ML/M.L			-0.1		
La^{3+}	ML/M.L $ML_2/M.L^2$	1.79 2.91				
Nd^{3+}	ML/M.L $ML_2/M.L^2$	1.85 3.02				
Sm^{3+}	ML/M.L $ML_2/M.L^2$	1.91 3.12				
Gd^{3+}	ML/M.L $ML_2/M.L^2$	1.84 3.03				
Er^{3+}	ML/M.L $ML_2/M.L^2$	1.74 2.89				
Co^{2+}	ML/M.L		0.51			
Ni^{2+}	ML/M.L		0.81			
Cu^{2+}	ML/M.L			1.35 ±0.03		

[t] 25-40°, 0

2-Furoic acid (continued)

Bibliography:

H^+	37GJ,60Lb,69RF,72LP	$La^{3+}-Er^{3+}$	69RF
Ca^{2+}	60Lb	$Co^{2+}-Cu^{2+}$	60Lb,72LP

$C_8H_8O_3$		2-Methoxybenzoic acid				HL
Metal ion	Equilibrium	Log K 25°, 0.1	Log K 25°, 0	ΔH 25°, 0	ΔS 25°, 0	
H^+	HL/H.L	3.89 +0.01	4.094	1.60	24.1	
Th^{4+}	$ML/M.L$	3.7				
	$ML_2/M.L_2$	6.8				
	$ML_3/M.L_3$	9.3				
	$ML_4/M.L_4$	11.2				
UO_2^{2+}	MOHL/M.OH.L	11.9				
Ni^{2+}	ML/M.L	0.8				
Cu^{2+}	ML/M.L	1.6				
Zn^{2+}	ML/M.L	0.9				
Cd^{2+}	ML/M.L	1.3				
Pb^{2+}	ML/M.L	1.9				

Bibliography:

H^+	37DLa,56Ha,59ZPL,60YY	UO_2^{2+}	56H
Th^{4+}	56Ha	$Ni^{2+}-Pb^{2+}$	60YY

$C_2H_4O_2S$		Mercaptoacetic acid (thioglycolic acid)					HL
Metal ion	Equilibrium	Log K 25°, 0.1	Log K 25°, 1.0	Log K 25°, 0	ΔH 25°, 0	ΔS 25°, 0	
H^+	HL/H.L	10.11 ±0.1	9.90[b]	10.55 ±0.01	-6.2 ±0.0 -6.52[b]	28 23.5[b]	
	$H_2L/HL.H$	3.43 ±0.04	3.42[b]±0.03	3.64 ±0.04	0.11[b]	16.0[b]	
Y^{3+}	MHL/M.HL	1.91[h]	1.49[d]				
	$M(HL)_2/M.(HL)^2$	3.19[h]	2.2[d]				
La^{3+}	MHL/M.HL	1.98[h]	1.42[d]		1.5[d]	12[d]	
	$M(HL)_2/M.(HL)^2$	2.98[h]	2.1[d]		3.9[d]	23[d]	
Ce^{3+}	MHL/M.HL	1.99[h]	1.42[d]				
	$M(HL)_2/M.(HL)^2$	3.03[h]	2.1[d]				

[b] 25°, 0.5;   [d] 25°, 2.0;   [h] 20°, 0.1

Mercaptoacetic acid (continued)

Metal ion	Equilibrium	Log K 25°, 0.1	Log K 25°, 1.0	Log K 25°, 0	ΔH 25°, 0	ΔS 25°, 0
Pr^{3+}	MHL/M.HL	2.03[h]				
	$M(HL)_2/M.(HL)^2$	3.07[h]				
Nd^{3+}	MHL/M.HL	2.07[h]	1.49[d]			
	$M(HL)_2/M.(HL)^2$	3.27[h]	2.3[d]			
Sm^{3+}	MHL/M.HL	2.11[h]	1.81[d]		1.3[d]	13[d]
	$M(HL)_2/M.(HL)^2$	3.47[h]	2.7[d]		3.3[d]	24[d]
Eu^{3+}	MHL/M.HL	2.07[h]	1.54[i]			
			1.75[d]			
	$M(HL)_2/M.(HL)^2$	3.41[h]	2.27[i]			
			2.6[d]			
Gd^{3+}	MHL/M.HL	2.01[h]	1.64[d]			
	$M(HL)_2/M.(HL)^2$	3.31[h]	2.4[d]			
Tb^{3+}	MHL/M.HL	1.96[h]	1.63[d]			
	$M(HL)_2/M.(HL)^2$	3.22[h]	2.4[d]			
Dy^{3+}	MHL/M.HL	1.94[h]				
	$M(HL)_2/M.(HL)^2$	3.26[h]				
Ho^{3+}	MHL/M.HL	1.92[h]	1.32[d]			
	$M(HL)_2/M.(HL)^2$	3.24[h]	2.1[d]			
Er^{3+}	MHL/M.HL	1.94[h]	1.28[d]		2.4[d]	14[d]
	$M(HL)_2/M.(HL)^2$	3.26[h]	2.2[d]		5.4[d]	28[d]
Tm^{3+}	MHL/M.HL	1.98[h]				
	$M(HL)_2/M.(HL)^2$	3.09[h]				
Yb^{3+}	MHL/M.HL	1.98[h]	1.32[d]			
	$M(HL)_2/M.(HL)^2$	3.30[h]	2.2[d]			
Lu^{3+}	MHL/M.HL	2.01[h]				
	$M(HL)_2/M.(HL)^2$	3.31[h]				
Am^{3+}	MHL/M.HL		1.55[i]			
	$M(HL)_2/M.(HL)^2$		2.6[i]			
Mn^{2+}	$ML/M.L$	4.38			(4)[t]	(30)[a]
	$ML_2/M.L^2$	7.56			(5)[t]	(50)[a]
Fe^{2+}	$ML_2/M.L^2$			10.92		
	ML/MOHL.H			1.62		
Co^{2+}	$ML/M.L$	5.84				
	$ML_2/M.L^2$	12.15				
Ni^{2+}	$ML_2/M.L^2$	13.01[h]±0.00	12.76[b]		-3[b]	50[b]
	$ML_3/M.L^3$	14.99[h]				
	$M_2L_3/M^2.L^3$	22.7[h]				
	$M_3L_4/M^3.L^4$	33.27[h]	32.22[b]		-21[b]	75[b]
	$M_4L_6.M^4.L^6$	49.85[h]	49.16[b]		-31[b]	120[b]
Zn^{2+}	$ML/M.L$	7.86				
		7.80[h]				
	$ML_2/M.L^2$	15.04				
		14.96[h]				
	$M_2L_3/M^2.L^3$	25.2[h]				
	$M_3L_4/M^3.L^4$	36.47[h]				

a 25°, 0.1; b 25°, 0.5; d 25°, 2.0; h 20°, 0.1; i 20°, 0.5; t 25-35°, 0.1

Mercaptoacetic acid (continued)

Metal ion	Equilibrium	Log K 25°, 0.1	Log K 25°, 1.0	Log K 25°, 0	ΔH 25°, 0	ΔS 25°, 0
Hg^{2+}	$ML_2/M.L^2$		43.8		$(-50)^u$	$(30)^c$
Sb(III)	$ML_2.(OH)^3/M(OH)_3.L^2$	-6.92^h				
	$ML_2/MOHL_2.H$	7.58^h				

^{c} 25°, 1.0; ^{h} 20°, 0.1; ^{u} 12-25°, 1.0

Bibliography:

H^+ 28L,53LK,55LM,58La,62CTC,62G,63C, 64IN,64PKP,64WI,65LJ,67PS,68MDK, 68PS,70AM,71RR,74DV

$Y^{3+}-Lu^{3+}$ 62BC,62G,64G,64PKP

Am^{3+} 62G

Mn^{2+} 58La,64SM

Fe^{2+} 53LK

Co^{2+} 58La

Ni^{2+} 60LL,67PS,74DV

Zn^{2+} 58La,67PS

Hg^{2+} 54SK

Sb(III)   70AM

Other references: 62CTG,65FCW,69RRV,69SCF, 70RB,72GP,73SM,75GS,75Y

$$\underset{\overset{|}{CH_3CHCO_2H}}{SH}$$

$C_3H_6O_2S$		DL-2-Mercaptopropanoic acid (thiolactic acid)				HL
Metal ion	Equilibrium	Log K 25°, 0.1	Log K 25°, 0.5	Log K 25°, 0	ΔH 25°, 0.5	ΔS 25°, 0.5
H^+	HL/H.L	10.08 ±0.02	9.93		-5.74	26.2
	$H_2L/HL.H$	3.48 ±0.05	3.38 ±0.02	3.69	0.49	17.1
Y^{3+}	MHL/M.HL		1.70^d		1.3^d	12^d
La^{3+}	MHL/M.HL		1.2^d		2.6^d	13^d
Ce^{3+}	MHL/M.HL		1.36^d		2.3^d	14^d
Pr^{3+}	MHL/M.HL		1.89^d		1.4^d	13^d
Nd^{3+}	MHL/M.HL		1.93^d		1.3^d	13^d
Sm^{3+}	MHL/M.HL		2.06^d		1.0^d	12^d
Eu^{3+}	MHL/M.HL		2.00^d		1.2^d	13^d
Gd^{3+}	MHL/M.HL		1.75^d		1.8^d	13^d
Tb^{3+}	MHL/M.HL		1.65^d		1.6^d	12^d
Dy^{3+}	MHL/M.HL		1.58^d		1.7^d	13^d
Ho^{3+}	MHL/M.HL		1.5^d		1.9^d	13^d
Er^{3+}	MHL/M.HL		1.5^d		2.5^d	15^d
Tm^{3+}	MHL/M.HL		1.5^d		2.6^d	15^d
Yb^{3+}	MHL/M.HL		1.4^d		3.2^d	17^d
Lu^{3+}	MHL/M.HL		1.51^d		1.7^d	12^d

^{d} 25°, 2.0; ^{z} optical isomerism not stated.

2-Mercaptopropanoic acid (continued)

Metal ion	Equilibrium	Log K 25°, 0.1	Log K 25°, 0.5	Log K 25°, 0	ΔH 25°, 0.5	ΔS 25°, 0.5
Ni^{2+}	ML/M.L		6.05			
	$ML_2/M.L^2$		$(13.14)^z$		(-0.2)	(59)
	$M_3L_4/M^3.L^4$		$(30.71)^z$		(-8)	(113)
Zn^{2+}	ML/M.L		6.85			
	$ML_2/M.L^2$		$(14.34)^z$		(-3)	(55)
	$M_2L_2/M^2.L_4^2$		$(17.16)^z$			
	$M_3L_4^2/M^3.L^4$		$(34.74)^z$		(-19)	(95)

^{z} optical isomerism not stated.

Bibliography:

H^+ 28L,63C,69RB,71RR,70SSa,75DG Ni^{2+},Zn^{2+} 75DG

Y^{3+}-Lu^{3+} 68CM

Other references: 63BC,66TF,70SS,72SS,73SM

$$HSCH_2CH_2CO_2H$$

$C_3H_6O_2S$ 3-Mercaptopropanoic acid H_2L

Metal ion	Equilibrium	Log K 30°, 0.1	Log K 25°, 2.0	Log K 25°, 0	ΔH 25°, 0	ΔS 25°, 0
H^+	HL/H.L			10.84	-6.10^s	29
	$H_2L/HL.H$	4.16		4.34		
Y^{3+}	MHL/M.HL		1.51		5.0^d	24^d
La^{3+}	MHL/M.HL		1.54		2.6^d	16^d
Ce^{3+}	MHL/M.HL		1.57		2.4^d	15^d
Pr^{3+}	MHL/M.HL		1.57		2.6^d	16^d
Nd^{3+}	MHL/M.HL		1.74		2.5^d	16^d
Sm^{3+}	MHL/M.HL		1.74		1.7^d	14^d
Eu^{3+}	MHL/M.HL		1.64		2.2^d	15^d
Gd^{3+}	MHL/M.HL		1.71		2.6^d	17^d
Tb^{3+}	MHL/M.HL		1.62		2.9^d	17^d
Dy^{3+}	MHL/M.HL		1.60		3.5^d	19^d
Ho^{3+}	MHL/M.HL		1.46		2.2^d	18^d
Er^{3+}	MHL/M.HL		1.42		4.8^d	23^d
Tm^{3+}	MHL/M.HL		1.40		4.7^d	22^d
Yb^{3+}	MHL/M.HL		1.43		4.0^d	20^d
Lu^{3+}	MHL/M.HL		1.49		4.9^d	23^d

^{d} 25°, 2.0; ^{s} 25°, 0.05

Bibliography:

H^+ 28L,62CTC,64IN

Y^{3+}-Lu^{3+} 68CM

Other references: 63BC,63SC,65FCW,66TF,
68SGc,68SGd,69SG,69SGa,69SGb,69SGc,
69SGM,72BD,72SM

$$CH_3SCH_2CO_2H$$

		(Methylthio)acetic acid			HL
$C_3H_6O_2S$					
Metal ion	Equilibrium	Log K 25°, 0.1	Log K 25°, 1.0	Log K 25°, 0	
H^+	HL/H.L	3.55	3.61	3.72	
Cu^{2+}	ML/M.L		2.40		
	$ML_2/M.L^2$		4.35		
Ag^+	ML/M.L	3.90^z			
	$ML_2/M.L^2$	7.4^z			
	MHL/M.HL	$3.16^{r,z}$			

^{r} 25°, 0.2; ^{z} acetate buffer.

Bibliography:
H^+ 30L,68PS,71S Ag^+ 68PS
Cu^{2+} 71Se

$$CH_3CH_2SCH_2CO_2H$$

		(Ethylthio)acetic acid			HL
$C_4H_8O_2S$					
Metal ion	Equilibrium	Log K 25°, 0.1	Log K 25°, 1.0	Log K 31°, 2.0	
H^+	HL/H.L	3.60	3.65	(3.62)	
Y^{3+}	ML/M.L			1.42	
	$ML_2/M.L^2$			2.1	
La^{3+}	ML/M.L			1.70	
	$ML_2/M.L^2$			2.5	
Nd^{3+}	ML/M.L			1.72	
	$ML_2/M.L^2$			2.5	
Sm^{3+}	ML/M.L			1.85	
	$ML_2/M.L^2$			2.8	
Eu^{3+}	ML/M.L			1.79	
	$ML_2/M.L^2$			2.7	
Gd^{3+}	ML/M.L			1.70	
	$ML_2/M.L^2$			2.6	
Tb^{3+}	ML/M.L			1.53	
	$ML_2/M.L^2$			2.5	
Ho^{3+}	ML/M.L			1.43	
	$ML_2/M.L^2$			2.4	
Er^{3+}	ML/M.L			1.42	
	$ML_2/M.L^2$			2.4	
Yb^{3+}	ML/M.L			1.40	
	$ML_2/M.L^2$			2.4	
Ni^{2+}	ML/M.L		1.04		
	$ML_2/M.L^2$		1.81		
	$ML_3/M.L^3$		2.28		

(Ethylthio)acetic acid (continued)

Metal ion	Equilibrium	Log K 25°, 0.1	Log K 25°, 1.0	Log K 31°, 2.0
Cu^{2+}	$ML/M.L$		2.56[j]	
	$ML_2/M.L^2$		4.76[j]	
	$ML_3/M.L^3$		4.85[j]	
Ag^+	$ML/M.L$	3.92[z]		
	$ML_2/M.L^2$	7.0[z]		
	$MHL/M.HL$	3.17[r,z]		
Zn^{2+}	$ML/M.L$		0.74	
	$ML_2/M.L^2$		1.20	
	$ML_3/M.L^3$		1.15	
Cd^{2+}	$ML/M.L$		1.27	
	$ML_2/M.L^2$		2.12	
	$ML_3/M.L^3$		2.51	
	$ML_4/M.L^4$		2.72	
Pb^{2+}	$ML/M.L$		1.72	
	$ML_2/M.L^2$		2.83	

[j] 20°, 1.0;   [r] 25°, 0.2;   [z] acetate buffer.

Bibliography:

H^+	61Sb,63BC,69S	Zn^{2+}	70Sa
$Y^{3+}-Yb^{3+}$	63BC	Cd^{2+}	69S
Ni^{2+}	70S	Pb^{2+}	71Sc
Cu^{2+}	61Sb		

Other references: 44L,72SC,72SCe,74MMa,74MMb, 74MMc

Ag^+ 68PS

$$Z-SCH_2CO_2H$$

$C_nH_mO_2S$			(Substitutedthio)acetic acid			HL
Z =	Metal ion	Equilibrium	Log K 25°, 0.1	Log K 25°, 0	ΔH 25°, 0.2	ΔS 25°, 0.2
Propyl	H^+	$HL/H.L$	3.62	3.77		
$(C_5H_{10}O_2S)$	Ag^+	$ML/M.L$	3.94[z]			
		$ML_2/M.L^2$	7.2[z]			
		$MHL/M.HL$		3.18[r,z]		
Butyl	H^+	$HL/H.L$	3.63	3.81		
$(C_6H_{12}O_2S)$	Cu^{2+}	$ML/M.L$	1.95			
	Ag^+	$ML/M.L$	4.02	3.92[r,z]	-6.8[z]	-5[z]
		$ML_2/M.L^2$	6.7[z]	6.70[r,z]	-7.2[z]	7[z]
		$MHL/M.HL$		3.15[r,z]		
		$M(HL)_2/M.(HL)^2$		5.31[r,z]		
	Cd^{2+}	$ML/M.L$	0.91			

[r] 25°, 0.2;   [z] actate buffer.

(Substitutedthio)acetic acid (continued)

$Z =$	Metal ion	Equilibrium	Log K 25°, 0.1	Log K 25°, 0	ΔH 25°, 0.2	ΔS 25°, 0.2
1-Methylpropyl	H^+	$HL/H.L$	3.66			
$(C_6H_{12}O_2S)$	Ag^+	$ML/M.L$	4.16^z			
		$ML_2/M.L^2$	7.0^z			
		$MHL/M.HL$		$3.34^{r,z}$		
Pentyl	H^+	$HL/H.L$	3.64			
$(C_7H_{14}O_2S)$	Ag^+	$ML/M.L$	3.92^z			
		$ML_2/M.L^2$	7.0^z			
		$MHL/M.HL$		$3.15^{r,z}$		
Hexyl	H^+	$HL/H.L$	3.65			
$(C_8H_{16}O_2S)$	Ag^+	$ML/M.L$	3.94^z			
		$ML_2/M.L^2$	7.1^z			
		$MHL/M.HL$		$3.16^{r,z}$		
Prop-2-enyl	H^+	$HL/H.L$	3.57			
$(C_5H_8O_2S)$	Ag^+	$ML/M.L$	3.78^z	$3.74^{r,z}$	-6.9^z	-6^z
		$ML_2/M.L^2$	6.8^z	$6.71^{r,z}$	-4.8^z	15^z
		$MHL/M.HL$		$3.04^{r,z}$		
		$M(HL)_2/M.(HL)^2$		$4.98^{r,z}$		
But-3-enyl	H^+	$HL/H.L$	3.61			
$(C_6H_{10}O_2S)$	Cu^{2+}	$ML/M.L$	2.13			
	Ag^+	$ML/M.L$	(4.70)	$4.77^{r,z}$	$(-11.0)^z$	$(-15)^z$
		$ML_2/M.L^2$	7.1^z	$7.02^{r,z}$	$(-20)^z$	$(-35)^z$
		$M_2L/ML.M$		$2.06^{r,z}$	2^z	16^z
		$MHL/M.HL$		$4.16^{r,z}$		
		$M(HL)_2/M.(HL)^2$		$6.37^{r,z}$		
	Cd^{2+}	$ML/M.L$	0.51			
Pent-4-enyl	H^+	$HL/H.L$	3.63			
$(C_7H_{12}O_2S)$	Ag^+	$ML/M.L$	4.21^z	$4.19^{r,z}$	-8.4^z	-9^z
		$ML_2/M.L^2$	7.0^z	$7.00^{r,z}$	-5^z	15^z
		$M_2L/ML.M$		$1.91^{r,z}$	-7^z	-15^z
		$MHL/M.HL$		$3.43^{r,z}$		
		$M(HL)_2/M.(HL)^2$		$5.85^{r,z}$		
Benzyl	H^+	$HL/H.L$	3.60	3.73		
$(C_9H_{10}O_2S)$	Ag^+	$ML/M.L$	3.69^z			
		$ML_2/M.L^2$	6.8^z			
		$MHL/M.HL$		$2.92^{r,z}$		

[r] 25°, 0.2; [z] acetate buffer.

Bibliography:
H^+ 30L, 68PS
Cu^{2+}, Cd^{2+} 72FG

Ag^+ 68PS, 71BFP, 72FG

Other reference: 44L

$$\begin{array}{c} H_3C \\ {}^{\diagdown}CHSCH_2CO_2H \\ H_3C \diagup \end{array}$$

$C_5H_{10}O_2S$		(2-Propylthio)acetic acid		HL
Metal ion	Equilibrium	Log K 25°, 1.0	Log K 25°, 0	
H^+	HL/H.L	3.65	3.72	
Cu^{2+}	ML/M.L	2.49		
	$ML_2/M.L^2$	4.77		

Bibliography:

H^+ 30L,71Sd Cu^{2+} 71Sd

$$\begin{array}{c} CH_3 \\ | \\ CH_3CSCH_2CO_2H \\ | \\ CH_3 \end{array}$$

$C_6H_{12}O_2S$		(2-Methyl-2-propylthio)acetic acid	HL
Metal ion	Equilibrium	Log K 25°, 1.0	
H^+	HL/H.L	3.64	
Cu^{2+}	ML/M.L	2.50	
	$ML_2/M.L^2$	4.92	

Bibliography: 71Se

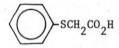

$C_8H_8O_2S$		(Phenylthio)acetic acid				HL
Metal ion	Equilibrium	Log K 25°, 0.1	Log K 20°, 0.2		ΔH 25°, 0.1	ΔS 25°, 0.1
H^+	HL/H.L	3.33 ±0.05			0.80	18.2
Mn^{2+}	ML/M.L	0.72				
Co^{2+}	ML/M.L	0.76				
Ni^{2+}	ML/M.L	0.7 -0.4				
Cu^{2+}	ML/M.L	1.43 +0.1				
Ag^+	ML/M.L	2.83 2.77[z]			-5.3[z]	-5[z]
	$ML_2/M.L^2$	4.9 [z]	4.8 [h,z]			
	MHL/M.HL		2.16[z]			
Zn^{2+}	ML/M.L	0.8 -0.5				

[h] 20°, 0.1;   [z] acetate buffer.

(Phenylthio)acetic acid (continued)

Metal ion	Equilibrium	Log K 25°, 0.1	Log K 20°, 0.2	ΔH 25°, 0.1	ΔS 25°, 0.1
Cd^{2+}	ML/M.L	1.2 −0.5			
Pb^{2+}	ML/M.L	1.8			

Bibliography:

H^+	62SY,68PRS,69BLP	Ag^+	68PRW,69BLP,72FG
Mn^{2+}-Cu^{2+},Zn^{2+}-Pb^{2+}	62SY,72FG	Other reference:	44L

$C_n H_m O_p S_q X_r$			(Substituted phenylthio)acetic acid			HL
Z =	Metal ion	Equilibrium	Log K 25°, 0.1	Log K 20°, 0.1	ΔH 25°, 0.1	ΔS 25°, 0.1
2-Methyl	H^+	HL/H.L		3.27		
$(C_9H_{10}O_2S)$	Ag^+	ML/M.L ML$_2$/M.L^2		2.73[z] 4.8 [z]		
		MHL/M.HL		2.16[r,z]		
3-Methyl	H^+	HL/H.L		3.28		
$(C_9H_{10}O_2S)$	Ag^+	ML/M.L ML$_2$/M.L^2		2.86[z] 5.1 [z]		
		MHL/M.HL		2.30[r,z]		
4-Methyl	H^+	HL/H.L		3.34		
$(C_9H_{10}O_2S)$	Ag^+	ML/M.L ML$_2$/M.L^2	2.91[z] 5.0 [z]	2.98[z] 5.3 [z]	−5.9[z]	−6[z]
		MHL/M.HL		2.34[r,z]		
2-Chloro	H^+	HL/H.L		3.12		
$(C_8H_7O_2ClS)$	Ag^+	ML/M.L ML$_2$/M.L^2		2.49[z] 4.6 [z]		
		MHL/M.HL		1.87[r,z]		
3-Chloro	H^+	HL/H.L		3.19		
$(C_8H_7O_2ClS)$	Ag^+	ML/M.L ML$_2$/M.L^2	2.43[z] 4.3 [z]	2.51[z] 4.6 [z]	−5.4[z]	−7[z]
		MHL/M.HL		2.04[r,z]		
4-Chloro	H^+	HL/H.L		3.22		
$(C_8H_7O_2ClS)$	Ag^+	ML/M.L ML$_2$/M.L^2	2.57[z] 4.4 [z]	2.64[z] 4.7 [z]	−5.4[z]	−6[z]
		MHL/M.HL		2.07[r,z]		

[r] 20°, 0.2; [z] acetate buffer.

(Substitutedphenylthio)acetic acid (continued)

Z =	Metal ion	Equilibrium	Log K 25°, 0.1	Log K 20°, 0.1	ΔH 25°, 0.1	ΔS 25°, 0.1
4-Bromo	H^+	HL/H.L		3.22		
$(C_8H_7O_2BrS)$	Ag^+	ML/M.L	2.53^z	2.61^z	-5.4^z	-7^z
		$ML_2/M.L^2$	4.6^z	4.6^z		
		MHL/M.HL		$2.21^{r,z}$		
3-Trifluoromethyl	H^+	HL/H.L		3.19		
$(C_9H_7O_2F_3S)$	Ag^+	ML/M.L		2.42^z		
		$ML_2/M.L^2$		4.0^z		
		MHL/M.HL		$1.76^{r,z}$		
4-Cyano	H^+	HL/H.L		3.01		
$(C_9H_7O_2NS)$	Ag^+	ML/M.L		2.09^z		
		$ML_2/M.L^2$		3.8^z		
		MHL/M.HL		$1.68^{r,z}$		
2-Nitro	H^+	HL/H.L		2.99		
$(C_8H_7O_4NS)$	Ag^+	ML/M.L		1.94^z		
		MHL/M.HL		$1.84^{r,z}$		
4-Nitro	H^+	HL/H.L		2.98		
$(C_8H_7O_4NS)$	Ag^+	ML/M.L	1.86^z	1.98^z	-4.5^z	-7^z
		MHL/M.HL		$1.58^{r,z}$		
2-Methoxy	H^+	HL/H.L		3.48		
$(C_9H_{10}O_3S)$	Mn^{2+}	ML/M.L	0.51			
	Co^{2+}	ML/M.L	0.74			
	Ni^{2+}	ML/M.L	(-0.1)			
	Cu^{2+}	ML/M.L	1.84			
	Ag^+	ML/M.L	2.95	3.02^z		
		$ML_2/M.L^2$		6.2^z		
		MHL/M.HL		$2.75^{r,z}$		
	Zn^{2+}	ML/M.L	1.00			
	Cd^{2+}	ML/M.L	0.86			
3-Methoxy	H^+	HL/H.L		3.28		
$(C_9H_{10}O_3S)$	Mn^{2+}	ML/M.L	0.59			
	Co^{2+}	ML/M.L	0.50			
	Ni^{2+}	ML/M.L	(0.23)			
	Cu^{2+}	ML/M.L	1.65			
	Ag^+	ML/M.L	2.70 2.65^z	2.73^z	-5.6^z	-7^z
		$ML_2/M.L^2$	4.6^z	4.9^z		
		MHL/M.HL		$2.12^{r,z}$		
	Zn^{2+}	ML/M.L	0.68			
	Cd^{2+}	ML/M.L	0.79			

^{r} 20°, 0.2; ^{z} acetate buffer

(Substitutedphenylthio)acetic acid (continued)

Z =	Metal ion	Equilibrium	Log K 25°, 0.1	Log K 20°, 0.1	ΔH 25°, 0.1	ΔS 25°, 0.1
4-Methoxy	H^+	HL/H.L		3.43		
$(C_9H_{10}O_3S)$	Ag^+	$ML/M.L$	2.97^z	3.07^z	-6.4^z	-8^z
		$ML_2/M.L^2$	5.2^z	5.5^z		
		MHL/M.HL		$2.45^{r,z}$		
2-Methylthio	H^+	HL/H.L		3.46		
$(C_9H_{10}O_2S_2)$	Ag^+	$ML/M.L$		3.95^z		
		$ML_2/M.L^2$		7.15^z		
		MHL/M.HL		$3.14^{r,z}$		
4-Amino	H^+	HL/H.L		4.48		
$(C_8H_9O_2NS)$		$H_2L/HL.H$		2.99		
	Ag^+	$ML/M.L$		3.22^z		
		$ML_2/M.L^2$		5.6^z		
		$MH_2L/M.H_2L$		$1.98^{r,z}$		

^{r} 20°, 0.2; ^{z} acetate buffer

Bibliography:

H^+ 68PRS Ag^+ 68PRW,69BLP,72FG

$Mn^{2+}-Cu^{2+},Zn^{2+},Cd^{2+}$ 72FG

$C_5H_4O_2S$		Thiole-2-carboxylic acid (2-thenoic acid)				HL
Metal ion	Equilibrium	Log K 25°, 0.1	Log K 25°, 1.0	Log K 25°, 0	ΔH 30°, 0	ΔS 30°, 0
H^+	HL/H.L	3.32	3.41	3.51	$(1)^t$	(20)
Ca^{2+}	ML/M.L			1.33		
La^{3+}	$ML/M.L$	1.77				
	$ML_2/M.L^2$	2.89				
Nd^{3+}	$ML/M.L$	2.01				
	$ML_2/M.L^2$	3.40				
Sm^{3+}	$ML/M.L$	2.02				
	$ML_2/M.L^2$	3.28				
Gd^{3+}	$ML/M.L$	1.92				
	$ML_2/M.L^2$	3.20				
Er^{3+}	$ML/M.L$	1.74				
	$ML_2/M.L^2$	2.79				
Co^{2+}	ML/M.L		0.83			
Ni^{2+}	ML/M.L		0.92			
Cu^{2+}	ML/M.L		1.31	1.90		

^{t} 30-40°, 0

Thiole-2-carboxylic acid (continued)

Bibliography:

H^+	60Lb,69RF,72LPa		La^{3+}-Er^{3+}	69RF
Ca^{2+}	60Lb		Co^{2+}-Cu^{2+}	60Lb,72LPN

$$Z-S-\langle\bigcirc\rangle-CO_2H$$

$C_nH_mO_2S$			4-(Substitutedthio)benzoic acid	HL
Z =	Metal ion	Equilibrium	Log K 25°, 0.2	
Methyl	H^+	HL/H.L	3.9	
$(C_8H_8O_2S)$	Ag^+	ML/M.L	2.79^z	
Ethyl $(C_9H_{10}O_2S)$	Ag^+	ML/M.L	2.82^z	
Propyl $(C_{10}H_{12}O_2S)$	Ag^+	ML/M.L	2.82^z	
Butyl $(C_{11}H_{14}O_2S)$	Ag^+	ML/M.L	2.82^z	
Pentyl $(C_{12}H_{16}O_2S)$	Ag^+	ML/M.L	2.84^z	
Hexyl $(C_{13}H_{18}O_2S)$	Ag^+	ML/M.L	2.84^z	
Prop-2-enyl $(C_{10}H_{10}O_2S)$	Ag^+	ML/M.L	2.76^z	
But-3-enyl $(C_{11}H_{12}O_2S)$	Ag^+	ML/M.L	3.84^z	
Pent-4-enyl $(C_{12}H_{14}O_2S)$	Ag^+	ML/M.L	3.05^z	
Benzyl $(C_{14}H_{12}O_2S)$	Ag^+	ML/M.L	2.76^z	
Phenyl $(C_{13}H_{10}O_2S)$	Ag^+	ML/M.L	2.44^z	

^{z} acetate buffer

Bibliography: 68PS

$$R-SeCH_2CO_2H$$

$C_nH_mO_2Se$			(Alkylseleno)acetic acid			HL
R =	Metal ion	Equilibrium	Log K 25°, 0.1	Log K 25°, 0.2	ΔH 25°, 0.2	ΔS 25°, 0.2
Butyl	H^+	HL/H.L	3.82			
$(C_6H_{12}O_2Se)$	Cu^{2+}	ML/M.L	1.95			
	Ag^+	ML/M.L	4.65	4.58^z	-7.8^z	-5^z
		$ML_2/M.L^2$		8.01^z	-6.5^z	15^z
		MHL/M.HL		3.81^z		
		$M(HL)_2/M.(HL)^2$		6.69^z		
	Cd^{2+}	ML/M.L	0.70			

^{z} acetate buffer

(Alkylseleno)acetic acid (continued)

R =	Metal ion	Equilibrium	Log K 25°, 0.1	Log K 25°, 0.2	ΔH 25°, 0.2	ΔS 25°, 0.2
But-3-enyl	H^+	HL/H.L	3.77			
$(C_6H_{10}O_2Se)$	Cu^{2+}	ML/M.L	2.02			
	Ag^+	ML/M.L	4.92	5.16^z	$(-13.0)^z$	$(-20)^z$
		$ML_2/M.L^2$		8.0^z	$(-24)^z$	$(-44)^z$
		$M_2L/ML.M$		1.97^z	-1^z	6^z
		MHL/M.HL		4.75^z		
		$M_2HL/MHL.M$		2.79^z		
	Cd^{2+}	ML/M.L	0.70			
Pent-4-enyl	H^+	HL/H.L	3.80			
$(C_7H_{12}O_2Se)$	Ag^+	ML/M.L		4.63^z	-9.1^z	-9^z
		$ML_2/M.L^2$		7.97^z	-8^z	10^z
		$M_2L/ML.M$		1.81^z	-4^z	-5^z
		MHL/M.HL		3.99^z		
		$M_2HL/MHL.M$		2.73^z		

---

^{z} acetate buffer.

Bibliography:

H^+ 71BFP Ag^+ 71BFP,72FG

Cu^{2+},Cd^{2+} 72FG

Z—⟨ ⟩—$SeCH_2CO_2H$

$C_nH_mO_pN_qSeX_r$		(Substitutedphenylseleno)acetic acid				HL
Z =	Metal ion	Equilibrium	Log K 25°, 0.1	Log K 20°, 0.1	ΔH 25°, 0.1	ΔS 25°, 0.1
Z = H	H^+	HL/H.L	3.64		0.22	17.9
$(C_8H_8O_2Se)$	Mn^{2+}	ML/M.L	0.32			
	Co^{2+}	ML/M.L	0.63			
	Cu^{2+}	ML/M.L	1.28			
	Ag^+	ML/M.L	3.68	3.70^z	-6.9^z	-7^z
			3.62^z			
		$ML_2/M.L^2$	6.1^z	6.42^z		
		MHL/M.HL		$2.96^{r,z}$		
	Zn^{2+}	ML/M.L	0.60			
	Cd^{2+}	ML/M.L	0.68			

---

^{r} 25°, 0.2; ^{z} acetate buffer.

(Substitutedphenylseleno)acetic acid (continued)

Z =	Metal ion	Equilibrium	Log K 25°, 0.1	Log K 20°, 0.1	ΔH 25°, 0.1	ΔS 25°, 0.1
2-Methyl	H^+	HL/H.L		3.65		
$(C_9H_{10}O_2Se)$	Ag^+	ML/M.L		3.53^z		
		$ML_2/M.L^2$		6.37^z		
		MHL/M.HL		$2.77^{r,z}$		
3-Methyl	H^+	HL/H.L		3.67		
$(C_9H_{10}O_2Se)$	Ag^+	ML/M.L		3.74^z		
		$ML_2/M.L^2$		6.64^z		
		MHL/M.HL		$2.99^{r,z}$		
4-Methyl	H^+	HL/H.L		3.72		
$(C_9H_{10}O_2Se)$	Ag^+	ML/M.L	3.73^z	3.82^z	-7.3^z	-7^z
		$ML_2/M.L^2$	6.5^z	6.72^z		
		MHL/M.HL		$3.02^{r,z}$		
2-Chloro	H^+	HL/H.L		3.46		
$(C_8H_7O_2ClSe)$	Ag^+	ML/M.L		3.27^r		
		$ML_2/M.L^2$		5.98^r		
		MHL/M.HL		$2.57^{r,z}$		
3-Chloro	H^+	HL/H.L		3.53		
$(C_8H_7O_2ClSe)$	Ag^+	ML/M.L	3.33^z	3.39^z	-6.9^z	-8^z
		$ML_2/M.L^2$	5.6^z	5.99^z		
		MHL/M.HL		$2.67^{r,z}$		
4-Chloro	H^+	HL/H.L		3.57		
$(C_8H_7O_2ClSe)$	Ag^+	ML/M.L	3.44^z	3.50^z	-6.9^z	-7^z
		$ML_2/M.L^2$	5.9^z	6.19^z		
		MHL/M.HL		$2.75^{r,z}$		
2-Bromo	H^+	HL/H.L		3.47		
$(C_8H_7O_2BrSe)$	Ag^+	ML/M.L		3.36^z		
		$ML_2/M.L^2$		6.28^z		
		MHL/M.HL		$2.69^{r,z}$		
4-Bromo	H^+	HL/H.L		3.59		
$(C_8H_7O_2BrSe)$	Ag^+	ML/M.L	3.42^z	3.49^z	-6.8^z	-7^z
		$ML_2/M.L^2$	6.0^z	6.31^z		
		MHL/M.HL		$2.73^{r,z}$		
2-Nitro	H^+	HL/H.L		3.31		
$(C_8H_7O_4NSe)$	Ag^+	ML/M.L		2.66^z		
		$ML_2/M.L^2$		4.9^z		
		MHL/M.HL		$2.18^{r,z}$		

[r] 25°, 0.2;   [z] acetate buffer.

(Substitutedphenylseleno)acetic acid (continued)

Z =	Metal ion	Equilibrium	Log K 25°, 0.1	Log K 20°, 0.1	ΔH 25°, 0.1	ΔS 25°, 0.1
3-Nitro	H^+	HL/H.L		3.44		
$(C_8H_7O_4NSe)$	Ag^+	ML/M.L		3.09^z		
		$ML_2/M.L^2$		5.6^z		
		MHL/M.HL		$2.35^{r,z}$		
4-Nitro	H^+	HL/H.L		3.34		
$(C_8H_7O_4NSe)$	Ag^+	ML/M.L	2.82^z	2.89^z	-6.0^z	-7^z
		$ML_2/M.L^2$	5.1^z	5.32^z		
		MHL/M.HL		$2.26^{r,z}$		
2-Methoxy	H^+	HL/H.L		3.76		
$(C_9H_{10}O_3Se)$	Mn^{2+}	ML/M.L	0.41			
	Co^{2+}	ML/M.L	0.65			
	Cu^{2+}	ML/M.L	1.65			
	Ag^+	ML/M.L		3.72^z		
		$ML_2/M.L^2$		6.83^z		
		MHL/M.HL		$3.02^{r,z}$		
	Zn^{2+}	ML/M.L	0.71			
	Cd^{2+}	ML/M.L	0.83			
3-Methoxy	H^+	HL/H.L		3.62		
$(C_9H_{10}O_3Se)$	Ag^+	ML/M.L		3.61^z		
		$ML_2/M.L^2$		6.53^z		
		MHL/M.HL		$2.90^{r,z}$		
4-Methoxy	H^+	HL/H.L		3.75		
$(C_9H_{10}O_3Se)$	Ag^+	ML/M.L	3.75^z	3.84^z	-7.8^z	-9^z
		$ML_2/M.L^2$	6.7^z	6.79^z		
		MHL/M.HL		$3.07^{r,z}$		
2-Ethoxy	H^+	HL/H.L		3.79		
$(C_{10}H_{12}O_3Se)$	Ag^+	ML/M.L		3.78^z		
		$ML_2/M.L^2$		7.07^z		
		MHL/M.HL		2.93^z		
4-Ethoxy	H^+	HL/H.L		3.75		
$(C_{10}H_{12}O_3Se)$	Ag^+	ML/M.L		3.85^z		
		$ML_2/M.L^2$		6.78^z		
		MHL/M.HL		$3.09^{r,z}$		
2-Methylthio	H^+	HL/H.L		3.69		
$(C_9H_{10}O_2SSe)$	Ag^+	ML/M.L		4.56^z		
		$ML_2/M.L^2$		8.30^z		
		MHL/M.HL		$3.86^{r,z}$		

[r] 25°, 0.2; [z] acetate buffer.

Bibliography:

H^+ 68PRS,69BLP Ag^+ 68PSW,69BLP,72FG

Mn^{2+}-Cu^{2+},Zn^{2+},Cd^{2+} 72FG

$C_5H_4O_4N_2$ 1,3-H-1,3-Diazine-2,4-dione-6-carboxylic acid H_2L
                                  (uracil-6-carboxylic acid) (orotic acid)

Metal ion	Equilibrium	Log K 25°, 0.1	Log K 25°, 0	ΔH 25°, 0	ΔS 25°, 0
H^+	HL/H.L	9.34	9.6	-8.7	14
	H_2L/HL.H	1.96	1.8	-0.5	6
Co^{2+}	MHL/M.HL	6.39^z			
Ni^{2+}	MHL/M.HL	6.82^z			
Zn^{2+}	MHL/M.HL	6.42			
Cd^{2+}	MHL/M.HL	5.87^z			

z $(CH_3)_4$NBr used as background electrolyte.

Bibliography:

H^+ 61TD,70WW Co^{2+}-Cd^{2+} 67TKL

$C_5H_4O_4N_2$ Uracil-5-carboxylic acid (iso-orotic acid) H_2L

Metal ion	Equilibrium	Log K 25°, 0.1
H^+	HL/H.L	8.78
	H_2L/HL.H	4.05
Mn^{2+}	MHL/M.HL	2.18 ±0.02
Co^{2+}	MHL/M.HL	2.48
Ni^{2+}	MHL/M.HL	2.95
Cu^{2+}	MHL/M.HL	4.12
Zn^{2+}	MHL/M.HL	2.67 ±0.02
Cd^{2+}	MHL/M.HL	2.02

Bibliography:

H^+,Ni^{2+}-Cd^{2+} 61TD Co^{2+} 66DT

Mn^{2+} 61TD,66DT

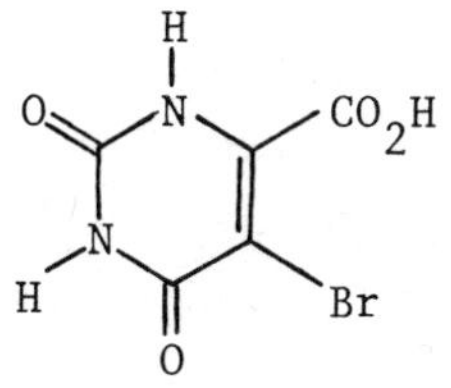

$C_5H_3O_4N_2Br$ 5-Bromo-orotic acid H_2L

Metal ion	Equilibrium	Log K 25°, 0.1
H^+	HL/H.L	7.22^z
	$H_2L/HL.H$	2.27^z
Mn^{2+}	MHL/M.HL	1.88^z
Co^{2+}	MHL/M.HL	3.27^z
Ni^{2+}	MHL/M.HL	4.19^z
Cu^{2+}	MHL/M.HL	5.58^z
Zn^{2+}	MHL/M.HL	3.26^z
Cd^{2+}	MHL/M.HL	2.43^z

^{z} $(CH_3)_4NBr$ used as background electrolyte.

Bibliography: 64TT

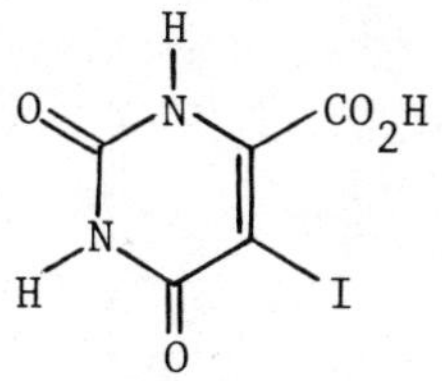

$C_5H_3O_4N_2I$ 5-Iodo-orotic acid H_2L

Metal ion	Equilibrium	Log K 25°, 0.1
H^+	HL/H.L	7.52^z
	$H_2L/HL.H$	1.77^z
Mn^{2+}	MHL/M.HL	2.25^z
Co^{2+}	MHL/M.HL	3.78^z
Ni^{2+}	MHL/M.HL	4.65^z
Cu^{2+}	MHL/M.HL	6.59^z
Zn^{2+}	MHL/M.HL	3.77^z
Cd^{2+}	MHL/M.HL	2.90^z

^{z} $(CH_3)_4NBr$ used as background electrolyte.

Bibliography: 64TT

$C_5H_3O_6N_3$ 5-Nitro-orotic acid H_2L

Metal ion	Equilibrium	Log K 25°, 0.1
H^+	HL/H.L	4.83^z
Mn^{2+}	MHL/M.HL	$1.77^z \pm 0.03$
Co^{2+}	MHL/M.HL	$2.43^z \pm 0.01$
Ni^{2+}	MHL/M.HL	3.04^z
Zn^{2+}	MHL/M.HL	$2.53^z \pm 0.02$
Cd^{2+}	MHL/M.HL	1.91^z

^{z} $(CH_3)_4NBr$ used as background electrolyte.

Bibliography:

$H^+, Ni^{2+}-Cd^{2+}$ 61TD Mn^{2+}, Co^{2+} 61TD,66DT

$C_5H_4O_3N_2S$ 2-Thio-iso-orotic acid H_2L

Metal ion	Equilibrium	Log K 25°, 0.1
Zn^{2+}	MHL/M.HL	3.94

Bibliography: 66DT

$C_7H_8O_3N_2S$ 2-(Ethylthio)-iso-orotic acid H_2L

Metal ion	Equilibrium	Log K 25°, 0.1
H^+	HL/H.L	10.41^z
	$H_2L/HL.H$	5.90^z

^{z} $(CH_3)_4NBr$ used as background electrolyte.

2-Ethylthio-iso-orotic acid (continued)

Metal ion	Equilibrium	Log K 25°, 0.1
Mn^{2+}	MHL/M.HL	2.07^z
Co^{2+}	MHL/M.HL	2.47^z
Ni^{2+}	MHL/M.HL	2.70^z
Cu^{2+}	MHL/M.HL	3.14^z
Zn^{2+}	MHL/M.HL	2.33^z
Cd^{2+}	MHL/M.HL	1.98^z

^{z} $(CH_3)_4NBr$ used as background electrolyte.

Bibliography: 61TD

$$\overset{O}{\underset{\|}{HSCNHCH_2CO_2H}}$$

$C_3H_5O_2NS_2$		N-(Dithiocarboxy)aminoacetic acid	H_2L

Metal ion	Equilibrium	Log K 25°, 0.1
H^+	HL/H.L	6.67
	$H_2L/HL.H$	3.75
Pb^{2+}	ML/M.L	7.3
	$ML_2/M.L^2$	13.0

Bibliography: 67BP

Other reference: 66BZ

$$HO_2CCO_2H$$

$C_2H_2O_4$ Metal ion	Equilibrium	Log K 25°, 0.1	Log K 25°, 1.0	Log K 25°, 0	ΔH 25°, 0	H_2L ΔS 25°, 0
		Ethanedioic acid (oxalic acid)				
H^+	HL/H.L	3.82 ±0.04	3.55 ±0.02	4.266 ±0.001	1.60 ±0.06	25.1
		3.55[b] ±0.05		3.80[e]	0.52[c]	18.0[c]
	H_2L/HL.H	1.04 ±0.10	1.04 ±0.04	1.252	0.9 ±0.1	
		1.02[b] ±0.03		1.26[e]		
K^+	ML/M.L			-0.8 [n]		
Be^{2+}	ML/M.L	4.08[h]	3.55			
	ML_2/M.L^2	5.38[h]	5.40			
Mg^{2+}	ML/M.L	2.76[h]		3.43[n]		
	ML_2/M.L^2	4.24				
Ca^{2+}	ML/M.L		1.66	3.00[n]		
	ML_2/M.L^2		2.69			
	MHL/M.HL	1.38		1.84		
	$M(HL)_2$/M.$(HL)^2$	1.8				
Sr^{2+}	ML/M.L		1.25	2.54[n]		
	ML_2/M.L^2		1.90			
	MHL/M.HL	1.11				
	$M(HL)_2$/M.$(HL)^2$	1.7		1.57		
Ba^{2+}	ML/M.L			2.31[n]		
Sc^{3+}	ML/M.L	8.74[r]	6.86[j]			
	ML_2/M.L^2		11.31[j]			
	ML_3/M.L^3		14.32[j]			
	ML_4/M.L^4		16.70[j]			
	MHL/M.HL	7.36[r]				
Y^{3+}	ML/M.L	5.46				
	ML_2/M.L^2	9.29				
La^{3+}	ML/M.L	4.71	4.3			
	ML_2/M.L^2	7.83	7.9			
	ML_3/M.L^3		10.3			
Ce^{3+}	ML/M.L	4.90	4.49[j]	6.52		
	ML_2/M.L^2	8.26	7.91[j]	10.48		
	ML_3/M.L^3		10.30[j]	11.30		
	ML_4/M.L^4		11.75[j]			
Pm^{3+}	ML/M.L	5.18				
	ML_2/M.L^2	8.78				
Eu^{3+}	ML/M.L	5.36	5.04[j] -0.3			
	ML_2/M.L^2	9.04	8.70[j] ±0.0			
	ML_3/M.L^3		11.45[j] -0.2			
	ML_4/M.L^4		13.09[j]			
Gd^{3+}	ML/M.L			7.01		

---

[b] 25°, 0.5;   [c] 25°, 1.0;   [e] 25°, 3.0;   [h] 20°, 0.1;   [j] 20°, 1.0;   [n] 18°, 0;   [r] 25°. 0.05

Oxalic acid (continued)

Metal ion	Equilibrium	Log K 25°, 0.1	Log K 25°, 1.0	Log K 25°, 0	ΔH 25°, 0	ΔS 25°, 0
Tb^{3+}	$ML/M.L$	5.45	5.08[j]			
	$ML_2/M.L^2$	9.25	8.86[j]			
	$ML_3/M.L^3$		11.85[j]			
	$ML_4/M.L^4$		13.41[j]			
Tm^{3+}	$ML/M.L$	5.60				
	$ML_2/M.L^2$	9.52				
Yb^{3+}	$ML/M.L$			7.30		
	$ML_2/M.L^2$			11.89		
Lu^{3+}	$ML/M.L$		5.28[j] -0.2			
	$ML_2/M.L^2$		9.53[j] -0.3			
	$ML_3/M.L^3$		12.74[j] ±0.0			
	$ML_4/M.L^4$		14.68[j]			
Ac^{3+}	$ML/M.L$	4.36	3.56			
	$ML_2/M.L^2$	7.08	6.16			
Am^{3+}	$ML/M.L$	5.25 4.82[b]	4.63			
	$ML_2/M.L^2$	8.85 8.60[b]	8.35			
	$ML_3/M.L^3$		11.15			
Cm^{3+}	$ML/M.L$	5.25				
	$ML_2/M.L^2$	8.85				
Bk^{3+}	$ML/M.L$	5.45				
	$ML_2/M.L^2$	9.14				
Cf^{3+}	$ML/M.L$	5.50				
	$ML_2/M.L^2$	9.37				
Th^{4+}	$ML/M.L$	8.8 [r]	8.23	10.6		
	$ML_2/M.L^2$		16.8	20.2		
	$ML_3/M.L^3$		22.8	26.4		
	$MHL/M.HL$	7.4 [r]				
NpO_2^+	$ML/M.L$	4.04[s]	3.74[j]			
	$ML_2/M.L^2$	7.36[s]	6.31[j]			
	$MHL/M.HL$	2.70[s]				
UO_2^{2+}	$ML/M.L$	6.36[h]	5.99[j]			
	$ML_2/M.L^2$	10.59[h]	10.64[j]			
	$ML_3/M.L^3$		11.0 [j]			
PuO_2^{2+}	$ML_2/M.L^2$		9.4 [j]			
Cr^{2+}	$ML/M.L$	3.85				
	$ML_2/M.L^2$	6.81				
Mn^{2+}	$ML/M.L$	3.2		3.95 ±0.03	1.4	23
	$ML_2/M.L^2$	4.4				
Fe^{2+}	$ML/M.L$		3.05			
	$ML_2/M.L^2$		5.15			
Co^{2+}	$ML/M.L$	3.84 ±0.03	3.25	4.72 ±0.08	0.6	24
	$ML_2/M.L^2$		5.60	7.0 ±0.1		
	$MHL/M.HL$	1.61				
	$M(HL)_2/M.(HL)^2$	2.89				

[b] 25°, 0.5; [h] 20°, 0.1; [j] 20°, 1.0; [r] 25°, 0.05; [s] 20°, 0.05

Oxalic acid (continued)

Metal ion	Equilibrium	Log K 25°, 0.1	Log K 25°, 1.0	Log K 25°, 0	ΔH 25°, 0	ΔS 25°, 0
Ni^{2+}	ML/M.L			5.16	0.2	24
Cu^{2+}	ML/M.L	4.84		6.23	-0.1	28
	$ML_2/M.L^2$	9.21		10.27		
	MHL/M.HL	2.49		3.18		
Mn^{3+}	ML/M.L		9.98[d]			
	$ML_2/M.L^2$		16.57[d]			
	$ML_3/M.L^3$		18.42[d]			
Fe^{3+}	ML/M.L	7.53[b] -0.1	7.59	7.74[e]		
	$ML_2/M.L^2$	13.64[b]				
	$ML_3/M.L^3$	18.49[b]				
	MHL/M.HL	4.35[b]				
VO^{2+}	ML/M.L		6.45[j]			
	$ML_2/M.L^2$		11.78[j]			
VO_2^+	ML/M.L	5.9[h]	5.0[j] ±0.1	6.64[o]		
	$ML_2/M.L^2$	9.54[h]	8.49[j]	10.20[o]		
Ag^+	ML/M.L		2.41			
Hg_2^{2+}	$ML_2/M.L^2$		6.98[u]			
	MOHL/M.OH.L		13.04[u]			
Zn^{2+}	ML/M.L	3.88[t]	3.43 ±0.01	4.87 ±0.02		
	$ML_2/M.L^2$	6.40[t]	6.16	7.65 ±0.05		
	MHL/M.HL	1.72				
	$M(HL)_2/M.(HL)^2$	3.12				
Cd^{2+}	ML/M.L		2.75	3.89[n]		
Hg^{2+}	ML/M.L	9.66				
Pb^{2+}	ML/M.L	4.00[t]	3.32	4.91	(-2)[w]	(15)[t]
	$ML_2/M.L^2$	5.82[t]	5.5	6.76	(-1)[w]	(30)[t]
Al^{3+}	ML/M.L		6.1			
	$ML_2/M.L^2$		11.09			
	$ML_3/M.L^3$		15.12			
Ga^{3+}	ML/M.L		6.45[j]			
	$ML_2/M.L^2$		12.38[j]			
	$ML_3/M.L^3$		17.86[j]			
In^{3+}	ML/M.L		5.30			
	$ML_2/M.L^2$		10.52			
	MHL/M.HL	3.08[v]				

---

[b] 25°, 0.5; [d] 25°, 2.0; [e] 25°, 3.0; [h] 20°, 0.1; [j] 20°, 1.0; [n] 18°, 0; [o] 20°, 0; [t] 25°, 0.16; [u] 27°, 2.5; [v] 25°, 0.3; [w] 15-40°, 0.3

Bibliography:

H^+ 39HF,48PB,60MN,61MN,65BC,65BSc,65NU, 65Sc,66MS,67RMa,68DM,69CMa,69GGR, 70AB,71VK,73VS,75FP,75VB

K^+ 31BR

Be^{2+} 67SS,70CF,75VB

Mg^{2+}-Ba^{2+} 27D,32MD,38CK,39P,57SAa,59TV, 62AMa,67HMS

Sc^{3+} 71Ga,73CS

Y^{3+}-Lu^{3+} 50CM,51CM,62YZ,65Sc,69GGR,71Sf

B. DI-CARBOXYLIC ACIDS

Oxalic acid (continued)

Bibliography:

Ac^{3+} 69SS,72MSS

Al^{3+} 68BC

$Am^{3+}-Cf^{3+}$ 65Sb,68AL,71S

Ga^{3+} 74KBZ

Th^{4+} 67ME,73CS

In^{3+} 60WT,66HS

NpO_2^+ 53GK,61ZM

UO_2^{2+} 69Hb

PuO_2^{2+} 73PDB

Cr^{2+} 70FK

$Mn^{2+}-Cu^{2+},Zn^{2+},Cd^{2+}$ 32MD,38MD,58SL,60MN, 60WT,61MM,61MN,65BC,65Ma,65SMa,66MN, 66RM,67KW,67W,69VP,70CG,70EM,74AR, 74MS,75B

Mn^{3+} 48T

Fe^{3+} 65BSc,66MS,68DM

VO^{2+} 70IVC

VO_2^+ 66Ia,69VI

Hg_2^{2+} 60YD

Hg^{2+} 73CS,73LU

Pb^{2+} 70K

Other references: 04K,05ASa,05SA,27S,29R, 34MD,34S,36BJ,36CE,37CV,40VB,42KP, 47L,49La,49Lb,50Ma,51BA,51MM,53GK, 54BC,54SLB,56FS,56KF,56VP,56YA,56Za, 57BD,57BDa,57DS,57GM,57L,57TT,58AO, 58MG,59BD,59BSR,59MZ,59PT,59TT,59TV, 59Z,60BD,60CI,60GS,60LP,60MK,60Sa, 61GA,61YB,61ZM,62AK,62BK,62IN,62MD, 62MR,62PLB,62YB,62YP,63FV,63KP,63PBa, 63Se,63Za,65BSa,64CK,64KS,65BL,65CV, 65HS,65NU,65PV,65SMc,66Ga,66KSG,66KFa, 66KZ,66LN ,66Me,66SSH,67EKG,67GP,67Ka, 67KH,67LN,67MNa,67NS,67SM,68FV,68GK, 68JG,68JK,68S,69KAa,69M ,69MKS,69SCF, 70GB,70SP,70ZP,71GM,71KC,71MMW,71NS, 71PK,71PL,71PSb,71SIa,72AD,72FD,72KG, 72MGb,72NK,73BB,73BR,73FS,73KG,74GM, 74SN,74VP,75BU,75KN,75PJ

$$HO_2CCH_2CO_2H$$

$C_3H_4O_4$ Metal ion	Equilibrium	Propanedioic acid (malonic acid)				H_2L
		Log K 25°, 0.1	Log K 25°, 1.0	Log K 25°, 0	ΔH 25°, 0	ΔS 25°, 0
H^+	HL/H.L	5.28 ±0.05	5.07 ±0.05	5.696 ±0.000	1.15 -0.01	30.0
		5.07[b]±0.06	5.14[d]±0.03	5.26[e]	0.48[c]±0.00	24.8[c]
	$H_2L/HL.H$	2.65 ±0.05	2.60 ±0.01	2.847 ±0.00	-0.04 ±0.02	12.9
		2.57[b]±0.05	2.68[d]±0.2	2.81[e]±0.01	-0.36[c]±0.01	10.7[c]
Na^+	ML/M.L			0.74		
Be^{2+}	ML/M.L	5.30[h]				
	$ML_2/M.L^2$	8.56[h]				
Mg^{2+}	ML/M.L	2.11		2.85 ±0.01	3.2	24
	MHL/M.HL	0.96				
Ca^{2+}	ML/M.L	1.51		2.35		
	MHL/M.HL	0.47[x]				
Sr^{2+}	ML/M.L	1.30				
	MHL/M.HL	0.41[x]				
Ba^{2+}	ML/M.L	1.22		2.07 ±0.07		
Sc^{3+}	ML/M.L		5.87		3.4[c]	38[c]
	$ML_2/M.L^2$		10.12		6.3[c]	68[c]
	$ML_3/M.L^3$		13.07		7.8[c]	86[c]

---

[b] 25°, 0.5;   [c] 25°, 1.0;   [d] 25°, 2.0;   [e] 25°, 3.0;   [h] 20°, 0.1;   [x] temperature not stated; 0.2

Malonic acid (continued)

Metal ion	Equilibrium	Log K 25°, 0.1	Log K 25°, 1.0	Log K 25°, 0	ΔH 25°, 0	ΔS 25°, 0
Y^{3+}	$ML/M.L$	4.40				
	$ML_2/M.L^2$	7.04				
La^{3+}	$ML/M.L$	3.69 +0.2	3.07	4.94 ±0.07	4.8	39
					2.9[c]	24[c]
	$ML_2/M.L^2$	5.90	5.1		4.9[c]	40[c]
	$MHL_2/ML_2.H$		4.1		-0.6[c]	17[c]
	$MHL/M.HL$		1.24		0.8[c]	8[c]
Ce^{3+}	$ML/M.L$	3.83 +0.3	3.25		2.9[c]	25[c]
	$ML_2/M.L^2$	6.17	5.2		5[c]	40[c]
	$MHL_2/ML_2.H$		4.1		0[c]	20[c]
	$MHL/M.HL$		1.32		0.8[c]	9[c]
Pr^{3+}	$ML/M.L$	3.91 +0.3	3.29		3.0[c]	25[c]
	$ML_2/M.L^2$	6.30	5.6		5.1[c]	43[c]
	$MHL_2/ML_2.H$		3.7		-0.6[c]	15[c]
	$MHL/M.HL$		1.42		0.8[c]	9[c]
Nd^{3+}	$ML/M.L$	3.95 +0.3	3.38		3.1[c]	26[c]
	$ML_2/M.L^2$	6.41	5.9		5.0[c]	44[c]
	$MHL_2/ML_2.H$		3.6		-0.4[c]	15[c]
	$MHL/M.HL$		1.42		0.7[c]	9[c]
Sm^{3+}	$ML/M.L$	4.19 +0.3	3.67		3.0[c]	27[c]
	$ML_2/M.L^2$	6.84	6.1		5.1[c]	45[c]
	$MHL_2/ML_2.H$		3.8		-0.7[c]	15[c]
	$MHL/M.HL$		1.53		0.9[c]	10[c]
Eu^{3+}	$ML/M.L$	4.30 +0.3	3.72		3.1[c]	27[c]
	$ML_2/M.L^2$	6.99	6.24		4.8[c]	45[c]
	$MHL_2/ML_2.H$		3.76		-0.8[c]	15[c]
	$MHL/M.HL$		1.42		0.9[c]	10[c]
Gd^{3+}	$ML/M.L$	4.32 +0.4	3.73		3.0[c]	27[c]
	$ML_2/M.L^2$	6.97	6.24		4.8[c]	45[c]
	$MHL_2/ML_2.H$		3.6		-0.5[c]	15[c]
	$MHL/M.HL$		1.45		1.1[c]	10[c]
Tb^{3+}	$ML/M.L$	4.44 +0.3	3.82		3.0[c]	28[c]
	$ML_2/M.L^2$	7.15	6.38		5.3[c]	47[c]
	$MHL_2/ML_2.H$		3.71		-1.2[c]	13[c]
	$MHL/M.HL$		1.30		1.3[c]	10[c]
Dy^{3+}	$ML/M.L$	4.47 +0.4	3.85		3.1[c]	28[c]
	$ML_2/M.L^2$	7.17	6.35		5.3[c]	47[c]
	$ML_3/M.L^3$		7.6		5[c]	50[c]
	$MHL_2/ML_2.H$		3.5		-1.0[c]	13[c]
	$MHL/M.HL$		1.24		1.5[c]	11[c]

[c] 25°, 1.0

Malonic acid (continued)

Metal ion	Equilibrium	Log K 25°, 0.1	Log K 25°, 1.0	Log K 25°, 0	ΔH 25°, 0	ΔS 25°, 0
Ho^{3+}	$ML/M.L$	4.39	3.83		3.2[c]	28[c]
	$ML_2/M.L^2$	6.97	6.37		5.3[c]	47[c]
	$ML_3/M.L^3$		7.7		7.2[c]	59[c]
	$MHL_2/ML_2.H$		3.58		-1.0[c]	13[c]
	$MHL/M.HL$		1.24		1.1[c]	9[c]
Er^{3+}	$ML/M.L$	4.42 +0.4	3.85		3.2[c]	28[c]
	$ML_2/M.L^2$	7.04	6.39		5.4[c]	47[c]
	$ML_3/M.L^3$		7.61		7[c]	60[c]
	$MHL_2/ML_2.H$		3.54		-0.5[c]	15[c]
	$MHL/M.HL$		1.28		1.0[c]	9[c]
Tm^{3+}	$ML/M.L$	4.42 +0.4	3.85		3.5[c]	29[c]
	$ML_2/M.L^2$	7.01	6.42		5.5[c]	48[c]
	$ML_3/M.L^3$		7.62		8.1[c]	62[c]
	$MHL_2/ML_2.H$		3.56		-1.4[c]	12[c]
	$MHL/M.HL$		1.21		0.9[c]	9[c]
Yb^{3+}	$ML/M.L$	4.53 +0.3	3.87	5.70	3.4[c]	29[c]
	$ML_2/M.L^2$	7.27	6.43	8.6	5.8[c]	49[c]
	$ML_3/M.L^3$		7.78		7.9[c]	62[c]
	$MHL_2/ML_2.H$		3.33		-0.6[c]	13[c]
	$MHL/M.HL$		1.14		1.4[c]	10[c]
Lu^{3+}	$ML/M.L$	4.45 +0.3	3.88		3.5[c]	30[c]
	$ML_2/M.L^2$	7.13	6.42		6.1[c]	50[c]
	$ML_3/M.L^3$		7.88		8.4[c]	64[c]
	$MHL_2/ML_2.H$		3.09		-0.2[c]	13[c]
	$MHL/M.HL$		1.20		1.4[c]	10[c]
Th^{4+}	$ML/M.L$		7.42[j]			
	$ML_2/M.L^2$		12.68[j]			
NpO_2^+	$ML/M.L$		2.75[j]			
UO_2^{2+}	$ML/M.L$	(5.43)[b]	(5.66)			
	$ML_2/M.L^2$	(9.31)[b]	(9.66)			
	$ML/MOHL.H$	(5.02)[b]				
PuO_2^{2+}	$ML/M.L$		4.84[j]			
	$ML_2/M.L^2$		8.45[j]			
Cr^{2+}	$ML/M.L$	3.92				
	$ML_2/M.L^2$	7.13				
Mn^{2+}	$ML/M.L$	2.30[s]		3.28 ±0.01	3.7	27
Co^{2+}	$ML/M.L$	2.97 ±0.02		3.74 ±0.03	2.9	27
	$ML_2/M.L^2$	4.4		5.1 ±0.0		
	$MHL/M.HL$	0.82				
Ni^{2+}	$ML/M.L$	3.24 ±0.06		4.05 ±0.05	2.0 ±0.2	25
	$ML_2/M.L^2$	4.9				
	$MHL/M.HL$	1.04				

---

[b] 25°, 0.5;   [c] 25°, 1.0;   [j] 20°, 1.0;   [s] 25°, 0.16

Malonic acid (continued)

Metal ion	Equilibrium	Log K 25°, 0.1	Log K 25°, 1.0	Log K 25°, 0	ΔH 25°, 0	ΔS 25°, 0
Cu^{2+}	$ML/M.L$	5.05 ±0.05	4.63	5.70 ±0.1	2.9	36
	$ML_2/M.L^2$	7.8 ±0.2	7.7	8.2		
	$MHL/M.HL$	2.15				
Cr^{3+}	$ML/M.L$		8.26^v			
Fe^{3+}	$ML/M.L$	7.52^b±0.06				
VO^{2+}	$ML/M.L$		5.23^j			
	$ML_2/M.L^2$		8.85^j			
	$MHL/M.HL$			1.63^e		
VO_2^+	$ML/M.L$		4.56^j			
	$ML_2/M.L^2$		6.53^j			
Zn^{2+}	$ML/M.L$	2.96 ±0.01	2.47	3.84 ±0.02	3.1 −0.1	28
	$ML_2/M.L^2$	4.4	3.8	5.4		
	$MHL/M.HL$	0.99				
Cd^{2+}	$ML/M.L$	2.64 ±0.1	1.92^j	3.22 ±0.04		
	$ML_2/M.L^2$		2.88^j			
	$MHL/M.HL$	1.49	0.69^j			
Pb^{2+}	$ML/M.L$	3.1	2.60^u			
	$ML_2/M.L^2$		3.62^u			
	$ML_3/M.L^3$		4.32^u			

^{b} 25°, 0.5; ^{e} 25°, 3.0; ^{j} 20°, 1.0; ^{u} 30°, 2.0; ^{v} 40°, 2.0

Bibliography:

H^+ 38CK,39A,40HB,49SD,60YY,61DI,62SS,
63Ca,65C,67RMa,68DM,68KK,68OV,68PFN,
69GH,69MBa,69MF,69VOR,70GSb,70IP,
70IPa,71CD,71DG,72DCa,73D,73IT,73PDB,
74OD,75VB

Na^+ 65AE

Be^{2+} 67AM,75VB

$Mg^{2+}-Ba^{2+}$ 38CK,49SD,51PJ,52EM,52SL,62JS,
63LM,68OV

Sc^{3+} 69GH

$Y^{3+}-Lu^{3+}$ 51PJ,56GNa,66AM,68PFN,71DG,
72DCa,73D

Th^{4+} 72TM

NpO_2^+ 72MB

UO_2^{2+} 67RMa,69VOR

PuO_2^{2+} 73PDB

Cr^{2+} 70FK

$Mn^{2+}-Cu^{2+},Zn^{2+}-Pb^{2+}$ 34FR,35D,38CK,49SD,
51PJ,56GNb,57LW,60YY,61NN,62BN,62JS,
62SS,63Ca,63MNa,65Ma,65N,65SMa,66MN,
66RM,68GP,68OV,69MBa,69PJ,70EM,70GSb,
74MS,74OD,74UY

Cr^{3+} 69MF

Fe^{3+} 68DM,71CD

VO^{2+} 70IVC,73IT

VO_2^+ 70IVa

Other references: 29RF,30R,31IR,35BJ,40CN,
52Sb,53BB,54SLB,58DB,61MM,61YR,62BK,
62YP,63YZ,66AMa,66KZ,66MTa,67Ka,68KK,
68S,69Ma,70BJ,70GNa,70NK,71GM,71MMW,
71WC,72KG,72S,73KG,73RM,74BSa,74GM,
74HB,74TG,74VP,75DNb,75KN,75TG,75W

$$\begin{array}{c} CH_3 \\ | \\ HO_2CCHCO_2H \end{array}$$

$C_4H_6O_4$		Methylpropanedioic acid (methylmalonic acid)			H_2L
Metal ion	Equilibrium	Log K 25°, 0.1	Log K 25°, 0.5	Log K 25°, 0	
H^+	HL/H.L	5.40	5.16 ±0.02	5.76	
	H_2L/HL.H	2.94	2.80 ±0.02	3.01	
Mg^{2+}	ML/M.L	1.73			
Ca^{2+}	ML/M.L	1.65			
Sr^{2+}	ML/M.L	1.43			
Ba^{2+}	ML/M.L	1.42			
UO_2^{2+}	ML/M.L		5.56		
	ML_2/M.L^2		9.53		
Co^{2+}	ML/M.L	2.45			
Ni^{2+}	ML/M.L	2.62			
Cu^{2+}	ML/M.L	4.89			
	ML_2/M.L^2	7.49			
	MHL/M.HL	1.66			
Fe^{3+}	ML/M.L		7.56		
Zn^{2+}	ML/M.L	2.55			
Cd^{2+}	ML/M.L	2.58			
	MHL/M.HL	1.27			

Bibliography:

H^+ 31GI,660Ca,69VOR,71CD Ni^{2+}-Cd^{2+} 660Ca

Mg^{2+}-Ba^{2+},Co^{2+} 680V Fe^{3+} 71CD

UO_2^{2+} 69VOR   Other references: 30R,31IR

$$\begin{array}{c} CH_2CH_3 \\ | \\ HO_2CCHCO_2H \end{array}$$

$C_5H_8O_4$		Ethylpropanedioic acid (ethylmalonic acid)		H_2L
Metal ion	Equilibrium	Log K 25°, 0.1	Log K 25°, 0	
H^+	HL/H.L	5.45 ±0.01	5.83	
	H_2L/HL.H	2.83 ±0.02	2.99	
Mg^{2+}	ML/M.L	1.62		
Ca^{2+}	ML/M.L	1.59		
Sr^{2+}	ML/M.L	1.40		
Ba^{2+}	ML/M.L	1.39		
Co^{2+}	ML/M.L	2.51		
Ni^{2+}	ML/M.L	2.57 ±0.04		

Ethylmalonic acid (continued)

Metal ion	Equilibrium	Log K 25°, 0.1	Log K 25°, 0
Cu^{2+}	$ML/M.L$	4.95	
	$ML_2/M.L^2$	7.77	
	$MHL/M.HL$	1.74	
Zn^{2+}	$ML/M.L$	2.53	
Cd^{2+}	$ML/M.L$	2.59	
	$MHL/M.HL$	1.28	

Bibliography:

H^+ 31GI,680V,740D Ni^{2+} 680V,740D

$Mg^{2+}-Co^{2+},Cu^{2+}-Cd^{2+}$ 680V   Other references: 30R,31IR

$$\begin{array}{c} CH_3CHCH_3 \\ | \\ HO_2CCHCO_2H \end{array}$$

$C_6H_{10}O_4$ 1-Methylethylpropanedioic acid (isopropylmalonic acid) H_2L

Metal ion	Equilibrium	Log K 25°, 0.1	Log K 25°, 0
H^+	$HL/H.L$	5.52	5.88
	$H_2L/HL.H$	2.81	2.94
Ni^{2+}	$ML/M.L$	2.60	
Cu^{2+}	$ML/M.L$	5.43	
	$ML_2/M.L^2$	9.09	
	$MHL/M.HL$	1.6	
Zn^{2+}	$ML/M.L$	2.90	
Cd^{2+}	$ML/M.L$	2.63	

Bibliography:

H^+ 31GI,74R   Other references: 30R,31IR

$Ni^{2+}-Cd^{2+}$ 74R

$$\begin{array}{c} CH_2CH_2CH_2CH_3 \\ | \\ HO_2CCHCO_2H \end{array}$$

$C_7H_{12}O_4$ Butylpropanedioic acid (butylmalonic acid) H_2L

Metal ion	Equilibrium	Log K 25°, 0.1	Log K 25°, 0.5	Log K 25°, 0
H^+	$HL/H.L$	5.50 ±0.00	5.20	5.96
	$H_2L/HL.H$	2.81 +0.1	2.77	3.02
Ni^{2+}	$ML/M.L$	2.57 ±0.08		
Cu^{2+}	$ML/M.L$	5.05		
	$ML_2/M.L^2$	8.13		
	$MHL/M.HL$	1.6		

B. DI-CARBOXYLIC ACIDS

Butylmalonic acid (continued)

Metal ion	Equilibrium	Log K 25°, 0.1	Log K 25°, 0.5	Log K 25°, 0
Fe^{3+}	ML/M.L		7.30	
Zn^{2+}	ML/M.L	2.70		
Cd^{2+}	ML/M.L	2.56		

Bibliography:

H^+	62BN,71CD,74R	$Cu^{2+}-Cd^{2+}$	74R
Ni^{2+}	62BN,74R	Fe^{3+}	71CD

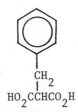

$$CH_2$$
$$HO_2CCHCO_2H$$

$C_{10}H_{10}O_4$		Benzylpropanedioic acid (benzylmalonic acid)			H_2L

Metal ion	Equilibrium	Log K 25°, 0.5	Log K 25°, 0	ΔH 25°, 0	ΔS 25°, 0
H^+	HL/H.L	5.02	5.87 −0.01		
	H_2L/HL.H	2.60	2.91 −0.01		
Mg^{2+}	ML/M.L		2.70		
Mn^{2+}	ML/M.L		2.98		
Co^{2+}	ML/M.L		3.35		
Ni^{2+}	ML/M.L		3.48		
Cu^{2+}	ML/M.L		5.43	0.5	27
Fe^{3+}	ML/M.L	7.26			
Zn^{2+}	ML/M.L		3.50	$(5)^t$	(30)
	$ML_2/M.L^2$		8.95		

^{t} 25-35°, 0

Bibliography:

H^+	70NP,71CD,72PA	Cu^{2+}	73PAc
Mg^{2+}	72PA	Fe^{3+}	71CD
$Mn^{2+}-Ni^{2+}$	70NP	Zn^{2+}	73PAb

$$HO_2CCHCO_2H$$

$C_9H_8O_4$ Phenylpropanedioic acid (phenylmalonic acid) H_2L

Metal ion	Equilibrium	Log K 25°, 0.1
H^+	HL/H.L	5.11
	H_2L/HL.H	2.40
Ni^{2+}	ML/M.L	2.35
Cu^{2+}	ML/M.L	4.57
	ML_2/M.L^2	7.23
	MHL/M.HL	1.8
Zn^{2+}	ML/M.L	2.37
Cd^{2+}	ML/M.L	2.23

Bibliography: 74R

$$HO_2CCCO_2H \quad \text{with } CH_3 \text{ above and } CH_3 \text{ below the central C}$$

$C_5H_8O_4$ Dimethylpropanedioic acid (dimethylmalonic acid) H_2L

Metal ion	Equilibrium	Log K 25°, 0.1	Log K 25°, 0.5	Log K 25°, 0
H^+	HL/H.L	5.68	5.46	6.06
	H_2L/HL.H	3.01	2.90	3.17
Mg^{2+}	ML/M.L	1.55		
Ca^{2+}	ML/M.L	1.52		
Sr^{2+}	ML/M.L	1.33		
Ba^{2+}	ML/M.L	1.35		
UO_2^{2+}	ML/M.L		5.55	
	ML_2/M.L^2		9.38	
Co^{2+}	ML/M.L	1.90		
Ni^{2+}	ML/M.L	1.95		
Cu^{2+}	ML/M.L	4.57		
	ML_2/M.L^2	7.09		
	MHL/M.HL	0.70		
Hg_2^{2+}	MOHL/M.OH.L		13.58[r]	
	ML_2/M.L^2		7.52[r]	
Zn^{2+}	ML/M.L	2.20		
Cd^{2+}	ML/M.L	2.54		
	MHL/M.HL	1.30		

[r] 27°, 0.75

B. DI-CARBOXYLIC ACIDS

Dimethylmalonic acid (continued)

Bibliography:

H^+ 31GI,60YD,66OCa,69VOR

$Mg^{2+}-Ba^{2+},Co^{2+}$ 680V

UO_2^{2+} 69VOR

$Ni^{2+},Cu^{2+},Zn^{2+},Cd^{2+}$ 66OCa

Hg_2^{2+} 60YD

Other references: 30R,31IR

$$\begin{array}{c} CH_2CH_3 \\ | \\ HO_2CCCO_2H \\ | \\ CH_2CH_3 \end{array}$$

$C_7H_{12}O_4$ Metal ion	Equilibrium	Log K 25°, 0.1	Log K 25°, 0.5	Log K 25°, 0	ΔH 25°, 0	H_2L ΔS 25°, 0
H^+	HL/H.L	6.96 ±0.02	6.80	7.417	0.83 ±0.01	36.8
	H_2L/HL.H	2.00 ±0.04	1.95	2.152	0.71	12.2
Y^{3+}	ML/M.L	4.60				
	ML_2/M.L^2	7.05				
La^{3+}	ML/M.L	3.61				
	ML_2/M.L^2	5.95				
Ce^{3+}	ML/M.L	3.78				
	ML_2/M.L^2	6.32				
Pr^{3+}	ML/M.L	3.91				
	ML_2/M.L^2	6.49				
Nd^{3+}	ML/M.L	4.01				
	ML_2/M.L^2	6.63				
Sm^{3+}	ML/M.L	4.33				
	ML_2/M.L^2	6.92				
Eu^{3+}	ML/M.L	4.46				
	ML_2/M.L^2	7.05				
Gd^{3+}	ML/M.L	4.49				
	ML_2/M.L^2	7.05				
Tb^{3+}	ML/M.L	4.63				
	ML_2/M.L^2	7.24				
Dy^{3+}	ML/M.L	4.61				
	ML_2/M.L^2	7.29				
Ho^{3+}	ML/M.L	4.63				
	ML_2/M.L^2	7.16				
Er^{3+}	ML/M.L	4.66				
	ML_2/M.L^2	7.26				
Tm^{3+}	ML/M.L	4.70				
	ML_2/M.L^2	7.31				
Yb^{3+}	ML/M.L	4.76				
	ML_2/M.L^2	7.43				
Lu^{3+}	ML/M.L	4.69				
	ML_2/M.L^2	7.40				

Diethylpropanedioic acid (continued)

Metal ion	Equilibrium	Log K 25°, 0.1	Log K 25°, 0.5	Log K 25°, 0	ΔH 25°, 0	ΔS 25°, 0
UO_2^{2+}	$ML/M.L$		6.36			
	$ML_2/M.L^2$		11.04			
Co^{2+}	$ML/M.L$	2.25				
Ni^{2+}	$ML/M.L$	2.37				
Cu^{2+}	$ML/M.L$	4.96				
	$ML_2/M.L^2$	7.72				
Zn^{2+}	$ML/M.L$	2.44				
Cd^{2+}	$ML/M.L$	2.54				

Bibliography:

H^+ 65ME,67CIH,68PFN,69VOR,70IPa,70OV Co^{2+}-Cd^{2+} 70OV

Y^{3+}-Lu^{3+} 68PFN   Other references: 30R,30IR

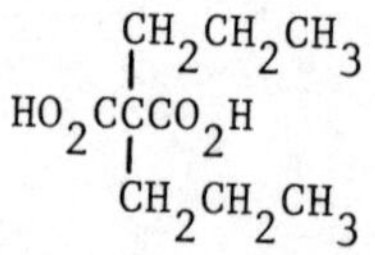

UO_2^{2+} 69VOR

$$CH_2CH_2CH_3$$
$$HO_2CCCO_2H$$
$$CH_2CH_2CH_3$$

$C_9H_{16}O_4$ Dipropylpropanedioic acid (dipropylmalonic acid) H_2L

Metal ion	Equilibrium	Log K 25°, 0.1	Log K 25°, 0
H^+	$HL/H.L$	7.19 ±0.04	7.51
	$H_2L/HL.H$	1.86 ±0.04	2.07
Y^{3+}	$ML/M.L$	4.74	
	$ML_2/M.L^2$	7.36	
La^{3+}	$ML/M.L$	3.66	
	$ML_2/M.L^2$	6.03	
Ce^{3+}	$ML/M.L$	3.96	
	$ML_2/M.L^2$	6.62	
Pr^{3+}	$ML/M.L$	4.01	
	$ML_2/M.L^2$	6.93	
Nd^{3+}	$ML/M.L$	4.06	
	$ML_2/M.L^2$	7.05	
Sm^{3+}	$ML/M.L$	4.38	
	$ML_2/M.L^2$	7.34	
Eu^{3+}	$ML/M.L$	4.57	
	$ML_2/M.L^2$	7.42	
Gd^{3+}	$ML/M.L$	4.58	
	$ML_2/M.L^2$	7.30	
Tb^{3+}	$ML/M.L$	4.73	
	$ML_2/M.L^2$	7.48	
Dy^{3+}	$ML/M.L$	4.78	
	$ML_2/M.L^2$	7.47	

Dipropylpropanedioic acid (continued)

Metal ion	Equilibrium	Log K 25°, 0.1	Log K 25°, 0
Ho^{3+}	ML/M.L	4.72	
	$ML_2/M.L^2$	7.28	
Er^{3+}	ML/M.L	4.73	
	$ML_2/M.L^2$	7.35	
Tm^{3+}	ML/M.L	4.76	
	$ML_2/M.L^2$	7.39	
Yb^{3+}	ML/M.L	4.81	
	$ML_2/M.L^2$	7.56	
Lu^{3+}	ML/M.L	4.78	
	$ML_2/M.L^2$	7.53	
Co^{2+}	ML/M.L	2.22	
Ni^{2+}	ML/M.L	2.48	
Cu^{2+}	ML/M.L	5.17	
	$ML_2/M.L^2$	8.27	
Zn^{2+}	ML/M.L	2.45	

Bibliography:

H^+ 31GI,65ME,68PFN,70OV $Co^{2+}-Zn^{2+}$ 70OV

$Y^{3+}-Lu^{3+}$ 68PFN   Other references: 30R,31IR

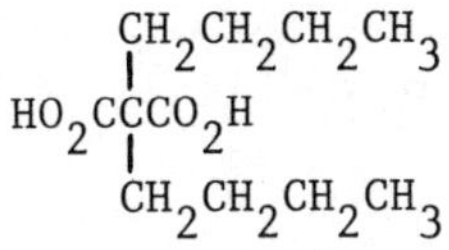

$$CH_2CH_2CH_2CH_3$$
$$|$$
$$HO_2CCCO_2H$$
$$|$$
$$CH_2CH_2CH_2CH_3$$

$C_{11}H_{20}O_4$ Dibutylpropanedioic acid (dibutylmalonic acid) H_2L

Metal ion	Equilibrium	Log K 25°, 0.1
H^+	HL/H.L	7.22 ±0.03
	$H_2L/HL.H$	1.95 ±0.06
Y^{3+}	ML/M.L	4.67
	$ML_2/M.L^2$	7.23
Sm^{3+}	ML/M.L	4.43
Eu^{3+}	ML/M.L	4.53
	$ML_2/M.L^2$	7.28
Gd^{3+}	ML/M.L	4.54
	$ML_2/M.L^2$	7.32
Tb^{3+}	ML/M.L	4.70
	$ML_2/M.L^2$	7.45
Dy^{3+}	ML/M.L	4.75
	$ML_2/M.L^2$	7.43
Ho^{3+}	ML/M.L	4.69
	$ML_2/M.L^2$	7.36

Dibutylpropanedioic acid (continued)

Metal ion	Equilibrium	Log K 25°, 0.1
Er³⁺	ML/M.L	4.74
	ML₂/M.L²	7.53
Tm³⁺	ML/M.L	4.73
	ML₂/M.L²	7.34
Yb³⁺	ML/M.L	4.80
	ML₂/M.L²	7.61
Lu³⁺	ML/M.L	4.78
	ML₂/M.L²	7.76
Co²⁺	ML/M.L	2.26
Ni²⁺	ML/M.L	2.35
Cu²⁺	ML/M.L	5.10
	ML₂/M.L²	8.15
Zn²⁺	ML/M.L	2.55

Bibliography:

H⁺   68PFN,700V   Co²⁺-Zn²⁺   700V

Y³⁺-Lu³⁺   68PFN

C₅H₆O₄		Cyclopropane-1,1-dicarboxylic acid		H₂L
Metal ion	Equilibrium	Log K 25°, 0.1	ΔH 25°, 0.1	ΔS 25°, 0.1
H⁺	HL/H.L	7.18 ±0.02	-0.39	31.6
	H₂L/HL.H	1.62 ±0.01	0.31	8.5
Ni²⁺	ML/M.L	3.75		

Bibliography:

H⁺   74OD,75CG   Ni²⁺   74OD

C₆H₈O₄		Cyclobutane-1,1-dicarboxylic acid			H₂L
Metal ion	Equilibrium	Log K 25°, 0.1	Log K 25°, 0.5	ΔH 25°, 0.1	ΔS 25°, 0.1
H⁺	HL/H.L	5.45 ±0.00	5.22	0.81	27.7
	H₂L/HL.H	2.94 ±0.02	2.86	0.26	14.3
Mg²⁺	ML/M.L	2.11			
	MHL/M.HL	0.95			
Ca²⁺	ML/M.L	1.54			

B. DI-CARBOXYLIC ACIDS

Cyclobutane-1,1-dicarboxylic acid (continued)

Metal ion	Equilibrium	Log K 25°, 0.1	Log K 25°, 0.5	ΔH 25°, 0.1	ΔS 25°, 0.1
Ba^{2+}	ML/M.L	1.46			
La^{3+}	ML/M.L	3.42			
	$ML_2/M.L^2$	5.54			
Ce^{3+}	ML/M.L	3.51			
	$ML_2/M.L^2$	5.77			
Pr^{3+}	ML/M.L	3.65			
	$ML_2/M.L^2$	5.99			
Nd^{3+}	ML/M.L	3.73			
	$ML_2/M.L^2$	6.01			
Sm^{3+}	ML/M.L	3.91			
	$ML_2/M.L^2$	6.14			
Eu^{3+}	ML/M.L	4.00			
	$ML_2/M.L^2$	6.18			
Gd^{3+}	ML/M.L	4.02			
	$ML_2/M.L^2$	6.13			
Tb^{3+}	ML/M.L	4.10			
	$ML_2/M.L^2$	6.33			
Dy^{3+}	ML/M.L	4.13			
	$ML_2/M.L^2$	6.36			
Ho^{3+}	ML/M.L	4.09			
	$ML_2/M.L^2$	6.29			
Er^{3+}	ML/M.L	4.07			
	$ML_2/M.L^2$	6.33			
Tm^{3+}	ML/M.L	4.08			
	$ML_2/M.L^2$	6.40			
Yb^{3+}	ML/M.L	4.10			
	$ML_2/M.L^2$	6.40			
Lu^{3+}	ML/M.L	4.05			
	$ML_2/M.L^2$	6.10			
Co^{2+}	ML/M.L	2.22 ±0.02			
	$ML_2/M.L^2$	3.2			
Ni^{2+}	ML/M.L	2.27 ±0.07			
	$ML_2/M.L^2$	3.3			
Cu^{2+}	ML/M.L	5.01 ±0.01			
	$ML_2/M.L^2$	8.3 ±0.2			
	MHL/M.HL	1.37			
Fe^{3+}	ML/M.L		7.46		
Zn^{2+}	ML/M.L	2.50 ±0.03			
	$ML_2/M.L^2$	4.0			
Cd^{2+}	ML/M.L	2.68			
	MHL/M.HL	1.30			

Bibliography:

H^+ 67PNF,71CD,75CG

Mg^{2+}-Ba^{2+} 66OC

La^{3+}-Lu^{3+} 67PNF

Co^{2+}-Cu^{2+},Zn^{2+},Cd^{2+} 66OC,69PJ

Fe^{3+} 71CD

$C_7H_{10}O_4$ Metal ion	Equilibrium	Cyclopentane-1,1-dicarboxylic acid Log K $25°, 0.1$	ΔH $25°, 0.1$	H_2L ΔS $25°, 0.1$
H^+	HL/H.L	5.79 ±0.02	0.90	29.6
	H_2L/HL.H	3.07 ±0.01	0.06	14.3
Eu^{3+}	ML/M.L	4.17		
	$ML_2/M.L^2$	6.70		
Gd^{3+}	ML/M.L	4.18		
	$ML_2/M.L^2$	6.67		
Tb^{3+}	ML/M.L	4.26		
	$ML_2/M.L^2$	6.78		
Dy^{3+}	ML/M.L	4.29		
	$ML_2/M.L^2$	6.81		
Ho^{3+}	ML/M.L	4.25		
	$ML_2/M.L^2$	6.74		
Er^{3+}	ML/M.L	4.24		
	$ML_2/M.L^2$	6.76		
Tm^{3+}	ML/M.L	4.25		
	$ML_2/M.L^2$	6.82		
Yb^{3+}	ML/M.L	4.26		
	$ML_2/M.L^2$	6.88		
Lu^{3+}	ML/M.L	4.22		
	$ML_2/M.L^2$	6.86		

Bibliography:

H^+ 71PJ,75CG Eu^{3+}-Lu^{3+} 71PJ

$$HO_2CCH_2CH_2CO_2H$$

$C_4H_6O_4$ Metal ion	Equilibrium	Butanedioic acid (succinic acid) Log K $25°, 0.1$	Log K $25°, 1.0$	Log K $25°, 0$	ΔH $25°, 0$	H_2L ΔS $25°, 0$
H^+	HL/H.L	5.24 ±0.04	5.12 ±0.02	5.636	0.11 -0.2	26.2
		5.10[b]±0.05	5.21[d]	5.40[e]		
	H_2L/HL.H	4.00 ±0.02	3.95 ±0.02	4.207	-0.7 ±0.1	16.9
		3.94[b] 0.02	4.07[d]	4.22[e]		
Na^+	ML/M.L			0.3		
Be^{2+}	ML/M.L		3.13			
	MHL/M.HL		1.44			
	ML/MOHL.H		5.59			
Mg^{2+}	ML/M.L	1.18[s]±0.02		2.05		
	MHL/M.HL	0.52[x]				

[b] $25°, 0.5$;   [d] $25°, 2.0$;   [e] $25°, 3.0$;   [s] $25°, 0.15$;   [x] temperature not stated, 0.2

B. DI-CARBOXYLIC ACIDS

Succinic acid (continued)

Metal ion	Equilibrium	Log K 25°, 0.1	Log K 25°, 1.0	Log K 25°, 0	ΔH 25°, 0	ΔS 25°, 0
Ca^{2+}	ML/M.L	1.20^h		2.00		
	MHL/M.HL	0.54^h				
Sr^{2+}	ML/M.L	1.06^x				
	MHL/M.HL	0.48^x				
Ba^{2+}	ML/M.L	$1.00^s \pm 0.03$		2.02 ± 0.06		
	MHL/M.HL	0.45^x				
La^{3+}	ML/M.L			3.96		
	MHL/M.HL	1.48^s				
	$M(HL)_2/M.(HL)^2$	2.7^s				
Pr^{3+}	MHL/M.HL	1.72^s				
	$M(HL)_2/M.(HL)^2$	3.0^s				
Sm^{3+}	MHL/M.HL	2.00^s				
	$M(HL)_2/M.(HL)^2$	3.4^s				
Eu^{3+}	MHL/M.HL	1.99^s				
	$M(HL)_2/M.(HL)^2$	3.3^s				
Gd^{3+}	MHL/M.HL	1.83^s				
	$M(HL)_2/M.(HL)^2$	3.1^s				
Tb^{3+}	MHL/M.HL	1.66^s				
	$M(HL)_2/M.(HL)^2$	3.2^s				
Dy^{3+}	MHL/M.HL	1.72^s				
	$M(HL)_2/M.(HL)^2$	3.1^s				
Er^{3+}	MHL/M.HL	1.71^s				
	$M(HL)_2/M.(HL)^2$	3.0^s				
Yb^{3+}	MHL/M.HL	1.72^s				
	$M(HL)_2/M.(HL)^2$	2.9^s				
Lu^{3+}	MHL/M.HL	1.76^s				
	$M(HL)_2/M.(HL)^2$	3.1^s				
Th^{4+}	ML/M.L		6.23^j			
NpO_2^+	ML/M.L		1.72^j			
UO_2^{2+}	ML/M.L	3.87^b	3.68			
	MHL/M.HL	2.13^b				
PuO_2^{2+}	ML/M.L		3.42^j			
Mn^{2+}	ML/M.L	1.48^r		2.26	3.0	20
	MHL/M.HL	0.7^r		1.2		
Fe^{2+}	ML/M.L	1.4^t				
Co^{2+}	ML/M.L	1.70^h		2.32 ± 0.10	3.2	21
	MHL/M.HL	0.99^h				
Ni^{2+}	ML/M.L	1.6		2.34	2.5	19
	MHL/M.HL			1.3		
Cu^{2+}	ML/M.L	2.6 ± 0.0		3.28 ± 0.06	4.6	30
		2.26^r				
Fe^{3+}	ML/M.L	6.88^b				

^{b} 25°, 0.5; ^{h} 20°, 0.1; ^{j} 20°, 1.0; ^{r} 25°, 0.2; ^{s} 25°, 0.16; ^{t} 37°, 0.15;
^{x} temperature not stated, 0.2

Succinic acid (continued)

Metal ion	Equilibrium	Log K 25°, 0.1	Log K 25°, 1.0	Log K 25°, 0	ΔH 25°, 0	ΔS 25°, 0
Hg_2^{2+}	$ML_2/M.L^2$		7.28[u]			
	MOHL/M.OH.L		13.45[u]			
Zn^{2+}	$ML/M.L$	1.76[h]	1.48	2.52 ±0.05	4.4	26
	$ML_2/M.L^2$		2.00			
	MHL/M.HL	0.96[h]				
Cd^{2+}	$ML/M.L$	2.1	1.67[j]	2.72 ±0.10		
	$ML_2/M.L^2$		2.79[j]			
	MHL/M.HL		0.99[j]			
Pb^{2+}	$ML/M.L$	2.8	2.40[v]			
	$ML_2/M.L^2$		3.73[v]			
	$ML_3/M.L^3$		4.11[v]			

[h] 20°, 0.1;   [j] 20°, 1.0;   [u] 27°, 2.5;   [v] 30°, 2.0

Bibliography:

H^+ 39A,48CW,50PB,60YY,62SS,67CIH,68DM, 69VOR,70MKS,74CC,74MS,75ST,75VG

Na^+ 65AE

Be^{2+} 75VG

Mg^{2+}-Ba^{2+} 38CK,40TD,46J,51PJ,52S,62JS, 63Ca

La^{3+}-Lu^{3+} 51PJ,68KK

Th^{4+} 72TM

NpO_2^+ 72MB

UO_2^{2+} 67RMa,69VOR

PuO_2^{2+} 73PDB

Mn^{2+},Co^{2+}-Cu^{2+},Zn^{2+}-Pb^{2+} 34FR,60YY,61MNa, 62JS,62SS,63Ca,65Ma,65SMa,67MNT,68GP, 70EM,74MS,75SJ

Fe^{2+} 74CC

Fe^{3+} 68DM

Hg_2^{2+} 60YD

Other references: 50SR,53BB,54S,57LW,61MM, 61YR,62BK,63FK,63YK,66KZ,66MTa,67Ka, 67MN,68Md,68S,70RSa,73KJ,73RM,74TG, 75KN,75W

$$\underset{HO_2CCH_2CHCO_2H}{\overset{\overset{\textstyle CH_3}{|}}{}}$$

			H_2L

$C_5H_8O_4$ <u>DL-Methylbutanedioic acid</u> (methylsuccinic acid)

Metal ion	Equilibrium	Log K 25°, 0.1	Log K 25°, 0
H^+	HL/H.L	5.35	5.79
	$H_2L/HL.H$	3.88	4.10
Y^{3+}	$ML/M.L$	3.12	
	$ML_2/M.L^2$	(4.92)[z]	
La^{3+}	$ML/M.L$	2.95	
	$ML_2/M.L^2$	(4.59)[z]	
Pr^{3+}	$ML/M.L$	3.21	
	$ML_2/M.L^2$	(4.89)[z]	
Nd^{3+}	$ML/M.L$	3.26	
	$ML_2/M.L^2$	(5.01)[z]	

[z] optical isomerism not stated.

Methylsuccinic acid (continued)

Metal ion	Equilibrium	Log K 25°, 0.1	Log K 25°, 0
Sm^{3+}	$ML/M.L$ $ML_2/M.L^2$	3.41 $(5.05)^z$	
Eu^{3+}	$ML/M.L$ $ML_2/M.L^2$	3.37 $(5.02)^z$	
Gd^{3+}	$ML/M.L$ $ML_2/M.L^2$	3.33 $(5.12)^z$	
Tb^{3+}	$ML/M.L$ $ML_2/M.L^2$	3.22 $(5.06)^z$	
Dy^{3+}	$ML/M.L$ $ML_2/M.L^2$	3.17 $(5.13)^z$	
Ho^{3+}	$ML/M.L$ $ML_2/M.L^2$	3.08 $(5.09)^z$	
Er^{3+}	$ML/M.L$ $ML_2/M.L^2$	3.08 $(5.15)^z$	
Tm^{3+}	$ML/M.L$ $ML_2/M.L^2$	3.08 $(5.14)^z$	
Yb^{3+}	$ML/M.L$ $ML_2/M.L^2$	3.07 $(5.16)^z$	
Lu^{3+}	$ML/M.L$ $ML_2/M.L^2$	3.07 $(5.12)^z$	

^{z} optical isomerism not stated.

Bibliography: 70RFV

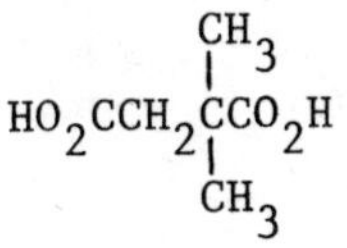

$$\underset{\displaystyle \overset{|}{CH_3}}{\overset{\displaystyle \overset{CH_3}{|}}{HO_2CCH_2CCO_2H}}$$

$C_6H_{10}O_4$	2,2-Dimethylbutanedioic acid (2,2-dimethylsuccinic acid)		H_2L

Metal ion	Equilibrium	Log K 25°, 1.0
H^+	$HL/H.L$ $H_2L/HL.H$	5.81 3.83
UO_2^{2+}	$ML/M.L$ $MHL/M.HL$ $ML/MOHL.H$	5.13 2.20 4.78

Bibliography: 70CVa

$$HO_2CHCHCO_2H$$
$$H_3C \quad CH_3$$

| $C_6H_{10}O_4$ | | meso-2,3-Dimethylbutanedioic acid | | | H_2L |
| | | (meso-2,3-dimethylsuccinic acid) | | | |

Metal ion	Equilibrium	Log K 25°, 1.0	Log K 25°, 0	ΔH 25°, 0	ΔS 25°, 0
H^+	HL/H.L	4.85	5.30	3.77	18.7
	H_2L/HL.H	3.50	3.67	0.30	17.8
UO_2^{2+}	ML/M.L	3.76			
	MHL/M.HL	1.89			

Bibliography:

H^+ 70CVa,73PT

UO_2^{2+} 70CVa

Other reference: 31GI

$$CH_3$$
$$HO_2CCHCHCO_2H$$
$$CH_3$$

| $C_6H_{10}O_4$ | | DL-2,3-Dimethylbutanedioic acid | | | H_2L |
| | | (DL-2,3-dimethylsuccinic acid) | | | |

Metal ion	Equilibrium	Log K 25°, 1.0	Log K 25°, 0	ΔH 25°, 0	ΔS 25°, 0
H^+	HL/H.L	5.42	5.93	2.25	24.7
	H_2L/HL.H	(2.65)	3.82	0.35	36.9
UO_2^{2+}	ML/M.L	4.24			
	MHL/M.HL	1.99			
	ML/MOHL.H	4.96			

Bibliography:

H^+ 70CVa,73PT

UO_2^{2+} 70CVa

Other reference: 31GI

$$CHCO_2H$$
$$CHCO_2H$$

| $C_4H_4O_4$ | | cis-Butenedioic acid (maleic acid) | | | | H_2L |

Metal ion	Equilibrium	Log K 25°, 0.1	Log K 25°, 1.0	Log K 25°, 0	ΔH 25°, 0	ΔS 25°, 0
H^+	HL/H.L	5.83 ±0.04	5.62 ±0.02	6.332	0.8	32
		5.61[b]			0.2[c]	26[c]
	H_2L/HL.H	1.75 ±0.05	1.63 ±0.03	1.910	-0.1	
		1.64[b]	1.71[d]	1.82[e]	0.1[c]	

[b] 25°, 0.5;   [c] 25°, 1.0;   [d] 25°, 2.0;   [e] 25°, 3.0

Maleic acid (continued)

Metal ion	Equilibrium	Log K 25°, 0.1	Log K 25°, 1.0	Log K 25°, 0	ΔH 25°, 0	ΔS 25°, 0
Na^+	ML/M.L			0.7		
Ca^{2+}	ML/M.L	1.10[r]		2.43		
Ba^{2+}	ML/M.L			2.26		
Y^{3+}	ML/M.L	3.61				
	$ML_2/M.L^2$	5.54				
La^{3+}	ML/M.L	3.45		4.55	3.1[a]	26[a]
	$ML_2/M.L^2$	5.43				
Pr^{3+}	ML/M.L	3.64			2.9[a]	26[a]
			2.81		2.5[c]	21[c]
	$ML_2/M.L^2$	5.80	4.70		4.8[c]	38[c]
Nd^{3+}	ML/M.L	3.66			2.9[a]	26[a]
	$ML_2/M.L^2$	5.81				
Sm^{3+}	ML/M.L	3.82			3.0[a]	28[a]
			3.00		2.5[c]	22[c]
	$ML_2/M.L^2$	6.01	4.90		4.7[c]	38[c]
Eu^{3+}	ML/M.L	3.83			3.4[a]	29[a]
	$ML_2/M.L^2$	5.98				
Gd^{3+}	ML/M.L	3.79			3.6[a]	29[a]
			2.97		3.0[c]	24[c]
	$ML_2/M.L^2$	5.90	4.79		4.7[c]	38[c]
Tb^{3+}	ML/M.L	3.75			4.1[a]	31[a]
	$ML_2/M.L^2$	5.83				
Dy^{3+}	ML/M.L	3.75			4.2[a]	31[a]
	$ML_2/M.L^2$	5.83				
Ho^{3+}	ML/M.L	3.67			4.4[a]	31[a]
			2.89		3.7[c]	26[c]
	$ML_2/M.L^2$	5.69	4.67		5.7[c]	41[c]
Er^{3+}	ML/M.L	3.64			4.3[a]	31[a]
	$ML_2/M.L^2$	5.70				
Tm^{3+}	ML/M.L	3.62			4.4[a]	31[a]
	$ML_2/M.L^2$	5.64				
Yb^{3+}	ML/M.L	3.64			4.4[a]	31[a]
			2.81		3.9[c]	26[c]
	$ML_2/M.L^2$	5.73	4.65		6.4[c]	43[c]
Lu^{3+}	ML/M.L	3.60			4.4[a]	31[a]
	$ML_2/M.L^2$	5.68				
Th^{4+}	ML/M.L		6.34[j]			
	$ML_2/M.L^2$		10.55[j]			
NpO_2^+	ML/M.L		2.20[j]			
UO_2^{2+}	ML/M.L		4.46			
PuO_2^{2+}	ML/M.L		4.25[j]			
Mn^{2+}	ML/M.L	1.68[r]				
Ni^{2+}	ML/M.L	2.0				

[a] 25°, 0.1;   [c] 25°, 1.0;   [j] 20°, 1.0;   [r] 25°, 0.16

Maleic acid (continued)

Metal ion	Equilibrium	Log K 25°, 0.1	Log K 25°, 1.0	Log K 25°, 0	ΔH 25°, 0	ΔS 25°, 0
Cu^{2+}	$ML/M.L$	3.4		3.90		
	$ML_2/M.L^2$	4.9[s]				
	$ML_3/M.L^3$	6.2[s]				
Cu^+	$ML/M.L$	3.05				
Zn^{2+}	$ML/M.L$	2.0				
Cd^{2+}	$ML/M.L$	2.4				
		2.2[s]				
	$ML_2/M.L^2$	3.6[s]				
	$ML_3/M.L^3$	3.8[s]				
Pb^{2+}	$ML/M.L$	3.2	2.75			
		3.0[s]				
	$ML_2/M.L^2$	4.5[s]	4.03			
	$ML_3/M.L^3$	5.4[s]	4.36			
	$MHL/M.HL$		0.58			
	$M(HL)_2/M.(HL)^2$		0.7			
In^{3+}	$ML/M.L$	5.0[s]				
	$ML_2/M.L^2$	7.1[s]				
	$ML_3/M.L^3$	6.2[s]				

[s] 25°, 0.2

Bibliography:

H^+ 39A,60DL,60YY,62BK,67CIH,67RMa,70RFV, 72MB,73DM,75OS

Na^+ 65AE

Ca^{2+},Ba^{2+} 40TD,52SL

Y^{3+}-Lu^{3+} 51PJ,70RFV,73CDS,73DM

Th^{4+} 72TM

NpO_2^+ 72MB

UO_2^{2+} 67RMa

PuO_2^{2+} 73PDB

Mn^{2+} 57LW

Ni^{2+},Cu^{2+},Zn^{2+}-Pb^{2+} 51PJ,60YY,67NMH,75OS

Cu^+ 49AK,49KAa

In^{3+} 67NMH

Other references: 48AK,68TKT,70VS,74BSa, 74HB,75KN,75ST,75W

$$CH_3CCO_2H$$
$$\underset{\|}{}$$
$$HCCO_2H$$

$C_5H_6O_4$	cis-Methylbutenedioic acid (citraconic acid)			H_2L
Metal ion	Equilibrium	Log K 25°, 0.1	Log K 18°, 0	
H^+	$HL/H.L$	5.60	6.17	
	$H_2L/HL.H$	2.2		
Ca^{2+}	$ML/M.L$	1.3[r]		
Sr^{2+}	$ML/M.L$	1.3[r]		
Mn^{2+}	$ML/M.L$	1.77[r]		

[r] 25°, 0.16

B. DI-CARBOXYLIC ACIDS

Citraconic acid (continued)

Metal ion	Equilibrium	Log K 25°, 0.1	Log K 18°, 0
Ni^{2+}	ML/M.L	1.8	
Cu^{2+}	ML/M.L	3.4	
Cu^{+}	ML/M.L	1.34	
Zn^{2+}	ML/M.L	1.8	
Cd^{2+}	ML/M.L	2.2	
Pb^{2+}	ML/M.L	3.3	

Bibliography:

H+ 24L,60YY

Ca^{2+},Sr^{2+} 52SL

Mn^{2+} 57LW

$Ni^{2+}-Cu^{2+},Zn^{2+}-Pb^{2+}$ 60YY

Cu^{+} 49KAa

Other references: 69JR,69SSa,72KS

$$HO_2CCH=CHCO_2H$$

$C_4H_4O_4$		trans-Butenedioic acid (fumaric acid)				H_2L
Metal ion	Equilibrium	Log K 25°, 0.1	Log K 25°, 1.0	Log K 25°, 0	ΔH 25°, 0	ΔS 25°, 0
H^{+}	HL/H.L	4.10 / 3.92 [b]	3.91 / 4.01 [d]	4.494 / 4.17 [e]	0.7	23
	H_2L/HL.H	2.85 / 2.78 [b]	2.80 / 2.92 [d]	3.053 / 3.08 [e]	-0.1	14
Ca^{2+}	ML/M.L	(0.48) [r]		2.00		
Sr^{2+}	ML/M.L	(0.54) [r]				
Ba^{2+}	ML/M.L			1.59		
La^{3+}	ML/M.L	2.74		3.01	2.7 [a]	22 [a]
Ce^{3+}	ML/M.L	2.80			3.2 [a]	23 [a]
Pr^{3+}	ML/M.L	2.84			3.2 [a]	24 [a]
Nd^{3+}	ML/M.L	2.74			3.7 [a]	25 [a]
Sm^{3+}	ML/M.L	2.83			3.5 [a]	25 [a]
Eu^{3+}	ML/M.L	2.86			3.4 [a]	25 [a]
Gd^{3+}	ML/M.L	2.88			3.8 [a]	26 [a]
Tb^{3+}	ML/M.L	2.77			3.6 [a]	25 [a]
Dy^{3+}	ML/M.L	2.80			3.8 [a]	25 [a]
Ho^{3+}	ML/M.L	2.80			3.6 [a]	25 [a]
Er^{3+}	ML/M.L	2.80			3.9 [a]	26 [a]
Tm^{3+}	ML/M.L	2.81			3.5 [a]	25 [a]
Yb^{3+}	ML/M.L	2.80			3.8 [a]	26 [a]
Lu^{3+}	ML/M.L	2.81			3.8 [a]	26 [a]
Mn^{2+}	ML/M.L	0.99 [r]				

---

[a] 25°, 0.1; [b] 25°, 0.5; [d] 25°, 2.0; [e] 25°, 3.0; [r] 25°, 0.16

Fumaric acid (continued)

Metal ion	Equilibrium	Log K 25°, 0.1	Log K 25°, 1.0	Log K 25°, 0	ΔH 25°, 0	ΔS 25°, 0
Cu^{2+}	ML/M.L			2.51		
Cu^{+}	ML/M.L	3.96				

Bibliography:

H^{+}	39A,67CIH,72KN	Cu^{2+}	51PJ
$Ca^{2+}-Ba^{2+}$	40TD,52SL	Cu^{+}	49KAa
$La^{3+}-Lu^{3+}$	51PJ,73CDS	Other references:	48KA,50SR,74BSa,74CC
Mn^{2+}	57LW		

$$\overset{\overset{\displaystyle CH_3}{|}}{HO_2CCH=CHCO_2H}$$

$C_5H_7O_4$	trans-Methylbutenedioic acid (mesaconic acid)	H_2L

Metal ion	Equilibrium	Log K 25°, 0.1	Log K 18°, 0
H^{+}	HL/H.L		4.82
Cu^{+}	ML/M.L	2.61	

Bibliography:

H^{+}	24L	Cu^{+}	49KAa

$$\overset{\overset{\displaystyle CH_2}{\|}}{HO_2CCH_2CCO_2H}$$

$C_5H_6O_4$	Methylenebutanedioic acid (itaconic acid)	H_2L

Metal ion	Equilibrium	Log K 25°, 0.1	Log K 25°, 1.0	Log K 18°, 0
H^{+}	HL/H.L	5.14	4.99	5.54
	$H_2L/HL.H$	3.68	3.63	
Ca^{2+}	ML/M.L	1.2		
Sr^{2+}	ML/M.L	0.96		
Nd^{3+}	ML/M.L		2.00	
	MHL/M.HL		1.38	
	$M(HL)_2/M.(HL)^2$		2.55	
Sm^{3+}	ML/M.L		2.07	
	MHL/M.HL		1.49	
	$M(HL)_2/M.(HL)^2$		2.74	
Gd^{3+}	ML/M.L		2.18	
	MHL/M.HL		1.50	
	$M(HL)_2/M.(HL)^2$		2.78	
Dy^{3+}	ML/M.L		2.22	
	MHL/M.HL		1.41	
	$M(HL)_2/M.(HL)^2$		2.64	

Itaconic acid (continued)

Metal ion	Equilibrium	Log K 25°, 0.1	Log K 25°, 1.0	Log K 18°, 0
Er^{3+}	ML/M.L		2.27	
	MHL/M.HL		1.25	
	$M(HL)_2/M.(HL)^2$		2.25	
	$M(HL)_2/MHL_2.H$		3.83	
Mn^{2+}	ML/M.L		1.13	
Co^{2+}	ML/M.L		1.36	
	MHL/M.HL		0.61	
	$M(HL)_2/M.(HL)^2$		1.20	
Ni^{2+}	ML/M.L	1.8	1.39	
	MHL/M.HL		0.40	
Cu^{2+}	ML/M.L	2.8	2.15	
	MHL/M.HL		1.03	
Cu^+	ML/M.L	3.34		
Zn^{2+}	ML/M.L	1.9	1.35	
	MHL/M.HL		0.25	
Cd^{2+}	ML/M.L	2.3	1.72	
	MHL/M.HL		0.85	
	$M(HL)_2/M.(HL)^2$		1.58	
Pb^{2+}	ML/M.L	3.1		

Bibliography:

H^+ 24L,60YY,70ST Cu^+ 49KAa

Ca^{2+},Sr^{2+} 52SL   Other references: 67LC,68RS,68RSb,70RSa,
                                                                     72STa
Nd^{3+}-Er^{3+} 72STb

Mn^{2+}-Cu^{2+},Zn^{2+}-Pb^{2+} 60YY,73RT

$$HO_2CCH_2CH_2CH_2CO_2H$$

$C_5H_8O_4$		Pentanedioic acid (glutaric acid)				H_2L
Metal ion	Equilibrium	Log K 25°, 0.1	Log K 25°, 1.0	Log K 25°, 0	ΔH 25°, 0	ΔS 25°, 0
H^+	HL/H.L	5.03 ±0.03	4.87	5.43 ±0.02	0.6	27
		4.87[b]±0.03	4.98[d]	5.15[d]		
	H_2L/HL.H	4.13 ±0.02	4.11	4.34 ±0.00	0.1	20
		4.11[b]±0.02	4.25[d]	4.43[e]		
Mg^{2+}	ML/M.L	1.08[x]				
	MHL/M.HL	0.52[x]				
Ca^{2+}	ML/M.L	1.06[x]				
	MHL/M.HL	0.50[x]				
Ba^{2+}	ML/M.L			2.04		
Sc^{3+}	ML/M.L	4.82[s]				
Y^{3+}	ML/M.L	3.25[s]				

[b] 25°, 0.5;   [d] 25°, 2.0;   [e] 25°, 3.0;   [s] 25°, 0.05;   [x] temperature not stated, 0.2

Glutaric acid (continued)

Metal ion	Equilibrium	Log K 25°, 0.1	Log K 25°, 1.0	Log K 25°, 0	ΔH 25°, 0	ΔS 25°, 0
La^{3+}	ML/M.L	3.02[s]		3.82		
Th^{4+}	ML/M.L	7.44[s]				
	MHL/M.HL		3.48[j]			
	$M(HL)_2/M.(HL)^2$		6.14[j]			
UO_2^{2+}	ML/M.L	3.53[b]				
	MHL/M.HL	2.30[b]	1.89[j]			
	$M(HL)_2/M.(HL)^2$		3.58[j]			
	$M(HL)_2/MHL_2.H$		9.5[j]			
Mn^{2+}	ML/M.L	1.13[r]				
Co^{2+}	ML/M.L			2.28 ±0.07		
Ni^{2+}	ML/M.L	1.6				
Cu^{2+}	ML/M.L	2.4	2.40	3.16		
Fe^{3+}	ML/M.L	6.78[b]				
Zn^{2+}	ML/M.L	1.6	1.25	2.45		
	$ML_2/M.L^2$		1.74			
	MHL/M.HL	0.84[x]				
Cd^{2+}	ML/M.L	2.0	1.60[u]			
Pb^{2+}	ML/M.L	2.8	2.48[u]			
	$ML_2/M.L^2$		3.45[u]			
	$ML_3/M.L^3$		3.90[u]			

[b] 25°, 0.5;   [j] 20°, 1.0;   [r] 25°, 0.16;   [s] 25°, 0.05;   [u] 30°, 2.0;   [x] temperature not stated, 0.2

Bibliography:

H^+	31GI,36JS,39A,60YY,67CIH,68DM,69VOR, 70MKS,73RM	Co^{2+}	65Ma,65SMa
Mg^{2+},Ca^{2+}	38CK	Ni^{2+}	60YY
Ba^{2+}	51PJ	Cu^{2+}	51PJ,55GL,60YY
Sc^{3+},Y^{3+},Th^{4+}	73CS	Fe^{3+}	68DM
La^{3+}	51PJ,73CS	Zn^{2+}	38CK,60YY,70EM,74MS
Th^{4+}	72TM,73CS	Cd^{2+}	60YY,75BC
UO_2^{2+}	69VOR,73CB	Pb^{2+}	60YY,68GP
Mn^{2+}	57LW	Other references:	52SL,66KZ,66RM,69Pa,74BSa, 75W

$$HO_2CCH_2CH_2CH_2CH_2CO_2H$$

$C_6H_{10}O_4$		Hexanedioic acid (adipic acid)				H_2L
Metal ion	Equilibrium	Log K 25°, 0.1	Log K 25°, 1.0	Log K 25°, 0	ΔH 25°, 0	ΔS 25°, 0
H^+	HL/H.L	5.03 ±0.03	4.92	5.42 ±0.01	0.6	27
		4.93[b]±0.00	5.05[d]	5.23[e]		
	$H_2L/HL.H$	4.26 ±0.03	4.21	4.42 ±0.01	0.3	21
		4.22[b]±0.04	4.34[d]	4.52[e]		

[b] 25°, 0.5;   [d] 25°, 2.0;   [e] 25°, 3.0

Adipic acid (continued)

Metal ion	Equilibrium	Log K 25°, 0.1	Log K 25°, 1.0	Log K 25°, 0	ΔH 25°, 0	ΔS 25°, 0
Ca^{2+}	ML/M.L			2.19		
Ba^{2+}	ML/M.L			1.89 ±0.04		
La^{3+}	ML/M.L			4.10		
UO_2^{2+}	ML/M.L	3.54[b]				
	MHL/M.HL	2.38[b]				
Co^{2+}	ML/M.L			2.19 ±0.04		
Ni^{2+}	ML/M.L	1.6				
Cu^{2+}	ML/M.L	2.3 ±0.0		3.35		
Zn^{2+}	ML/M.L	1.8	1.23	2.40		
	$ML_2/M.L^2$		1.78			
Cd^{2+}	ML/M.L	2.1				
Pb^{2+}	ML/M.L	2.8	2.38[r]			
	$ML_2/M.L^2$		3.20[r]			
	$ML_3/M.L^3$		3.69[r]			

---

[b] 25°, 0.5; [r] 30°, 2.0

Bibliography:

H^+ 31GI,36JS,39A,60YY,67CIH,69LP,69VO,
       70MKS,74MS,75ST

Ca^{2+}-La^{3+} 40TD,51PJ

UO_2^{2+} 69VO

Co^{2+} 65Ma,65SMa

Ni^{2+},Cd^{2+} 60YY

Cu^{2+} 51PJ,60YY,75SJ

Zn^{2+} 60YY,74MS,70EM

Pb^{2+} 60YY,68GP

Other references: 61MM,62JS,66KZ,66RM,68GGa,
                   68GGb,69GGa,69RS,73KJ,75BC

$$HO_2CCH_2CH_2CH_2CH_2CH_2CO_2H$$

$C_7H_{12}O_4$		Heptanedioic acid (pimelic acid)				H_2L
Metal ion	Equilibrium	Log K 25°, 0.1	Log K 25°, 1.0	Log K 25°, 0	ΔH 25°, 0	ΔS 25°, 0
H^+	HL/H.L	5.08 ±0.03	4.97	5.43 ±0.01	0.9	28
		4.96[b]±0.00	5.09[d]	5.25[e]		
	$H_2L/HL.H$	4.31 +0.01	4.28	4.49 ±0.01	0.3	22
		4.28[b]±0.02	4.42[d]	4.62[e]		
UO_2^{2+}	ML/M.L	3.68[b]				
	MHL/M.HL	2.45[b]				
Mn^{2+}	ML/M.L	(1.33)[r]				
		1.08				
Co^{2+}	ML/M.L	1.50				
Ni^{2+}	ML/M.L	(1.20)				
Cu^{2+}	ML/M.L	2.21				

---

[b] 25°, 0.5; [d] 25°, 2.0; [e] 25°, 3.0; [r] 25°, 0.16

Pimelic acid (continued)

Metal ion	Equilibrium	Log K 25°, 0.1	Log K 25°, 1.0	Log K 25°, 0	ΔH 25°, 0	ΔS 25°, 0
Zn²⁺	ML/M.L	1.3				
Cd²⁺	ML/M.L	1.76				
Pb²⁺	ML/M.L	2.62				

Bibliography:

H⁺	31GI,39A,67CIH,69VO,75LP	Mn²⁺	57LW,75LP
UO₂²⁺	69VO	Co²⁺-Pb²⁺	75LP

$$HO_2CCH_2CH_2CH_2CH_2CH_2CH_2CO_2H$$

$C_9H_{16}O_4$ Nonanedioic acid (azelaic acid) H_2L

Metal ion	Equilibrium	Log K 25°, 0.1	Log K 25°, 1.0	Log K 25°, 0
H⁺	HL/H.L	5.12 ±0.03	4.99	5.42 ±0.01
		4.98[b]	4.12[d]	5.31[d]
	H₂L/HL.H	4.39 ±0.03	4.33	4.55 ±0.01
		4.30[b]	4.48[d]	4.66[e]
Mn²⁺	ML/M.L	1.03[r]		

---
[b] 25°, 0.5;   [d] 25°, 2.0;   [e] 25°, 3.0;   [r] 25°, 0.16

Bibliography:

H⁺	31GI,39A,72KMK	Other reference: 70MKS
Mn²⁺	57LW	

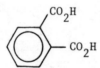

$C_8H_6O_4$ Benzene-1,2-dicarboxylic acid (phthalic acid) H_2L

Metal ion	Equilibrium	Log K 25°, 0.1	Log K 25°, 1.0	Log K 25°, 0	ΔH 25°, 0	ΔS 25°, 0
H⁺	HL/H.L	4.93 -0.01	4.71 ±0.03	5.408	0.50 +0.04	26.4
		4.73[b]±0.01	4.73[d]	4.87[e]±0.03		
	H₂L/HL.H	2.75 ±0.02	2.66 ±0.03	2.950	0.64 +0.01	15.6
		2.66[b]±0.01	2.80[d]	2.99[e]±0.07		
Na⁺	ML/M.L			0.7		
Be²⁺	ML/M.L	3.97				
	ML₂/M.L²	5.69				
Ca²⁺	ML/M.L	1.07		2.42 ±0.01		
Ba²⁺	ML/M.L	0.92		2.33		

---
[b] 25°, 0.5;   [d] 25°, 2.0;   [e] 25°, 3.0

Phthalic acid (continued)

Metal ion	Equilibrium	Log K 25°, 0.1	Log K 25°, 1.0	Log K 25°, 0	ΔH 25°, 0	ΔS 25°, 0
Y^{3+}	ML/M.L	3.46				
La^{3+}	ML/M.L	3.44		4.74		
Pr^{3+}	ML/M.L	3.56				
Nd^{3+}	ML/M.L	(3.88)				
Sm^{3+}	ML/M.L	3.70				
Eu^{3+}	ML/M.L	3.70				
Gd^{3+}	ML/M.L	3.63				
Tb^{3+}	ML/M.L	3.46				
Dy^{3+}	ML/M.L	3.48				
Ho^{3+}	ML/M.L	3.55				
Er^{3+}	ML/M.L	3.76				
Tm^{3+}	ML/M.L	3.53				
Yb^{3+}	ML/M.L	3.48				
Lu^{3+}	ML/M.L	(3.65)				
Th^{4+}	ML/M.L		5.92^j			
	$ML_2/M.L^2$		10.05^j			
NpO_2^+	ML/M.L		2.22^j			
UO_2^{2+}	ML/M.L	4.81^r	4.38			
PuO_2^{2+}	ML/M.L		4.11^j			
Mn^{2+}	ML/M.L			2.74	2.2	20
Co^{2+}	ML/M.L	2.03_b 1.53^b	1.42_d 1.50^d	2.83 ±0.07	1.9	19
	MHL/M.HL	1.28^b				
Ni^{2+}	ML/M.L	2.17 ±0.03 1.70^b	1.57 1.63^d	2.95	1.8	19
	MHL/M.HL	0.7^b				
Cu^{2+}	ML/M.L	3.15 ±0.05 2.81^b	2.69 2.64^d	4.04 ±0.03	2.0	25
	$ML_2/M.L^2$		3.73_d 4.14^d		5.3	
	MHL/M.HL	1.2^b				
Hg_2^{2+}	ML/M.L	4.90^h				
Zn^{2+}	ML/M.L $ML_2/M.L^2$	2.2		2.91 ±0.02 4.2	3.2	24
Cd^{2+}	ML/M.L	2.5				
Al^{3+}	ML/M.L $ML_2/M.L^2$	3.18^b 6.32^b				
Ga^{3+}	ML/M.L	5.15				

---

^{b} 25°, 0.5; ^{d} 25°, 2.0; ^{h} 18°, 0.1; ^{j} 20°, 1.0; ^{r} 31°, 0.1

Phthalic acid (continued)

Bibliography:

H^+ 45HP,56YS,64PK,67RMa,70NL,75LK,75ST $Mn^{2+}-Cu^{2+}$ 51PJ,56YS,62DN,65Ma,65SMa,75LKa

Na^+ 65AE Hg^{2+} 66GC

Be^{2+} 62BK Zn^{2+},Cd^{2+} 60YY,65N,70EM

Ca^{2+},Ba^{2+} 40TD,46J,70GNa Al^{3+} 70NL

$Y^{3+}-Lu^{3+}$ 51PJ,73SA Ga^{3+} 68ZK

Th^{4+} 72TM   Other references: 29Ra,53BB,55VG,58G,66KPa,
                                                            67HH,70KC,70NPa,73KJ,73NK,73NKa,74PJ,
NpO_2^+ 72MB   75PA,75PAa,75SJ

UO_2^{2+} 67RMa,67SP

PuO_2^{2+} 73PDB

4-Methylphthalic acid

$C_9H_8O_4$ H_2L

Metal ion	Equilibrium	Log K 30°, 0.1	Log K 25°, 0
H^+	HL/H.L	5.26	5.68
	H_2L/HL.H	(3.04)	(3.06)
Mn^{2+}	ML/M.L		2.82
Co^{2+}	ML/M.L		2.88
Ni^{2+}	ML/M.L	2.15	2.97
Cu^{2+}	ML/M.L	3.56	
	$ML_2/M.L^2$	5.47	
Zn^{2+}	ML/M.L		3.05
Cd^{2+}	ML/M.L	2.16	
	$ML_2/M.L^2$	3.96	

Bibliography:

H^+,Ni^{2+} 70NPa,71NP Cu^{2+},Cd^{2+} 70NPa

Mn^{2+},Co^{2+},Zn^{2+} 71NP

Benzene-1,3-dicarboxylic acid (isophthalic acid)

$C_8H_6O_4$ H_2L

Metal ion	Equilibrium	Log K 25°, 0	ΔH 25°, 0	ΔS 25°, 0
H^+	HL/H.L	4.50	0.4	22
	H_2L/HL.H	3.50	0.0	16
Ca^{2+}	ML/M.L	2.00		
Ba^{2+}	ML/M.L	1.55		

Isophthalic acid (continued)

Bibliography:

H^+ 72PTR   Other reference: 74KG

Ca^{2+}, Ba^{2+} 40TD

$C_mH_nO_qX_r$		Substitutedphthalic acid			H_2L
Z =	Metal ion	Equilibrium	Log K 25°, 0.1	Log K 25°, 0	
3-Chloro	H^+	HL/H.L	4.49		
$(C_8H_5O_4Cl)$	Ni^{2+}	ML/M.L	2.02		
	Cu^{2+}	ML/M.L	2.86		
3-Bromo	H^+	HL/H.L	4.35		
		$H_2L/HL.H$	2.37		
$(C_8H_5O_4Br)$	Ni^{2+}	ML/M.L	2.10		
	Cu^{2+}	ML/M.L	2.95		
4-Bromo	H^+	HL/H.L	4.60		
		$H_2L/HL.H$	2.50		
$(C_8H_5O_4Br)$	Ni^{2+}	ML/M.L	2.05		
	Cu^{2+}	ML/M.L	2.84		
3-Nitro	H^+	HL/H.L	3.93	4.48[s]	
		$H_2L/HL.H$		2.11[s]	
$(C_8H_5O_6N)$	UO_2^{2+}	ML/M.L	3.82[P]		
	Mn^{2+}	ML/M.L		3.16[s]	
	Co^{2+}	ML/M.L		3.16[s]	
	Ni^{2+}	ML/M.L	1.72	3.24[s]	
	Cu^{2+}	ML/M.L	2.42	3.78[s]	
	Zn^{2+}	ML/M.L		3.33[s]	
4-Nitro	H^+	HL/H.L	4.12	4.44	
		$H_2L/HL.H$		2.02	
$(C_8H_5O_6N)$	UO_2^{2+}	ML/M.L	4.02[P]		
	Mn^{2+}	ML/M.L		2.80	
	Co^{2+}	ML/M.L		2.87	
	Ni^{2+}	ML/M.L	1.65	3.02	
	Cu^{2+}	ML/M.L	2.42		

[P] 31°, 0.1;   [s] 35°, 0

Substitutedphthalic acid (continued)

Bibliography:

$H^+, Mn^{2+} - Zn^{2+}$ 56YS,70NPb,71NP   Other references: 73NK,73NKa,74PJ

UO_2^{2+} 67SP

$$\begin{array}{c} \text{OH} \\ | \\ \text{HO}_2\text{CCHCO}_2\text{H} \end{array}$$

$C_3H_4O_5$	Hydroxypropanedioic acid (tartronic acid)	H_2L

Metal ion	Equilibrium	Log K 20°, 0.1
H^+	HL/H.L	4.24
	H_2L/HL.H	2.02
Mg^{2+}	ML/M.L	2.17
	MHL/M.HL	1.23
Ca^{2+}	ML/M.L	2.27
	MHL/M.HL	1.30
Ba^{2+}	ML/M.L	1.80
	MHL/M.HL	0.87
Co^{2+}	ML/M.L	3.25
	MHL/M.HL	1.91
Ni^{2+}	ML/M.L	3.45
	MHL/M.HL	2.10
Cu^{2+}	ML/M.L	5.34
	MHL/M.HL	3.62
	$ML/MH_{-1}L.H$	4.03
Zn^{2+}	ML/M.L	3.22
	MHL/M.HL	1.91
Cd^{2+}	ML/M.L	2.85
	MHL/M.HL	1.61

Bibliography: 63Ca

Other references: 64ZT,67Me

$$\begin{array}{c} \text{OH} \\ | \\ \text{HO}_2\text{CCH}_2\text{CHCO}_2\text{H} \end{array}$$

$C_4H_6O_5$	L-Hydroxybutanedioic acid (malic acid)	H_2L

Metal ion	Equilibrium	Log K 25°, 0.1	Log K 25°, 1.0	Log K 25°, 0	ΔH 25°, 0	ΔS 25°, 0
H^+	HL/H.L	4.71 +0.01	4.45 ±0.01	5.097	0.28	24.3
	H_2L/HL.H	3.24 ±0.03	3.11	3.459	-0.71	13.5
Li^+	ML/M.L	0.38				
Na^+	ML/M.L	0.28				
K^+	ML/M.L	0.18				

Malic acid (continued)

Metal ion	Equilibrium	Log K 25°, 0.1	Log K 25°, 1.0	Log K 25°, 0	ΔH 25°, 0	ΔS 25°, 0
Rb^+	ML/M.L	0.04				
Cs^+	ML/M.L	-0.15				
Be^{2+}	ML/M.L		2.70			
	MHL/M.HL		1.21			
	$M_2L_2/M^2.L^2$		8.48			
	$M_2(OH)_2L.H/M^2.L^2$		3.05			
	$M_2(OH)_2L_2.H^2/M^2.L^2$		1.35			
	$M_4(OH)_4L_2.H^4/M^4.L^2$		-1.74			
Mg^{2+}	ML/M.L	1.70^h	1.55^x			
	MHL/M.HL	0.90^h	0.77^x			
Ca^{2+}	ML/M.L	1.96^h	1.80^x	2.66		
	MHL/M.HL	1.06^h	1.02^x			
Sr^{2+}	ML/M.L		1.45^x			
	MHL/M.HL		0.72^x			
Ba^{2+}	ML/M.L	1.45^h	1.30^x	2.20		
	MHL/M.HL	$(0.67)^h$	0.67^x			
Ra^{2+}	ML/M.L		0.95^s			
Y^{3+}	ML/M.L	4.91^h				
	$ML_2/M.L^2$	$(8.18)^{h,z}$				
La^{3+}	ML/M.L	$4.37^h+0.01$				
	$ML_2/M.L^2$	$(7.16)^{h,z}$				
Pr^{3+}	ML/M.L	4.65^h				
	$ML_2/M.L^2$	$(7.74)^{h,z}$				
Nd^{3+}	ML/M.L	4.77^h				
	$ML_2/M.L^2$	$(7.94)^{h,z}$				
Sm^{3+}	ML/M.L	4.89^h				
	$ML_2/M.L^2$	$(8.16)^{h,z}$				
Eu^{3+}	ML/M.L	4.85^h				
	$ML_2/M.L^2$	$(8.11)^{h,z}$				
Gd^{3+}	ML/M.L	4.76^h				
	$ML_2/M.L^2$	$(7.00)^{h,z}$				
Tb^{3+}	ML/M.L	4.77^h				
	$ML_2/M.L^2$	$(8.03)^{h,z}$				
Dy^{3+}	ML/M.L	4.78^h				
	$ML_2/M.L^2$	$(8.10)^{h,z}$				
Ho^{3+}	ML/M.L	4.90^h				
	$ML_2/M.L^2$	$(8.25)^{h,z}$				
Er^{3+}	ML/M.L	4.96^h				
	$ML_2/M.L^2$	$(8.35)^{h,z}$				
Tm^{3+}	ML/M.L	5.00^h				
	$ML_2/M.L^2$	$(8.50)^{h,z}$				
Yb^{3+}	ML/M.L	$5.05^h-0.1$				
	$ML_2/M.L^2$	$(8.58)^{h,z}$				

^{h} 20°, 0.1; ^{s} 25°, 0.16; ^{z} optical isomerism not stated; ^{x} temperature not stated, 0.2

Malic acid (continued)

Metal ion	Equilibrium	Log K 25°, 0.1	Log K 25°, 1.0	Log K 25°, 0	ΔH 25°, 0	ΔS 25°, 0
Lu^{3+}	$ML/M.L$	5.08 [h]				
	$ML_2/M.L^2$	(8.67) [h,z]				
UO_2^{2+}	$MH_{-1}L.H^3/M.H_2L$		−5.6			
	$M_2(H_{-1}L)_2.H^6/M^2.(H_2L)^2$	−7.90 [r]	(−7.75) [w]			
Mn^{2+}	$ML/M.L$	2.24 [r]				
Fe^{2+}	$ML/M.L$	2.6 [h]				
Co^{2+}	$ML/M.L$	2.86 [h]				
	$MHL/M.HL$	1.64 [h]				
Ni^{2+}	$ML/M.L$	3.17 [h]				
	$MHL/M.HL$	1.83 [h]				
Cu^{2+}	$ML/M.L$	3.42 [h]				
	$MHL/M.HL$	2.00 [h]				
	$ML/MH_{-1}L.H$	4.54 [h]				
	$M_2L_2/M^2.L^2$		(8.0) [w]			
	$M_2L_2/M_2(H_{-1}L)_2.H^2$		(7.8) [w]			
Fe^{3+}	$ML/M.L$	7.1 [h]				
	$M_2(H_{-1}L)_2.H^2/M^2.L^2$	12.85 [h]				
	$M_2(H_{-1}L)_2L/M_2(H_{-1}L)_2.L$	5.0 [h]				
	$M_3(H_{-1}L)_4L.H^4/M^3.L^5$	26.0 [h]				
Zn^{2+}	$ML/M.L$	2.93 [h] −0.1		3.32		
	$MHL/M.HL$	1.66 [h] −0.1		2.00		
Cd^{2+}	$ML/M.L$	2.36 [h]				
	$MHL/M.HL$	1.34 [h]				
Pb^{2+}	$MHL/M.HL$		2.45			
	$M(HL)_2/M.(HL)^2$		3.70			

[h] 20°, 0.1;　[r] 25°, 0.16;　[z] optical isomerism not stated;　[w] DL-mixture

Bibliography:

H^+　59EB,63Ca,64RMa,64Ta,68BL

Li^+-Cs^+　64RZ

Be^{2+}　75VG

Mg^{2+}-Ra^{2+}　38CK,40TD,52S,54S,63Ca

Y^{3+}-Lu^{3+}　62D,70RFV

UO_2^{2+}　60FN,64RMa

Mn^{2+}　57LW

Fe^{2+},Fe^{3+}　64Ta

Co^{2+},Ni^{2+},Cd^{2+}　63Ca

Cu^{2+}　63Ca,67RM

Zn^{2+}　38CK,51BW,63Ca

Pb^{2+}　68BL

Other references:　51BW,57Lb,57V,58NP,60RE,
61MM,62BA,62CTC,62CTG,62GL,62GLa,63ED,
63YZ,64Da,65MN,66Ga,67Ka,68GPa,68RS,
69LA,69PV,70AKN,70AL,70CV,70Z,71WC,
72MK,73KGa,73KGG,73SMa,74VPa,75F,75PMa,
75PSa,75SP,75TK,75W

$$\begin{array}{c} \text{OH} \\ | \\ \text{HO}_2\text{CCHCHCO}_2\text{H} \\ | \\ \text{OH} \end{array}$$

$C_4H_6O_6$		D-2,3-Dihydroxybutanedioic acid (D-tartaric acid)				H_2L
Metal ion	Equilibrium	Log K 25°, 0.1	Log K 25°, 1.0	Log K 25°, 0	ΔH 25°, 0	ΔS 25°, 0
H^+	HL/H.L	3.95 ±0.02	3.73 ±0.04	4.366	−0.20	19.3
		(3.67)[b]	3.81[d]	3.93[e]		
	H_2L/HL.H	2.82 ±0.03	2.69 ±0.09	3.036	−0.74	11.4
		(2.62)[b]	2.83[d]	2.98[e]		
Na^+	ML/M.L	0.73[r,y] ±0.07		0.83 ±0.05		
	MHL/M.HL	−0.05[r,y]		0.22 ±0.02		
K^+	ML/M.L	0.0 [r,y]				
	MHL/M.HL	−0.5 [r,y]				
Be^{2+}	ML/M.L		1.74			
	ML/MOHL.H		4.40			
	$M_2(OH)_2L_2.H^2/M_4^2.L_2^2$		−1.46			
	$M_4(OH)_6L_2.H^6/M_4^2.L_4$		−15.27			
	$M_4(OH)_6L_4.H^6/M^4.L^4$		−9.83			
Mg^{2+}	ML/M.L	1.36[x]				
	MHL/M.HL	0.92[x]				
Ca^{2+}	ML/M.L	1.80[x]−0.08		2.80		
	MHL/M.HL	1.11[x]				
Sr^{2+}	ML/M.L	1.65[x]±0.1				
	MHL/M.HL	0.91[x]				
Ba^{2+}	ML/M.L	1.62[x]+0.1		2.95 ±0.05		
	MHL/M.HL	0.88[x]				
Ra^{2+}	ML/M.L	1.24[s]				
Sc^{3+}	$ML_2/M.L^2$	(12.5)[h,z]				
Y^{3+}	ML/M.L	4.03[u]				
La^{3+}	ML/M.L	3.74[u]				
		3.68[y]		4.61	(0)[t]	(20)
	$ML_2/M.L^2$	6.13[y]		7.59	(11)[t]	(70)
	MHL/M.HL	2.44[y]		2.48	(1)[t]	(15)
Pr^{3+}	ML/M.L	4.08[u]				
Nd^{3+}	ML/M.L	4.20[u]				
Sm^{3+}	ML/M.L	4.27[u]				
Eu^{3+}	ML/M.L	4.33[u]				
Gd^{3+}	ML/M.L	4.15[u]				
Tb^{3+}	ML/M.L	4.06[u]				
Dy^{3+}	ML/M.L	4.13[u]				

[b] 25°, 0.5; [d] 25°, 2.0; [e] 25°, 3.0; [h] 20°, 0.1; [r] 25°, 0.2; [s] 25°, 0.16; [t] 15-35°, 0;

[u] 25°, 0.05; [x] temperature not stated, 0.2; [y] $(CH_3)_4$NCl used as background electrolyte;

[z] optical isomerism not stated.

D-Tartaric acid (continued)

Metal ion	Equilibrium	Log K 25°, 0.1	Log K 25°, 1.0	Log K 25°, 0	ΔH 25°, 0	ΔS 25°, 0
Ho^{3+}	$ML/M.L$	4.14^u				
Er^{3+}	$ML/M.L$	4.14^u				
Tm^{3+}	$ML/M.L$	4.18^u				
Yb^{3+}	$ML/M.L$	4.26^u				
Lu^{3+}	$ML/M.L$	4.32^u				
UO_2^{2+}	$MH_{-1}L.H^3/M.H_2L$		-5.6			
	$M_2(H_{-1}L)_2.H^6/M^2.(H_2L)^2$	-7.98	-8.00			
Mn^{2+}	$ML/M.L$	2.49				
Fe^{2+}	$ML/M.L$ $ML_2/M.L^2$	2.2^h	1.43 $(2.50)^w$			
Co^{2+}	$ML/M.L$ $ML_2/M.L^2$	2.19		3.05 ± 0.03 $(4.0)^w \pm 0.02$		
Ni^{2+}	$ML/M.L$		2.06			
	$MHL/M.HL$		0.8			
Cu^{2+}	$ML/M.L_2$	3.39^b	2.90 ± 0.3			
	$M_2L_2/M^2.L^2$		8.24 ± 0.0			
	$M_2L_2/M_2H_{-1}L_2.H$		4.42			
	$M_2H_{-1}L_2/M_2(H_{-1}L)_2.H$		4.14			
Fe^{3+}	$ML/M.L$	6.49^h	5.73			
	$M_2(H_{-1}L)_2.H^2/M^2.L^2$	$(11.87)^{h,w}$	$(10.9)^w$			
	$M_2(H_{-1}L)_2/M_2(H_{-1}L)(H_{-2}L).H$	$(2.8)^{h,w}$				
	$M_2(H_{-1}L)_2/M_2(H_{-2}L)_2.H^2$		$(4.9)^w$			
	$M_3(H_{-2}L)_3.H^6/M^3.L^3$	$(9.5)^{h,w}$				
VO^{2+}	$ML.H/M.HL$	1.10^v				
	$ML_2.H^2/M.(HL)^2$	1.40^v				
Zn^{2+}	$ML/M.L$	2.68^x 2.2^b		3.82		
	$ML_2/M.L^2$			$(5.0)^w$		
Hg^{2+}	$ML/M.L$	7.0				
Sn^{2+}	$ML/M.L$ $ML_2/M.L^2$	5.2^h $(9.91)^{h,z}$				
Pb^{2+}	$ML/M.L_2$ $ML_2/M.L^2$	3.12 ± 0.03	2.60 $(4.0)^w$			
	$MHL/M.HL$ $MHL_2/ML_2.H$		1.76 $(3.5)^w$			
	$ML.OH/M.L_2$		1.1			
	$M_2L_2.(OH)^2/M^2.L^2$		$(5.5)^w$			
	$M_2L_3.(OH)^3/M^2.L^3$		$(4.7)^w$			
Al^{3+}	$ML/M.L_2$ $ML_2/M.L^2$		5.32 $(9.77)^w$			

---

[b] $25°$, 0.5; [h] $20°$, 0.1; [u] $25°$, 0.05; [v] $25°$, 0.25; [w] DL-mixture; [x] temperature not stated, 0.2; [z] optical isomerism not stated.

D-Tartaric acid (continued)

Metal ion	Equilibrium	Log K 25°, 0.1	Log K 25°, 1.0	Log K 25°, 0	ΔH 25°, 0	ΔS 25°, 0
In^{3+}	$ML/M.L$		4.44			
	$ML_2/M.L^2$		$(8.46)^w$			
$Sb(III)$	$M_2L_2.(OH)^2/[M(OH)_3]^2.(H_2L)^2$					
			-6.07^h			
	$M_2L_2.(OH)^2/(MOHL)^2$	-10.16^h				

^{h} 20°, 0.1; ^{w} DL-mixture

Bibliography:

H^+	51BC,62FC,64RM,64T,68BR,69OC,70AM, 70KS,71SK,72DM,72PT,73GK,75MH	Sn^{2+}	65Sb
Na^+,K^+	71DM,72DM	Pb^{2+}	70KS,72BVa,72RM
Be^{2+}	75VG	Al^{3+}	72RM
$Mg^{2+}-Ra^{2+}$	38CK,40TD,48SR,50S,50SR,52SL, 54S,62BA,63LM	In^{3+}	63Se,72RM
Sc^{3+}	63Se	$Sb(III)$	70AM

$Y^{3+}-Lu^{3+}$ 72DMa,72SSa

UO_2^{2+} 60FN,64RMa

Mn^{2+} 70KS

Fe^{2+} 64T,68BR

Co^{2+} 61MM,65Ma,65SM

Ni^{2+} 74KC

Cu^{2+} 48F,64L,67RM,69BLR,69OC,69SV,71BVc, 72PB

Fe^{3+} 64T,74BV

VO^{2+} 70KP

Zn^{2+} 38CK,62FC,70EM

Hg^{2+} 75LB

Other references: 46J,49M,51H,51MM,51S,53Sb, 54Sa,55Sa,55SR,56P,57Lc,57V,58T,59RE, 60BC,60Sa,61MMZ,61YR,62BA,62BTS,62GL, 62MM,62YP,62ZR,63FL,63GM,63LJ,63YZ, 63ZA,64RME,65BR,65BRP,65F,65FA,65HS, 65KB,65SB,65SSR,65TP,66DD,66Ga,66KS, 66KZ,66PB,66RM,67F,67Ka,66NA,67NG,67SB, 67TP,67TT,68PKa,68PKb,68PSa,69F,69LA 69PPa,69PS,69SC,69SKa,70AKN,70AL,70CV, 70PPa,70TP,71SSP,71WC,72GM,72MK,72TPP, 73BPS,73H,73KK,73TPa,73ZG,74GK,74Ia, 74KKb,74KPD,75KL,75KPT,75W,75YB

$$HO_2CCHCHCO_2H$$
$$||$$
$$HOOH$$

$C_4H_6O_6$		meso-2,3-Dihydroxybutanedioic acid (meso-tartaric acid)				H_2L
Metal ion	Equilibrium	Log K 25°, 0.1	Log K 25°, 1.0	Log K 25°, 0	ΔH 25°, 0	ΔS 25°, 0
H^+	$HL/H.L$	4.44 ±0.2	4.10	4.91	-1.48	17.5
	$H_2L/HL.H$	2.99 ±0.2	2.86	3.17	-0.81	11.8
Be^{2+}	$ML/M.L$		1.74			
	$ML/M(OH)L.H$		4.23			
	$M_2(OH)_3L.H^3/M^2.L$		-9.26			
	$M_2(OH)_2L_2.H^2/M^2.L^2$		0.31			
	$M_4(OH)_6L_2.H^6/M^4.L^2$		-14.70			

meso-Tartaric acid (continued)

Metal ion	Equilibrium	Log K 25°, 0.1	Log K 25°, 1.0	Log K 25°, 0	ΔH 25°, 0	ΔS 25°, 0
La^{3+}	$ML/M.L$	4.39				
	$ML_2/M.L^2$	7.40				
Nd^{3+}	$ML/M.L$	5.08				
	$ML_2/M.L^2$	8.53				
Cu^{2+}	$ML/M.L$	3.52	3.25			
	$ML_2/M.L^2$	6.04				
	$M_2(H_{-1}L)_2.H^2/M^2.L^2$		0.95			
Fe^{3+}	$ML/M.L$	6.66				
	$ML_2/M.L^2$	12.30				
Pb^{2+}	$ML/M.L$	3.59				
Al^{3+}	$ML/M.L$	5.62				
	$ML_2/M.L^2$	9.95				
In^{3+}	$ML/M.L$	4.97				
	$ML_2/M.L^2$	9.74				

Bibliography:

H^+ 60FN,69OC,70AM,72RM,73PT $La^{3+},Nd^{3+},Fe^{3+}-In^{3+}$ 72RM

Be^{2+} 75VG Cu^{2+} 69OC,72RM

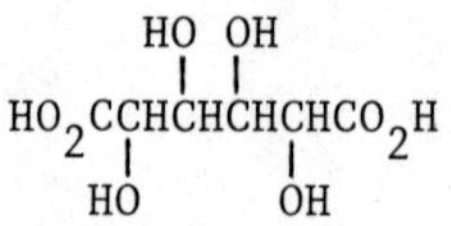

$C_6H_{10}O_8$			H_2L

Mucic acid

Metal ion	Equilibrium	Log K 25°, 1.0
H^+	$HL/H.L$	3.63
	$H_2L/HL.H$	3.08
Co^{2+}	$MH_{-1}L.H/M.L$	-9.36
	$MH_{-1}L/MH_{-2}L.H$	8.75
Ni^{2+}	$MH_{-1}L.H/M.L$	-9.34
	$MH_{-1}L/MH_{-2}L.H$	8.74

Bibliography: 68B

Other references: 70KK,75F

$$\underset{\displaystyle HO_2CCH_2CCO_2H}{\overset{\displaystyle O}{\overset{\displaystyle \|}{}}}$$

$C_4H_4O_5$ Metal ion	Equilibrium	Oxobutanedioic acid (oxaloacetic acid)				H_2L
		Log K 25°, 0.1	Log K 25°, 0.2	Log K 25°, 0	ΔH 25°, 0	ΔS 25°, 0
H^+	$L/H_{-1}L.H$	13.06				
	$HL/H.L$	3.89 ±0.00	3.78	4.37	(0)[r]	(20)
	$H_2L/HL.H$	2.27 ±0.05	2.25	2.56	(−4)[r]	(0)
Be^{2+}	$ML/M.L$		3.1			
Mg^{2+}	$ML/M.L$	1.96				
	$MH_{-1}L/M.H_{-1}L$	6.27				
	$M(H_{-1}L)_2/M.(H_{-1}L)^2$	11.09				
Ca^{2+}	$ML/M.L$		1.60[s]	2.6		
	$M_2L/ML.M$			2		
Y^{3+}	$ML/M.L$			5.63		
	$ML_2/M.L^2$			9.8		
La^{3+}	$ML/M.L$			5.26		
	$ML_2/M.L^2$			9		
Gd^{3+}	$ML/M.L$			5.54		
	$ML_2/M.L^2$			10.1		
Dy^{3+}	$ML/M.L$			5.66		
	$ML_2/M.L^2$			10.2		
Lu^{3+}	$ML/M.L$			5.88		
	$ML_2/M.L^2$			10.6		
Mn^{2+}	$ML/M.L$			2.8		
	$M_2L/ML.M$			2		
Co^{2+}	$ML/M.L$			3.1		
	$M_2L/ML.M$			2.3		
Ni^{2+}	$ML/M.L$			3.5		
	$M_2L/ML.M$			2.2		
Cu^{2+}	$ML/M.L$	4.10[t]	3.9 3.38[u]	4.9		
Zn^{2+}	$ML/M.L$			3.2		
	$M_2L/ML.M$			2.3		

[r] 25-37°, 0;   [s] 3°, 0.16;   [t] 37°, 0.1;   [u] 37°, 0.2

Bibliography:

H^+ 52P,64TG

Be^{2+} 72DT

Mg^{2+} 64TGa

Ca^{2+} 52SL,58GS

Ra^{2+} 50SR

Y^{3+}-Lu^{3+} 56GN

Mn^{2+}-Ni^{2+},Zn^{2+} 58GS

Cu^{2+} 52Pa,58GS,72DT

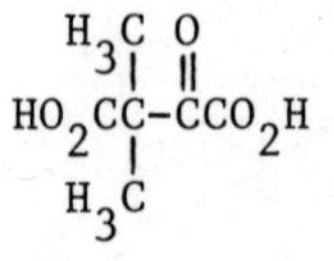

$$
\begin{array}{c}
H_3C\ O \\
|\ \ || \\
HO_2CC-CCO_2H \\
| \\
H_3C
\end{array}
$$

$C_6H_8O_5$		3,3-Dimethyl-2-oxobutanedioic acid	H_2L

Metal ion	Equilibrium	Log K 25°, 0
H^+	HL/H.L	4.62
	H_2L/HL.H	1.77
Cu^{2+}	ML/M.L	4.7

Bibliography: 58GH

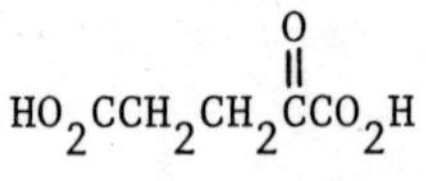

$$
\begin{array}{c}
O \\
|| \\
HO_2CCH_2CH_2CCO_2H
\end{array}
$$

$C_5H_6O_5$		2-Oxopentanedioic acid (α-ketoglutaric acid)		H_2L

Metal ion	Equilibrium	Log K 25°, 0.16	Log K 25°, 0.5
H^+	HL/H.L		4.44
	H_2L/HL.H		1.85
Ca^{2+}	ML/M.L	1.29	
Sr^{2+}	ML/M.L	1.14	
Zn^{2+}	ML/M.L		1.13
	ML_2/M.L^2		1.7

Bibliography:
H^+,Zn^{2+} 70SF   Other reference: 75SD
Ca^{2+},Sr^{2+} 52SL

$$
\begin{array}{c}
CH_2CH_2CO_2H \\
| \\
Z-NHCCO_2H \\
| \\
CH_3
\end{array}
$$

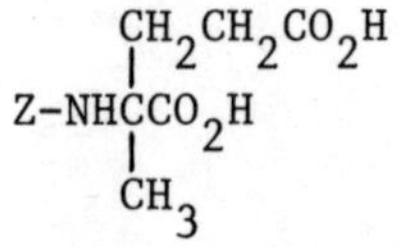

$C_nH_mO_pN_r$		N-Substituted-2-methylglutamic acid		H_2L

Z =	Metal ion	Equilibrium	Log K 25°, 0.1
N-Acetyl	H^+	HL/H.L	4.60
		H_2L/HL.H	3.29
$(C_8H_{13}O_5N)$	Cu^{2+}	ML/M.L	2.2
N-Benzoyl	H^+	HL/H.L	4.63
		H_2L/HL.H	3.51
$(C_{13}H_{15}O_5N)$	Cu^{2+}	ML/M.L	2.2

B. DI-CARBOXYLIC ACIDS

N-Substituted-2-methylglutamic acid (continued)

Z =	Metal ion	Equilibrium	Log K 25°, 0.1
N-(4-Nitrobenzoyl)	H^+	HL/H.L	4.57
		H_2L/HL.H	3.23
$(C_{13}H_{14}O_7N_2)$	Cu^{2+}	ML/M.L	2.1

Bibliography: 65NC

$$\text{HO}_2\text{CCH}_2\text{NHC}\overset{\overset{\text{O}}{\|}}{-}\overset{\overset{\text{O}}{\|}}{\text{C}}\text{NHCH}_2\text{CO}_2\text{H}$$

$C_6H_8O_6N_2$ Oxamide-N,N'-diacetic acid H_2L

Metal ion	Equilibrium	Log K 20°, 0.1
H^+	HL/H.L	3.49
	H_2L/HL.H	2.95
Ni^{2+}	$ML/MH_{-1}L.H$	7.8
	$M(H_{-1}L)_2.H/MH_{-1}L.L$	-7.6
Cu^{2+}	$MH_{-1}L.H/M.L$	-1.9
	$M(H_{-2}L)_2.H/MH_{-1}L.L$	-6.6
	$M_2H_{-2}L.H/MH_{-1}L.M$	1.6

Bibliography: 74SB

$$\text{HO}_2\text{CCH}_2\text{OCH}_2\text{CO}_2\text{H}$$

$C_4H_6O_5$ Oxydiacetic acid (diglycolic acid) H_2L

Metal ion	Equilibrium	Log K 25°, 0.1	Log K 25°, 1.0	Log K 25°, 0	ΔH 25°, 1.0	ΔS 25°, 1.0
H^+	HL/H.L	3.93 ±0.02	3.76 ±0.00 3.96[d]	4.37	0.76	19.8
	H_2L/HL.H	2.79 ±0.03	2.75 ±0.06 2.91[d]	2.97	-0.38	11.3
Mg^{2+}	ML/M.L	1.8 ±0.2				
	MHL/M.HL	-0.3				
Ca^{2+}	ML/M.L	3.38 ±0.02				
	MHL/M.HL	1.36				
Sr^{2+}	ML/M.L	2.50 ±0.03				
	MHL/M.HL	1.0				
Ba^{2+}	ML/M.L	2.15				
Sc^{3+}	ML/M.L		8.3		-0.5	36
	$ML_2/M.L^2$		12.8		-3.0	48

[d] 25°, 2.0

Oxydiacetic acid (continued)

Metal ion	Equilibrium	Log K 25°, 0.1	Log K 25°, 1.0	Log K 25°, 0	ΔH 25°, 1.0	ΔS 25°, 1.0
Y^{3+}	$ML/M.L$		5.24		1.7	30
	$ML_2/M.L^2$		9.76		0.5	46
	$ML_3/M.L^3$		13.03		-3.7	47
La^{3+}	$ML/M.L$		4.93		-0.1	22
	$ML_2/M.L^2$		8.41		-0.8	36
	$ML_3/M.L^3$		10.25		-0.5	45
Ce^{3+}	$ML/M.L$		5.16		-0.4	22
	$ML_2/M.L^2$		8.92		-1.3	36
	$ML_3/M.L^3$		11.23		-1.8	45
Pr^{3+}	$ML/M.L$		5.33		-0.7	22
	$ML_2/M.L^2$		9.23		-1.7	36
	$ML_3/M.L^3$		11.63		-2.5	45
Nd^{3+}	$ML/M.L$		5.45		-0.9	22
	$ML_2/M.L^2$		9.50		-2.1	36
	$ML_3/M.L^3$		12.16		-3.0	45
Sm^{3+}	$ML/M.L$		5.55		-1.1	22
	$ML_2/M.L^2$		9.89		-2.9	36
	$ML_3/M.L^3$		12.79		-4.3	44
Eu^{3+}	$ML/M.L$		5.53		-0.8	23
	$ML_2/M.L^2$		10.04		-2.9	36
	$ML_3/M.L^3$		13.20		-4.5	45
Gd^{3+}	$ML/M.L$		5.40		-0.4	24
	$ML_2/M.L^2$		9.93		-2.7	36
	$ML_3/M.L^3$		13.04		-4.6	44
Tb^{3+}	$ML/M.L$		5.23		0.8	27
	$ML_2/M.L^2$		9.98		-1.9	39
	$ML_3/M.L^3$		13.25		-4.5	45
Dy^{3+}	$ML/M.L$		5.31		1.3	29
	$ML_2/M.L^2$		9.98		-1.1	42
	$ML_3/M.L^3$		13.36		-4.4	46
Ho^{3+}	$ML/M.L$		5.28		1.6	30
	$ML_2/M.L^2$		9.95		-0.3	45
	$ML_3/M.L^3$		13.31		-4.4	46
Er^{3+}	$ML/M.L$		5.34		1.7	30
	$ML_2/M.L^2$		10.02		0.7	48
	$ML_3/M.L^3$		13.23		-4.2	46
Tm^{3+}	$ML/M.L$		5.49		1.6	30
	$ML_2/M.L^2$		10.22		1.2	51
	$ML_3/M.L^3$		13.29		-3.8	48
Yb^{3+}	$ML/M.L$		5.55		1.4	30
	$ML_2/M.L^2$		10.36		1.1	51
	$ML_3/M.L^3$		13.17		-3.9	47
Lu^{3+}	$ML/M.L$		5.64		1.2	30
	$ML_2/M.L^2$		10.55		0.8	51
	$ML_3/M.L^3$		13.16		-3.8	47
UO_2^{2+}	$ML/M.L$		5.11[j]			
NpO_2^{2+}	$ML/M.L$		5.16[j]			
PuO_2^{2+}	$ML/M.L$		4.97[j]			

[j] 20°, 1.0

Oxydiacetic acid (continued)

Metal ion	Equilibrium	Log K 25°, 0.1	Log K 25°, 1.0	Log K 25°, 0	ΔH 25°, 1.0	ΔS 25°, 1.0
Mn^{2+}	ML/M.L	2.53 ±0.01				
Fe^{2+}	ML/M.L	2.56 ±0.08				
	MHL/M.HL	0.64				
Co^{2+}	ML/M.L	2.70 ±0.05				
	MHL/M.HL	(1.08)				
Ni^{2+}	$ML/M.L$	3.80 ±0.02	2.25[d]			
	$ML_2/M.L^2$		3.45[d]			
	MHL/M.HL	0.84				
Cu^{2+}	ML/M.L	3.95 ±0.02				
	MHL/M.HL	1.42				
Fe^{3+}	ML/M.L	5.04[b]				
VO^{2+}	ML/M.L	5.01[b]				
Zn^{2+}	ML/M.L	3.61 ±0.04				
	MHL/M.HL	(2.08)				
Cd^{2+}	ML/M.L	3.21 +0.1				
	MHL/M.HL	1.02				
Sn^{2+}	ML/M.L	5.56				
Pb^{2+}	ML/M.L	4.41 +0.1				
	MHL/M.HL	1.92				
Al^{3+}	$ML/M.L$	3.16[b]				
	$ML_2/M.L^2$	5.25[b]				

[b] 25°, 0.5;   [d] 25°, 2.0

Bibliography:

H^+ 57TB,60YY,63GT,66S,69GH,69RS,72AN, 72GO,72Na,73CB,74MSM,75FC,76MM

Mg^{2+}-Ba^{2+} 57TB,74MSM,75FC,76MM

Sc^{3+} 69GH

Y^{3+}-Lu^{3+} 63Gb,63GT,72GO

UO_2^{2+}-PuO_2^{2+} 73CB

Fe^{2+} 72Nb

Co^{2+}-Cu^{2+},Zn^{2+}-Pb^{2+} 57TB,60YY,64KK,72AN, 75FC,75MT,76MM

Fe^{3+} 72N

VO^{2+} 73N

Al^{3+} 72Na

Other references: 61COC,66PRK,74BSa

$$HO_2CCHOCHCO_2H$$
$$H_3C \quad CH_3$$

$C_6H_{10}O_5$		DL-2,2'-Oxydipropanoic acid (dilactic acid)	H_2L

Metal ion	Equilibrium	Log K 25°, 0.1
Y^{3+}	$ML/M.L$	5.44
	$ML_2/M.L^2$	(9.90)[z]
La^{3+}	$ML/M.L$	4.84
	$ML_2/M.L^2$	(7.88)[z]

[z] optical isomerism not stated.

Dilactic acid (continued)

Metal ion	Equilibrium	Log K 25°, 0.1
Ce^{3+}	$ML/M.L$ $ML_2/M.L^2$	5.09 $(8.49)^z$
Pr^{3+}	$ML/M.L$ $ML_2/M.L^2$	5.25 $(8.85)^z$
Nd^{3+}	$ML/M.L$ $ML_2/M.L^2$	5.35 $(9.15)^z$
Sm^{3+}	$ML/M.L$ $ML_2/M.L^2$	5.49 $(9.57)^z$
Eu^{3+}	$ML/M.L$ $ML_2/M.L^2$	5.45 $(9.67)^z$
Gd^{3+}	$ML/M.L$ $ML_2/M.L^2$	5.38 $(9.65)^z$
Tb^{3+}	$ML/M.L$ $ML_2/M.L^2$	5.47 $(9.85)^z$
Dy^{3+}	$ML/M.L$ $ML_2/M.L^2$	5.57 $(10.04)^z$
Ho^{3+}	$ML/M.L$ $ML_2/M.L^2$	5.59 $(10.16)^z$
Er^{3+}	$ML/M.L$ $ML_2/M.L^2$	5.59 $(10.32)^z$
Tm^{3+}	$ML/M.L$ $ML_2/M.L^2$	5.68 $(10.52)^z$
Yb^{3+}	$ML/M.L$ $ML_2/M.L^2$	5.67 $(10.65)^z$
Lu^{3+}	$ML/M.L$ $ML_2/M.L^2$	5.62 $(10.75)^z$

^{z} optical isomerism not stated.

Bibliography: 66SPa

$$HO_2CCH_2CH_2OCH_2CH_2CO_2H$$

$C_6H_{10}O_5$ 3,3'-Oxydipropanoic acid H_2L

Metal ion	Equilibrium	Log K 25°, 0.1
H^+	$HL/H.L$ $H_2L/HL.H$	4.62 3.77
Mn^{2+}	$ML/M.L$	2.0
Co^{2+}	$ML/M.L$	1.69
Ni^{2+}	$ML/M.L$	1.39
Cu^{2+}	$ML/M.L$ $MHL/M.HL$	2.52 1.4
Zn^{2+}	$ML/M.L$	2.13
Cd^{2+}	$ML/M.L$	1.66
Pb^{2+}	$ML/M.L$ $MHL/M.HL$	2.66 1.73

Bibliography: 75LP

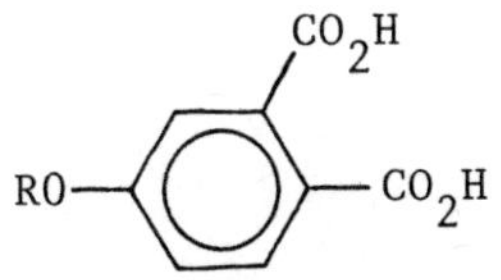

$$RO-\underset{}{\bigcirc}\begin{array}{c}CO_2H\\CO_2H\end{array}$$

4-(Alkyloxy)phthalic acid H_2L

$C_nH_mO_5$

R =	Metal ion	Equilibrium	Log K 25°, 0.1
Methyl	H^+	HL/H.L	5.16
		H_2L/HL.H	2.67
$(C_9H_8O_5)$	Ni^{2+}	ML/M.L	2.27
	Cu^{2+}	ML/M.L	3.32
Ethyl	H^+	HL/H.L	5.12
		H_2L/HL.H	2.74
$(C_{10}H_{10}O_5)$	Ni^{2+}	ML/M.L	2.18
	Cu^{2+}	ML/M.L	3.25

Bibliography: 56YS

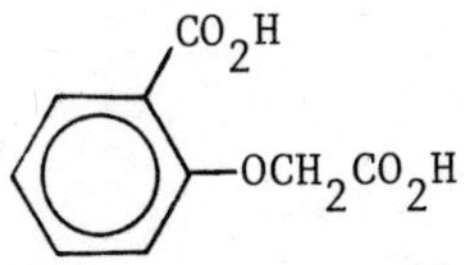

$$\underset{}{\bigcirc}\begin{array}{c}CO_2H\\OCH_2CO_2H\end{array}$$

$C_9H_8O_5$ **(2-Carboxyphenyloxy)acetic acid [2-(carboxymethoxy)benzoic acid]** H_2L

Metal ion	Equilibrium	Log K 25°, 0.1
H^+	HL/H.L	4.36
	H_2L/HL.H	2.51
Ni^{2+}	ML/M.L	2.0
Cu^{2+}	ML/M.L	3.1
Zn^{2+}	ML/M.L	(2.6)
Cd^{2+}	ML/M.L	2.0
Pb^{2+}	ML/M.L	2.6

Bibliography: 62SY

$$HO_2CCH_2OCH_2CH_2OCH_2CO_2H$$

(Ethylenedioxy)diacetic acid H_2L

$C_6H_{10}O_6$

Metal ion	Equilibrium	Log K 25°, 0.1	Log K 25°, 1.0	ΔH 25°, 1.0	ΔS 25°, 1.0
H^+	HL/H.L	3.83	3.68	0.26	17.7
	H_2L/HL.H	2.99	(3.05)	0.24	(14.8)
Mg^{2+}	ML/M.L	1.9			
Ca^{2+}	ML/M.L	3.15			
Sr^{2+}	ML/M.L	2.40			

(Ethylenedioxy)diacetic acid (continued)

Metal ion	Equilibrium	Log K 25°, 0.1	Log K 25°, 1.0	ΔH 25°, 1.0	ΔS 25°, 1.0
Ba^{2+}	ML/M.L	2.29			
La^{3+}	$ML/M.L$		4.35	1.6	25
	$ML_2/M.L^2$		7.71	-1.0	32
	$ML_3/M.L^3$		8.6		
	$MHL_2/ML_2.H$		1.71	-2.9	-2
Ce^{3+}	$ML/M.L$		4.64	1.3	26
	$ML_2/M.L^2$		7.89	-0.8	33
	$ML_3/M.L^3$		8.8		
	$MHL_2/ML_2.H$		1.61	-3.5	-4
Pr^{3+}	$ML/M.L$		4.81	1.0	25
	$ML_2/M.L^2$		7.89	-0.5	34
	$ML_3/M.L^3$		8.59		
	$MHL_2/ML_2.H$		1.65	-9	-20
Nd^{3+}	$ML/M.L$		4.92	0.8	25
	$ML_2/M.L^2$		7.95	0.1	37
	$ML_3/M.L^3$		8.63		
	$MHL_2/ML_2.H$		1.91	1	10
Sm^{3+}	$ML/M.L$		5.08	0.4	25
	$ML_2/M.L^2$		7.98	1.9	43
	$ML_3/M.L^3$		8.95		
	$MHL_2/ML_2.H$		2.35	-2	4
Gd^{3+}	$ML/M.L$		4.89	0.9	25
	$ML_2/M.L^2$		7.87	2.9	46
	$ML_3/M.L^3$		8.83		
	$MHL_2/ML_2.H$		2.34	-3	1
Tb^{3+}	$ML/M.L$		4.75	1.7	27
	$ML_2/M.L^2$		7.90	3.4	48
	$ML_3/M.L^3$		8.69		
	$MHL_2/ML_2.H$		1.94	-3	-1
Dy^{3+}	$ML/M.L$		4.67	2.6	30
	$ML_2/M.L^2$		7.93	4.2	50
	$ML_3/M.L^3$		8.78		
	$MHL_2/ML_2.H$		2.02	-2	3
Ho^{3+}	$ML/M.L$			3.1	
	$ML_2/M.L^2$			4.6	
	$MHL_2/ML_2.H$			-3	
Er^{3+}	$ML/M.L$		4.61	3.7	34
	$ML_2/M.L^2$		8.04	6.9	60
	$ML_3/M.L^3$		8.94		
	$MHL_2/ML_2.H$		1.4	-4	-7
Tm^{3+}	$ML/M.L$		4.64	4.0	35
	$ML_2/M.L^2$		8.23	7.6	63
	$ML_3/M.L^3$		8.46		
	$MHL_2/ML_2.H$		1.28	1	9

(Ethylenedioxy)diacetic acid (continued)

Metal ion	Equilibrium	Log K 25°, 0.1	Log K 25°, 1.0	ΔH 25°, 1.0	ΔS 25°, 1.0
Yb^{3+}	$ML/M.L$		4.85	3.8	35
	$ML_2/M.L_2$		8.68	7.5	65
	$ML_3/M.L_3$		8.52		
	$MHL_2/ML_2.H$		1.1	3	15
Lu^{3+}	$ML/M.L$			3.6	
	$ML_2/M.L_2$			7.0	
	$MHL_2/ML_2.H$			-2.0	
Mn^{2+}	$ML/M.L$	2.79			
Fe^{2+}	$ML/M.L$	2.35			
Co^{2+}	$ML/M.L$	1.69			
Ni^{2+}	$ML/M.L$	1.79			
Cu^{2+}	$ML/M.L$	3.39			
Zn^{2+}	$ML/M.L$	2.65			

Bibliography:

H^+ 74GG,74MSM $La^{3+}-Lu^{3+}$ 74GG

$Mg^{2+}-Ba^{2+}$ 74MSM $Mn^{2+}-Zn^{2+}$ 75MT

$$HO_2CCH_2OCH_2CH_2OCH_2CH_2OCH_2CO_2H$$

$C_8H_{14}O_7$	Oxybis(ethyleneoxyacetic acid) (diethylenetrioxydiacetic acid)		H_2L

Metal ion	Equilibrium	Log K 25°, 0.1
H^+	$HL/H.L$	3.80
	$H_2L/HL.H$	3.03
Mg^{2+}	$ML/M.L$	1.8
Ca^{2+}	$ML/M.L$	2.40
Sr^{2+}	$ML/M.L$	2.29
Ba^{2+}	$ML/M.L$	2.29
Mn^{2+}	$ML/M.L$	2.90
Fe^{2+}	$ML/M.L$	2.71
Co^{2+}	$ML/M.L$	2.29
Ni^{2+}	$ML/M.L$	2.39
Cu^{2+}	$ML/M.L$	2.85
Zn^{2+}	$ML/M.L$	2.60

Bibliography:

H^+-Ba^{2+} 74MSM $Mn^{2+}-Zn^{2+}$ 75MT

$$HO_2CCH_2OCH_2CH_2OCH_2CH_2OCH_2CH_2OCH_2CO_2H$$

$C_{10}H_{18}O_8$ Ethylenebis(oxyethyleneoxyacetic acid) H_2L
                        (triethylenetetraoxydiacetic acid)

Metal ion	Equilibrium	Log K 25°, 0.1
H^+	HL/H.L	3.80
	H_2L/HL.H	3.05
Mg^{2+}	ML/M.L	1.4
Ca^{2+}	ML/M.L	2.14
Sr^{2+}	ML/M.L	2.29
Ba^{2+}	ML/M.L	2.29
Mn^{2+}	ML/M.L	2.18
Fe^{2+}	ML/M.L	2.46
Co^{2+}	ML/M.L	1.92
Ni^{2+}	ML/M.L	1.94
Cu^{2+}	ML/M.L	2.65
Zn^{2+}	ML/M.L	2.18

Bibliography:

H^+-Ba^{2+} 74MSM $Mn^{2+}-Zn^{2+}$ 75MT

$$\overset{\displaystyle SH}{\underset{\displaystyle |}{HO_2CCH_2CHCO_2H}}$$

$C_4H_6O_4S$ DL-Mercaptobutanedioic acid (thiomalic acid) H_2L

Metal ion	Equilibrium	Log K 25°, 0.1	Log K 25°, 1.0	ΔH 25°, 1.0	ΔS 25°, 1.0
H^+	HL/H.L	10.38 ±0.01	9.79 ±0.06		
		10.45[h]±0.06	10.02[e]		
	H_2L/HL.H	4.60 ±0.04	4.42 ±0.02		
		4.60[h]±0.02	4.88[e]		
	H_3L/H_2L.H	3.3 ±0.3	2.7 ±0.2		
		3.2 [h]±0.2	3.50[e]		
Co^{2+}	ML/M.L	6.88			
Ni^{2+}	ML/M.L	7.67			
		7.97 [h]			
	$ML_2/M.L^2$	(13.88)[z]			
		(12.3)[h,z]			
	MHL/ML.H	4.59 [h]			
Fe^{3+}	MHL.H^2/M.H_3L		0.41	(5)[t]	(20)
Ag^+	ML/M.L	7.85			

[e] 25°, 3.0;   [h] 20°, 0.1;   [t] 10-25°, 1.0;   [z] optical isomerism not stated.

B. DI-CARBOXYLIC ACIDS

Thiomalic acid (continued)

Metal ion	Equilibrium	Log K 25°, 0.1	Log K 25°, 1.0	ΔH 25°, 1.0	ΔS 25°, 1.0
Zn^{2+}	ML/M.L	8.24	7.52		
		8.36^h+0.5			
	$ML_2/M.L^2$	$(14.56)^z$			
		$(14.88)^h$-0.2			
	MHL/ML.H	2.4^h			
	ML/MOHL.H	8.36	8.01		
Hg^{2+}	ML/M.L	9.94			
	$ML_2/M.L^2$	$(18.07)^z$			
Sb(III)	$ML_2.(OH)^3/M(OH)_3.L^2$	-5.90^h			
	$MHL_2/ML_2.H$	3.46^h			
	$MH_2L_2/MHL_2.H$	2.4^h			

^{h} 20°, 0.1; ^{z} optical isomerism not stated.

Biliography:

H^+ 62CTC,63C,65LM,68RSb,69PP,69RBa,70AM, 71BCP,72CA,74LD

Co^{2+},Ag^+,Hg^{2+} 65LM

Ni^{2+} 65LM,69PP

Fe^{3+} 73EM

Zn^{2+} 65LM,69PP,70AM

Sb(III) 70AM

Other references: 59CF,62CTG,63MNS,65NKK, 65SN,66SN,67MS,68SG,68SGa,68SGb,69SGe, 70RSa,72PR,72SMa,75BS,75PMa

$$\underset{\underset{SH}{|}}{\overset{\overset{SH}{|}}{HO_2CCHCHCO_2H}}$$

$C_4H_6O_4S_2$		2,3-Dimercaptobutanedioic acid (dithiotartaric acid)			H_4L

Isomer	Metal ion	Equilibrium	Log K 25°, 0.1	Log K 20°, 0.1
meso-	H^+	HL/H.L	10.79	11.82
		$H_2L/HL.H$	8.89	9.44
		$H_3L/H_2L.H$	3.48	3.46
		$H_4L/H_3L.H$	2.71	2.40
	Ni^{2+}	ML/M.L	11.69	
		$ML_2/M.L^2$	13.33	
		MHL/ML.H	8.7	
		$MH_2L/MHL.H$	3.3	
		ML/MOHL.H	11.14	
		$ML_2/MOHL_2.H$	10.4	
	Zn^{2+}	ML/M.L	14.42	15.82
		$ML_2/M.L^2$	17.74	19.39
		MHL/ML.H	5.7	5.6
		$MH_2L/MHL.H$	3.4	2.7
		$M_2L/M.ML$	4.07	3.85
		ML/MOHL.H	10.11	12.28
		$ML_2/MOHL_2.H$	9.4	9.7
	Sb(III)	$M_2L_2.(OH)^6/[M(OH)_3]^2.L^2$		-9.2
		$M_2L_2/(MOHL)^2.H^2$		13.17

Dithiotartaric acid (continued)

Isomer	Metal ion	Equilibrium	Log K 25°, 0.1	Log K 20°, 0.1
DL-	H^+	$H_2L/HL.H$		9.61
		$H_3L/H_2L.H$		3.96
		$H_4L/H_3L.H$		2.25
	Sb(III)	$ML_2.M(OH)_3/(MOHL)^2.OH$		-10.7
		$MH_4L_2/MH_3L_2.H$		2.57
		$MH_3L_2/MH_2L_2.H$		3.60
		$MH_2L_2/MHL_2.H$		4.61
		$MHL_2/ML_2.H$		6.82
		$ML_2/MOHL_2.H$		4.90

Bibliography:

H^+, Zn^{2+} 55AS,65LM,70AM   Sb(III)   70AM

Ni^{2+} 65LM   Other references: 72Ea,73ERa

$$HO_2CCH_2SCH_2CO_2H$$

$C_4H_6O_4S$ Thiodiacetic acid H_2L

Metal ion	Equilibrium	Log K 25°, 0.1	Log K 25°, 1.0	Log K 25°, 0	ΔH 25°, 1.0	ΔS 25°, 1.0
H^+	$HL/H.L$	4.15 ±0.02	4.01 ±0.04	4.54 ±0.03	0.43	19.8
		4.01[b]±0.03	4.12[d]±0.02	4.56[f]		
	$H_2L/HL.H$	3.14 ±0.02	3.11 ±0.04	3.27 ±0.03	-0.11	14.0
		3.08[b]±0.03	3.21[d]±0.02	3.66[f]		
Ca^{2+}	$ML/M.L$	1.4[P]				
Ce^{3+}	$ML/M.L$		2.66		3.1	23
	$ML_2/M.L^2$		4.49		4.9	37
	$MHL/M.HL$		1.37		2	13
	$MHL_2/ML_2.H$		3.2		-1	11
Pr^{3+}	$ML/M.L$		2.74		2.9	22
	$ML_2/M.L^2$		4.39		4.8	36
	$MHL/M.HL$		1.46		1	10
	$MHL_2/ML_2.H$		3.2		-1	11
Sm^{3+}	$ML/M.L$		2.90		2.7	22
	$ML_2/M.L^2$		4.68		4.9	38
	$MHL/M.HL$		1.60		1	11
	$MHL_2/ML_2.H$		3.1		-1	11
Tb^{3+}	$ML/M.L$		2.52		4.1	25
	$ML_2/M.L^2$		4.12		6.3	40
	$MHL/M.HL$		1.22		3	16
	$MHL_2/ML_2.H$		3.2		-1	11

[b] 25°, 0.5; [d] 25°, 2.0; [f] 25°, 4.0; [P] 30°, 0.1

B. DI-CARBOXYLIC ACIDS

Thiodiacetic acid (continued)

Metal ion	Equilibrium	Log K 25°, 0.1	Log K 25°, 1.0	Log K 25°, 0	ΔH 25°, 1.0	ΔS 25°, 1.0
Er^{3+}	ML/M.L		2.36		5.0	28
	$ML_2/M.L^2$		3.85		7.9	44
	MHL/M.HL		1.15		3	15
	$MHL_2/ML_2.H$		3.4		−2	9
Yb^{3+}	ML/M.L		2.36		5.4	29
	$ML_2/M.L^2$		3.76		8.6	46
	MHL/M.HL		1.11		3	15
	$MHL_2/ML_2.H$		3.5		−2	9
UO_2^{2+}	ML/M.L		3.16[j]			
	$MHL_2/M.HL.L$		4.38[j]			
V^{2+}	ML/M.L	1.73				
Cr^{2+}	ML/M.L	3.00				
	$ML_2/M.L^2$	5.39				
Mn^{2+}	ML/M.L	1.72				
	MHL/M.HL	0.6				
Fe^{2+}	ML/M.L	2.88	2.45[b]			
	$ML_2/M.L^2$	5.24	3.82[b]			
	MHL/M.HL	1.61				
Co^{2+}	ML/M.L	3.51 −0.1				
	$ML_2/M.L^2$	6.19 −0.7				
	MHL/M.HL	1.72				
Ni^{2+}	ML/M.L	4.20 −0.1	3.93[d]			
	$ML_2/M.L^2$	7.01 −0.3	7.03[d]			
	$ML_3/M.L^3$		8.55[d]			
	MHL/M.HL	2.15	1.70[d]			
Cu^{2+}	ML/M.L	4.65 −0.1	4.18[j]			
	$ML_2/M.L^2$	7.50 −0.2	7.08[j]			
	$ML_3/M.L^3$		(8.6)[j]			
	$ML_4/M.L^4$		(11.9)[j]			
	MHL/M.HL	2.63				
Cr^{3+}	MHL/M.HL	4.9				
Fe^{3+}	MHL/M.HL	3.6				
VO^{2+}	ML/M.L		3.14[b]			
Ag^+	ML/M.L	4.04				
	MHL/M.HL	3.06				
	$MH_2L/M.H_2L$	2.35				
Zn^{2+}	ML/M.L	3.30 −0.3				
	$ML_2/M.L^2$	5.85				
Cd^{2+}	ML/M.L	3.14 −0.8				
	$ML_2/M.L^2$	5.57				
Pb^{2+}	ML/M.L	3.6				
Al^{3+}	ML/M.L		1.93[b]			
	MOHL/ML.OH		10.41[b]			

[b] 25°, 0.5; [d] 25°, 2.0; [j] 20°, 1.0

Thiodiacetic acid (continued)

H^+	39A,57TB,60YY,66S,70PP,72AN,72Na, 73CB,73DG	$Fe^{2+}-Cu^{2+},Zn^{2+}-Pb^{2+}$	57TB,60YY,61COC,61Sb, 70PP,72AN,72Nc,73PP,75AH
Ca^{2+}	57TB	VO^{2+}	73N
$Ce^{3+}-Yb^{3+}$	73DG	Ag^+	75LP
UO_2^{2+}	73CB	Al^{3+}	72Na
$V^{2+}-Mn^{2+},Cr^{3+},Fe^{3+}$	70PP	Other references:	44L,69SD,71KK,74MMa

$$HO_2CCHSCHCO_2H$$
$$H_3C \quad CH_3$$

$C_6H_{10}O_4S$ · · · · · · · · · · DL-2,2'-Thiodipropanoic acid · · · · · · · · · · H_2L

Metal ion	Equilibrium	Log K 25°, 0.1	Log K 25°, 0.2	Log K 18°, 0
H^+	HL/H.L	4.19		4.62
	H_2L/HL.H	3.12		
Mn^{2+}	ML/M.L	(2.1)		
Co^{2+}	ML/M.L	2.14		
	MHL/M.HL	(2.2)		
Ni^{2+}	ML/M.L	3.59		
	MHL/M.HL	1.7		
Cu^{2+}	ML/M.L	3.97		
	MHL/M.HL	1.8		
Ag^+	ML/M.L		3.81 [z]	
	$ML_2/M.L^2$		(6.02) [y,z]	
	MHL/M.HL	3.15		
	$MH_2L/M.H_2L$	2.74	2.81 [z]	
	$M(H_2L)_2/M.(H_2L)^2$		(4.32) [y,z]	
Zn^{2+}	ML/M.L	2.13		
Cd^{2+}	ML/M.L	2.25		
Pb^{2+}	ML/M.L	3.12		
	MHL/M.HL	1.8		

[y] optical isomerism not stated;   [z] acetate buffer

Bibliography:

H^+	24L,75LP	$Mn^{2+}-Pb^{2+}$	75LP

$$HO_2CCH_2SSCH_2CO_2H$$

$C_4H_6O_4S_2$ · · · · · · · · · · · Dithiodiacetic acid · · · · · · · · · · · H_2L

Metal ion	Equilibrium	Log K 25°, 0.1	Log K 18°, 1.0	Log K 18°, 0
H^+	HL/H.L	3.81 -0.01	3.66	4.20
		3.65 [i]	3.75 [k] +0.1	3.92 [l]
	H_2L/HL.H	2.91 ±0.03	2.89	3.08
		2.88 [i]	3.01 [d] +0.01	3.18 [l]

[d] 25°, 2.0;   [i] 18°, 0.5;   [k] 18°, 2.0;   [l] 18°, 3.0

Dithiodiacetic acid (continued)

Metal ion	Equilibrium	Log K 25°, 0.1	Log K 18°, 1.0	Log K 18°, 0
Mn^{2+}	ML/M.L	1.7		
Co^{2+}	ML/M.L	1.5		
Ni^{2+}	ML/M.L	1.8		
Zn^{2+}	ML/M.L	1.6		
Cd^{2+}	ML/M.L	1.9		
Pb^{2+}	ML/M.L	2.4		

Bibliography:

H^+ 39A,68SKM,72AN Mn^{2+}-Pb^{2+} 68SKM

$$HO_2CCH_2CH_2SCH_2CO_2H$$

$C_5H_8O_4S$		(2-Carboxyethylthio)acetic acid		H_2L

Metal ion	Equilibrium	Log K 25°, 0.1	Log K 25°, 0.2	Log K 25°, 2.0
H^+	HL/H.L	4.34		4.52
	$H_2L/HL.H$	3.27		3.57
Mn^{2+}	ML/M.L	1.70		
Co^{2+}	ML/M.L	2.11		
Ni^{2+}	ML/M.L	2.30		2.48
	$ML_2/M.L^2$			4.15
	MHL/M.HL			1.18
Cu^{2+}	ML/M.L	3.75		
	MHL/M.HL	1.5		
Ag^+	ML/M.L		4.40[z]	
	$ML_2/M.L^2$		7.46[z]	
	MHL/M.HL	3.63		
	$MH_2L/M.H_2L$	3.41	3.50[z]	
	$M(H_2L)_2/M.(H_2L)^2$		6.01[z]	
Zn^{2+}	ML/M.L	1.70		
Cd^{2+}	ML/M.L	3.09		
	MHL/M.HL	1.05		
Pb^{2+}	ML/M.L	3.59		
	MHL/M.HL	2.01		

[z] acetate buffer.

Bibliography:

H^+,Ni^{2+} 75AH,75LP   Other reference: 61COC

Mn^{2+},Co^{2+},Cu^{2+}-Pb^{2+} 75LP

$$HO_2CCH_2CH_2SCH_2CH_2CO_2H$$

$C_6H_{10}O_4S$ 3,3'-Thiodipropanoic acid H_2L

Metal ion	Equilibrium	Log K 25°, 0.1	Log K 18°, 1.0	Log K 18°, 0
H^+	HL/H.L	4.68 ±0.02	4.54	5.08
		4.55[i]	4.67[k]	4.84[l]
	$H_2L/HL.H$	3.87 ±0.03	3.88	4.09
		3.84[i]	4.00[k]	4.18[l]
Ni^{2+}	ML/M.L		(1.2)[d]	
	MHL/M.HL		(0.67)[d]	
Cu^{2+}	ML/M.L	2.97 ±0.0		
	MHL/M.HL	1.59		
Ag^+	ML/M.L	3.85	3.95[r]	
	$ML_2/M.L^2$		6.48[r]	
	MHL/M.HL	3.34		
	$MH_2L/M.H_2L$	3.22	3.18[r]	
	$M(H_2L)_2/M.(H_2L)^2$		5.59[r]	
Zn^{2+}	ML/M.L	1.72 −0.1	1.49[s]	
	MHL/M.HL	1.28		
Cd^{2+}	ML/M.L	2.31 −0.3	1.80[s]	
	MHL/M.HL	1.77		
Pb^{2+}	ML/M.L	2.57 +0.1	2.08[s]	
	MHL/M.HL	1.82		

[i] 18°, 0.5; [k] 18°, 2.0; [l] 18°, 3.0; [r] 25°, 0.2 acetate buffer; [s] 30°, 1.2

Bibliography:

H^+ 39A,68SKM,75LP Ag^+ 75LP

Mn^{2+}-Cu^{2+},Zn^{2+}-Pb^{2+} 68SKM,72AN,72RG,72RGa, Other references: 44L,61COC,72RGb,74MMc
        75AH,75LP

$$HO_2CCH_2CH_2SSCH_2CH_2CO_2H$$

$C_6H_{10}O_4S_2$ 3,3'-Dithiodipropanoic acid H_2L

Metal ion	Equilibrium	Log K 20°, 0.15
H^+	HL/H.L	4.47
	$H_2L/HL.H$	3.88
Cu^{2+}	ML/M.L	3.02
	MHL/M.HL	2.54

Bibliography: 63HP

Other reference: 61OC

$$HO_2C \overset{\diagup}{\underset{\diagdown}{\bigcirc}} -SCH_2CO_2H$$

			Log K	Log K
C₉H₈O₄S		(Carboxyphenylthio)acetic acid [(carboxymethylthio)benzoic acid]		H₂L

$C_9H_8O_4S$ — (Carboxyphenylthio)acetic acid [(carboxymethylthio)benzoic acid] — H_2L

Isomer	Metal ion	Equilibrium	Log K 25°, 0.1	Log K 20°, 0.1
2-Carboxy	H^+	HL/H.L	4.01	3.88
		H_2L/HL.H	2.93	2.90
	Ni^{2+}	ML/M.L	1.9	
	Cu^{2+}	ML/M.L	3.0	
	Ag^+	ML/M.L		3.13[z]
		MH_2L/M.H_2L		1.77[r,z]
	Zn^{2+}	ML/M.L	1.8	
	Cd^{2+}	ML/M.L	2.0	
	Pb^{2+}	ML/M.L	2.5	
3-Carboxy	H^+	HL/H.L		4.04
		H_2L/HL.H		3.17
	Ag^+	ML/M.L		2.90[z]
		ML_2/M.L^2		5.1[z]
		MH_2L/M.H_2L		1.94[r,z]
4-Carboxy	H^+	HL/H.L		4.18
		H_2L/HL.H		3.08
	Ag^+	ML/M.L		2.80[z]
		ML_2/M.L^2		5.3[z]
		MH_2L/M.H_2L		2.47[r,z]

[r] 25°, 0.2;   [z] acetate buffer

Bibliography:

H^+ 62SY, 68PRS Ag^+ 68PRW

$Ni^{2+}, Cu^{2+}, Zn^{2+}-Pb^{2+}$ 62SY

$$HO_2CCH_2SCH_2SCH_2CO_2H$$

$C_5H_8O_4S_2$ — (Methylenedithio)diacetic acid — H_2L

Metal ion	Equilibrium	Log K 18°, 0.1	Log K 18°, 1.0	Log K 18°, 0
H^+	HL/H.L	3.96[s]	3.82[u]	4.34[v]
		3.82[s]	3.92[u]+0.1	4.08[v]
	H_2L/HL.H	3.16[s]	3.10[u]	3.31[v]
		3.09[s]	3.23[u]+0.1	3.40[v]
Ni^{2+}	$ML/M.L_3$		1.31[d]	
	$ML_3/M.L^3$		6.6[d]	
	MHL/M.HL		0.85[d]	
Cu^{2+}	ML/M.L	3.17[h]		
	MHL/M.HL	1.98[h]		

[d] 25°, 2.0;   [h] 20°, 0.1;   [s] 18°, 0.5;   [u] 18°, 2.0;   [v] 18°, 3.0

(Methylenedithio)diacetic acid (continued)

Bibliography:

H^+ 39A,74AH Cu^{2+} 610C

Ni^{2+} 74AH   Other reference: 44L

$$HO_2CCH_2CH_2SCH_2SCH_2CH_2CO_2H$$

$C_7H_{12}O_4S_2$	3,3'-(Methylenedithio)dipropanoic acid		H_2L

Metal ion	Equilibrium	Log K 20°, 0.1
H^+	HL/H.L	4.7
	H_2L/HL.H	3.9
Cu^{2+}	ML/M.L	2.06
	MHL/M.HL	1.38

Bibliography: 610C

$$HO_2CCH_2SCH_2CH_2SCH_2CO_2H$$

$C_6H_{10}O_4S_2$		(Ethylenedithio)diacetic acid				H_2L
Metal ion	Equilibrium	Log K 25°, 0.1	Log K 25°, 1.0	Log K 18°, 0	ΔH 25°, 1.0	ΔS 25°, 1.0
H^+	HL/H.L	4.00 ±0.1	3.85 ±0.02	4.35		
		3.83[r]	3.94[k]+0.2			
	H_2L/HL.H	3.20 ±0.04	3.21 ±0.05	3.38		
		3.15[r]	3.29[k]+0.2			
Ca^{2+}	ML/M.L	1.74[h]				
	MHL/M.HL	1.16[h]				
Sm^{3+}	ML/M.L		2.32		2.9	20
	$ML_2/M.L^2$		3.49		5.1	33
	MHL/M.HL		1.49			
Dy^{3+}	ML/M.L		2.09		4.9	26
	$ML_2/M.L^2$		3.48		5.8	35
	MHL/M.HL		1.41			
Er^{3+}	ML/M.L		2.01		4.7	25
	$ML_2/M.L^2$		3.15		7.9	41
	MHL/M.HL		1.28			
V^{2+}	ML/M.L	1.39				
Cr^{2+}	ML/M.L	1.99				
Mn^{2+}	ML/M.L	1.04				
	MHL/M.HL	0.7				
Fe^{2+}	ML/M.L	2.73				
	MHL/M.HL	2.03				
Co^{2+}	ML/M.L	3.13				
	MHL/M.HL	1.95				

[h] 20°, 0.1;   [k] 18°, 2.0;   [r] 18°, 0.5

(Ethylenedithio)diacetic acid (continued)

Metal ion	Equilibrium	Log K 25°, 0.1	Log K 25°, 1.0	Log K 18°, 0	ΔH 25°, 1.0	ΔS 25°, 1.0
Ni^{2+}	ML/M.L	4.49 ±0.0	4.56[d]			
	MHL/M.HL	3.00	2.95[d]			
	M_2L/ML.M		1.44[d]			
Cu^{2+}	ML/M.L	5.68 ±0.0				
	MHL/M.HL	3.94 ±0.0				
	M_2L/ML.M		2.1[d]			
Cr^{3+}	MHL/M.HL	5.4				
Fe^{3+}	MHL/M.HL	4.4				
Cu^+	ML/M.L	7.3[s]				
	ML_2/M.L^2	10.2[s]				
Ag^+	ML/M.L	4.95				
	MH_2L/M.H_2L	4.45				
	$M(H_2L)_2$/M.$(H_2L)^2$	8.24				
	M_2L/ML.M	2.06				
	M_2H_2L/MH_2L.M	1.71				
Zn^{2+}	ML/M.L	2.68 ±0.0				
	MHL/M.HL	1.74[h]				
Cd^{2+}	ML/M.L	2.82 ±0.0				
	MHL/M.HL	1.93[h]				
Pb^{2+}	ML/M.L	3.8				

---

[d] 25°, 2.0;   [h] 20°, 0.1;   [s] 25°, 0.3

Bibliography:

H^+ 39A,62SY,74AH,71FP,71PP,74GG

Ca^{2+} 61SO

Sm^{3+}-Er^{3+} 74GG

V^{2+}-Co^{2+},Cr^{3+},Fe^{3+} 71PP

Ni^{2+},Cu^{2+},Zn^{2+},Cd^{2+},Pb^{2+} 61SO,62SY,71FP, 71PP,74AH

Cu^+ 61JW

Ag^+ 71FP

Other references: 44L,71KKK,72SR

$$HO_2CH_2SCH_2CH_2CH_2SCH_2CO_2H$$

$C_7H_{12}O_4S_2$ (Trimethylenedithio)diacetic acid H_2L

Metal ion	Equilibrium	Log K 18°, 0.1	Log K 18°, 1.0	Log K 18°, 0
H^+	HL/H.L	4.03 3.90[i]	3.90 4.02[k]+0.2	4.38
	H_2L/HL.H	3.28 3.22[i]	3.22 3.35[k]+0.2	3.44
Ni^{2+}	ML/M.L		2.70[d]	
	MHL/M.HL		1.55[d]	
Cu^{2+}	ML/M.L	4.41[h]		
	MHL/M.HL	3.05[h]		

---

[d] 25°, 2.0;   [h] 20°, 0.1;   [i] 18°, 0.5;   [k] 18°, 1.0

Bibliography:

H^+ 39A,74AH

Ni^{2+} 74AH

Cu^{2+} 61OC

Other reference: 44L

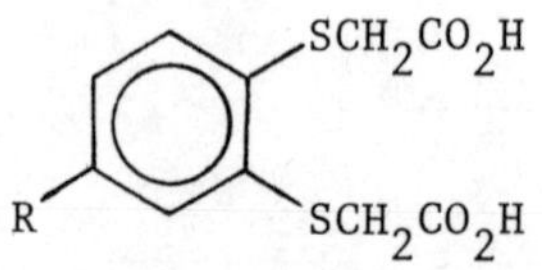

$$SCH_2CO_2H$$

R ———⟨benzene ring⟩——— SCH_2CO_2H

$C_nH_mO_4S_2$		4-Alkyl-1,2-bis(carboxymethylthio)benzene [(4-alkyl-1,2-phenylenedithio)diacetic acid]		H_2L
R =	Metal ion	Equilibrium	Log K 25°, 0.1	
R = H	H^+	HL/H.L	3.79	
		H_2L/HL.H	3.10	
($C_{10}H_{10}O_4S_2$)	Ag^+	ML/M.L	3.97	
4-Methyl	H^+	HL/H.L	3.78	
		H_2L/HL.H	3.08	
($C_{11}H_{12}O_4S_2$)	Ni^{2+}	ML/M.L	2.97	
	Cu^{2+}	ML/M.L	3.90	
	Ag^+	ML/M.L	4.17	
	Zn^{2+}	ML/M.L	2.20	
	Cd^{2+}	ML/M.L	2.40	

Bibliography: 71FP

$$HO_2CCH_2SCH_2CH_2OCH_2CH_2SCH_2CO_2H$$

$C_8H_{14}O_5S_2$		Oxybis(ethylenethioacetic acid)	H_2L
Metal ion	Equilibrium	Log K 20°, 0.1	
H^+	HL/H.L	3.84	
	H_2L/HL.H	3.60	
Co^{2+}	ML/M.L	2.60	
Ni^{2+}	ML/M.L	3.61	
Cu^{2+}	ML/M.L	5.87	
Ag^+	ML/M.L	4.53	
Pd^{2+}	ML/M.L	6.22	
Pt^{2+}	ML/M.L	3.80	
Zn^{2+}	ML/M.L	1.90	
Cd^{2+}	ML/M.L	2.4	

Bibliography: 72CAa

Other reference: 73Ca

$$HO_2CCH_2SCH_2CH_2SCH_2CH_2SCH_2CO_2H$$

$C_8H_{14}O_4S_3$		Thiobis(ethylenethioacetic acid) (diethylenetrithiodiacetic acid)	H_2L

Metal ion	Equilibrium	Log K 25°, 0.1
H^+	HL/H.L	3.93 ±0.01
	H_2L/HL.H	3.18 ±0.03
V^{2+}	ML/M.L	1.2
Cr^{2+}	ML/M.L	2.33
Mn^{2+}	ML/M.L	1.7
	MHL/M.HL	0.6
Fe^{2+}	ML/M.L	1.88
	MHL/M.HL	0.7
Co^{2+}	ML/M.L	2.28
	MHL/M.HL	1.58
Ni^{2+}	ML/M.L	4.70 −0.3
	MHL/M.HL	2.42
Cu^{2+}	ML/M.L	5.90 −0.2
	MHL/M.HL	3.36
Fe^{3+}	MHL/M.HL	5.3
Ag^+	ML/M.L	8.52
	MH_2L/M.H_2L	5.47
	M_2L/ML.M	2.47
	M_3L/M_2L.M	1.2
	M_2H_2L/MH_2L.M	2.01
Zn^{2+}	ML/M.L	2.07 +0.2
Cd^{2+}	ML/M.L	2.43 +0.1

Bibliography:

H^+,Ni^{2+},Cu^{2+}	71FP,71PPa	Ag^+	71FP
V^{2+}-Co^{2+},Fe^{3+}	71PPa	Other reference:	71KKK
Zn^{2+},Cd^{2+}	71FP,73PP		

$$HO_2CCH_2SeCH_2CO_2H$$

$C_4H_6O_4Se$		Selenodiacetic acid		H_2L

Metal ion	Equilibrium	Log K 25°, 0.1	Log K 25°, 0.2
H^+	HL/H.L	4.35	
	H_2L/HL.H	3.27	
Mn^{2+}	ML/M.L	(2.02)	
	MHL/M.HL	0.88	
Co^{2+}	ML/M.L	2.47	
	MHL/M.HL	1.37	
Ni^{2+}	ML/M.L	2.96	
	MHL/M.HL	1.76	
Cu^{2+}	ML/M.L	3.55	
	MHL/M.HL	2.50	

Selenodiacetic acid (continued)

Metal ion	Equilibrium	Log K 25°, 0.1	Log K 25°, 0.2
Ag^+	$ML/M.L$	4.46	4.20^z
	$ML_2/M.L^2$		6.76^z
	$MHL/M.HL$	3.42	
	$MH_2L/M.H_2L$	3.02	2.77^z
	$M(H_2L)_2/M.(H_2L)^2$		4.61^z
Zn^{2+}	$ML/M.L$	2.18	
	$MHL/M.HL$	1.05	
Cd^{2+}	$ML/M.L$	2.57	
	$MHL/M.HL$	1.82	
Pb^{2+}	$ML/M.L$	3.22	
	$MHL/M.HL$	2.16	

^{z} acetate buffer.

Bibliography: 75LP

$$HO_2CCHSeCHCO_2H$$
$$H_3C \quad CH_3$$

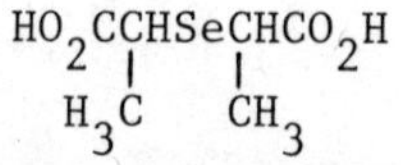

$C_6H_{10}O_4Se$		DL-2,2'-Selenodipropanoic acid		H_2L

Metal ion	Equilibrium	Log K 25°, 0.1	Log K 25°, 0.2
H^+	$HL/H.L$	4.66	
	$H_2L/HL.H$	3.35	
Mn^{2+}	$ML/M.L$	2.02	
Co^{2+}	$ML/M.L$	2.20	
	$MHL/M.HL$	1.1	
Ni^{2+}	$ML/M.L$	2.73	
	$MHL/M.HL$	1.8	
Cu^{2+}	$ML/M.L$	3.21	
	$MHL/M.HL$	2.1	
Ag^+	$ML/M.L$		4.55^z
	$ML_2/M.L^2$		$(6.92)^{y,z}$
	$MHL/M.HL$	3.72	
	$MH_2L/M.H_2L$	3.01	2.90^z
	$M(H_2L)_2/M.(H_2L)^2$		$(4.78)^{y,z}$
Zn^{2+}	$ML/M.L$	1.80	
Cd^{2+}	$ML/M.L$	2.09	
	$MHL/M.HL$	0.9	
Pb^{2+}	$ML/M.L$	2.80	
	$MHL/M.HL$	1.7	

^{y} optical isomerism not stated; ^{z} acetate buffer.

Bibliography: 75LP

$$HO_2CCH_2CH_2SeCH_2CH_2CO_2H$$

$C_6H_{10}O_4Se$ 3,3'-Selenodipropanoic acid H_2L

Metal ion	Equilibrium	Log K 25°, 0.1	Log K 25°, 0.2
H^+	HL/H.L	4.70	
	H_2L/HL.H	3.90	
Mn^{2+}	ML/M.L	1.50	
Co^{2+}	ML/M.L	1.82	
	MHL/M.HL	1.43	
Ni^{2+}	ML/M.L	1.83	
	MHL/M.HL	1.56	
Cu^{2+}	ML/M.L	2.60	
	MHL/M.HL	1.59	
Ag^+	ML/M.L		4.97[z]
	ML_2/M.L^2		7.36[z]
	MHL/M.HL	4.32	
	MH_2L/M.H_2L	3.96	4.05[z]
	$M(H_2L)_2$/M.$(H_2L)^2$		5.62[z]
Zn^{2+}	ML/M.L	1.77	
	MHL/M.HL	1.05	
Cd^{2+}	ML/M.L	2.07	
	MHL/M.HL	1.52	
Pb^{2+}	ML/M.L	2.58	
	MHL/M.HL	1.95	

[z] acetate buffer.

Bibliography: 75LP

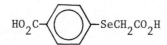

$C_9H_8O_4Se$ (4-Carboxyphenylseleno)acetic acid H_2L

Metal ion	Equilibrium	Log K 20°, 0.1	Log K 20°, 0.2
H^+	HL/H.L	4.29	
	H_2L/HL.H	3.38	
Ag^+	ML/M.L	3.52[z]	
	ML_2/M.L^2	6.02[z]	
	MH_2L/M.H_2L		2.8[z]

[z] acetate buffer.

Bibliography:

H^+ 68PRS Ag^+ 68PSW

$$HO_2CCH_2CH_2TeCH_2CH_2CO_2H$$

$C_6H_{10}O_4Te$ 3,3'-Tellurodipropanoic acid H_2L

Metal ion	Equilibrium	Log K 25°, 0.1	Log K 25°, 0.2
H^+	HL/H.L	4.77	
	H_2L/HL.H	3.92	
Mn^{2+}	ML/M.L	1.2	
Co^{2+}	ML/M.L	2.36	
	MHL/M.HL	1.9	
Ni^{2+}	ML/M.L	3.0	
	MHL/M.HL	2.8	
Cu^{2+}	ML/M.L	3.2	
	MHL/M.HL	2.7	
Ag^+	ML/M.L		6.28^z
	ML_2/M.L^2		9.18^z
	MHL/M.HL	5.35	
	MH_2L/M.H_2L	4.97	5.03^z
	$M(H_2L)_2$/M.$(H_2L)^2$		7.77^z
Zn^{2+}	ML/M.L	1.4	
Cd^{2+}	ML/M.L	2.85	
	MHL/M.HL	2.4	
Pb^{2+}	ML/M.L	2.94	
	MHL/M.HL	2.3	

^{z} acetate buffer.

Bibliography: 75LP

$C_{10}H_{11}O_4As$ As-Phenylarsinodiacetic acid [bis(carboxymethyl)phenylarsine] H_2L

Metal ion	Equilibrium	Log K 20°, 0.1	Log K 25°, 0.3
H^+	HL/H.L	5.03 -0.01	4.93
	H_2L/HL.H	3.61 ±0.01	3.61
Ni^{2+}	ML/M.L	1.5	
Cu^{2+}	ML/M.L	2.51	
Cu^+	ML/M.L		5.7
	ML_2/M.L^2		7.4
	MHL/ML.H	4.0	3.96
	MH_2L/MHL.H		2.70
Ag^+	ML/M.L	6.13	
	MHL/M.HL	5.13	
	MH_2L/M.H_2L	4.64	
Zn^{2+}	ML/M.L	1.4	
Cd^{2+}	ML/M.L	1.0	

As-Phenylarsinodiacetic acid (continued)

Metal ion	Equilibrium	Log K 20°, 0.1	Log K 25°, 0.3
Hg^{2+}	$ML/M.L$	14.7	
	$ML_2/M.L^2$	19.92	

Bibliography:

H^+ 61JW,64PI,69PR Cu^+ 61JW,64PI

Ni^{2+},Cu^{2+},Zn^{2+}-Hg^{2+} 64PI Ag^+ 64PI,69PR,72FG

$C_m H_n O_q As X_r$		As-(Substitutedphenyl)arsinodiacetic acid		H_2L
Z =	Metal ion	Equilibrium	Log K 20°, 0.1	
2-Chloro	H^+	HL/H.L	5.12	
		H_2L/HL.H	3.59	
$(C_{10}H_{10}O_4AsCl)$	Ag^+	ML/M.L	5.20	
		MHL/M.HL	4.13	
		$MH_2L/M.H_2L$	4.05	
4-Chloro	H^+	HL/H.L	4.94	
		H_2L/HL.H	3.59	
$(C_{10}H_{10}O_4AsCl)$	Ag^+	ML/M.L	5.96	
		MHL/M.HL	4.96	
		$MH_2L/M.H_2L$	4.47	
2-Methyl	H^+	HL/H.L	5.09	
		H_2L/HL.H	3.71	
$(C_{11}H_{13}O_4As)$	Ag^+	ML/M.L	5.93	
		MHL/M.HL	4.98	
		$MH_2L/M.H_2L$	4.41	
3-Methyl	H^+	HL/H.L	5.08	
		H_2L/HL.H	3.68	
$(C_{11}H_{13}O_4As)$	Ag^+	ML/M.L	6.06	
		MHL/M.HL	4.97	
		$MH_2L/M.H_2L$	4.40	
4-Methyl	H^+	HL/H.L	5.10	
		H_2L/HL.H	3.70	
$(C_{11}H_{13}O_4As)$	Ag^+	ML/M.L	6.44	
		MHL/M.HL	5.58	
		$MH_2L/M.H_2L$	4.83	
2-Methoxy	H^+	HL/H.L	5.29	
		H_2L/HL.H	3.68	
$(C_{11}H_{13}O_5As)$	Ag^+	ML/M.L	6.14	
		MHL/M.HL	4.67	
		$MH_2L/M.H_2L$	4.54	

As-(Substitutedphenyl)arsinodiacetic acid (continued)

Z =	Metal ion	Equilibrium	Log K 20°, 0.1
3-Methoxy	H^+	HL/H.L	4.99
		H_2L/HL.H	3.58
$(C_{11}H_{13}O_5As)$	Ag^+	ML/M.L	6.20
		MHL/M.HL	4.80
		MH_2L/M.H_2L	4.56
4-Methoxy	H^+	HL/H.L	5.03
		H_2L/HL.H	3.64
$(C_{11}H_{13}O_5As)$	Ag^+	ML/M.L	6.30
		MHL/M.HL	5.50
		MH_2L/M.H_2L	4.76

Bibliography: 69PR,72FG

$C_{11}H_{13}O_4AsS$		As-(Methylthiophenyl)arsenodiacetic acid		H_2L
Isomer	Metal ion	Equilibrium	Log K 25°, 0.1	
2-Methylthio	H^+	HL/H.L	5.24	
		H_2L/HL.H	3.78	
	Mn^{2+}	ML/M.L	2.86	
		MHL/M.HL	2.35	
	Fe^{2+}	ML/M.L	3.80	
		MHL/M.HL	2.32	
	Co^{2+}	ML/M.L	2.93	
		MHL/M.HL	2.37	
	Ni^{2+}	ML/M.L	3.27	
		MHL/M.HL	2.64	
	Cu^{2+}	ML/M.L	3.68	
		MHL/M.HL	2.19	
	Ag^+	ML/M.L	5.14	
		MH_2L/M.H_2L	4.64	
		$M(H_2L)_2$/M.$(H_2L)^2$	8.73	
		M_2L/ML.M	2.74	
		M_2H_2L/MH_2L.M	2.18	
	Zn^{2+}	ML/M.L	3.20	
		MHL/M.HL	2.52	
	Cd^{2+}	ML/M.L	3.22	
		MHL/M.HL	2.51	

As-(Methylthiophenyl)arsenodiacetic acid (continued)

Isomer	Metal ion	Equilibrium	Log K 25°, 0.1
4-Methylthio	H^+	HL/H.L	4.95
		$H_2L/HL.H$	3.56
	Ni^{2+}	ML/M.L	2.92
	Cu^{2+}	ML/M.L	2.83
	Cd^{2+}	ML/M.L	1.5

Bibliography: 71FP

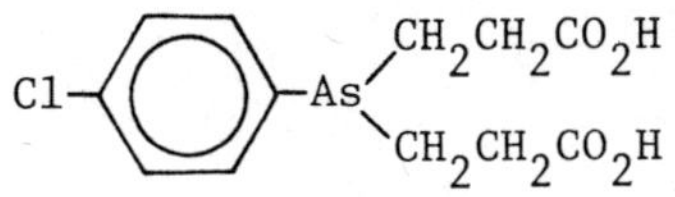

$C_{12}H_{14}O_4AsCl$	As-(4-Chlorophenyl)-3,3'-arsinodipropanoic acid	H_2L

Metal ion	Equilibrium	Log K 20°, 0.1
H^+	HL/H.L	5.08
	$H_2L/HL.H$	4.17
Cu^{2+}	ML/M.L	1.5
Ag^+	ML/M.L	5.00
	MHL/M.HL	3.98

Bibliography: 64PI

$$\begin{array}{c} CO_2H \\ | \\ HO_2CCH_2CHCH_2CO_2H \end{array}$$

$C_6H_8O_6$	Propane-1,2,3-tricarboxylic acid (tricarballylic acid)					H_3L
Metal ion	Equilibrium	Log K 25°, 0.1	Log K 20°, 0.1	Log K 25°, 0	ΔH 25°, 0	ΔS 25°, 0
H^+	HL/H.L	5.82 ±0.05	5.89	6.38	-1.81	35.2
	H_2L/HL.H	4.50 ±0.03	4.54	4.87	-1.49	25.3
	H_3L/H_2L.H	3.48 ±0.02	3.47	3.67	-0.56	14.9
Mg^{2+}	ML/M.L	2.00[s]	2.06			
	MHL/M.HL	0.91[s]	1.20			
	MH_2L/M.H_2L		0.77			
Ca^{2+}	ML/M.L	1.78[s] ±0.04	2.17			
	MHL/M.HL		1.46			
	MH_2L/M.H_2L		0.88			
Sr^{2+}	ML/M.L	1.68[s]				
Ba^{2+}	ML/M.L		1.95			
	MHL/M.HL		1.15			
	MH_2L/M.H_2L		0.73			
La^{3+}	ML/M.L	3.71				
	MHL/M.HL	2.50				
Sm^{3+}	ML/M.L	4.23				
	MHL/M.HL	2.94				
Gd^{3+}	ML/M.L	4.21				
	MHL/M.HL	2.88				
Eu^{3+}	ML/M.L	4.02				
	MHL/M.HL	2.72				
Am^{3+}	ML/M.L	5.61				
	MHL/M.HL	4.96				
Mn^{2+}	ML/M.L	1.99[s]				
Co^{2+}	ML/M.L		2.44			
	MHL/M.HL		1.60			
	MH_2L/M.H_2L		0.95			
Ni^{2+}	ML/M.L	2.70[s]	2.65			
	MHL/M.HL	1.56[s]	1.66			
	MH_2L/M.H_2L		1.07			
Cu^{2+}	ML/M.L		3.70			
	MHL/M.HL		2.57			
	MH_2L/M.H_2L		1.40			
	M_2L/ML.M		1.60			
Zn^{2+}	ML/M.L		2.43			
	MHL/M.HL		1.61			
	MH_2L/M.H_2L		0.94			

[s] 25°, 0.15

Bibliography:

H^+ 59LL,64COM,72PTR Mn^{2+} 57LW
Mg^{2+}-Ba^{2+} 36ML,52SL,59LL,64COM Co^{2+}-Zn^{2+} 59LL,64COM
La^{3+}-Er^{3+} 69BB   Other reference: 70Lb
Am^{3+} 72EM

C. TRI-CARBOXYLIC ACIDS

$$HO_2CCH_2CH_2CHCH_2CH_2CO_2H$$
$$\underset{CO_2H}{|}$$

$C_8H_{12}O_6$		Pentane-1,3,5-tricarboxylic acid (γ-carboxypimelic acid)	H_3L

Metal ion	Equilibrium	Log K 25°, 0.16
Ca^{2+}	ML/M.L	1.59
Sr^{2+}	ML/M.L	1.54

Bibliography: 52SL

$$HO_2CCCH_2CO_2H$$
$$\underset{HO_2CCH}{\|}$$

$C_6H_6O_6$		cis-Propene-1,2,3-tricarboxylic acid (cis-aconitic acid)	H_3L

Metal ion	Equilibrium	Log K 25°, 0.16
Mn^{2+}	ML/M.L	2.47

Bibliography: 57LW

$$HO_2CCCH_2CO_2H$$
$$\underset{CHCO_2H}{\|}$$

$C_6H_6O_6$		trans-Propene-1,2,3-tricarboxylic acid (trans-aconitic acid)	H_3L

Metal ion	Equilibrium	Log K 25°, 0.16
Ca^{2+}	ML/M.L	1.50
Sr^{2+}	ML/M.L	1.51
Mn^{2+}	ML/M.L	2.27

Bibliography:
Ca^{2+}, Sr^{2+} 52SL Mn^{2+} 57LW

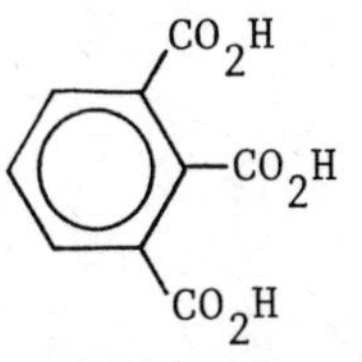

$C_9H_6O_6$		Benzene-1,2,3-tricarboxylic acid (hemimellitic acid)			H_3L

Metal ion	Equilibrium	Log K 25°, 0.1	Log K 25°, 0	ΔH 25°, 0	ΔS 25°, 0
H^+	HL/H.L	5.51	7.13	-0.37	31.4
	$H_2L/HL.H$	3.82	4.75	-0.21	21.4
	$H_3L/H_2L.H$	2.62	2.88	1.06	16.7

Hemimellitic acid (continued)

Metal ion	Equilibrium	Log K 25°, 0.1	Log K 25°, 0	ΔH 25°, 0	ΔS 25°, 0
Ni^{2+}	ML/M.L	2.86	4.18		
Cu^{2+}	ML/M.L	4.40	5.72		
	MHL/M.HL	2.7	3.6		

Bibliography:

H$^+$ 61Y,72PTR   Ni^{2+},Cu^{2+} 61Y

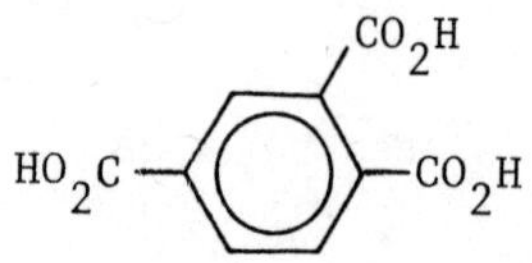

$C_9H_6O_6$ Benzene-1,2,4-tricarboxylic acid (trimellitic acid) H_3L

Metal ion	Equilibrium	Log K 25°, 0.1	Log K 25°, 0	ΔH 25°, 0	ΔS 25°, 0
H$^+$	HL/H.L	5.01	5.54	0.95	28.6
	H$_2$L/HL.H	3.71	4.04	0.09	18.5
	H$_3$L/H$_2$L.H	2.4	2.48	1.24	25.5
Ni^{2+}	ML/M.L	1.86	3.18		
Cu^{2+}	ML/M.L	3.17	4.49		

Bibliography:

H$^+$ 61Y,72PTR   Ni^{2+},Cu^{2+} 61Y

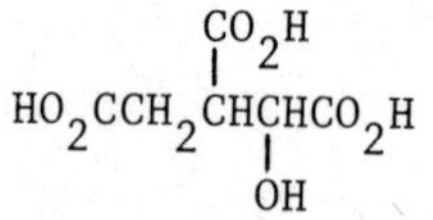

$C_6H_8O_7$ DL-1-Hydroxypropane-1,2,3-tricarboxylic acid (isocitric acid) H_3L

Metal ion	Equilibrium	Log K 25°, 0.1
H$^+$	HL/H.L	5.75[y]
	H$_2$L/HL.H	4.28[y]
	H$_3$L/H$_2$L.H	3.02[y]
Mg^{2+}	ML/M.L	2.32
		2.72[y]
	MHL/M.HL	1.43[y]
Ca^{2+}	ML/M.L	2.47[s]
Sr^{2+}	ML/M.L	2.02[s]
Mn^{2+}	ML/M.L	2.55[s]
		3.06[y]
	MHL/M.HL	1.76[y]

[s] 25°, 0.16;   [y] (CH$_3$)$_4$NCl used as background electrolyte and corrected for Cl$^-$.

C. TRI-CARBOXYLIC ACIDS

Isocitric acid (continued)

Bibliography:

H^+ 70GT

Mg^{2+} 69Ba,70GT,71LDb

Ca^{2+},Sr^{2+} 52SL

Mn^{2+} 57LW,70GT

Other reference: 72MGa

$$\begin{array}{c} CO_2H \\ | \\ HO_2CCH_2CCH_2CO_2H \\ | \\ OH \end{array}$$

$C_6H_8O_7$	2-Hydroxypropane-1,2,3-tricarboxylic acid (citric acid)		H_3L

Metal ion	Equilibrium	Log K 25°, 0.1	Log K 25°, 1.0	Log K 25°, 0	ΔH 25°, 0	ΔS 25°, 0
H^+	HL/H.L	5.69 ±0.05	5.33 ±0.06	6.396 +0.004	0.80	32.0
		5.83[y]±0.01	5.18[d]			
	H_2L/HL.H	4.35 ±0.05	4.08 ±0.03	4.761 −0.002	−0.58	19.8
		4.35[y]±0.01	4.16[d]			
	H_3L/H_2L.H	2.87 ±0.08	2.80 ±0.05	3.128 −0.07	−1.00	11.0
		2.87[y]±0.02	2.90[d]			
Li^+	ML/M.L	0.83				
Na^+	ML/M.L	0.70 ±0.0				
K^+	ML/M.L	0.59 −0.2				
Rb^+	ML/M.L	0.49				
Cs^+	ML/M.L	0.32				
Be^{2+}	ML/M.L	4.5[r]				
	MHL/M.HL	2.2[r]				
	$MH_2L/M.H_2L$	1.4[r]				
Mg^{2+}	ML/M.L	3.37 ±0.02	3.25[s]±0.05			
		3.87[y]	3.40[h]			
	MHL/M.HL	1.92[y]	1.60[s]			
			1.84[h]			
	$MH_2L/M.H_2L$		0.84[h]			
Ca^{2+}	ML/M.L	3.50	3.18[s]±0.03	4.68		
			3.55[h]			
	MHL/M.HL		2.10[h]	3.09		
	$MH_2L/M.H_2L$		1.05[h]	1.10		
Sr^{2+}	ML/M.L	3.05	2.81[s]±0.1	4.11		
Ba^{2+}	ML/M.L	2.76	2.55[s]±0.01			
			2.89[h]			
	MHL/M.HL		1.75[h]			
	$MH_2L/M.H_2L$		0.79[h]			
Ra^{2+}	ML/M.L		2.36[s]			
Y^{3+}	ML/M.L	7.87[t]		9.42		

[d] 25°, 2.0; [h] 20°, 0.1; [r] 34°, 0.15; [s] 25°, 0.16; [t] 25°, 0.05; [y] $(CH_3)_4N^+$ used as background electrolyte or a correction was made for the background electrolyte.

Citric acid (continued)

Metal ion	Equilibrium	Log K 25°, 0.1	Log K 25°, 1.0	Log K 25°, 0	ΔH 25°, 0	ΔS 25°, 0
La^{3+}	ML/M.L	7.63[t]	6.65[j]	9.18		
		7.17				
	$ML_2/M.L^2$	10.2				
	$MHL_2/ML.HL$	2.2				
Ce^{3+}	$ML/M.L$	7.39 ±0.01				
	$ML_2/M.L^2$	10.4				
	$MHL_2/ML.HL$	2.4				
Pr^{3+}	ML/M.L	7.95[t]		9.50		
Nd^{3+}	ML/M.L	7.96[t]		9.51		
Pm^{3+}	$ML/M.L$	7.75				
	$ML_2/M.L^2$	11.0				
	$MHL_2/ML.HL$	2.4				
Sm^{3+}	ML/M.L	8.04[t]		9.59		
Eu^{3+}	ML/M.L	7.91[t]		9.46		
Gd^{3+}	ML/M.L	7.83[t]		9.38		
Tb^{3+}	ML/M.L	7.75[t]		9.30		
Dy^{3+}	ML/M.L	7.79[t]		9.34		
		7.58				
Ho^{3+}	ML/M.L	7.84[t]		9.39		
Er^{3+}	ML/M.L	7.86[t]		9.41		
Tm^{3+}	ML/M.L	8.00[t]		9.55		
Yb^{3+}	ML/M.L	8.10[t]		9.65		
Lu^{3+}	ML/M.L	8.12[t]		9.67		
Am^{3+}	$ML/M.L$	7.74	6.96			
	$ML_2/M.L^2$	10.9	10.3			
	MHL/M.HL		4.53			
	$ML/MH_{-1}L.H$		5.61			
	$MHL_2/ML.HL$	2.5				
Cm^{3+}	$ML/M.L$	7.74				
	$ML_2/M.L^2$	10.9				
	$MHL_2/ML.HL$	2.5				
Bk^{3+}	$ML/M.L$	7.89				
	$ML_2/M.L^2$	11.2				
	$MHL_2/ML.HL$	2.6				
Cf^{3+}	$ML/M.L$	7.93				
	$ML_2/M.L^2$	11.2				
	$MHL_2/ML.HL$	2.6				
UO_2^{2+}	$ML/M.L$	7.4	6.9			
	$M_2L_2/M^2.L^2$	18.87	17.70 -0.1			
Mn^{2+}	ML/M.L	4.15[y]	3.70[s] ±0.03			
	MHL/M.HL	2.16[y]	2.08[s]			

[j] 20°, 1.0; [s] 25°, 0.16; [t] 25°, 0.05; [y] $(CH_3)_4NCl$ used as background electrolyte.

Citric acid (continued)

Metal ion	Equilibrium	Log K 25°, 0.1	Log K 25°, 1.0	Log K 25°, 0	ΔH 25°, 0	ΔS 25°, 0
Fe^{2+}	ML/M.L	4.4[h]				
	MHL/M.HL	2.65[h]				
Co^{2+}	ML/M.L	5.00[h]	4.83[s]			
	MHL/M.HL	3.02[h]	3.19[s]			
	$MH_2L/M.H_2L$	1.25[h]				
Ni^{2+}	ML/M.L	5.40[h]	5.11[s]			
	MHL/M.HL	3.30[h]	3.19[s]			
	$MH_2L/M.H_2L$	1.75[h]				
Cu^{2+}	ML/M.L	5.90[h]				
	MHL/M.HL	3.42[h]				
	$MH_2L/M.H_2L$	2.26[h]				
	$ML/M(H_{-1}L).H$	4.34[h]				
	$M_2L/ML.M$	2.20[h]				
	$M_2L_2/M^2.L^2$		13.2			
	$M_2L_2/M_2(H_{-1}L)_2.H^2$		8.03			
Fe^{3+}	ML/M.L	11.50[h]				
	$(ML)^2/M_2(H_{-1}L)_2.H^2$	1.6[h]				
Tl^+	ML/M.L	1.04		1.48		
		0.65[b]				
Zn^{2+}	ML/M.L	4.98[h]	4.70[s]±0.1			
		4.27[b]				
	$ML_2/M.L^2$	5.90[b]				
	MHL/M.HL	2.98[h]	2.96[s]			
		2.94[b]				
	$MH_2L/M.H_2L$	1.25[h]				
Cd^{2+}	ML/M.L	3.75[h]		5.36		
		3.15[b]				
	$ML_2/M.L^2$	4.54[b]				
	MHL/M.HL	2.20[h]				
	$MH_2L/M.H_2L$	0.97[h]				
Hg^{2+}	ML/M.L	10.9				
Pb^{2+}	ML/M.L		4.08[d]	4.34[e]		
	$ML_2/M.L^2$		6.1[d]	6.08[e]		
	$ML_3/M.L^3$			6.97[e]		
	MHL/M.HL		2.97[d]			
	$MH_2L/M.H_2L$		1.51[d]			
	$MH_2L_2/ML_2.H^2$		8.9[d]			
	$MH_4L_2/MH_2L_2.H^2$		6.7[d]			
	$M(OH)_2L.OH/M(OH)_3L$		-0.94[d]			
	$M(OH)_2L_2.OH/M(OH)_3.L^2$		-0.47[d]			
	$M_2(OH)_4L.(OH)^3/[M(OH)_3]^2.L$		-0.7[d]			
Ga^{3+}	ML/M.L	10.02				
	ML/MOHL.H	2.9				
As(III)	$M(OH)_2L/M(OH)_2.L$	9.3				
	$M(OH)_2HL/M(OH)_2.HL$	8.5				

[b] 25°, 0.5;   [d] 25°, 2.0;   [e] 25°, 3.0;   [h] 20°, 0.1;   [s] 25°, 0.16

Citric acid (continued)

Bibliography:

H^+ 29BU,49BP,53WW,59LL,59OK,63Mb,64COM,
          64Ta,65RM,65TG,68SM,70BB,70GT,71SK,
          73BVa,73GK,73RM,74FM,75BSa,75PC,76HM

Li^+-Cs^+ 61W,64RZ

Be^{2+} 55FT

Mg^{2+}-Ra^{2+} 34HM,36ML,38N,46J,48SR,52S,52SL,
          53DH,54S,59LL,61W,63Mb,64COM,69Ba,
          70GT,71R,75FC

Y^{3+}-Lu^{3+} 70BB,72BK

Am^{3+} 72EM

UO_2^{2+} 65RM,72MK

Mn^{2+},Co^{2+}-Cu^{2+},Zn^{2+},Cd^{2+} 58SL,58TF,58Wa,
          59LL,59OK,62FC,63Mb,64COM,67RMG,70GT

Fe^{2+},Fe^{3+} 64Ta

Tl^+ 56SD

Hg^{2+} 67SKS

Pb^{2+} 63DG,73BV

Gd^{2+} 76HM

As(III) 70ET

Other references: 40A,42K,47T,47TM,48LQ,
          48SRa,50M,50SR,51H,51M,51Sa,52PD,52Sa,
          53Sa,53Sb,54HS,54Sa,55Sa,56HE,57HE,
          57KP,57Lb,57LW,57P,57PP,58HF,58MS,59KV,
          60AS,60DP,60FN,60RE,61MMZ,61PP,62GLa,
          62KB,62RM,62SBb,62Ta,62ZR,63TM,64MZ,
          64RME,64TM,65HS,65KB,65PPa,65Sb,65SB,
          65SKP,66Ga,66KS,66KZ,66N,66NU,66SS,
          66SSa,66TP,67Ka,67NN,68Ga,68PK,68PSa,
          69SK,70AKN,70BC,70Lb,71GB,71K,71MMW,
          71OO,71TZ,72KC,72MG,73BPS,73HH,73TPb,
          74HHG,74PKa,74PKb,74PV,74RM,75GG,75KB,
          75NA,75OO

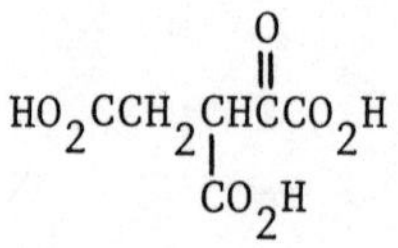

$$HO_2CCH_2\underset{\underset{\displaystyle CO_2H}{|}}{CH}\overset{\overset{\displaystyle O}{\|}}{C}CO_2H$$

$C_6H_6O_6$	1-Oxopropane-1,2,3-tricarboxylic acid (oxalosuccinic acid)		H_3L
Metal ion	Equilibrium	Log K 25°, 0.16	
Ca^{2+}	ML/M.L	1.5	

Bibliography: 52SL

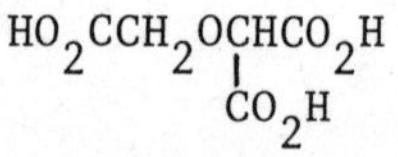

$$HO_2CCH_2\underset{\underset{\displaystyle CO_2H}{|}}{O}CHCO_2H$$

$C_5H_6O_7$	Carboxymethoxypropanedioic acid		H_3L
Metal ion	Equilibrium	Log K 25°, 0.1	
H^+	HL/H.L	4.41	
	$H_2L/HL.H$	3.25	
	$H_3L/H_2L.H$	1.72	
Mg^{2+}	ML/M.L	2.77	
	MHL/M.HL	1.43	
Ca^{2+}	ML/M.L	4.62	
	MHL/M.HL	2.86	
Sr^{2+}	ML/M.L	3.79	
	MHL/M.HL	2.29	
Mn^{2+}	ML/M.L	3.76	
	MHL/M.HL	2.19	
Fe^{2+}	ML/M.L	3.67	
	MHL/M.HL	1.96	

Carboxymethoxypropanedioic acid (continued)

Metal ion	Equilibrium	Log K 25°, 0.1
Co^{2+}	ML/M.L	3.87
	MHL/M.HL	2.27
Ni^{2+}	ML/M.L	3.84
	MHL/M.HL	2.32
Cu^{2+}	ML/M.L	5.20
	MHL/M.HL	3.80
Zn^{2+}	ML/M.L	4.87
	MHL/M.HL	2.88
Cd^{2+}	ML/M.L	4.49
	MHL/M.HL	2.75
Sn^{2+}	ML/M.L	7.00
	MHL/M.HL	4.15
Pb^{2+}	ML/M.L	6.04
	MHL/M.HL	3.65

Bibliography: 76MM

$$HO_2CCH_2SCHCO_2H$$
$$CH_2CO_2H$$

$C_6H_8O_6S$	Carboxymethylthiobutanedioic acid		H_3L

Metal ion	Equilibrium	Log K 20°, 0.1
H^+	HL/H.L	5.12
	$H_2L/HL.H$	3.79
	$H_3L/H_2L.H$	3.26
Mn^{2+}	ML/M.L	2.1
Co^{2+}	ML/M.L	3.45
	$ML_2/M.L^2$	6.08
Ni^{2+}	ML/M.L	4.32
	$ML_2/M.L^2$	7.31
Cu^{2+}	ML/M.L	4.80
	$ML_2/M.L^2$	8.05
Ag^+	ML/M.L	3.22
Pb^{2+}	ML/M.L	5.20
Pt^{2+}	ML/M.L	4.6
Zn^{2+}	ML/M.L	3.19
	$ML_2/M.L^2$	5.61
Cd^{2+}	ML/M.L	2.71
	$ML_2/M.L^2$	4.47

Bibliography: 72CA

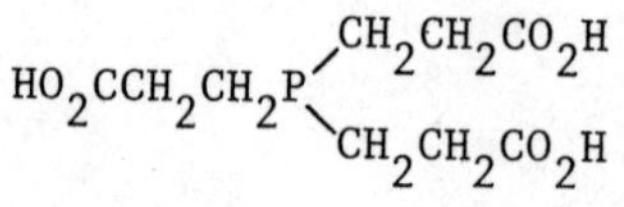

$$HO_2CCH_2CH_2P \Big\langle {}^{CH_2CH_2CO_2H}_{CH_2CH_2CO_2H}$$

$C_9H_{15}O_6P$

<u>3,3',3"-Phosphinotripropanoic acid</u>
[tris(2-carboxyethyl)phosphine]

H_3L

Metal ion	Equilibrium	Log K 25°, 0.1
H^+	HL/H.L	7.66
	$H_2L/HL.H$	4.36
	$H_3L/H_2L.H$	3.67
	$H_4L/H_3L.H$	2.99
Co^{2+}	ML/M.L	2.23
Ni^{2+}	ML/M.L	3.80
Zn^{2+}	ML/M.L	2.92

Bibliography: 73PPa

Other reference: 69LP

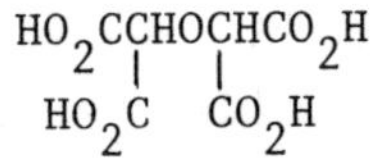

D. TETRA-CARBOXYLIC ACIDS

167

$$HO_2CCHOCHCO_2H$$
$$HO_2C \quad CO_2H$$

$C_6H_6O_9$		Oxybis(propanedioic acid) (ditartronic acid)	H_4L
Metal ion	Equilibrium	Log K 25°, 0.1	
H^+	HL/H.L	4.71	
	$H_2L/HL.H$	3.66	
	$H_3L/H_2L.H$	2.33	
	$H_4L/H_3L.H$	1.27	
Mg^{2+}	ML/M.L	3.43	
	MHL/M.HL	2.05	
Ca^{2+}	ML/M.L	5.17	
	MHL/M.HL	3.69	
Sr^{2+}	ML/M.L	4.69	
	MHL/M.HL	3.32	
Mn^{2+}	ML/M.L	4.51	
	MHL/M.HL	3.03	
Fe^{2+}	ML/M.L	4.41	
	MHL/M.HL	2.92	
Co^{2+}	ML/M.L	4.55	
	MHL/M.HL	3.04	
Ni^{2+}	ML/M.L	4.40	
	MHL/M.HL	3.08	
Cu^{2+}	ML/M.L	5.54	
	MHL/M.HL	4.39	
Zn^{2+}	ML/M.L	5.62	
	MHL/M.HL	3.85	
Cd^{2+}	ML/M.L	5.44	
	MHL/M.HL	3.83	
Sn^{2+}	ML/M.L	7.90	
	MHL/M.HL	5.51	

Bibliography: 76MM

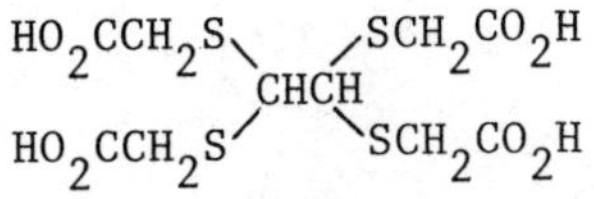

$$HO_2CCH_2S \quad SCH_2CO_2H$$
$$CHCH$$
$$HO_2CCH_2S \quad SCH_2CO_2H$$

$C_{10}H_{14}O_8S_4$		(Ethanediylidenetetrathio)tetraacetic acid		H_4L	
Metal ion	Equilibrium	Log K 25°, 0.1	Log K 20°, 0.1	ΔH 25°, 0.1	ΔS 25°, 0.1
H^+	HL/H.L	4.60 ±0.03	(4.93)		
	$H_2L/HL.H$	3.96 +0.01	3.99		
	$H_3L/H_2L.H$	3.55 ±0.01	3.56		
	$H_4L/H_3L.H$	2.89 +0.01	(3.24)		
Mn^{2+}	ML/M.L	2.32			
	MHL/M.HL	1.79			
	$MH_2L/M.H_2L$	0.9			

(Ethylidenetetrathio)tetraacetic acid (continued)

Metal ion	Equilibrium	Log K 25°, 0.1	Log K 20°, 0.1	ΔH 25°, 0.1	ΔS 25°, 0.1
Fe^{2+}	ML/M.L	2.68			
	MHL/M.HL	1.97			
	MH$_2$L/M.H$_2$L	1.63			
	M$_2$L/ML.M	1.50			
Co^{2+}	ML/M.L	2.86			
	MHL/M.HL	2.15			
	MH$_2$L/M.H$_2$L	1.64			
	M$_2$L/ML.M	1.2			
Ni^{2+}	ML/M.L	4.08 +0.09			
	MHL/M.HL	3.09 −0.2			
	MH$_2$L/M.H$_2$L	2.22			
	M$_2$L/ML.M	1.57			
Cu^{2+}	ML/M.L	5.35 +0.4	5.00	1.7	29
	MHL/M.HL	4.45 −1.0	4.08		
	MH$_2$L/M.H$_2$L	3.43	3.24		
	MH$_3$L/M.H$_3$L		2.64		
	M$_2$L/ML.M	2.40	2.33	0.8	13

Bibliography:

H$^+$ 61SOa,70GM,73PPb

Mn^{2+}-Co^{2+} 73PPb

Ni^{2+} 70GM,73PPb

Cu^{2+} 61SOa,70CL,70GM,73PPb

Other reference: 72PP

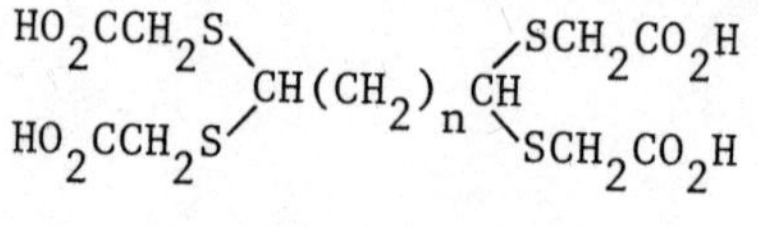

$$\begin{array}{c} HO_2CCH_2S \\ HO_2CCH_2S \end{array} CH(CH_2)_n CH \begin{array}{c} SCH_2CO_2H \\ SCH_2CO_2H \end{array}$$

C$_m$H$_n$O$_8$S$_4$	Metal ion	Equilibrium	Log K 25°, 0.1	
n =				(Alkyldiylidenetetrathio)tetraacetic acid
n = 1	H$^+$	HL/H.L	4.71	
1,3-Propane		H$_2$L/HL.H	4.23	
(C$_{11}$H$_{16}$O$_8$S$_4$)		H$_3$L/H$_2$L.H	3.44	
		H$_4$L/H$_3$L.H	3.40	
	Ca^{2+}	ML/M.L	(4.0)	
		MHL/M.HL	(3.8)	
	Ni^{2+}	ML/M.L	2.8	
	Cu^{2+}	ML/M.L	4.9	
	Hg^{2+}	ML/M.L	16.6	

H$_4$L

(Alkyldiylidenetetrathio)tetraacetic acid (continued)

n =	Metal ion	Equilibrium	Log K 25°, 0.1
n = 2 1,4-Butane ($C_{12}H_{18}O_8S_4$)	H^+	HL/H.L	5.05
		$H_2L/HL.H$	4.04
		$H_3L/H_2L.H$	3.70
		$H_4L/H_3L.H$	3.55
	Mg^{2+}	ML/M.L	3.2
		MHL/M.HL	2.5
	Ca^{2+}	ML/M.L	3.6
		MHL/M.HL	(3.96)
	Ni^{2+}	ML/M.L	2.4
	Cu^{2+}	ML/M.L	4.9
	Cd^{2+}	ML/M.L	3.6
	Hg^{2+}	ML/M.L	16.9
n = 3 1,5-Pentane ($C_{13}H_{20}O_8S_4$)	H^+	HL/H.L	4.90
		$H_2L/HL.H$	4.30
		$H_3L/H_2L.H$	3.75
		$H_4L/H_3L.H$	3.60
	Mg^{2+}	ML/M.L	3.3
	Ca^{2+}	ML/M.L	3.7
		MHL/M.HL	(4.3)
	Ni^{2+}	ML/M.L	2.7
	Cu^{2+}	ML/M.L	5.0
	Cd^{2+}	ML/M.L	3.4
	Hg^{2+}	ML/M.L	19.0

Bibliography: 75JB

$$\begin{array}{c} HO_2CCH_2CH_2S \\ HO_2CCH_2CH_2S \end{array} CH(CH_2)_n CH \begin{array}{c} SCH_2CH_2CO_2H \\ SCH_2CH_2CO_2H \end{array}$$

$C_nH_mO_8S_4$	3,3',3'',3''' - (Alkanediylidenetetrathio)tetrapropanoic acid			H_4L

n =	Metal ion	Equilibrium	Log K 25°, 0.1
n = 0 Ethane ($C_{14}H_{22}O_8S_4$)	Mg^{2+}	ML/M.L	1.92
	Ca^{2+}	ML/M.L	2.59
	Ni^{2+}	ML/M.L	2.79
	Zn^{2+}	ML/M.L	3.00
	Hg^{2+}	ML/M.L	17.3

3,3',3",3"', -(Alkanediylidenetetrathio)tetrapropanoic acid (continued)

n =	Metal ion	Equilibrium	Log K 25°, 0.1
n = 1	H^+	HL/H.L	5.63
1,3-Propane		$H_2L/HL.H$	4.94
$(C_{15}H_{24}O_8S_4)$		$H_3L/H_2L.H$	4.34
		$H_4L/H_3L.H$	4.19
	Mg^{2+}	ML/M.L	2.20
	Ca^{2+}	ML/M.L	2.79
	Ni^{2+}	ML/M.L	3.83
	Zn^{2+}	ML/M.L	2.92
	Hg^{2+}	ML/M.L	17.5
n = 2	H^+	HL/H.L	5.38
1,4-Butane		$H_2L/HL.H$	4.82
$(C_{16}H_{26}O_8S_4)$		$H_3L/H_2L.H$	4.14
		$H_4L/H_3L.H$	3.91
	Mg^{2+}	ML/M.L	2.20
	Ca^{2+}	ML/M.L	2.55
	Ni^{2+}	ML/M.L	3.87
	Zn^{2+}	ML/M.L	3.67
	Hg^{2+}	ML/M.L	17.7
n = 3	H^+	HL/H.L	5.45
1,5-Pentane		$H_2L/HL.H$	4.80
$(C_{17}H_{28}O_8S_4)$		$H_3L/H_2L.H$	4.29
		$H_4L/H_3L.H$	3.88
	Mg^{2+}	ML/M.L	2.21
	Ca^{2+}	ML/M.L	3.11
	Ni^{2+}	ML/M.L	3.65
	Zn^{2+}	ML/M.L	2.62
	Hg^{2+}	ML/M.L	17.9

Bibliography:

H^+ 74JP $Mg^{2+}-Hg^{2+}$ 75PJ

$C_{16}H_{18}O_8S_4$	1,2-Bis[bis(carboxymethylthio)methyl]benzene [(1,2-phthalylidenetetrathio)tetraacetic acid]	H_4L

Metal ion	Equilibrium	Log K 25°, 0.1
H^+	HL/H.L	4.97
	$H_2L/HL.H$	4.23
	$H_3L/H_2L.H$	3.71
	$H_4L/H_3L.H$	3.24

1,2-Phthalylidenetetrathiotetraacetic acid (continued)

Metal ion	Equilibrium	Log K 25°, 0.1
Mg^{2+}	ML/M.L	3.8
	MHL/M.HL	3.5
Ca^{2+}	ML/M.L	3.5
Ni^{2+}	ML/M.L	3.2
Cu^{2+}	ML/M.L	6.5
Zn^{2+}	ML/M.L	2.2
Hg^{2+}	$MH_2L/M.H_2L$	18.8
	$M(OH)_2L/M.(OH)^2.L$	45
Pb^{2+}	ML/M.L	6.0

Bibliography:

H^+,Ca^{2+},Hg^{2+},Pb^{2+} 74JB Mg^{2+},Ni^{2+}-Zn^{2+} 75JB

$$CH_3OPO_3HPO_3HPO_3H_2$$

$CH_7O_{10}P_3$ Triphosphoric acid methylester (methyltriphosphate) H_4L

Metal ion	Equilibrium	Log K 20°, 0.1
H^+	HL/H.L	6.45
Cu^{2+}	ML/M.L	6.17
	MHL/M.HL	3.65

Bibliography: 64SBE

$$\overset{\displaystyle OH}{\underset{\displaystyle}{|}}$$
$$HOCH_2CHCH_2OPO_3H_2$$

$C_3H_9O_6P$ Propane-1,2,3-triol-1-(dihydrogen phosphate) H_2L
                             (glycerol-1-phosphate)

Metal ion	Equilibrium	Log K 20°, 0.1	Log K 25°, 0	ΔH 25°, 0	ΔS 25°, 0
H^+	HL/H.L	6.08	6.656	0.7	33
Mg^{2+}	ML/M.L	1.80			
Ca^{2+}	ML/M.L	1.66			

Bibliography:

H^+ 57SA, 58DG   Other reference: 71MM

Mg^{2+}, Ca^{2+} 57SA

$$\overset{\displaystyle OPO_3H_2}{\underset{\displaystyle}{|}}$$
$$HOCH_2CHCH_2OH$$

$C_3H_9O_6P$ Propane-1,2,3-triol-2-(dihydrogen phosphate) H_2L
                             (glycerol-2-phosphate)

Metal ion	Equilibrium	Log K 25°, 0	ΔH 25°, 0	ΔS 25°, 0
H^+	HL/H.L	6.650	0.4	32
	H_2L/HL.H	1.335	2.9	16
Mg^{2+}	ML/M.L	2.49	3.4	23

Bibliography:

H^+ 54AC

Mg^{2+} 54CC   Other reference: 71MM

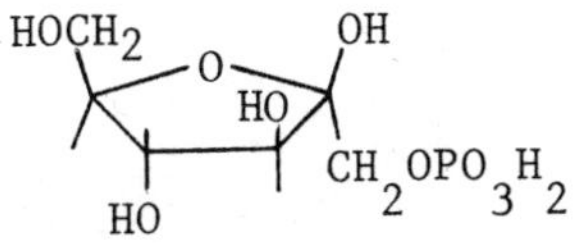

$$HOCH_2 \quad O \quad OH$$
$$HO$$
$$CH_2OPO_3H_2$$
$$HO$$

$C_6H_{13}O_9P$		D(-)-Fructose-1-(dihydrogen phosphate)	H_2L
Metal ion	Equilibrium	Log K 20°, 0.1	
H^+	HL/H.L	5.84	
Mg^{2+}	ML/M.L	1.59	
Ca^{2+}	ML/M.L	1.47	

Bibliography: 57SA

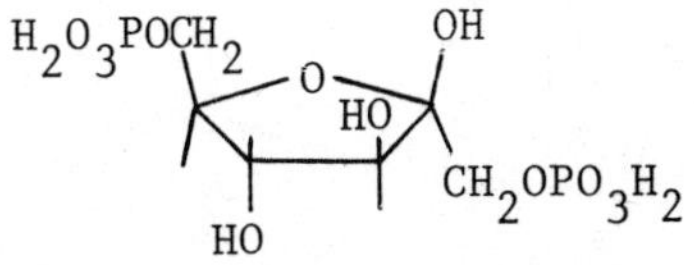

$$H_2O_3POCH_2 \quad O \quad OH$$
$$HO$$
$$CH_2OPO_3H_2$$
$$HO$$

$C_6H_{14}O_{12}P_2$		D(-)-Fructose-1,6-bis(dihydrogen phosphate)		H_4L
Metal ion	Equilibrium	Log K 25°, 0.08	Log K 25°, 0	
H^+	HL/H.L	6.66[y]	7.28	
	$H_2L/HL.H$	5.86[y]	6.43	
Mg^{2+}	ML/M.L	2.7 [y]		
	MHL/M.HL	2.12[y]		

[y] NaCl used as background electrolyte.

Bibliography: 65M

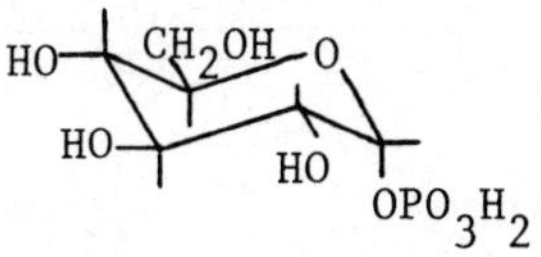

$$HO \quad CH_2OH \quad O$$
$$HO$$
$$HO$$
$$OPO_3H_2$$

$C_6H_{13}O_9P$		α-D(+)-Glucose-1-(dihydrogen phosphate)			H_2L
Metal ion	Equilibrium	Log K 25°, 0.1	Log K 25°, 0	ΔH 25°, 0	ΔS 25°, 0
H^+	HL/H.L		6.504 +0.01	0.4	31
	$H_2L/HL.H$	1.46			
Mg^{2+}	ML/M.L		2.47	2.9	21
Ca^{2+}	ML/M.L		2.50	2.4	20
Mn^{2+}	ML/M.L	2.19			
Co^{2+}	ML/M.L	2.18			
Zn^{2+}	ML/M.L	2.37			

α-D(+)-Glucose-1-(dihydrogen phosphate) (continued)

Bibliography:

H^+ 52TM,55AC,66MT Mn^{2+}-Zn^{2+} 66DT

Mg^{2+} 54CC   Other references: 66IR,71MM

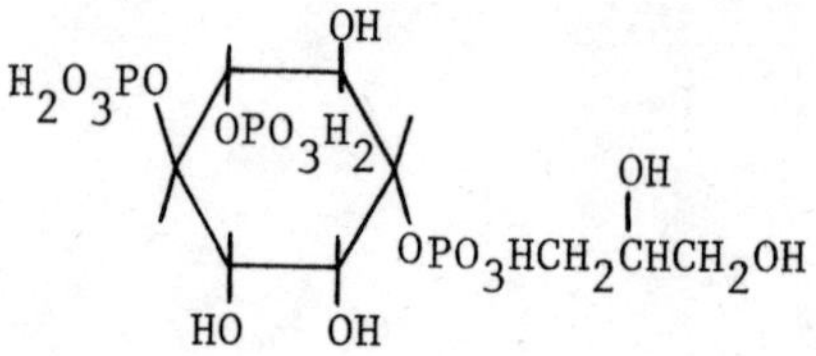

Ca^{2+} 56CD

Metal ion	Equilibrium	Log K 25°, 0.1

$C_9H_{21}O_{17}P_3$ 1-(Glycerylphosphoryl)inositol-3,4-bis(dihydrogen phosphate) H_5L

Metal ion	Equilibrium	Log K 25°, 0.1
H^+	HL/H.L	8.05[x]
	H_2L/HL.H	5.70[x]
Mg^{2+}	ML/M.L	3.5 [x]
	MHL/M.HL	2.4 [x]
Ca^{2+}	ML/M.L	3.3 [x]
	MHL/M.HL	2.2 [x]

[x] $(C_3H_7)_4$NI used as background electrolyte.

Bibliography: 65HF

$C_8H_{19}O_2S_2P$ Phosphorodithioic acid 0,0'-dialkyl ester   HL
                          (dialkyldithiophosphoric acid)

R =	Metal ion	Equilibrium	Log K 25°, 1.0
dibutyl	H^+	HL/H.L	0.08[y]
	Zn^{2+}	ML_2/M.L^2	3.81[y]
bis(2-methyl-propyl)	H^+	HL/H.L	−0.04[y]
	Zn^{2+}	ML_2/M.L^2	4.00[y]

[y] NaCl used as background electrolyte.

Bibliography: 63HZ

B. PHOSPHONIC ACIDS

$$HOCH_2PO_3H_2$$

Hydroxymethanephosphonic acid $\quad H_2L$

CH₅O₄P → CH_5O_4P

Metal ion	Equilibrium	Log K 25°, 0.1
H^+	HL/H.L	7.01^x
	H_2L/HL.H	1.7^x
Li^+	ML/M.L	0.72^x
Na^+	ML/M.L	0.61^x
K^+	ML/M.L	0.34^x
Mg^{2+}	ML/M.L	1.92^x
Ca^{2+}	ML/M.L	1.87^x

[x] $(CH_3)_4NCl$ used as background electrolyte.

Bibliography: 72WF

Other reference: 72EZc

$$H_2O_3PCH_2PO_3H_2$$

Methylenediphosphonic acid $\quad H_4L$

$CH_6O_6P_2$

Metal ion	Equilibrium	Log K 25°, 0.1	Log K 25°, 0.5	Log K 25°, 0	ΔH 25°, 0.5	ΔS 25°, 0.5
H^+	HL/H.L	10.57^x-0.1 10.33^h	10.54^x	10.96 -0.3	$(-3)^{t,x}$	$(40)^x$
	H_2L/HL.H	7.00^x+0.3 6.87^h	6.87^x	7.35 +0.1	$(2)^{t,x}$	$(40)^x$
	H_3L/H_2L.H	2.77^x±0.02 2.5^h 1.7^x	2.49^x $2.67^{c,x}$ $1.7^{c,x}$	3.05 -0.1 2.2	$(3)^{t,x}$ $(-3)^{t,x}$	$(20)^x$
	H_4L/H_3L.H					
Li^+	ML/M.L		2.48^x		0.7^x	14^x
	MHL/M.HL		$(0.20)^x$			
Na^+	ML/M.L		1.13^x		1.6^x	11^x
	MHL/M.HL		0.39^x			
K^+	ML/M.L		1.02^x		1.4^x	10^x
	MHL/M.HL		0.20^x			
Cs^+	ML/M.L		0.84^x			
	MHL/M.HL		0.04^x			
Mg^{2+}	ML/M.L		5.8^x			
	MHL/M.HL		2.92^x			
Ca^{2+}	ML/M.L		4.7^x		2.6^x	30^x
	MHL/M.HL		2.46^x			

[c] 25°, 1.0;   [h] 20°, 0.1;   [t] 25-50°, 0.1;   [x] $(CH_3)_4N^+$ salt used as background electrolyte.

Bibliography:

H^+ 50SZ,62IM,67CI,67GQ

Li^+-Cs^+ 67CI,68CI

Mg^{2+},Ca^{2+} 68CI

Other references: 63KE,67KL,68DM

$$\overset{\displaystyle CH_3}{\underset{\displaystyle H_2O_3PCHPO_3H_2}{|}}$$

$C_2H_8O_6P_2$		Ethane-1,1-diphosphonic acid				H_4L
Metal ion	Equilibrium	Log K 25°, 0.1	Log K 25°, 0.5	Log K 25°, 0	ΔH 25°, 0.5	ΔS 25°, 0.5
H^+	HL/H.L	11.7 [x]	11.5 [x]	12.0		
	$H_2L/HL.H$	7.22 [x]	7.18 [x]	7.49		
	$H_3L/H_2L.H$	2.88 [x]	2.66 [x]	3.14		
Li^+	ML/M.L		3.1 [x]		−0.4 [x]	13 [x]
	MHL/M.HL		0.99 [x]			
Na^+	ML/M.L		1.5 [x]		1.7 [x]	(13) [x]
	MHL/M.HL		0.50 [x]			
K^+	ML/M.L		1.2 [x]		0.8 [x]	8 [x]
	MHL/M.HL		0.28 [x]			
Cs^+	ML/M.L		1.0 [x]			
	MHL/M.HL		0.09 [x]			
Mg^{2+}	ML/M.L		6.3 [x]			
	MHL/M.HL		2.99 [x]			
Ca^{2+}	ML/M.L		5.2 [x]		2.2 [x]	31 [x]
	MHL/M.HL		2.74 [x]			

[x] $(CH_3)_4NCl$ used as background electrolyte.

Bibliography:

H^+ 67CI, 67GQ Mg^{2+}, Ca^{2+} 68CI

Li^+-Cs^+ 67CI, 68CI

$$\overset{\displaystyle CH_3}{\underset{\displaystyle \underset{\displaystyle CH_3}{\overset{\displaystyle |}{H_2O_3PCPO_3H_2}}}{|}}$$

$C_3H_{10}O_6P_2$		Propane-2,2-diphosphonic acid				H_4L
Metal ion	Equilibrium	Log K 25°, 0.1	Log K 25°, 0.5	Log K 25°, 0	ΔH 25°, 0.5	ΔS 25°, 0.5
H^+	HL/H.L	11.8 [a]	(12.4) [x]	12.1		
	$H_2L/HL.H$	7.78 [a]	7.75 [x]	8.04		
	$H_3L/H_2L.H$	2.98 [a]	2.94 [x]	3.16		
Li^+	ML/M.L		3.8 [x]		−1.3 [x]	13 [x]
	MHL/M.HL		1.38 [x]			
Na^+	ML/M.L		2.08 [x]		0.3 [x]	11 [x]
	MHL/M.HL		0.57 [x]			
K^+	ML/M.L		1.60 [x]		0.0 [x]	8 [x]
	MHL/M.HL		0.35 [x]			
Cs^+	ML/M.L		1.40 [x]			
	MHL/M.HL		(0.35) [x]			
Mg^{2+}	ML/M.L		6.8 [x]			
	MHL/M.HL		3.33 [x]			

[x] $(CH_3)_4NCl$ used as background electrolyte.

Propane-2,2-diphosphonic acid (continued)

Metal ion	Equilibrium	Log K 25°, 0.1	Log K 25°, 0.5	Log K 25°, 0	ΔH 25°, 0.5	ΔS 25°, 0.5
Ca^{2+}	ML/M.L		6.3[x]		-1.0[x]	26[x]
	MHL/M.HL		3.14[x]			

---

[x] $(CH_3)_4NCl$ used as background electrolyte.

Bibliography:

H^+ 67CI,67GQ

Mg^{2+},Ca^{2+} 68CI

Li^+-Cs^+ 67CI,68CI

$$\begin{array}{c} CH_3 \\ | \\ H_2O_3PCPO_3H_2 \\ | \\ OH \end{array}$$

$C_2H_8O_7P_2$		1-Hydroxyethane-1,1-diphosphonic acid				H_4L
Metal ion	Equilibrium	Log K 25°, 0.1	Log K 25°, 0.5	Log K 25°, 0	ΔH 25°, 0.5	ΔS 25°, 0.5
H^+	HL/H.L	11.1[x]±0.1	(11.4)[x]	11.5		
	$H_2L/HL.H$	7.01[x]±0.02	6.97[x]	7.31		
	$H_3L/H_2L.H$	2.80[x]-0.5	2.54[x]	3.03		
Li^+	ML/M.L		3.35[x]		-0.6[x]	13[x]
	MHL/M.HL	1.33[x]	1.08[x]			
	$M_2L/M^2.L$	4.78[x]				
Na^+	ML/M.L		2.07[x]		0.4[x]	11[x]
	MHL/M.HL	0.81[x]	0.54[x]			
	$M_2L/M^2.L$	2.66[x]				
K^+	ML/M.L		1.79[x]		0.4[x]	9[x]
	MHL/M.HL	0.62[x]	0.36[x]			
	$M_2L/M^2.L$	2.04[x]				
Cs^+	ML/M.L		1.6[x]			
	MHL/M.HL		0.24[x]			
Mg^{2+}	ML/M.L	7.28[x]	(6.4)[x]			
	MHL/M.HL	3.81[x]	(3.32)[x]			
	$M_2L/M^2.L$	10.8[x]				
Ca^{2+}	ML/M.L	6.84[x]	(5.7)[x]		-0.5[x]	25[x]
	MHL/M.HL	3.54[x]	(3.6)[x]			
	$M_2L/M^2.L$	12.2[x]				
Cu^{2+}	ML/M.L	11.84[x]				
	MHL/M.HL	7.47[x]				
	$MH_2L/M.H_2L$	4.80[x]				

---

[x] $(CH_3)_4N^+$ salt used as background electrolyte.

Bibliography:

H^+ 67CI,67GQ,71WF,72WF

Cu^{2+} 71WF

Li^+-Cs^+ 67CI,68CI,72WF

Other references: 67KL,71CA,71GC

Mg^{2+},Ca^{2+} 68CI,72WF

$$H_2O_3PCH_2CH_2PO_3H_2$$

$C_2H_8O_6P_2$		Ethylenediphosphonic acid			H_4L
Metal ion	Equilibrium	Log K 25°, 0.1	Log K 25°, 1.0	Log K 25°, 0	
H^+	HL/H.L	9.08 [x]	8.96 [x]	9.28	
	H_2L/HL.H	(7.50) [x]	(7.42) [x]	(7.62)	
	H_3L/H_2L.H	(2.96) [x]	(2.74) [x]	(3.18)	
	H_4L/H_3L.H	1.5 [x]	1.5 [x]	1.5	
Mg^{2+}	ML/M.L		2.85 [x]		
	MHL/M.HL		2.67 [x]		
Ca^{2+}	ML/M.L		2.80 [x]		
	MHL/M.HL		2.60 [x]		

[x] $(CH_3)_4$NBr used as background electrolyte.

Bibliography: 62IM

$$\overset{\textstyle CH_3}{\underset{\textstyle }{|}}$$
$$H_2O_3PCHCH_2PO_3H_2$$

$C_3H_{10}O_6P_2$		Propane-1,2-diphosphonic acid	H_4L
Metal ion	Equilibrium	Log K 20°, 0.1	
H^+	HL/H.L	9.27	
	H_2L/HL.H	7.00	
	H_3L/H_2L.H	2.6	
Mg^{2+}	ML/M.L	3.04	
	MHL/M.HL	2.08	
Ca^{2+}	ML/M.L	2.65	
	MHL/M.HL	1.7	
Ba^{2+}	ML/M.L	2.20	
	MHL/M.HL	1.3	

Bibliography: 51SR

$$H_2O_3PCH_2CH_2CH_2PO_3H_2$$

$C_3H_{10}O_6P_2$		Trimethylenediphosphonic acid			H_4L
Metal ion	Equilibrium	Log K 25°, 0.1	Log K 25°, 1.0	Log K 25°, 0	
H^+	HL/H.L	8.43 [x] 8.35 [h]	8.33 [x]	8.63	
	H_2L/HL.H	7.50 [x] 7.34 [h]	7.41 [x]	7.65	
	H_3L/H_2L.H	2.81 [x] 2.6 [h]	2.69 [x]	3.06	
	H_4L/H_3L.H	1.6 [x]	1.7 [x]	1.6	

[h] 20°, 0.1;   [x] $(CH_3)_4$NBr used as background electrolyte.

Trimethylenediphosphonic acid (continued)

Metal ion	Equilibrium	Log K 25°, 0.1	Log K 25°, 1.0	Log K 25°, 0
Mg^{2+}	ML/M.L	2.84^h		
	MHL/M.HL	2.08^h		
Ca^{2+}	ML/M.L	2.58^h		
	MHL/M.HL	1.8^h		
Ba^{2+}	ML/M.L	2.34^h		
	MHL/M.HL	1.6^h		

^{h} 20°, 0.1

Bibliography:

H^+ 50SZ,62IM $Mg^{2+}-Ba^{2+}$ 51SR

$$H_2O_3PCH_2CH_2CH_2CH_2PO_3H_2$$

$C_4H_{12}O_6P_2$	Tetramethylenediphosphonic acid			H_4L
Metal ion	Equilibrium	Log K 25°, 0.1	Log K 25°, 1.0	Log K 25°, 0
H^+	HL/H.L	8.38^x 8.38^h	8.30^x	8.58
	H_2L/HL.H	7.58^x 7.54^h	7.47^x	7.78
	H_3L/H_2L.H	2.85^x 2.7^h	2.70^x	3.19
	H_4L/H_3L.H	1.7^x	1.7^x	1.7
Mg^{2+}	ML/M.L	2.77^h		
	MHL/M.HL	2.05^h		
Ca^{2+}	ML/M.L	2.54^h		
	MHL/M.HL	1.7^h		
Ba^{2+}	ML/M.L	2.28^h		
	MHL/M.HL	1.5^h		

^{h} 20°, 0.1; ^{x} $(CH_3)_4NBr$ used as background electrolyte.

Bibliography:

H^+ 50SZ,62IM $Mg^{2+}-Ba^{2+}$ 51SR

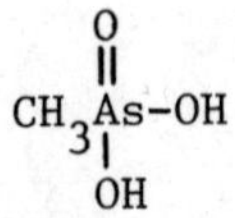

$$\begin{array}{c} O \\ \parallel \\ CH_3As\!-\!OH \\ \mid \\ OH \end{array}$$

CH$_5$O$_3$As   Methylarsonic acid   H$_2$L

Metal ion	Equilibrium	Log K 25°, 0.1
H$^+$	HL/H.L	8.66
	H$_2$L/HL.H	3.96
CH$_3$Hg$^+$	ML/MOH.HL	2.0

Bibliography: 66GD

Other references: 35BB,71GH

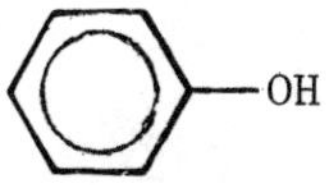

C_6H_6O		Hydroxybenzene (phenol)				HL
Metal ion	Equilibrium	Log K 25°, 0,1	Log K 25°, 1.0	Log K 25°, 0	ΔH 25°, 0	ΔS 25°, 0
H^+	HL/H.L	9.82 ±0.04	$(9.64)^b$ 9.52^e	9.98 ±0.01	-5.5 -6.1^a	27 24^a
Sc^{3+}	ML/M.L	7.17				
Y^{3+}	ML/M.L	2.40				
La^{3+}	ML/M.L	1.51				
UO_2^{2+}	ML/M.L	5.8^h				
Fe^{3+}	ML/M.L	7.78	$(7.81)^b$	8.20	-0.5^a	33^a
Ag^+	ML/M.L	0.18	0.34		$(-1)^r$	$(-2)^c$

^{a} 25°, 0.1; ^{b} 25°, 0.5; ^{c} 25°, 1.0; ^{e} 25°, 3.0; ^{h} 20°, 0.1; ^{r} 0-25°, 1.0

Bibliography:

H^+ 48WB,55Aa,59B,59FF,59FH, 67Z,68Mc,69DM

Sc^{3+} 71UK

Y^{3+},La^{3+} 66PM

UO_2^{2+} 65BS,66BS

Fe^{3+} 55M,67Mc,68Mc,69DM

Ag^+ 38WL,49KAb

Other reference: 69LR

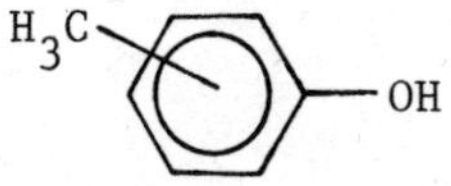

C_7H_8O			Methylphenol (cresol)			HL
Isomer	Metal ion	Equilibrium	Log K 25°, 0.5	Log K 25°, 0	ΔH 25°, 0	ΔS 25°, 0
3-methyl	H^+	HL/H.L		10.09 ±0.01	-5.5	28
	Fe^{3+}	ML/M.L		8.51		
4-methyl	H^+	HL/H.L	9.87	10.26 ±0.01	-5.5	29
	Fe^{3+}	ML/M.L	8.33	9.26		

Bibliography:

H^+ 56B,57HK,62CL,68Mc,69P   Other reference: 75GJ

Fe^{3+} 55M,65EH,68Mc

C_6H_5OX				Halogenophenol			HL
X =	Metal ion	Equilibrium	Log K 25°, 0.1	Log K 25°, 0	ΔH 25°, 0	ΔS 25°, 0	
2-Fluoro	H^+	HL/H.L	8.49	8.71			
	Fe^{3+}	ML/M.L	7.19	7.33			
3-Fluoro	H^+	HL/H.L	8.81	9.21			
	Fe^{3+}	ML/M.L	(7.77)	(7.77)			
4-Fluoro	H^+	HL/H.L	9.46	9.91			
	Fe^{3+}	ML/M.L	8.29	8.38			
2-Chloro	H^+	HL/H.L	8.29 ±0.04	8.53 ±0.00	-4.5 -0.1	24	
					-4.8[a]	22[a]	
	Fe^{3+}	ML/M.L	6.09 +1	7.32	0.2[a]	28[a]	
3-Chloro	H^+	HL/H.L	8.78 ±0.02	9.13	-6.1[a]	19[a]	
	Fe^{3+}	ML/M.L	6.90 +0.6	7.89	-0.2[a]	31[a]	
4-Chloro	H^+	HL/H.L	9.14 ±0.04	9.42	-5.6	24	
					-6.1[a]	21[a]	
	Fe^{3+}	ML/M.L	6.97 +1	7.92	-1.2[a]	28[a]	
2-Bromo	H^+	HL/H.L	8.22	8.44			
	Fe^{3+}	ML/M.L	6.98	7.19			
3-Bromo	H^+	HL/H.L	8.75	9.03			
	Fe^{3+}	ML/M.L	7.65				
4-Bromo	H^+	HL/H.L	9.06 9.02[b]	9.35 ±0.01			
	Fe^{3+}	ML/M.L	8.00 7.72[b]	8.10			
2-Iodo	H^+	HL/H.L		8.51			
	Fe^{3+}	ML/M.L		7.43			
3-Iodo	H^+	HL/H.L	8.74	9.06			
	Fe^{3+}	ML/M.L	7.57				
4-Iodo	H^+	HL/H.L		9.31			
	Fe^{3+}	ML/M.L		8.63			

[a] 25°, 0.1; [b] 25°, 0.5

Halogenophenol (continued)

Bibliography:

H^+ 48CD,59FH,61BR,62CL,66JM,68Mc,69DM   Other reference: 69NA

Fe^{3+} 55M,65EH,66JM,67Mc,68Mc,69DM

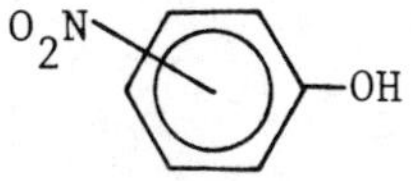

$C_6H_5O_qN_r$			Nitrophenol			HL
	Metal		Log K	Log K	ΔH	ΔS
Isomer	ion	Equilibrium	25°, 0.1	25°, 0	25°, 0	25°, 0
2-Nitro	H^+	HL/H.L	7.05 ±0.01	7.21 ±0.01	−4.5 −0.1	18
($C_6H_5O_3N$)	La^{3+}	ML/M.L	2.20			
	Fe^{3+}	ML/M.L	5.99			
3-Nitro	H^+	HL/H.L	8.09 ±0.05	8.39 ±0.01	−4.7	23
($C_6H_5O_3N$)			8.03[b]		−4.8[a]	21[a]
	Fe^{3+}	ML/M.L _	6.09 +0.9	7.22	1.0[a]	31[a]
			6.69[b]			
4-Nitro	H^+	HL/H.L	6.90 −0.01	7.15 ±0.01	−4.5 −0.1	18
($C_6H_5O_3N$)			6.85[b]		−4.8[a]	15[a]
	UO_2^{2+}	ML/M.L	4.40[h]			
	Fe^{3+}	ML/M.L	5.06 +0.7	5.74	0.5[a]	24[a]
			5.60[b]			
2,4-Dinitro	H^+	HL/H.L	3.93	4.11		
($C_6H_4O_5N_2$)	La^{3+}	ML/M.L	1.0			

[a] 25°, 0.1;   [b] 25°, 0.5;   [h] 20°, 0.1

Bibliography:

H^+ 48CD,48WB,54BS,55RB,56B,59FF,59FH, UO_2^{2+} 67Ba
          61BR,63RP,66JM,66PM,68Mc,69DM
                                                    Fe^{3+} 55M,66JM,67Mc,68Mc,69DM
La^{3+} 66PM
                                                    Other reference: 72PG

C_7H_5ON Cyanophenol   HL

Isomer	Metal ion	Equilibrium	Log K 25°, 0.1	Log K 25°, 0	ΔH 25°, 0	ΔS 25°, 0
2-Cyano	H^+	HL/H.L	6.86		-4.0[a]	18[a]
	Fe^{3+}	ML/M.L	5.54		0.7[a]	27[a]
3-Cyano	H^+	HL/H.L	8.36	8.59 ±0.02	-5.0 / -5.3[a]	22 / 20[a]
	Fe^{3+}	ML/M.L	6.31		0.9[a]	31[a]
4-Cyano	H^+	HL/H.L	7.70	7.96 ±0.01	-4.8 / -4.8[a]	20 / 19[a]
	Fe^{3+}	ML/M.L	5.80		-0.2[a]	26[a]

[a] 25°, 0.1

Bibliography:

H^+ 48WB,59FF,64KO,69DM Fe^{3+} 69DM

$C_6H_6O_4S$ 4-Hydroxybenzenesulfonic acid (4-sulfophenol) H_2L

Metal ion	Equilibrium	Log K 25°, 0.1	Log K 25°, 0.5	Log K 25°, 0	ΔH 25°, 0	ΔS 25°, 0
H^+	HL/H.L	8.66	8.41	9.053	-4.0	28
Fe^{3+}	ML/M.L		6.72			

Bibliography:

H^+ 43BS,68Mc,73YB Fe^{3+} 68Mc

$C_6H_7O_4P$ 2-Hydroxybenzenephosphonic acid (2-phosphonophenol) H_3L

Metal ion	Equilibrium	Log K 25°, 0.1	Log K 25°, 0.6	Log K 25°, 0
H^+	HL/H.L			15.40
	$H_2L/HL.H$	6.23		6.48
	$H_3L/H_2L.H$	1.54		1.66
UO_2^{2+}	MHL/M.HL	5.81		5.92
Fe^{3+}	MHL/M.HL	8.69	8.19	

2-Phosphonophenol (continued)

Bibliography:

H^+ 740,740a UO_2^{2+},Fe^{3+} 740a

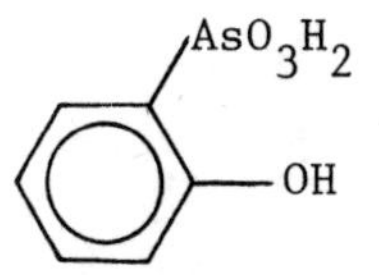

$C_6H_7O_4As$		2-Hydroxybenzenearsonic acid (2-arsonophenol)		H_3L
Metal ion	Equilibrium	Log K 25°, 0.1	Log K 25°, 0	
H^+	HL/H.L		13.27	
	$H_2L/HL.H$	7.50	7.92 −0.2	
	$H_3L/H_2L.H$	3.83	4.02 ±0.02	
UO_2^{2+}	MHL/M.HL	8.64	8.75	
	$M(HL)_2/M.(HL)^2$		13.86	
Fe^{3+}	MHL/M.HL	11.74		

Bibliography:

H^+ 6800,6900,7300 Fe^{3+} 70Ka

UO_2^{2+} 6900   Other reference: 73VG

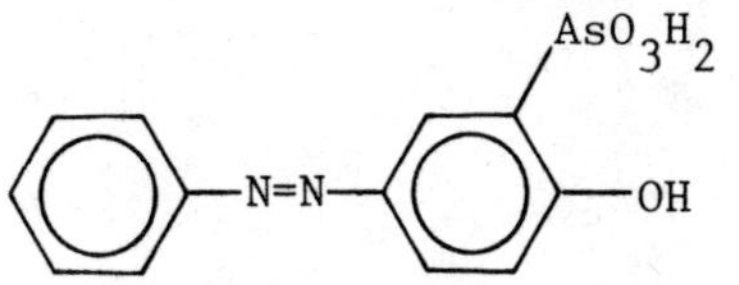

$C_{12}H_{11}O_4N_2As$		2-Hydroxy-5-(phenylazo)benzenearsonic acid (3-arsono-4-hydroxyazobenzene)		H_3L
Metal ion	Equilibrium	Log K 25°, 0.08	Log K 25°, 0	
H^+	HL/H.L		11.82	
	$H_2L/HL.H$	6.88	7.29	
	$H_3L/H_2L.H$	3.43	3.59	
Fe^{3+}	MHL/M.HL	11.31		

Bibliography: 740b

$$\text{[benzene ring with } CO_2H \text{ and } OH \text{ substituents]}$$

$C_7H_6O_3$	2-Hydroxybenzoic acid (salicylic acid)				H_2L
Metal ion	Equilibrium	Log K 25°, 0.1	Log K 25°, 1.0	Log K 25°, 0	ΔH 25°, 0	ΔS 25°, 0
H^+	HL/H.L	13.4 ±0.0 13.0 [t]	13.15 13.12 [e]	13.74	-8.56 ±0.05	34.2
	H_2L/HL.H	2.81 ±0.01 2.78 [b] ±0.04 2.81 [t]	2.78 ±0.02 3.16 [e] ±0.02	2.97 ±0.00	-0.77 ±0.04 -0.9 [a]	11.0
Be^{2+}	ML/M.L $ML_2/M.L^2$	12.37 [h] 22.02 [h]				
Ca^{2+}	MHL/M.HL	0.15 [r] -0.01		0.36		
Ba^{2+}	MHL/M.HL			0.21		
La^{3+}	MHL/M.HL			2.08		
Th^{4+}	MHL/M.HL $M(HL)_2/M.(HL)^2$ $M(HL)_3/M.(HL)^3$ $M(HL)_4/M.(HL)^4$	4.25 7.60 10.05 11.60				
UO_2^{2+}	ML/M.L $ML_2/M.L^2$ MHL/M.HL	12.08 [h] 20.83 [h] 2.2				
Cr^{2+}	ML/M.L $ML_2/M.L^2$	8.41 15.36				
Mn^{2+}	ML/M.L $ML_2/M.L^2$	5.90 [s] 9.8 [s]				
Fe^{2+}	ML/M.L $ML_2/M.L^2$	6.55 [s] 11.2 [s]				
Co^{2+}	ML/M.L $ML_2/M.L^2$	6.72 [s] 11.4 [s]				
Ni^{2+}	ML/M.L $ML_2/M.L^2$	6.95 [s] 11.7 [s]				
Cu^{2+}	ML/M.L $ML_2/M.L^2$	10.62 ±0.02 10.60 [s] 18.45 [s]	10.13 [t] 18.2 [t]		-4.4 [b]	34 [a]
Fe^{3+}	ML/M.L $ML_2/M.L^2$ $ML_3/M.L^3$ MHL/M.HL	16.3 16.35 [s] 16.1 [b] 28.25 [s] 4.4	15.81 [e] 27.49 [e] 35.31 [e]	17.44	(-5) [u]	(55) [e]
VO^{2+}	ML/M.L $ML_2/M.L^2$	13.38 (12.7) [h] (22.4) [h]			(-2) [v]	(50) [a]

[a] 25°, 0.1;   [b] 25°, 0.5;   [e] 25°, 3.0;   [h] 20°, 0.1;   [r] 25°, 0.16;   [s] 20°, 0.15;   [t] 37°, 0.15;
[u] 15-35°, 3.0;   [v] 25-35°, 0.1

Salicylic acid (continued)

Metal ion	Equilibrium	Log K 25°, 0.1	Log K 25°, 1.0	Log K 25°, 0	ΔH 25°, 0	ΔS 25°, 0
Zn^{2+}	$ML/M.L$	6.85^s				
Cd^{2+}	$ML/M.L$	5.55^s				
$B(III)$	$M(OH)_2L.H/M(OH)_3.H_2L$	-1.51				
	$ML_2.H/M(OH)_3.(H_2L)^2$	0.8				
Al^{3+}	$ML/M.L$	12.9^h				
	$ML_2/M.L^2$	23.2^h				
	$ML_3/M.L^3$	29.8^h				

^{h} 20°, 0.1; ^{s} 20°, 0.15

Bibliography:

H^+ 31LAb,38D,39EW,51BW,53H,54Aa,55Aa, 56Aa,58P,61PS,61TS,62BK,63EM,64EI, 66LM,66P,67PSS,69CMa,73L,70KC,73A, 75DI

Be^{2+} 67BZ

$Ca^{2+}-Ba^{2+}$ 46J,51BW,54S

La^{3+} 68EN

Th^{4+} 56Ha,60R

UO_2^{2+} 56H,67B

Cr^{2+} 61PS

$Mn^{2+}-Ni^{2+},Zn^{2+},Cd^{2+}$ 58P

Cu^{2+} 58P,66LM,67PSS,69CMa,70GMa

Fe^{3+} 58P,61JS, 63EMa,66P,70MR,70SMb

VO^{2+} 66MM,71ZB

$B(III)$ 69HH

Al^{3+} 69HBa

Other references: 42S,45B,47B,52B,53BB,54SL, 56CF,57BDH,57GL,57PC,59DA,59DG,59DGa, 59FE,60BSa,60C,61CO,61DA,62BK,62BV, 62BVG,62CTC,62LZ,63DN,63I,63ZG,64DA, 64E,64PPB,64PPD,65DA,65KB,65KJ,66GGJ, 66OTa,66RS,67AS,67SBa,69H,69KM,70KKM, 70KM,71H,71ZB,72L,72PB,72PS,72RMG,73JK, 73PL,73RM,73VV,74CS,74KKb,75BSb,75DJ, 75JK,75MJa,75ST,75SV

$C_mH_nO_3$				H_2L

Alkylsalicylic acid

R =	Metal ion	Equilibrium	Log K 25°, 0.1	Log K 25°, 0
3-Methyl	H^+	$HL/H.L$		14.59
$(C_8H_8O_3)$		$H_2L/HL.H$	2.82	2.95
	Fe^{3+}	$ML/M.L$		18.13
		$ML.H/M.HL$	2.58	2.53
		$ML_2.H/ML.HL$	0.5	
		$ML_3.H/ML_2.HL$	-3.7	
		$MHL/M.HL$	4.60	

Alkylsalicylic acid (continued)

R =	Metal ion	Equilibrium	Log K 25°, 0.1	Log K 25°, 0
4-Methyl	H^+	HL/H.L		14.26
		H_2L/HL.H	2.97	3.04
$(C_8H_8O_3)$	Fe^{3+}	ML.H/M.HL	2.99 3.02[h]	
		ML_2.H/ML.HL	1.3	
		MHL/M.HL	4.7	
5-Methyl	H^+	HL/H.L		14.57
		H_2L/HL.H	2.90	2.88
$(C_8H_8O_3)$	Fe^{3+}	ML.H/M.HL	2.98	
		ML_2.H/ML.HL	1.20	
		MHL/M.HL	4.4	
6-Methyl	H^+	H_2L/HL.H	3.16	3.32
	Fe^{3+}	ML.H/M.HL	2.58	
$(C_8H_8O_3)$		ML_2.H/ML.HL	0.60	
		ML_3.H/ML_2.HL	-3,0	
3-(2-Propyl)	H^+	H_2L/HL.H	2.76	
$(C_{10}H_{12}O_3)$	Fe^{3+}	ML.H/M.HL	2.56	

[h] 20°, 0.1

Bibliography:

H^+ 57BDH,62JSa,63EMb,66P,71CS   Other reference: 70KCa

Fe^{3+} 62JSb,63EMc,66P

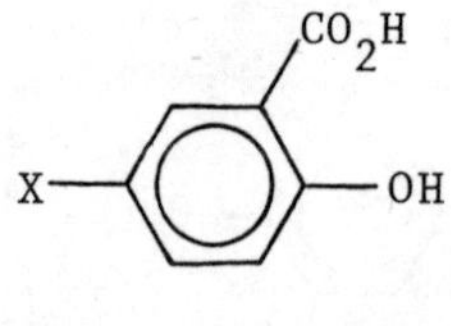

$C_7H_5O_3X$ 5-Haligenosalicylic acid H_2L

X =	Metal ion	Equilibrium	Log K 25°, 0
5-Chloro	H^+	HL/H.L	12.95
		H_2L/HL.H	2.64 ±0.01
	Fe^{3+}	ML/M.L	16.84
5-Bromo	H^+	HL/H.L	12.84
		H_2L/HL.H	2.64 ±0.03
	Fe^{3+}	ML/M.L	16.76

Bibliography:

H^+ 57BDH,63EM   Other references: 70HS,73JK,74J,75JK

Fe^{3+} 63EMc

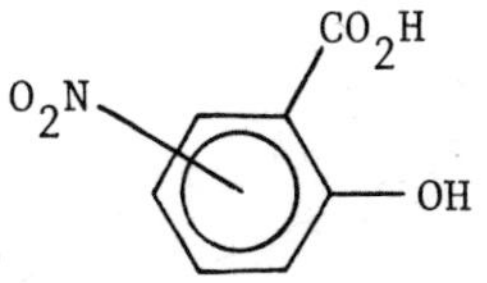

$$O_2N \overset{CO_2H}{\underset{OH}{\bigcirc}}$$

$C_7H_nO_qN_r$ Nitrosalicylic acid H_2L

Isomer	Metal ion	Equilibrium	Log K 25°, 0.1	Log K 25°, 0	ΔH 25°, 0.1	ΔS 25°, 0.1
3-Nitro	H^+	HL/H.L		10.33		
($C_7H_5O_5N$)		H_2L/HL.H		1.87		
	Fe^{3+}	ML/M.L		14.19		
5-Nitro	H^+	HL/H.L	10.11	10.34		
($C_7H_5O_5N$)		H_2L/HL.H	(2.20)	2.12		
	Mn^{2+}	ML/M.L	4.99	5.83	(4)[r]	(40)
	Co^{2+}	ML/M.L	5.57	6.41	(5)[r]	(45)
	Ni^{2+}	ML/M.L	5.87	6.71	(5)[r]	(50)
	Cu^{2+}	ML/M.L	8.40	9.24	(5)[r]	(60)
		ML_2/M.L^2	14.59	15.43	(9)[r]	(100)
	Fe^{3+}	ML/M.L		14.34		
3,5-Dinitro	H^+	HL/H.L	7.22 ±0.02 7.29[h]			
($C_7H_4O_7N_2$)		H_2L/HL.H	2.14 2.14[h][h]			
	UO_2^{2+}	ML/M.L	7.55[h]			
	Mn^{2+}	ML/M.L	3.52	4.36	(3)[r]	(30)
	Co^{2+}	ML/M.L	4.05	4.89	(4)[r]	(35)
	Ni^{2+}	ML/M.L	4.45	5.29	(4)[r]	(40)
	Cu^{2+}	ML/M.L	6.85	7.69	(5)[r]	(50)

[h] 20°, 0.1;   [r] 25-45°, 0.1

Bibliography:

H^+ 57BDH,63EM,63EMb,67B,75CT,75DNa,75DNc

UO_2^{2+} 67B

Mn^{2+}-Cu^{2+} 75DNa,75DNc

Fe^{3+} 63EMc

Other references: 66RS,68DD,69DD,69DDa, 69DDb,69DDc,70DD,70DDa,70DDb,70KA, 70KAa,71DD,71KG,72CH,72DN,72JK,72PSL, 73JK,74CS,75JK,75SJ,75PSL

$$HO_3S \overset{CO_2H}{\underset{OH}{\bigcirc}}$$

$C_7H_6O_6S$ <u>2-Hydroxy-5-sulfobenzoic acid</u> (5-sulfosalicylic acid) H_3L

Metal ion	Equilibrium	Log K 25°, 0.1	Log K 25°, 1.0	Log K 25°, 0	ΔH 25°, 0.5	ΔS 25°, 0.5
H^+	HL/H.L	11.72 ±0.08 11.80[h] ±0.1	11.40[b] ±0.08 11.51[b] ±0.06	12.53 11.74[e]	-9.3 (-7)[s]	26 (30)[e]
	H_2L/HL.H	2.49 ±0.02 2.49[h] ±0.01	2.32[b] ±0.02 2.35[b] ±0.05	2.84 2.67[e]		
	H_3L/H_2L.H		-0.75			
Be^{2+}	ML/M.L	11.54[h]	11.0 [b]			
	ML_2/M.L^2	20.43[h]	19.8 [b]			
La^{3+}	ML/M.L		5.92[j]			
	ML_2/M.L^2		10.73[j]			
Ce^{3+}	ML/M.L		6.03[j]			
	ML_2/M.L^2		10.91[j]			
Pr^{3+}	ML/M.L		6.23[j]			
	ML_2/M.L^2		11.24[j]			
Nd^{3+}	ML/M.L		6.35[j]			
	ML_2/M.L^2		11.85[j]			
Sm^{3+}	ML/M.L		6.77[j]			
	ML_2/M.L^2		15.56[j]			
Eu^{3+}	ML/M.L		6.79[j]			
	ML_2/M.L^2		12.46[j]			
Gd^{3+}	ML/M.L		6.93[j]			
	ML_2/M.L^2		12.83[j]			
Tb^{3+}	ML/M.L		6.95[j]			
	ML_2/M.L^2		12.86[j]			
Dy^{3+}	ML/M.L		7.15[j]			
	ML_2/M.L^2		13.01[j]			
Ho^{3+}	ML/M.L		7.18[j]			
	ML_2/M.L^2		12.98[j]			
Er^{3+}	ML/M.L		7.20[j]			
	ML_2/M.L^2		13.02[j]			
Tm^{3+}	ML/M.L		7.28[j]			
	ML_2/M.L^2		13.15[j]			
Yb^{3+}	ML/M.L		7.32[j]			
	ML_2/M.L^2		13.24[j]			
Lu^{3+}	ML/M.L		7.33[j]			
	ML_2/M.L^2		13.30[j]			
Th^{4+}	ML/M.L		12.30			
UO_2^{2+}	ML/M.L	11.14[h] 11.25[h]	10.44			
	ML_2/M.L^2	19.20[h] 18.75[h]				

[b] 25°, 0.5;   [e] 25°, 3.0;   [h] 20°, 0.1;   [j] 20°, 1.0;   [s] 15-35°, 3.0

A. MONO-HYDROXY PHENOLS

5-Sulfosalicylic acid (continued)

Metal ion	Equilibrium	Log K 25°, 0.1	Log K 25°, 1.0	Log K 25°, 0	ΔH 25°, 0.5	ΔS 25°, 0.5
Cr^{2+}	$ML/M.L$	7.14				
	$ML_2/M.L^2$	12.88				
Mn^{2+}	$ML/M.L$	5.24 −0.01	4.77			
		5.10[r]				
	$ML_2/M.L^2$	8.24 +0.4	8.19			
		8.00[r]				
Fe^{2+}	$ML/M.L$	5.90[r]				
	$ML_2/M.L^2$	9.9 [r]				
Co^{2+}	$ML/M.L$	6.13 +0.3				
		6.00[r]				
	$ML_2/M.L^2$	9.82 +1				
		9.60[r]				
Ni^{2+}	$ML/M.L$	6.42 +0.2				
		6.30[r]				
	$ML_2/M.L^2$	10.24 +0.6				
		10.20[r]				
Cu^{2+}	$ML/M.L$	9.43 ±0.09	8.91	10.74	−4.0	28
		9.50[r]	8.97[b]			
	$ML_2/M.L^2$	16.3 ±0.1	15.86	17.17		
		16.3 [r]	15.91[b]			
Cr^{3+}	$ML/M.L$	9.56				
Fe^{3+}	$ML/M.L$	14.60[r]	14.40[b]	14.42[e]	(−3)[s]	(60)[e]
	$ML_2/M.L^2$	25.15[r]	(22.2)[b]	25.2 [e]		
	$ML_3/M.L^3$		(30.6)[b]	32.2 [e]		
Ti(IV)	$ML_3/MO.H^2.L^3$	42.2 [h]				
	$MO(HL)/MO.HL$	3.1 [h]				
	$MO(HL)_2/MO.(HL)^2$	5.4 [h]				
VO^{2+}	$ML/M.L$	11.71[h]				
		12.0 [h]				
	$ML_2/M.L^2$	20.6				
	$ML/MOHL.H$	7.22				
	$(MOHL)_2/(MOHL)^2$	5.33				
Zn^{2+}	$ML/M.L$	6.05[r]				
	$ML_2/M.L^2$	10.7 [r]				
Cd^{2+}	$ML/M.L$	4.64[r]				
Al^{3+}	$ML/M.L$	12.3 [h]				
	$ML_2/M.L^2$	20.0 [h]				
	$ML_3/M.L^3$	25.8 [h]				

[b] 25°, 0.5; [e] 25°, 3.0; [h] 20°, 0.1; [r] 20°, 0.15; [s] 15-35°, 3.0

Bibliography:

H^+ 54A,55Aa,56Aa,58P,59BS,61NP,61PS,
64MP,64RM,66LM,66MM,66TJ,69Pb,70MR,
70SMa,71MP,71ZB,72BT,75MG

La^{3+}-Th^{4+} 72CDP

UO_2^{2+} 67B

Be^{2+} 67BZ,72BT

Cr^{2+} 61PS

5-Sulfosalicylic acid (continued)

$Mn^{2+}-Ni^{2+}$ 58P,60BS,62N

Cu^{2+} 59NH,58P,60BS,61NP,62N,62NM,64MP,
66LM,69Pb,75MG

Cr^{3+} 60BS

Fe^{3+} 54A,56Aa,58P,70MR

Ti(IV) 63Sc

VO^{2+} 66MM,71ZB

Zn^{2+},Cd^{2+} 58P

Al^{3+} 63NA,69HBa

Other references: 48FA,49FA,49TA,50SR,51BP,
51MB,54SL,55V,56IT,57GL,57NA,58RC,59CG,
60C,61Da,61N,61JS, 62SBa,62SSB,63NA,
64DA,64H,65DN,66BP,66KSG,66KZ,66OT,66RS,
67AM,67AS,68AS,68KT,69KM,69MN,70FK,70GI,
70KC,70KKM,70NPC,70PSa,72CS,72PSa,73JK,
73KJ,74CS,74J,74KKb,74PL,75SJ,75ST

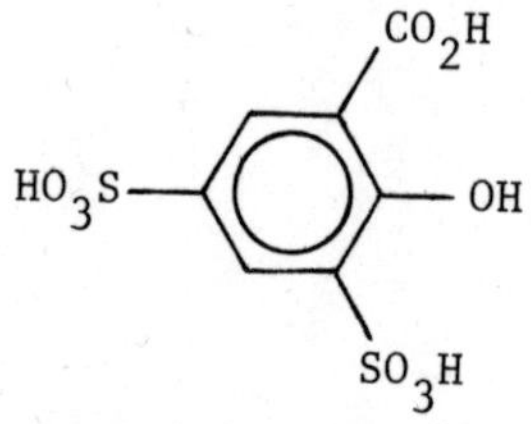

$C_7H_6O_9S_2$ 2-Hydroxy-3,5-disulfobenzoic acid H_4L

Metal ion	Equilibrium	Log K 25°, 0.1	Log K 25°, 1.0	Log K 25°, 0
H^+	HL/H.L	11.55 11.04[b]	10.95	12.50
	$H_2L/HL.H$	2.03 1.70[b]	1.71	2.69
Cu^{2+}	ML/M.L	9.88 9.13[b]	8.89	11.49
	$ML_2/M.L^2$	16.13 16.13[b]	16.15	

[b] 25°, 0.5

Bibliography: 75L

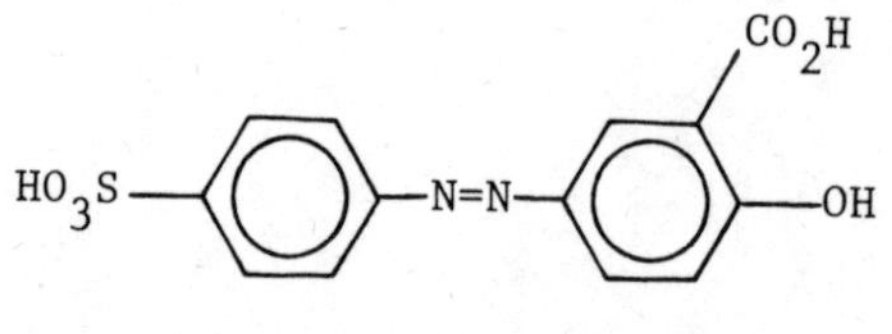

$C_{13}H_{10}O_6N_2S$ 5-(4-Sulfophenylazo)salicylic acid
(3-carboxy-4-hydroxy-4'-sulfoazobenzene) H_3L

Metal ion	Equilibrium	Log K 25°, 0.1
H^+	HL/H.L	11.04
	$H_2L/HL.H$	2.38
Mg^{2+}	ML/M.L	4.45
	$ML_2/M.L^2$	7.49
Ca^{2+}	ML/M.L	3.10
Mn^{2+}	ML/M.L	4.94
	$ML_2/M.L^2$	8.4
Co^{2+}	ML/M.L	5.84
	$ML_2/M.L^2$	9.77

5-(4-Sulfophenylazo)salicylic acid (continued)

Metal ion	Equilibrium	Log K 25°, 0.1
Ni^{2+}	$ML/M.L$	6.17
	$ML_2/M.L^2$	10.22
Fe^{3+}	$ML_2/ML.L$	10.70

Bibliography: 64MT

Other reference: 68AVa

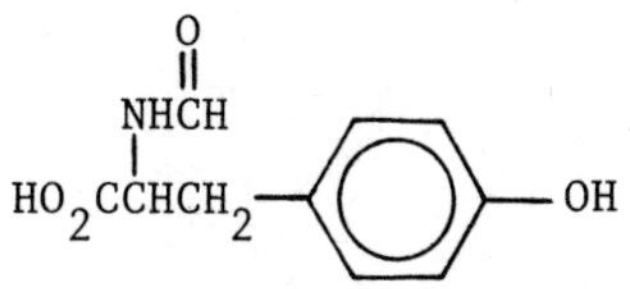

$C_{10}H_{10}O_4N$	L-N-Formyltyrosine	H_2L

Metal ion	Equilibrium	Log K 20°, 0.37
H^+	$HL/H.L$	9.63
	$H_2L/HL.H$	3.11
Cu^{2+}	$ML/M.L$	7.24
	$ML_2/M.L^2$	11.67
	$ML/M(H_{-1}L).H$	4.00
	$M(H_{-1}L)L/M(H_{-1}L).L$	2.71
	$M(H_{-1}L)L/M(H_{-1}L)_2.H$	7.28

Bibliography: 74W

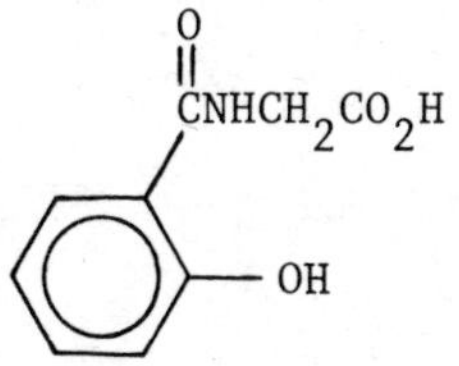

$C_9H_9O_4N$	N-(2-Hydroxybenzoyl)glycine (salicyluric acid)	H_2L

Metal ion	Equilibrium	Log K 25°, 0.1
H^+	$H_2L/HL.H$	3.41
Fe^{3+}	$ML.H/M.HL$	2.09
	$ML_2.H/ML.HL$	0.57
	$ML_3.H/ML_2.HL$	-4.1
	$MHL/M.HL$	3.90

Bibliography: 66P

$C_7H_7O_3N$ 4-Amino-2-hydroxybenzoic acid H_2L

Metal ion	Equilibrium	Log K 25°, 0.1	Log K 25°, 1.0	ΔH 25°, 1.0	ΔS 25°, 1.0
H^+	HL/H.L		13.7 [e]	(-11) [s]	(30) [e]
	H_2L/HL.H	3.63 -0.01	3.68 4.08 [e]	(0) [t]	(15)
	H_3L/H_2L.H	1.78 ±0.1	1.95 2.08 [e]	(0) [t]	(10)
Fe^{3+}	MHL/M.HL	6.00 [h] (6.18) [b]	5.93 6.20 [e]	(0) [t]	(25)
	ML.H/M.HL	3.64 [h] 4.06 [b]	3.17 [e]		
	ML_2.H/ML.HL	-0.97 [b]	-1.64 [e]		
	ML_3.H/ML_2.HL	-6.33 [h] -5.70 [b]	-5.0 [e]		

[b] 25°, 0.5;   [e] 25°, 3.0;   [h] 20°, 0.1;   [s] 15-35°, 3.0;   [t] 20-30°, 1.0

Bibliography:

H^+ 54Aa,54WS,55Aa,62JSa,67LR,72MSb

Fe^{3+} 54Aa,62JSa,70MR,72MSb

Other references: 55LR,60C,60CA,61GK,66NV,
                        66RS,70HS

$C_7H_6O_2$ 2-Hydroxybenzaldehyde (salicylaldehyde)   HL

Metal ion	Equilibrium	Log K 25°, 0.1	Log K 25°, 0.5	Log K 25°, 0	ΔH 25°, 3.0	ΔS 25°, 3.0
H^+	HL/H.L	8.13 ±0.01 8.22 [r]	8.07 8.80 [e]	8.37 ±0.00	(-6) [s]	(20)
Ca^{2+}	ML/M.L	1.1				
Y^{3+}	ML/M.L	4.34				
La^{3+}	ML/M.L	3.40				
Mn^{2+}	ML/M.L ML_2/M.L^2		2.15 4.0			
Ni^{2+}	ML/M.L ML_2/M.L^2		3.58 6.5			
Cu^{2+}	ML/M.L	5.56 5.75 [r]	5.36			
	ML_2/M.L^2		10.11			

[e] 25°, 3.0;   [r] 20°, 0.15;   [s] 15-35°, 3.0

A. MONO-HYDROXY PHENOLS

Salicylaldehyde (continued)

Metal ion	Equilibrium	Log K 25°, 0.1	Log K 25°, 0.5	Log K 25°, 0	ΔH 25°, 3.0	ΔS 25°, 3.0
Fe^{3+}	$ML/M.L$		8.75^e		$(2)^s$	(50)
	$ML_2/M.L^2$		15.55^e			
Zn^{2+}	$ML/M.L$		2.87			
	$ML_2/M.L^2$		5.00			

----

^{e} 25°, 3.0; ^{s} 15-35°, 3.0

Bibliography:

H^+	55Aa,56Aa,56RK,58P,65GA,66PM,68LB		Cu^{2+}	58P,66PM,68LB
Ca^{2+},Y^{3+},La^{3+}	66PM		Fe^{3+}	55A,56Aa
Mn^{2+},Ni^{2+},Zn^{2+}	68LB			

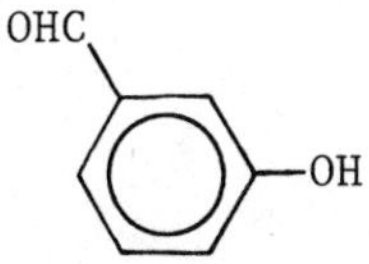

$C_7H_6O_2$		3-Hydroxybenzaldehyde		HL
Metal ion	Equilibrium	Log K 25°, 0	ΔH 25°, 0	ΔS 25°, 0
H^+	$HL/H.L$	9.01 ± 0.02	-5.3	23
Fe^{3+}	$ML/M.L$	8.11		

Bibliography:

H^+	56RK,67LL	Fe^{3+}	65EH

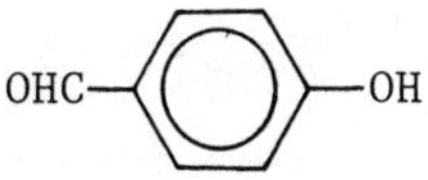

$C_7H_6O_2$		4-Hydroxybenzaldehyde				HL
Metal ion	Equilibrium	Log K 25°, 0.1	Log K 25°, 0	ΔH 25°, 0	ΔS 25°, 0
H^+	$HL/H.L$	7.41	7.62 ± 0.00	-4.4 / -4.0^a	20 / 20^a
Fe^{3+}	$ML/M.L$	5.36	7.56	0.5^a	26^a

----

^{a} 25°, 0.1

Bibliography:

H^+	56RK,67LL,69DM	Fe^{3+}	65EH,69DM

$C_7H_nO_qN_rCl_s$		Substitutedsalicylaldehyde		HL
Z =	Metal ion	Equilibrium	Log K 25°, 0.1	
3-Chloro	H^+	HL/H.L	6.61	
$(C_7H_5O_2Cl)$	Ca^{2+}	ML/M.L	0.7	
	Y^{3+}	ML/M.L	3.77	
	La^{3+}	ML/M.L	3.16	
	Cu^{2+}	ML/M.L	4.74	
	Zn^{2+}	ML/M.L	2.39	
4-Chloro	H^+	HL/H.L	7.18	
$(C_7H_5O_2Cl)$	Ca^{2+}	ML/M.L	1.1	
	Y^{3+}	ML/M.L	4.09	
	La^{3+}	ML/M.L	3.38	
	Cu^{2+}	ML/M.L	5.11	
5-Chloro	H^+	HL/H.L	7.41	
$(C_7H_5O_2Cl)$	Ca^{2+}	ML/M.L	0.9	
	Y^{3+}	ML/M.L	3.94	
	La^{3+}	ML/M.L	3.20	
	Cu^{2+}	ML/M.L	4.96	
6-Chloro	H^+	HL/H.L	8.26	
$(C_7H_5O_2Cl)$	Ca^{2+}	ML/M.L	1.74	
	Y^{3+}	ML/M.L	4.87	
	La^{3+}	ML/M.L	4.08	
	Cu^{2+}	ML/M.L	5.88	
3-Nitro	H^+	HL/H.L	5.21	
$(C_7H_5O_4N)$	Ca^{2+}	ML/M.L	1.0	
	Y^{3+}	ML/M.L	3.27	
	La^{3+}	ML/M.L	3.03	
	Cu^{2+}	ML/M.L	3.61	
5-Nitro	H^+	HL/H.L	5.32	
$(C_7H_5O_4N)$	Ca^{2+}	ML/M.L	0.8	
	Y^{3+}	ML/M.L	3.17	
	La^{3+}	ML/M.L	2.73	
	Cu^{2+}	ML/M.L	3.83	
	Zn^{2+}	ML/M.L	2.01	

Substitutedsalicylaldehyde (continued)

Z =	Metal ion	Equilibrium	Log K 25°, 0.1
3,5-Dinitro	H^+	HL/H.L	2.09
$(C_7H_4O_6N_2)$	Y^{3+}	ML/M.L	1.75
	La^{3+}	ML/M.L	1.84
	Cu^{2+}	ML/M.L	1.68
	Zn^{2+}	ML/M.L	0.8

Bibliography: 66PM

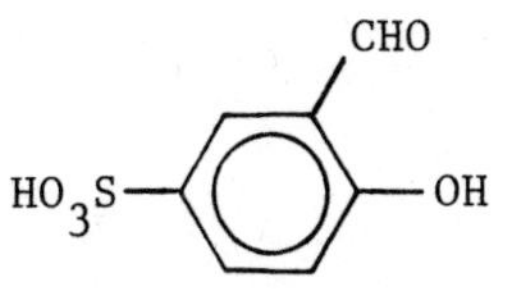

$C_7H_6O_5S$ 5-Sulfosalicylaldehyde H_2L

Metal ion	Equilibrium	Log K 25°, 0.1	Log K 25°, 1.0	Log K 25°, 0
H^+	HL/H.L	6.93	6.89	7.32
		$6.82^b \pm 0.00$	7.15^d	
	$H_2L/HL.H$	0.24^b		
Be^{2+}	ML/M.L	3.40^b		
	$M_2(OH)L/M_2(OH).L$	7.98^b		
	$M_3(OH)_3L/M_3(OH)_3.L$	3.26^b		
	$M_3(OH)_3L_2/M_3(OH)_3.L^2$	6.56^b		
	$M_3(OH)_3L_3/M_3(OH)_3.L^3$	8.15^b		
Cu^{2+}	ML/M.L	5.01	4.75	5.75
		4.76^b	4.92^d	
	$ML_2/M.L^2$	8.78	8.50	9.61
		8.46^b	8.87^d	

---

[b] 25°, 0.5;   [d] 25°, 2.0

Bibliography:

H^+ 71S,72BT Cu^{2+} 71S

Be^{2+} 72BT   Other reference: 72KEa

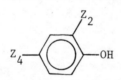

$C_8H_8O_2$		Acetylphenol (hydroxyacetophenone)				HL
Isomer	Metal ion	Equilibrium	Log K 30°, 0.1	Log K 25°, 0	ΔH 25°, 3.0	ΔS 25°, 3.0
2-Acetyl	H^+	HL/H.L	9.87[h] 9.94[h]	10.82[e]	(-9)[s]	(20)
	Co^{2+}	ML/M.L ML$_2$/M.L^2	4.20 7.04			
	Ni^{2+}	ML/M.L ML$_2$/M.L^2	4.36 7.2			
	Cu^{2+}	ML/M.L ML$_2$/M.L^2	6.49[h] 6.75[h] 11.74[h] 12.45[h]			
	Fe^{3+}	ML/M.L		10.52[e]	(1)[s]	(50)
3-Acetyl	H^+	HL/H.L	8.85[c]	9.25	-5.2[c]	23[c]
	Fe^{3+}	ML/M.L		8.36		
4-Acetyl	H^+	HL/H.L		8.05		
	Fe^{3+}	ML/M.L		7.20		

[c] 25°, 1.0;   [e] 25°, 3.0;   [h] 20°, 0.15;   [s] 15-35°, 3.0

Bibliography:

H^+ 55Aa,56Aa,58P,64EH,71PPb,73A Cu^{2+} 58P,71PPb

Co^{2+},Ni^{2+} 71PPb Fe^{3+} 55A,56Aa,65EH

$C_nH_mO_pN_r$		Substituted phenol				HL
Z =	Metal ion	Equilibrium	Log K 20°, 0.15	Log K 25°, 3.0	ΔH 25°, 3.0	ΔS 25°, 3.0
Z_2=CH_3CO_2,Z_4=H (2-hydroxyben- zoic acid methyl ester) (methyl sali- cylate) ($C_8H_8O_3$)	H^+	HL/H.L	9.75	10.19	(-8)[s]	(20)
	Cu^{2+}	ML/M.L	5.90			
	Fe^{3+}	ML/M.L		9.73	(2)[s]	(50)

[s] 15-35°, 3.0

A. MONO-HYDROXY PHENOLS

Substituted phenol (continued)

Z =	Metal ion	Equilibrium	Log K 20°, 0.15	Log K 25°, 3.0	ΔH 25°, 3.0	ΔS 25°, 3.0
$Z_2=H_2NCO, Z_4=H$	H^+	HL/H.L		8.89	$(-7)^s$	(20)
(salicylamide)	Cu^{2+}	ML.H/M.HL	−3.40			
$(C_7H_7O_2N)$		$ML_2.H^2/M.(HL)^2$	−7.90			
	Fe^{3+}	$ML/M.L_2$		10.02	$(0)^s$	(50)
		$ML_2/M.L^2$		17.3		
$Z_2=H_2NCO,$	H^+	HL/H.L		9.11	$(-5)^s$	(25)
$Z_4=NH_2$		$H_2L/HL.H$		3.00		
(4-aminosali-	Fe^{3+}	$ML/M.L_2$		10.7	$(0)^s$	(50)
cylamide)		$ML_2/M.L^2$		18.0		
$(C_7H_8O_2N_2)$		MHL/M.HL		1.80		

s 15-35°, 3.0

Bibliography:

H^+ 55Aa,56Aa,58P Fe^{3+} 55A,56A,56Aa

Cu^{2+} 58P

$C_6H_6O_2$		1,2-Dihydroxybenzene (catechol)				H_2L
Metal ion	Equilibrium	Log K 25°, 0.1	Log K 25°, 1.0	Log K 25°, 0	ΔH 25°, 0.1	ΔS 25°, 0.1
H^+	HL/H.L	13.0 ±0.1	13.0	(12.8)	(−5)[t]	(40)
	H_2L/HL.H	9.23 ±0.04	9.23	9.40 ±0.05	−6.0	22
Be^{2+}	ML/M.L	13.52[h]				
	$ML_2/M.L^2$	23.35[h]				
	MHL/M.HL	5.0[h]				
	MHL_2/ML.HL	2.8[h]				
Mg^{2+}	ML/M.L	5.7[p]				
Sc^{3+}	ML/M.L	17.04				
La^{3+}	ML/M.L	9.46				
Pr^{3+}	ML/M.L	10.31				
Nd^{3+}	ML/M.L	10.50				
Sm^{3+}	ML/M.L	11.06				
Eu^{3+}	ML/M.L	11.17				
Gd^{3+}	ML/M.L	11.20				
Dy^{3+}	ML/M.L	11.34				
Ho^{3+}	ML/M.L	11.42				
Er^{3+}	ML/M.L	11.43				
Tm^{3+}	ML/M.L	11.56				
Yb^{3+}	ML/M.L	11.67				
Lu^{3+}	ML/M.L	11.31				
UO_2^{2+}	ML/M.L	15.9[h]				
	MHL/M.HL	6.3[h]				
	MHL_2/ML.HL	4.9[h]				
	MHL_3/MHL_2.HL	3.7[h]				
Mn^{2+}	ML/M.L	7.72	7.47			
	$ML_2/M.L^2$	13.6	12.8			
Fe^{2+}	ML/M.L		7.95			
	$ML_2/M.L^2$		13.5			
Co^{2+}	ML/M.L	8.60 ±0.01	8.32			
	$ML_2/M.L^2$	15.0 ±0.4	14.7			
Ni^{2+}	ML/M.L	8.92 ±0.03	8.77			
	$ML_2/M.L^2$	14.4 ±0.6				
Cu^{2+}	ML/M.L	13.90 ±0.06	13.62 ±0.03	(14.1) ±0.0	(−10)	(30)
	$ML_2/M.L^2$	24.9 ±0.1	24.94 ±0.02	(24.6) −0.1		
	ML.H/M.HL	0.85	0.55		−5.0	−13

[h] 20°, 0.1;   [p] 30°, 0.1;   [t] 15−25°, 0.1

Catechol (continued)

Metal ion	Equilibrium	Log K 25°, 0.1	Log K 25°, 1.0	Log K 25°, 0	ΔH 25°, 0.1	ΔS 25°, 0.1
V^{3+}	$ML/M.L$	18.3 [i]				
	$M(OH)_2L/ML.(OH)^2$	19.4 [i]				
	$M(OH)_2L_2/M(OH)_2L.L$	10.1 [i]				
$Ti(IV)$	$MO(HL)_2/MO.(HL)^2$	24.3 [h]				
	$ML_3/MO(HL)_2.L$	21.5 [h]				
VO^{2+}	$ML/M.L_2$	17.7 [h]				
	$ML_2/M.L^2$	33.5 [h]				
$Mo(VI)$	$ML_2L_2/MO_4.(H_2L)^2$	5.1 [h] ±0.2				
	$MO_2L_2.H/HMO_3L.H_2L$	-3.30				
	$HMO_2L_2/MO_2L_2.H$	3.65				
$W(VI)$	$MO_2L_2/MO_4.(H_2L)^2$	6.7 [h] ±0.2				
	$MO_2L_2.H/HMO_3L.H_2L$	2.70				
Zn^{2+}	$ML/M.L$	9.90 / 9.80 [P]	9.54			
	$ML_2/M.L^2$	17.4 ±0.3	17.5			
Cd^{2+}	$ML/M.L$	8.2 [P]				
$B(III)$	$M(OH)_2L.H/M(OH)_3.H_2L$	-5.13±0.00				
		-5.17 [h]				
	$ML_2.H/M(OH)_3.(H_2L)^2$	-4.84				
Al^{3+}	$ML/M.L_2$	16.3 [h]				
	$ML_2/M.L^2_3$	29.3 [h]				
	$ML_3/M.L$	37.6 [h]				
Ga^{3+}	$ML/M.L$	18.9 [a]				
$As(III)$	$M(OH)_2L.H/M(OH)_3.H_2L$	-6.89				
	$ML_2.H/M(OH)_3.(H_2L)^2$	-6.4				
$Sb(III)$	$ML_2.OH/M(OH)_3.(HL)^2$	5.44 [h]				
	$ML_2.H/MOHL.H_2^3L$	-2.37 [h]				
$Ge(IV)$	$ML_3.H^2/M(OH)_4.(H_2L)^3$	-1.2 ±0.2				

[a] 22°, 0.1; [h] 20°, 0.1; [i] 21°, 0.5; [P] 30°, 0.1

Bibliography:

H^+ 56NM,57T,58P,63MNT,65SSc,66LM,68TM, 69CMa,69HG,71PB,72JW,73AW

Be^{2+} 67BZ

Mg^{2+},Ca^{2+} 66AP

Sc^{3+} 71UK

La^{3+}-Lu^{3+} 69PK

UO_2^{2+} 65BS,65SSc,66BR

Mn^{2+}-Cu^{2+},Zn^{2+} 56NM,57T,63MNT,65JNa,66AP, 66JN,67OH,68TM,69CMa,69HG,71GS,71PB, 72JWa

V^{3+} 74ZB

Ti(IV) 63Sa

VO^{2+} 71ZB

Mo(VI),W(VI) 59H,64SM,69HB,71SB,73B

B(III) 59AK,68HBa,69MB

Al^{3+} 69HBa

Ga^{3+} 68ZK

As(III) 59ARa

Sb(III) 70AM

Ge(IV) 59AM,63AN,67PB,73SB

Catechol (continued)

Other references: 42Sa,55VGa,57GL,57RL,   68OOH,68VA,69CS,69Ha,69HBb,69SKK,70E,70H,
    60HP,61SK,62AM,63SG,64DM,64DMa,65JN,   71AO,71RS,72JM,72MJ,73KL,73KLa,74KA,75ZB
    66DM,66SC,66TB,67CBa,67EKM,67LAa,68AP,

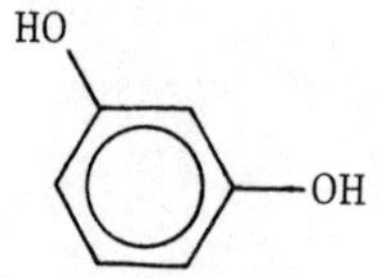

$C_6H_6O_2$		1,3-Dihydroxybenzene (resorcinol)	H_2L
Metal ion	Equilibrium	Log K 20°, 0.1	
H^+	HL/H.L	11.06	
	H_2L/HL.H	9.30	
UO_2^{2+}	ML/M.L	16.9	
	MHL/M.HL	5.9	

Bibliography: 65BS,66BR

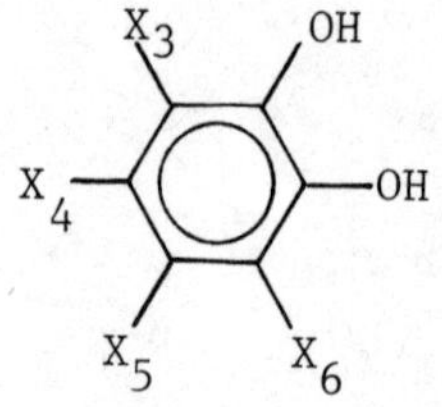

$C_6H_nO_2Cl_s$			Chloro-1,2-dihydroxybenzene			H_2L
X =	Metal ion	Equilibrium	Log K 25°, 0.1	Log K 30°, 0.1	ΔH 25°, 0.1	ΔS 25°, 0.1
X_4=Cl,	H^+	HL/H.L	11.97	(11.54)		
$X_3=X_5=X_6$=H		H_2L/HL.H	8.51 ±0.02	8.42	-6.9	16
(4-Chloro)	Mn^{2+}	ML/M.L		6.82		
		$ML_2/M.L^2$		11.48		
($C_6H_5O_2Cl$)	Co^{2+}	ML/M.L		7.64		
		$ML_2/M.L^2$		14.01		
		$ML_3/M.L^3$		18.24		
	Ni^{2+}	ML/M.L	8.38	7.90		
		$ML_2/M.L^2$	13.84	12.90		
		$ML_3/M.L^3$		17.06		
	Cu^{2+}	ML/M.L	12.89	12.56		
		$ML_2/M.L^2$	23.05	22.39		
		ML.H/M.HL	0.92		-4.0	-9
	Zn^{2+}	ML/M.L		8.63		
		$ML_2/M.L^2$		15.45		
	B(III)	$M(OH)_2L.H/M(OH)_3.H_2L$	-4.87			
	Ge(IV)	$ML_3.H^2/M(OH)_4.(H_2L)^3$	0.65			

B. DI-HYDROXYPHENOLS

Chloro-1,2-dihydroxybenzene (continued)

X =	Metal ion	Equilibrium	Log K 25°, 0.1	Log K 30°, 0.1	ΔH 25°, 0.1	ΔS 25°, 0.1
$X_3=X_4=X_5=X_6=Cl$	H^+	HL/H.L	10.10			
(tetrachloro)		H_2L/HL.H	5.80			
$(C_6H_2O_2Cl_4)$	B(III)	$M(OH)_2L.H/M(OH)_3.H_2L$				
			-3.10			
	Ge(IV)	$ML_3.H^2/M(OH)_4.(H_2L)^3$				
			6.67			

Bibliography:

H^+ 64MTa,67PB,69MB,72JWb

Mn^{2+},Co^{2+},Zn^{2+} 64MTa

Ni^{2+},Cu^{2+} 64MTa,72JWb

B(III)   69MB

Ge(IV)   67PB,69MB

Other references: 62Hb,63H,74NV

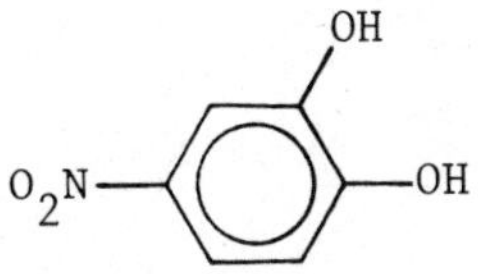

4-Nitro-1,2-dihydroxybenzene

$C_6H_5O_4N$

H_2L

Metal ion	Equilibrium	Log K 25°, 0.1	Log K 30°, 0.1	ΔH 25°, 0.1	ΔS 25°, 0.1
H^+	HL/H.L	10.85 ±0.05	10.75		
	H_2L/HL.H	6.69 ±0.03	6.62	-5.7	12
UO_2^{2+}	ML/M.L		12.9 [h]		
	$ML_2/M.L^2$		22.7 [h]		
	MHL/M.HL		4.5 [h]		
Mn^{2+}	ML/M.L		6.51		
	$ML_2/M.L^2$		11.25		
Co^{2+}	ML/M.L		7.48		
	$ML_2/M.L^2$		12.79		
	$ML_3/M.L^3$		15.93		
Ni^{2+}	ML/M.L	7.89	7.82		
	$ML_2/M.L^2$	13.39	13.09		
	$ML_3/M.L^3$		16.90		
Cu^{2+}	ML/M.L	11.67	11.65		
	$ML_2/M.L^2$	20.95	20.93		
	ML.H/M.HL	0.81		-4.0	-10
Zn^{2+}	ML/M.L		8.20		
	$ML_2/M.L^2$		15.00		
B(III)	$M(OH)_2L.H/M(OH)_3.H_2L$				
		-3.91	-4.0 [h]		
Ge(IV)	$ML_3.H^2/M(OH)_4.(H_2L)^3$	3.90			

[h] 20°, 0.1

4-Nitro-1,2-dihydroxybenzene (continued)

Bibliography:

H^+ 64MTa,67Ba,67PB,68HB,72JWb,75VH   B(III)   68HB,69MB

UO_2^{2+} 67Ba   Ge(IV)   67PB

Mn^{2+},Co^{2+},Zn^{2+} 64MTa   Other references: 72HK,75VH

Ni^{2+},Cu^{2+} 64MTa,72JWb

$C_6H_6O_5S$		1,2-Dihydroxybenzene-4-sulfonic acid		H_3L
Metal ion	Equilibrium	Log K 30°, 0.1	Log K 20°, 0.1	
H^+	HL/H.L	(12.16)	12.8	
	H_2L/HL.H	8.26	8.50	
Mg^{2+}	ML/M.L	6.27		
	ML_2/M.L^2	10.41		
Ca^{2+}	ML/M.L	4.40		
	ML_2/M.L^2	7.99		
Sr^{2+}	ML/M.L	3.61		
UO_2^{2+}	MHL/M.HL		6.4	
	$M_2OHL_2.H^3$/(MHL)^2		-9.0	
Mn^{2+}	ML/M.L	7.87		
	ML_2/M.L^2	12.53		
Co^{2+}	ML/M.L	8.54		
	ML_2/M.L^2	14.40		
	ML_3/M.L^3	17.48		
Ni^{2+}	ML/M.L	8.85		
	ML_2/M.L^2	14.41		
	ML_3/M.L^3	19.14		
Cu^{2+}	ML/M.L	13.29		
	ML_2/M.L^2	23.52		
Fe^{3+}	ML_2/ML.L	14.0		
	ML_3/ML_2.L	9.1		
VO^{2+}	ML/M.L		16.7	
	ML_2/M.L^2		31.2	
Mo(VI)	MO_2L_2/MO_4.$(H_2L)^2$		5.28	
Zn^{2+}	ML/M.L	9.40		
	ML_2/M.L^2	16.60		
B(III)	$M(OH)_2L.H$/$M(OH)_3.H_2L$		-4.72	
Al^{3+}	ML/M.L		16.6	
	ML_2/M.L^2		29.9	
	ML_3/M.L^3		39.2	

B. DI-HYDROXYPHENOLS

1,2-Dihydroxybenzene-4-sulfonic acid (continued)

Metal ion	Equilibrium	Log K 30°, 0.1	Log K 20°, 0.1
Ge(IV)	$ML_3 \cdot H^2/M(OH)_4 \cdot (H_2L)^3$		0.9

Bibliography:

H^+	63MNT,65BS,68HGa	B(III)	68HGa
Mg^{2+}-Sr^{2+},Mn^{2+}-Cu^{2+},Zn^{2+}	63MNT	Al^{3+}	69HBa
UO_2^{2+}	65BS,65BSa	Ge(IV)	73SB
Fe^{3+}	63MN	Mo(VI)	71SB
VO^{2+}	71ZB		

$C_6H_6O_8S_2$		1,2-Dihydroxybenzene-3,5-disulfonic acid (tiron)				H_4L
Metal ion	Equilibrium	Log K 25°, 0.1	Log K 25°, 1.0	Log K 25°, 0	ΔH 25°, 0.1	ΔS 25°, 0.1
H^+	HL/H.L	12.5 ±0.1 [b] 12.0 ±0.1	11.8 ±0.1	13.3	(-20)[r]	(-10)
	H_2L/HL.H	7.61 ±0.05 [b] 7.28 ±0.03	7.20 ±0.05	8.31	-5.0	18
Be^{2+}	ML/M.L $ML_2/M.L^2$	12.88[h] 22.25[h]				
	MHL/M.HL $MHL_2/ML.HL$	4.2[h] 2.3[h]				
Mg^{2+}	ML/M.L MHL/M.HL	6.86[h] 1.98[h]				
Ca^{2+}	ML/M.L MHL/M.HL	5.80[h] 2.18[h]				
Sr^{2+}	ML/M.L MHL/M.HL	4.55[h] 1.88[h]				
Ba^{2+}	ML/M.L MHL/M.HL	4.10[h] 2.0[h]				
Sc^{3+}	ML/M.L MHL/M.HL	18.96 8.94				
Y^{3+}	ML/M.L MHL/M.HL	13.72 5.13				
La^{3+}	ML/M.L	12.87				
Pr^{3+}	ML/M.L	13.47				
Nd^{3+}	ML/M.L MHL/M.HL	13.69 5.61				

[b] 25°, 0.5; [h] 20°, 0.1; [r] 25-35°, 0.1

Tiron (continued)

Metal ion	Equilibrium	Log K 25°, 0.1	Log K 25°, 1.0	Log K 25°, 0	ΔH 25°, 0.1	ΔS 25°, 0.1
Sm^{3+}	ML/M.L	13.92				
	MHL/M.HL	5.72				
Gd^{3+}	ML/M.L	14.10				
	MHL/M.HL	5.92				
Tb^{3+}	ML/M.L	14.14				
	MHL/M.HL	5.71				
Dy^{3+}	ML/M.L	14.36				
	MHL/M.HL	5.59				
Ho^{3+}	ML/M.L	14.39				
	MHL/M.HL	5.42				
Er^{3+}	ML/M.L	14.48				
	MHL/M.HL	5.45				
Tm^{3+}	ML/M.L	14.36				
	MHL/M.HL	5.67				
Yb^{3+}	ML/M.L	14.43				
	MHL/M.HL	5.65				
Th^{4+}	$(ML)^2.H_2L/M_2(OH)_2L_3.H^4$	11.9				
	$M_2L_3/M_2(OH)_2L_3.H^2$	12.8				
UO_2^{2+}	ML/M.L	(15.9)				
	MHL/M.HL	(6.4)[h]				
	$M_2(OH)L_2.H^3/(MHL)^2$	(−8.9)[h]				
	$M_3(OH)_2L_3.H^2/(ML)^3$	(2.8)				
	$M_3(OH)_2L_4/M_3(OH)_2L_3.L$	(8.96)				
Mn^{2+}	$ML/M.L$	8.6	7.20			
	$ML_2/M.L^2$		12.75			
	$ML_3/M.L^3$		16.28			
	MHL/M.HL		2.05			
Co^{2+}	ML/M.L	9.49[h]	8.19	10.78		
	$ML_2/M.L^2$		14.41			
	MHL/M.HL	3.08[h]				
Ni^{2+}	ML/M.L	9.96[h]	8.56	11.24		
	$ML_2/M.L^2$		14.90			
	MHL/M.HL	3.00[h]				
Cu^{2+}	ML/M.L	14.27 ±0.04 14.53[h]	12.76	15.62	(−21)[s]	(−5)
	$ML_2/M.L^2$	25.46 ±0.04 25.16[p]	23.73			
	MHL/M.HL	5.48[h]				
Fe^{3+}	ML/M.L	20.4 20.7[h]	19.5[b] 19.0[b]			
	$ML_2/M.L^2$	35.4 35.9[h]	34.6[b] 33.9[b]			
	$ML_3/M.L^3$	45.8 46.9[h]	46.0[b] 45.3[b]			
	MHL/M.HL	9.7 10.0[h]	9.1[b] 9.1			

[b] 25°, 0.5; [h] 20°, 0.1; [P] 30°, 0.1; [s] 20−30°, 0.1

B. DI-HYDROXYPHENOLS

Tiron (continued)

Metal ion	Equilibrium	Log K 25°, 0.1	Log K 25°, 1.0	Log K 25°, 0	ΔH 25°, 0.1	ΔS 25°, 0.1
VO^{2+}	$ML/M.L$	16.74^h 16.8			$(-3)^r$	(70)
	$ML_2/M.L^2$	32.79^h 31.2				
	$ML/MOHL.H$	6.3				
	$(MOHL)_2/(MOHL)^2$	4.3				
Ti(IV)	$ML_3/MO.H^2.L^3$	57.6^h				
	$MO(HL)_2.H^2/MO.(H_2L)^2$	-0.3^h				
	$ML_3.H^2/MO(HL)_2.H_2L$	-2.9^h				
Mo(VI)	$MO_2L_2/MO_4.(H_2L)^2$	6.44^h				
Zn^{2+}	$ML/M.L$	10.41^h 10.19^p	9.00	11.68	$(-9)^s$	(20)
	$ML_2/M.L^2$	18.52^p	16.91			
	$MHL/M.HL$	3.30^h				
Cd^{2+}	$ML/M.L$		7.69	10.29		
	$ML_2/M.L^2$		13.29			
Hg^{2+}	$ML/M.L$	19.86				
Pb^{2+}	$ML/M.L$		11.95	14.77		
	$ML_2/M.L^2$		18.28			
B(III)	$M(OH)_2L.H/M(OH)_3.H_2L$	-3.70 ± 0.03 -3.30^b	-3.17 -3.14^d	-4.34		
Al^{3+}	$ML/M.L$	16.6^h 16.7^h	14.9^d 14.6^d	19.06		
	$ML_2/M.L^2$	30.0^h 30.3^h	28.4^d 28.1^d	31.10		
	$ML_3/M.L^3$	40.0^h		33.5		
Ga^{3+}	$ML/M.L$	19.24				
In^{3+}	$ML/M.L$	16.34				
As(III)	$ML_2.H/M(OH)_3.(H_2L)^2$			-8.19		
Sb(III)	$ML_2.OH/M(OH)_3.(HL)^2$	3.95^h				
	$ML_2.H/MOHL.H_2L$	-1.23^h				
	$ML_2/ML.L$	14.4	13.0			
	$MHL_2/ML_2.H$	2.00	2.04			
Ge(IV)	$ML_3.H^2/M(OH)_4.(H_2L)^3$	2.27 ± 0.04 3.50^b	3.70			

^{b} 25°, 0.5; ^{d} 25°, 2.0; ^{h} 20°, 0.1; ^{p} 30°, 0.1; ^{r} 25-35°, 0.1; ^{s} 20-30°, 0.1

Bibliography:

H^+ 51WS,57MC,57N,63MNT,64Ma,66LM,66MM, 69CMa,70KC,72MSa,73SHG,75L,75SG,75VH

Be^{2+} 67BZ

Mg^{2+}-Ba^{2+} 64SM

Sc^{3+},Hg^{2+},Ga^{3+},In^{3+} 72GKa

Y^{3+}-Yb^{3+} 70SSK

Th^{4+} 66MMa

UO_2^{2+} 60GR,65BS,65BSa,65SSc

Mn^{2+} 58CG,75SG

Tiron (continued)

Co^{2+}-Cu^{2+},Zn^{2+}-Pb^{2+} 56Na,58Na,59N,60Na,
     63MNT,64SM,66LM,68OOH,69CMa,73SHG

Fe^{3+} 51WS,64Ma

VO^{2+} 66MM,71ZB

Mo(VI) 71SB

Ti(IV) 63Sa

B(III) 60N,68HBa

Al^{3+} 57Na,69HBa

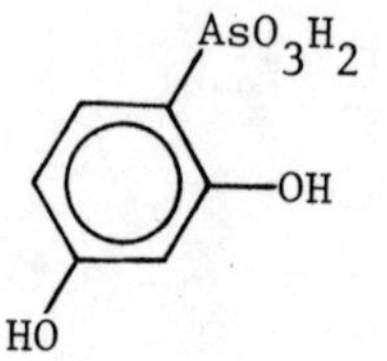

As(III)   64ATa

Sb(III)   70AM,71OB

Ge(IV)   66AT,67PB,73SB

Other references: 54S,55Va,56NV,56S,57N,58N,
     59AM,59CG,65Da,65DM,65DMa,65ND,65ON,
     66KE,67EKM,67LA,68AS,70AK,70E,71CN,
     71N,72MGS,74KA,75VH

$C_6H_7O_5As$ 2,4-Dihydroxybenzenearsonic acid H_4L

Metal ion	Equilibrium	Log K 25°, 0.1	Log K 25°, 0
H^+	HL/H.L		14.27
	H_2L/HL.H		9.82 +0.4
	H_3L/H_2L.H	7.56	7.98 −0.3
	H_4L/H_3L.H	4.01	4.20 ±0.02
UO_2^{2+}	$MH_2L/M.H_2L$	8.76	8.83
	$M(H_2L)_2/M.(H_2L)^2$		14.12

Bibliography:

H^+ 6800,6900,7300

UO_2^{2+} 6900

Other reference: 73VG

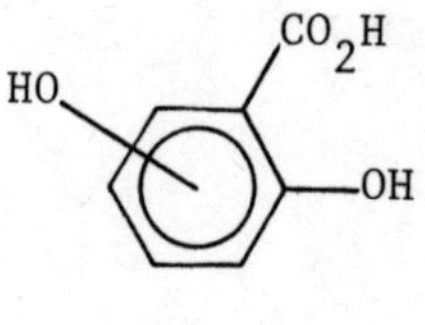

$C_7H_6O_4$ Hydroxysalicylic acid H_3L

Isomer	Metal ion	Equilibrium	Log K 25°, 0.1	Log K 25°, 0
3-Hydroxy (2-pyrocate-chuic acid)	H^+	H_2L/HL.H	8.83[h]	
		H_3L/H_2L.H	4.33[h]	
	Sb(III)	$ML.(OH)^2/M(OH)_3$.HL	4.17[h]	
		ML_2.H/MOHL.H_2L^3	−2.8[h]	

[h] 20°, 0.1

Hydroxysalicylic acid (continued)

Isomer	Metal ion	Equilibrium	Log K 25°, 0.1	Log K 25°, 0
4-Hydroxy (β-resorcylic acid)	H^+	$H_2L/HL.H$	8.60[r]	8.69[c]
		$H_3L/H_2L.H$	3.13 ±0.03	3.19[c]
	UO_2^{2+}	$MH_2L/M.H_2L$	2.10[r]	
		$MHL.H/M.H_2L$	-0.66[r]	
		$M_2HL_2.H^3/M^2.(H_2L)^2$	-4.17[r]	
	Fe^{3+}	$MHL.H/M.H_2L$	3.14[h] 3.19	
		$M(HL)_2.H/MHL.H_2L$	1.75[h] 1.61	
		$M(HL)_3.H/M(HL)_2.H_2L$	-3.7	
		$MH_2L/M.H_2L$	4.80	
5-Hydroxy (gentisic acid)	H^+	$H_3L/H_2L.H$	2.70[r]	
	UO_2^{2+}	$MH_2L/M.H_2L$	1.51[r]	
		$MHL.H/M.H_2L$	-1.00[r]	
	Fe^{3+}	$M(HL)_2.H/MHL.H_2L$	2.07[h]	
		$M(HL)_3.H/M(HL)_2.H_2L$	-3.94[h]	
6-Hydroxy (γ-resorcylic acid)	H^+	$H_3L/H_2L.H$	1.08	1.23
	Fe^{3+}	$MHL.H/M.H_2L$	2.76	
		$M(HL)_2.H/MHL.H_2L$	1.2	

[c] 25°, 1.0; [h] 20°, 0.1; [r] 25°, 0.2

Bibliography:

H^+ 62JSa,62JSb,66P,68OC,70AM,73LS,75SG   Sb(III)   70AM

UO_2^{2+} 68OC   Other references: 64RSa,65DD,65GJS,66GS,66RS,
                                                        69GS,69JJ,71Hb,71Sb,71SG,72PSb,72PSc,
Fe^{3+} 62JSc,66P   72PSd,73JK,73PS,73SJ,74J,75JK,75SJ

$C_7H_6O_4$		3,4-Dihydroxybenzoic acid (protocatechuic acid)		H_3L

Metal ion	Equilibrium	Log K 25°, 0.1	Log K 30°, 0.1	Log K 25°, 1.0
H^+	$HL/H.L$		12.2 ±0.3	(12.80)
	$H_2L/HL.H$	8.84	8.70 ±0.03	8.68
	$H_3L/H_2L.H$	4.35	4.34 ±0.05	4.34
Mg^{2+}	$ML/M.L$		5.67	
	$ML_2/M.L^2$		9.84	

3,4-Dihydroxybenzoic acid (continued)

Metal ion	Equilibrium	Log K 25°, 0.1	Log K 30°, 0.1	Log K 25°, 1.0
Ca^{2+}	$ML/M.L$		3.71	
	$ML_2/M.L^2$		6.36	
Mn^{2+}	$ML/M.L$		7.22	(7.43)
	$ML_2/M.L^2$		12.28	(12.64)
Co^{2+}	$ML/M.L$		7.96	
	$ML_2/M.L^2$		13.36	
	$ML_3/M.L^3$		17.42	
Ni^{2+}	$ML/M.L$		8.27	
	$ML_2/M.L^2$		12.98	
	$ML_3/M.L^3$		16.87	
Cu^{2+}	$ML/M.L$		12.79	
	$ML_2/M.L^2$		22.60	
$Ti(IV)$	$ML_3/MO.H^2.L^3$	58.6 [h]		
	$MO(H_2L)_2.H^2/MO.(H_3L)^2$	-1.35 [h]		
	$ML_3.H^5/MO(H_2L)_2.H_3L$	-3.9 [h]		
$Nb(V)$	$M(HL)_3/MO_2.(HL)^3.H^4$	63.1 [h]		
	$MO_2(H_2L).H/MO_2.H_3L$	2.3 [h]		
	$MO(HL)_2.H/MO_2(H_2L).H_3L$	1.3 [h]		
Zn^{2+}	$ML/M.L$		8.91	
	$ML_2/M.L^2$		15.62	
$B(III)$	$M(OH)_2L.H/M(OH)_3.H_2L$	-5.01		

[h] 20°, 0.1

Bibliography:

H^+ 63MNT,66AP,68AV,75SG

Mg^{2+}-Ca^{2+},Co^{2+},Cu^{2+},Zn^{2+} 63MNT

Mn^{2+} 63MNT,75SG

$Ti(IV)$ 63Sa

Nb(V)   64SH

B(III) 68AV

Other references: 59H,64SB,68JH,69Ha,69HBb, 69HO,70PB,72IM,72JM,72MJ,73KL,73MI

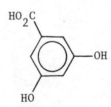

$C_7H_6O_4$	3,5-Dihydroxybenzoic acid (α-resorcylic acid)				H_3L
Metal ion	Equilibrium	Log K 25°, 0.2	Log K 25°, 0	ΔH 25°, 0	ΔS 25°, 0
H^+	$HL/H.L$	10.54			
	$H_2L/HL.H$	9.00	9.08	-2.1	
	$H_3L/H_2L.H$	3.84	4.12	0.8	
UO_2^{2+}	$MH_2L/M.H_2L$	2.13			
	$M(OH)H_2L.H/M.H_2L$	-2.02			

Bibliography:

H^+ 68OC,71PS

UO_2^{2+} 68OC

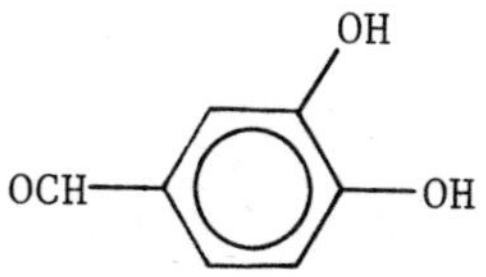

C$_7$H$_6$O$_3$ 3,4-Dihydroxybenzaldehyde (protocatechualdehyde)   H$_2$L

Metal ion	Equilibrium	Log K 25°, 0.1
H$^+$	H$_2$L/HL.H	7.22
B(III)	M(OH)$_2$L.H/M(OH)$_3$.H$_2$L	-3.95
As(III)	M(OH)$_2$L.H/M(OH)$_3$.H$_2$L	-6.18
Ge(IV)	ML$_3$.H^2/M(OH)$_4$.(H$_2$L)^3	2.78

Bibliography: 68AO

Other references: 59H,69Ha,69HO

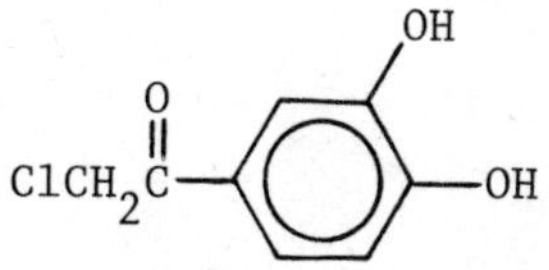

C$_8$H$_7$O$_3$Cl   4-(Chloroacetyl)-1,2-dihydroxybenzene   H$_2$L

Metal ion	Equilibrium	Log K 25°, 0.1
H$^+$	HL/H.L	11.54
	H$_2$L/HL.H	7.45
B(III)	M(OH)$_2$L.H/M(OH)$_3$.H$_2$L	-4.35
Ge(IV)	ML$_3$.H^2/M(OH)$_4$.(H$_2$L)^3	2.20

Bibliography: 69MB

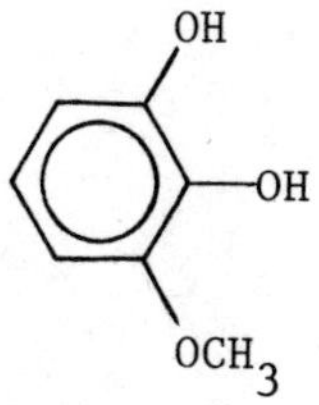

C$_7$H$_8$O$_3$ 1,2-Dihydroxy-3-methoxybenzene (pyrogallol 1-methyl ester)   H$_2$L

Metal ion	Equilibrium	Log K 25°, 0.1
H$^+$	HL/H.L	11.99
	H$_2$L/HL.H	9.20
Cu^{2+}	ML/M.L	13.00
	ML$_2$/M.L^2	23.06

Bibliography: 73SHG

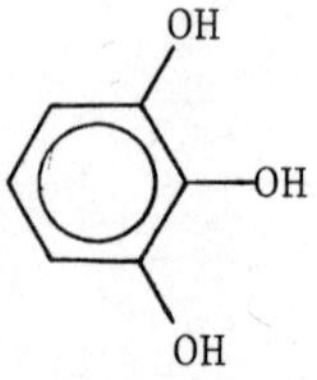

$C_6H_6O_3$		1,2,3-Trihydroxybenzene (pyrogallol)		H_3L

Metal ion	Equilibrium	Log K 25°, 0.1	Log K 20°, 0.1
H^+	$HL/H.L$		14
	$H_2L/HL.H$		11.08
	$H_3L/H_2L.H$	8.98	8.94
Be^{2+}	$MHL/M.HL$		13.5
	$MH_2L/M.H_2L$		4.6
UO_2^{2+}	$M_2L.H^3/M^2.H_3L$	−6.84	
	$MH_2L_2.H/ML.H_3L$	−4.69	
V^{3+}	$ML/M.L$		18.1^i
	$M(OH)_2L/ML.(OH)^2$		19.9^i
	$M(OH)_2L_2/M(OH)_2L.L$		11.0^i
VO^{2+}	$MHL/M.HL$		15.0
	$ML_2/M(HL)^2$		28.7
$Mo(VI)$	$MO_2(HL)_2/MO_4.(H_3L)^2$		5.5 ±0.2
	$MO_2(HL)_2.H/HMO_3L.H_3L$ −3.23		
	$HMO_2(HL)_2/MO_2(HL)_2.H$ 3.70		
$W(VI)$	$MO_2(HL)_2/MO_2(H_3L)^2$		7.0
	$MO_2(HL)_2.H/HMO_3(HL).H_3L$ −2.38		
$B(III)$	$M(OH)_2HL.H/M(OH)_3.H_3L$ −5.05		−4.98
	$M(OH)(HL)_2.H^2/M(OH)_3.(H_3L)^2$ −4.40		
$As(III)$	$M(OH)_2HL.H/M(OH)_3.H_3L$ −6.32		
	$M(HL)_2.H/M(OH)_3.(H_3L)^2$ −6.05		
$Ge(IV)$	$M(HL)_3.H^2/H_2MO_3.(H_3L)^3$ −0.6 ±0.2		

---

^{i} 21°, 0.5

Bibliography:

H^+	67BZ,71AK,75VH	$B(III)$	68HBa,71AK
Be^{2+}	67BZ	$As(III)$	59ARa
V^{3+}	74ZB	$Ge(IV)$	59AM,63AN,67BP,73SB
VO^{2+}	71ZB		
UO_2^{2+}	65Bb		
$Mo(VI),W(VI)$	59H,71SB,73B		

Other references: 56LP,58Pa,59AM,59H,65AM, 66DM,66PRS,66SC,67DM,69HBb,69HE, 70CSa,70E,70GO,71PB,72JM,72MJ,73KL, 74KA,75VH

$C_6H_6O_6S$ 2,3,4-Trihydroxybenzenesulfonic acid (pyrogallolsulfate) H_4L

Metal ion	Equilibrium	Log K 20°, 0.1
H^+	$H_2L/HL.H$	11.3
	$H_3L/H_2L.H$	8.28 ±0.01
Ge(IV)	$ML_3.H^2/M(OH)_4.(H_2L)^3$	1.3

Bibliography:

H^+	67Ba,75VH	Other references: 63HO,66TK,75VH
Ge(IV)	73SB	

$C_7H_6O_5$ 2,4,6-Trihydroxybenzoic acid (pyrogallolcarboxylic acid) H_4L

Metal ion	Equilibrium	Log K 20°, 0.1
H^+	$H_2L/HL.H$	12.3
	$H_3L/H_2L.H$	8.73
	$H_4L/H_3L.H$	3.02
Mo(VI)	$MO_2L_2/MO_4.(H_2L)^2$	5.2

Bibliography:

H^+	67Ba	Other references: 69HK,71AKB
Mo(VI)	71SB	

$C_7H_6O_5$ 3,4,5-Trihydroxybenzoic acid (gallic acid) H_2L

Metal ion	Equilibrium	Log K 25°, 0.1	Log K 20°, 0.1	Log K 25°, 0
H^+	$H_2L/HL.H$		11.45	
	$H_3L/H_2L.H$	8.68	8.70	9.11
	$H_4L/H_3L.H$	4.27	4.26	4.43
UO_2^{2+}	$MH_3L/M.H_3L$		2.3	
Mo(VI)	$MO_2(HL)_2/MO_4.(H_2L)^2$		5.4	
B(III)	$M(OH)_2HL.H/M(OH)_3H_3L$			
		-4.86		

Gallic acid (continued)

Bibliography:

H^+ 69AV,74KKa   B(III)   69AV

UO_2^{2+} 69HS   Other references: 59H,64SB,69HBb,70AK,70PB,
                                                                     71LM,73KL,74KK,75KA
Mo(VI) 71SB

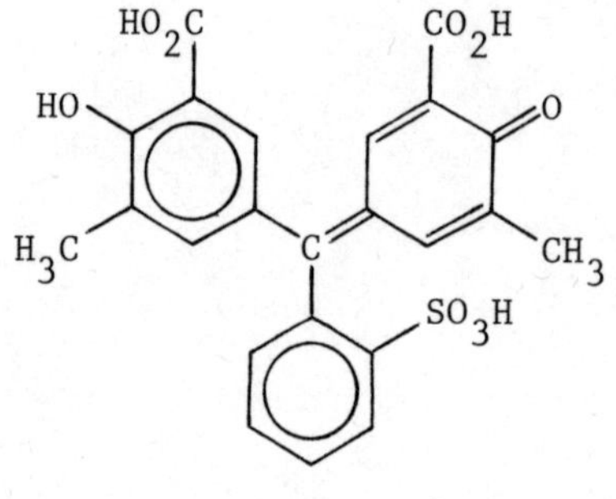

$C_{23}H_{18}O_9S$ 4'-Hydroxy-3,3'-dimethyl-2"-sulfofuchsone-5,5'-dicarboxylic acid H_4L
                              (Eriochrome Cyanine R)

Metal ion	Equilibrium	Log K 20°, 0.1
H^+	HL/H.L	11.85
	$H_2L/HL.H$	5.47
	$H_3L/H_2L.H$	2.23
Be^{2+}	MHL/M.HL	5.45
	$M_2L_2/M^2.L^2$	28.3
Fe^{3+}	ML/M.L	17.9
	$M_2L/M^2.L$	22.5
	$M_2L_2/M^2.L^2$	37.9

Bibliography:

H^+,Be^{2+} 67SK   Other references: 59SM,72GMa,73BSb

Fe^{3+} 65LS

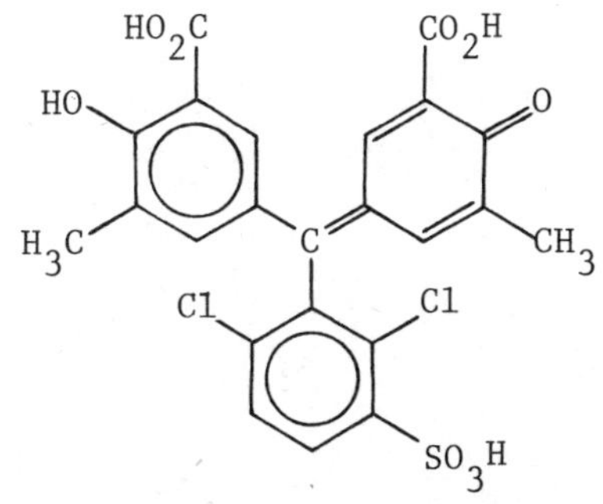

$C_{23}H_{16}O_9Cl_2S$ 2",6"-Dichloro-4'-hydroxy-3,3'-dimethyl-3"-sulfo- fuchsone-5,5'-dicarboxylic acid (Chrome Azurol S) H_4L

Metal ion	Equilibrium	Log K 25°, 0.1	Log K 20°, 0.1
H^+	HL/H.L	11.75	11.80 ±0.01
	$H_2L/HL.H$	4.88	4.80 ±0.09
	$H_3L/H_2L.H$	2.25	2.25 +0.1
Be^{2+}	MHL/M.HL	4.66	4.83
	$M_2L/M^2.L$	15.8	
	$M_2L_2/M^2.L^2$		26.8
UO_2^{2+}	MHL/M.HL		5.35
	$M_2L/M^2.L$		18.3
Cu^{2+}	MHL/M.HL	4.02	
	$M_2L/M^2.L$	13.7	
Fe^{3+}	ML/M.L		15.6
	$M_2L/M^2.L$		20.2
	$M_2L_2/M^2.L^2$		36.2

Bibliography:

H^+ 63LK,67SK,68BS

Be^{2+} 67SK,68BS

UO_2^{2+} 70CSb

Cu^{2+} 66SL

Fe^{3+} 63LK

Other references: 60Sc,62AMY,62SDa,63SD, 63SDb,63SSa,63SSc,64MDa,64SS,66DMD,66Sa, 67IH,67S,67SMa,67SPa,67SSD,68MPS,71N, 72Ba,72MSD,74Ib,75Ba,75P

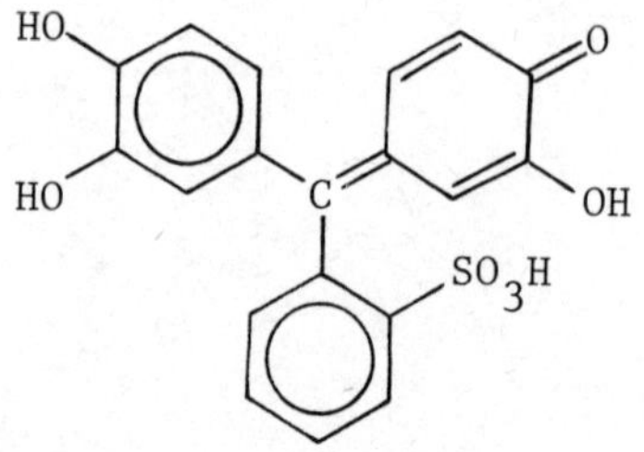

$C_{19}H_{14}O_7S$ 3,3',4'-Trihydroxyfuchsone-2"-sulfonic acid H_4L
                                  (pyrocatechol violet)

Metal ion	Equilibrium	Log K 25°, 0.1
H^+	HL/H.L	12.8
	$H_2L/HL.H$	9.76
	$H_3L/H_2L.H$	7.80
	$H_4L/H_3L.H$	0.8
Th^{4+}	$MH_2L/M.H_2L$	9.83
	$MH_3L_2/MH_2L.HL$	15.38
	$MH_4L_3/MH_3L_2.HL$	13.62
	$M_2HL/M^2.HL$	20.04
UO_2^{2+}	$MH_2L/M.H_2L$	7.05
	$M(H_2L)_2/M.(H_2L)^2$	12.65
	$M_2HL/M_2.HL$	14.7
Cr^{3+}	$MH_2L/M.H_2L$	8.70
	$MH_3L_2/MH_2L.HL$	11.15
	$MH_4L_3/MH_3L_2.HL$	8.96

Bibliography:

H^+-UO_2^{2+} 73CG

Cr^{3+} 73CDC

Other references: 56RC,67MP,68BRN,68TL,70BR, 70BRa,72WV,72YV

$C_{10}H_7O_2N$ 2-Nitroso-1-naphthol   HL

Metal ion	Equilibrium	Log K 25°, 0.1	Log K 25°, 0.5	Log K 25°, 0	ΔH 25°, 0	ΔS 25°, 0
H^+	HL/H.L	7.25 -0.01	7.21	7.47 ±0.01	-4.8	18
			7.22[a]	7.33[b]		
Th^{4+}	$ML/M.L$	8.30				
	$ML_2/M.L^2$	15.54				
	$ML_3/M.L^3$	23.04				
	$ML_4/M.L^4$	29.26				
Zn^{2+}	$ML/M.L$	3.91				
Cd^{2+}	$ML/M.L$	3.33				

[a] 25°, 1.0;   [b] 25°, 2.0

Bibliography:

H^+ 55DJ,70BTL,70SM Zn^{2+},Cd^{2+} 70SM

Th^{4+} 56DD,60R   Other references: 71LG,75Lb

$C_{10}H_7O_5NS$ 2-Nitroso-1-naphthol-4-sulfonic acid (nitroso-NW acid) H_2L

Metal ion	Equilibrium	Log K 25°, 0.1	Log K 25°, 0	ΔH 25°, 0	ΔS 25°, 0
H^+	HL/H.L	6.20	6.62 ±0.01	-4.6	15
	$H_2L/HL.H$	2.63	1.94		
Y^{3+}	$ML/M.L$	2.87	3.97		
La^{3+}	$ML/M.L$	3.46	4.56		
	$ML_3/M.L^3$	9.9			
Ce^{3+}	$ML/M.L$	(2.70)	(3.79)		
Nd^{3+}	$ML/M.L$	3.47	4.57		
Sm^{3+}	$ML/M.L$	3.48	4.57		
Gd^{3+}	$ML/M.L$	3.41	4.50		
Dy^{3+}	$ML/M.L$	3.18	4.27		
Ho^{3+}	$ML/M.L$	3.13	4.22		
Er^{3+}	$ML/M.L$	3.09	4.19		
Yb^{3+}	$ML/M.L$	3.09	4.18		
Mn^{2+}	$ML/M.L$		2.07		

Nitroso-NW acid (continued)

Metal ion	Equilibrium	Log K 25°, 0.1	Log K 25°, 0	ΔH 25°, 0	ΔS 25°, 0
Co^{2+}	ML/M.L	4.34			
Ni^{2+}	ML/M.L	5.50	6.28		
	$ML_2/M.L^2$	10.42	11.26		
	$ML_3/M.L^2$	17.25	14.55		
Cu^{2+}	ML/M.L		7.8		
	$ML_2/M.L^2$		13.0		
VO^{2+}	ML/M.L	5.96			
Zn^{2+}	ML/M.L	3.06	3.86		
Cd^{2+}	ML/M.L	2.36	3.12		
Pb^{2+}	ML/M.L	3.84	4.74		

Bibliography:

H^+ 70BTL,73Sb

Y^{3+}-Yb^{3+} 67Ma

Mn^{2+},Cu^{2+}-Cd^{2+} 61M

Co^{2+} 72BTL

Ni^{2+} 61M,73S

VO^{2+} 75ML

Pb^{2+} 66Md

Other references: 69MS,70SM,71LS,72LS,72LPS,
74LS,75BSP,75Lb

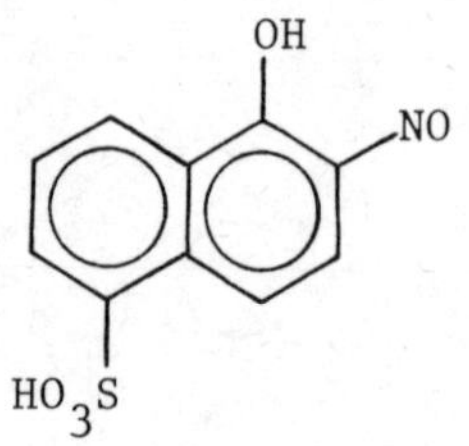

| $C_{10}H_7O_5NS$ | 2-Nitroso-1-naphthol-5-sulfonic acid (nitroso-C acid) | | | H_2L |

Metal ion	Equilibrium	Log K 25°, 0.1	Log K 25°, 0.5	Log K 25°, 0
H^+	HL/H.L	6.89	6.79	7.32
Pr^{3+}	ML/M.L	3.50		
	$ML_2/M.L^2$	6.3		
Nd^{3+}	ML/M.L	3.83		
	$ML_2/M.L^2$	6.9		
Eu^{3+}	ML/M.L	3.86	3.39 [c] 3.31 [c]	5.01
	$ML_2/M.L^2$	6.71	6.23 [c] 6.31 [c]	8.21
Ni^{2+}	ML/M.L	5.76	5.45	6.50
	$ML_2/M.L^2$	10.88	10.53	11.65
	$ML_3/M.L^3$	15.26	15.11	15.30
Zr^{4+}	ML/M.L	8.27	7.98	9.71
VO^{2+}	ML/M.L	6.19		6.94
Zn^{2+}	ML/M.L	3.58		
	$ML_2/M.L^2$	6.29		

[c] 25°, 1.0

Nitroso-C acid (continued)

Metal ion	Equilibrium	Log K 25°, 0.1	Log K 25°, 0.5	Log K 25°, 0
Cd^{2+}	$ML/M.L$	3.18	2.60	
	$ML_2/M.L^2$	6.14	5.81	

Bibliography:

H^+	69MSb	VO^{2+}	75ML
Pr^{3+},Nd^{3+}	70MSL	Zn^{2+}	69MSb
Eu^{3+}	72MS	Cd^{2+}	71MSa
Ni^{2+}	71Sa	Other reference:	70SM
Zr^{4+}	72MP		

$C_{10}H_7O_5NS$	2-Nitroso-1-naphthol-6-sulfonic acid	H_2L

Metal ion	Equilibrium	Log K 25°, 0.1	Log K 25°, 0
H^+	$HL/H.L$	7.02	7.39
Ni^{2+}	$ML/M.L$	5.65	6.41
	$ML_2/M.L^2$	10.77	11.57
	$ML_3/M.L^3$	15.2	15.4
Cu^{2+}	$ML/M.L$	7.30	8.01
Zn^{2+}	$ML/M.L$	3.71	
	$ML_2/M.L^2$	6.58	
Cd^{2+}	$ML/M.L$	3.02	
	$ML_2/M.L^2$	5.46	

Bibliography:

H^+	70MSM	Zn^{2+}	70MSM
Ni^{2+}	71MS	Cd^{2+}	71MSa
Cu^{2+}	72ML		

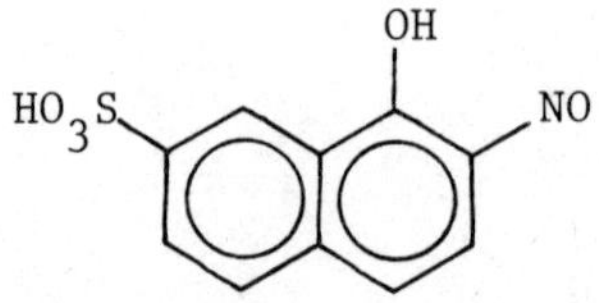

$C_{10}H_7O_5NS$ 2-Nitroso-1-naphthol-7-sulfonic acid H_2L

Metal ion	Equilibrium	Log K 25°, 0.1	Log K 25°, 0.5	Log K 25°, 0
H^+	HL/H.L	6.97	6.81	7.40
Ni^{2+}	ML/M.L	5.94		6.75
	$ML_2/M.L^2$	10.98		11.92
	$ML_3/M.L^3$	18.05		15.03
Zn^{2+}	ML/M.L	3.76		
	$ML_2/M.L^2$	6.45		
Cd^{2+}	ML/M.L	2.96		
	$ML_2/M.L^2$	5.29		

Bibliography:

H^+ 70MS Zn^{2+}, Cd^{2+} 70MS

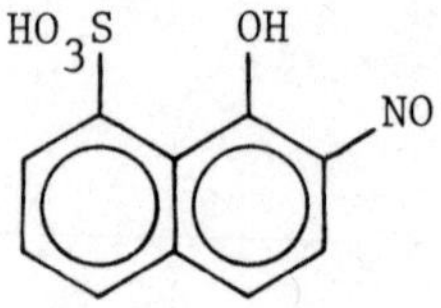

Ni^{2+} 73Sb

$C_{10}H_7O_5NS$ 2-Nitroso-1-naphthol-8-sulfonic acid (nitroso-S acid) H_2L

Metal ion	Equilibrium	Log K 25°, 0.1	Log K 25°, 0.5	Log K 25°, 0
H^+	HL/H.L	7.74	7.50 / 7.50[c]	8.19
Eu^{3+}	ML/M.L	5.46		6.78
	$ML_2/M.L^2$	8.42		10.33
Ni^{2+}	ML/M.L	5.93	5.31[c] / 5.13[c]	6.92
	$ML_2/M.L^2$	10.42	9.81[c] / 9.68[c]	11.41
	$ML_3/M.L^3$	13.87	13.81[c] / 13.73[c]	13.88
Cu^{2+}	ML/M.L	7.67		8.64
	$ML_2/M.L^2$	11.77		12.42
VO^{2+}	ML/M.L	7.19		
Zn^{2+}	ML/M.L	4.04		
	$ML_2/M.L^2$	6.34		
Cd^{2+}	ML/M.L	3.41		

[c] 25°, 1.0

Nitroso-S acid (continued)

Bibliography:

H^+ 69MSa,70SM Cu^{2+} 72RL

Eu^{3+} 72MS VO^{2+} 75ML

Ni^{2+} 70MMT Zn^{2+},Cd^{2+} 69MSa,70SM

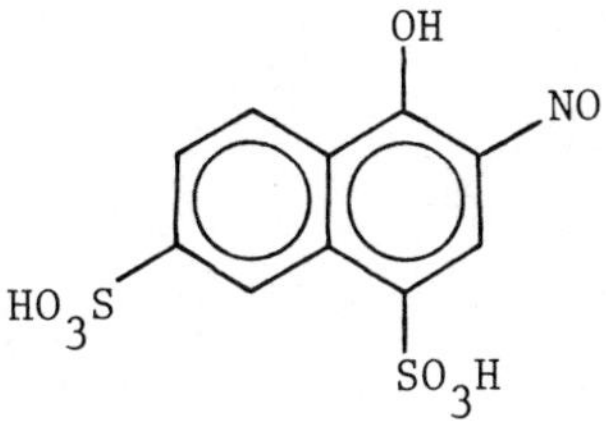

$C_{10}H_7O_8NS_2$		2-Nitroso-1-naphthol-4,6-disulfonic acid			H_3L
Metal ion	Equilibrium	Log K 25°, 0.1	Log K 25°, 0.5	Log K 25°, 0	
H^+	HL/H.L	5.87	5.65	6.51	
Pr^{3+}	ML/M.L $ML_2/M.L^2$	3.67 5.89			
Nd^{3+}	ML/M.L $ML_2/M.L^2$	3.76 6.02			
Sm^{3+}	ML/M.L $ML_2/M.L^2$	3.89 6.38			
Eu^{3+}	ML/M.L $ML_2/M.L^2$	3.82 6.35			
Gd^{3+}	ML/M.L $ML_2/M.L^2$	3.70 6.26			
Tb^{3+}	ML/M.L $ML_2/M.L^2$	3.48 6.14			
Dy^{3+}	ML/M.L $ML_2/M.L^2$	3.39 6.02			
Ni^{2+}	ML/M.L $ML_2/M.L^2$ $ML_3/M.L^3$	5.45 10.07 13.60	4.95 9.58 13.55	6.58 10.79 12.3	
VO^{2+}	ML/M.L	5.68			
Zn^{2+}	ML/M.L $ML_2/M.L^2$	3.51 5.83		4.65 7.25	
Cd^{2+}	ML/M.L $ML_2/M.L^2$	2.82 4.93	2.18	4.16	

Bibliography:

H^+,Ni^{2+} 73Sb Zn^{2+} 73MS

$Pr^{3+}-Dy^{3+}$ 74S Cd^{2+} 74SJ

VO^{2+} 75ML

$C_{10}H_7O_8NS_2$ 2-Nitroso-1-naphthol-4,7-disulfonic acid H_3L

Metal ion	Equilibrium	Log K 25°, 0.1	Log K 25°, 0.5	Log K 25°, 0
H^+	HL/H.L	5.83	5.58 [c] 5.55	6.45
Pr^{3+}	ML/M.L	3.80		
	$ML_2/M.L^2$	5.86		
Nd^{3+}	ML/M.L	3.88		
	$ML_2/M.L^2$	6.00		
Sm^{3+}	ML/M.L	4.02		
	$ML_2/M.L^2$	6.27		
Eu^{3+}	ML/M.L	3.96		
	$ML_2/M.L^2$	6.24		
Gd^{3+}	ML/M.L	3.83		
	$ML_2/M.L^2$	6.23		
Tb^{3+}	ML/M.L	3.65		
	$ML_2/M.L^2$	6.19		
Dy^{3+}	ML/M.L	3.59		
	$ML_2/M.L^2$	6.05		
Ni^{2+}	ML/M.L	5.55	9.98	6.74
	$ML_2/M.L^2$	9.98	9.47	10.71
	$ML_3/M.L^3$	13.00	13.03	11.7
Zn^{2+}	ML/M.L	3.63		4.79
	$ML_2/M.L^2$	5.73		6.69
Cd^{2+}	ML/M.L	2.92	2.28	4.22
	$ML_2/M.L^2$	4.92		6.2

[c] 25°, 1.0

Bibliography:

H^+,Ni^{2+}	73Sb		Zn^{2+}	73MS
$Pr^{3+}-Dy^{3+}$	74S		Cd^{2+}	74SJ

$$HO_3S \quad OH$$

$$NO$$

$$SO_3H$$

$C_{10}H_7O_8NS_2$		2-Nitroso-1-naphthol-4,8-disulfonic acid (nitroso-Schollkopf's acid)			H_3L
Metal ion	Equilibrium	Log K 25°, 0.1	Log K 25°, 0.5	Log K 25°, 0	
H^+	HL/H.L	6.66	6.32 6.24[c]	7.32	
Sc^{3+}	ML/M.L $ML_2/M.L^2$	6.18 10.21		7.96 12.24	
Y^{3+}	ML/M.L $ML_2/M.L^2$	4.59 7.66		6.49 9.79	
La^{3+}	ML/M.L $ML_2/M.L^2$	4.65 7.50		6.50 9.54	
Ce^{3+}	ML/M.L $ML_2/M.L^2$	4.95 7.73		6.79 9.76	
Pr^{3+}	ML/M.L $ML_2/M.L^2$	5.21 7.86		7.05 9.88	
Nd^{3+}	ML/M.L $ML_2/M.L^2$	5.33 8.00		7.17 10.00	
Sm^{3+}	ML/M.L $ML_2/M.L^2$	5.49 8.34		7.33 10.35	
Eu^{3+}	ML/M.L $ML_2/M.L^2$	5.49 8.36		7.34 10.40	
Gd^{3+}	ML/M.L $ML_2/M.L^2$	5.38 8.31		7.23 10.39	
Tb^{3+}	ML/M.L $ML_2/M.L^2$	5.19 8.26		7.04 10.33	
Dy^{3+}	ML/M.L $ML_2/M.L^2$	5.07 8.30		6.93 10.37	
Ho^{3+}	ML/M.L $ML_2/M.L^2$	4.91 8.22		6.75 10.30	
Er^{3+}	ML/M.L $ML_2/M.L^2$	4.74 7.96		6.59 10.03	
Tm^{3+}	ML/M.L $ML_2/M.L^2$	4.70 7.91		6.54 9.96	
Yb^{3+}	ML/M.L $ML_2/M.L^2$	4.81 7.88		6.66 9.91	
Lu^{3+}	ML/M.L $ML_2/M.L^2$	4.81 7.84		6.67 9.91	

[c] 25°, 1.0

Nitroso-Schollkopf's acid (continued)

Metal ion	Equilibrium	Log K 25°, 0.1	Log K 25°, 0.5	Log K 25°, 0
Ni^{2+}	ML/M.L	5.74	4.96 4.66[c]	7.05
	$ML_2/M.L^2$	9.51	8.89 8.69[c]	10.29
	$ML_3/M.L^3$	12.6	12.6 12.5 [c]	11.6
Zn^{2+}	ML/M.L	3.78		
	$ML_2/M.L^2$	5.79		
Cd^{2+}	ML/M.L	3.26		
	$ML_2/M.L^2$	5.37		

[c] 25°, 1.0

Bibliography:

H^+ 69MSa

Sc^{3+},Y^{3+} 73Sa

$La^{3+}-Lu^{3+}$ 72MS,73S

Ni^{2+} 70MM

Zn^{2+},Cd^{2+} 69MSa,70SM

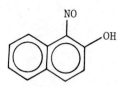

$C_{10}H_7O_2N$ HL

1-Nitroso-2-naphthol

Metal ion	Equilibrium	Log K 25°, 0.1	Log K 25°, 0.5	Log K 25°, 0	ΔH 25°, 0	ΔS 25°, 0
H^+	HL/H.L	7.65 ±0.02	7.59 7.62[c]	7.87 ±0.01	-4.5	21
Gd^{3+}	ML/M.L	4.70				
Th^{4+}	ML/M.L	8.50				
	$ML_2/M.L^2$	16.13				
	$ML_3/M.L^3$	24.02				
	$ML_4/M.L^4$	30.28				
Zn^{2+}	ML/M.L	4.63				

[c] 25°, 1.0

Bibliography:

H^+ 55DJ,69MSc,70BTL

Gd^{3+} 69MSc

Th^{4+} 55DD,60R

Zn^{2+} 71MSa

Other references: 64AS,69BFa,75Lb

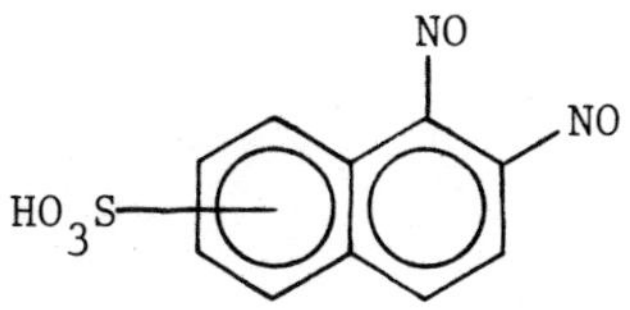

$C_{10}H_7O_5NS$		1-Nitroso-2-naphthol-sulfonic acid		H_2L

Isomer	Metal ion	Equilibrium	Log K 25°, 0.1	Log K 25°, 0
4-Sulfonic acid	H^+	HL/H.L	7.15 7.00[b]	7.56 7.00[c]
	Ni^{2+}	ML/M.L	6.43 6.05[b]	7.20 5.92[c]
		$ML_2/M.L^2$	12.28 11.83[b]	13.15 11.73[c]
		$ML_3/M.L^3$	17.25 16.94[b]	17.32 16.97[c]
	Zn^{2+}	ML/M.L	4.30 3.89[b]	5.08 3.72[c]
		$ML_2/M.L^2$	7.49 7.00[b]	8.39 6.88[d]
5-Sulfonic acid	H^+	HL/H.L	7.23 7.08[b]	7.65 7.10[c]
	Ni^{2+}	ML/M.L	6.56 6.20[b]	7.35 6.10[c]
		$ML_2/M.L^2$	12.63 12.21[b]	13.48 12.11[c]
		$ML_3/M.L^3$	17.90 17.50[b]	18.01 17.56[c]
	Zn^{2+}	ML/M.L	4.46 4.05[b]	5.26 3.89[c]
		$ML_2/M.L^2$	7.99 7.46[b]	8.92 7.32[c]
	Cd^{2+}	ML/M.L	3.46 2.95[b]	4.38 2.81[c]
		$ML_2/M.L^2$	6.1	7.6

[b] 25°, 0.5;   [c] 25°, 1.0

Bibliography:

H^+, Ni^{2+} 73Sb Cd^{2+} 74SJ

Zn^{2+} 73MS

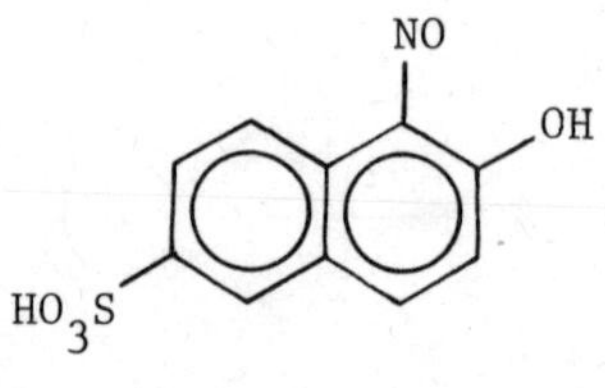

$C_{10}H_7O_5NS$ 1-Nitroso-2-naphthol-6-sulfonic acid (nitroso-Schaffer's acid) H_2L

Metal ion	Equilibrium	Log K 25°, 0.1	Log K 25°, 0.5	Log K 25°, 0
H^+	HL/H.L	7.22	7.14 7.13[c]	7.60
Y^{3+}	ML/M.L	4.24		
La^{3+}	ML/M.L	3.94		
Ce^{3+}	ML/M.L	4.17		
Pr^{3+}	ML/M.L	4.39 -0.1		
Nd^{3+}	ML/M.L	4.47 -0.1		
Sm^{3+}	ML/M.L	4.69		
Eu^{3+}	ML/M.L	4.64 -0.3		
Gd^{3+}	ML/M.L	4.57 +0.4		
Tb^{3+}	ML/M.L	4.42		
Dy^{3+}	ML/M.L	4.38		
Ho^{3+}	ML/M.L	4.40		
Er^{3+}	ML/M.L	4.41		
Tm^{3+}	ML/M.L	4.47		
Yb^{3+}	ML/M.L	4.53		
Lu^{3+}	ML/M.L	4.58		
Ni^{2+}	ML/M.L	6.64	6.29	7.41
	$ML_2/M.L^2$	12.77	12.39	13.58
	$ML_3/M.L^3$	18.21	17.99	18.26
Cu^{2+}	ML/M.L	8.66	8.45	9.45
Zn^{2+}	ML/M.L	4.19	3.70	5.05
	$ML_2/M.L^2$	7.73	7.31	8.56
Cd^{2+}	ML/M.L	3.35		
	$ML_2/M.L^2$	6.45		

---

[c] 25°, 1.0

Bibliography:

H^+	69MSb,69MSc	Cu^{2+}	72ML
Y^{3+}-Lu^{3+}	69MSc,70MSL,71MSP,73PM	Zn^{2+}	69MSb
Ni^{2+}	71Sa	Cd^{2+}	71MSa

$C_{10}H_7O_5NS$		1-Nitroso-2-naphthol-7-sulfonic acid		H_2L
Metal ion	Equilibrium	Log K 25°, 0.1	Log K 25°, 0	
H^+	HL/H.L	7.31		
Eu^{3+}	$ML/M.L$	4.80	5.86	
	$ML_2/M.L^2$	8.29	9.98	
Ni^{2+}	$ML/M.L$	6.73	7.52	
	$ML_2/M.L^2$	12.75	13.59	
	$ML_3/M.L^3$	18.05	18.13	
Zn^{2+}	$ML/M.L$	4.13		
	$ML_2/M.L^2$	7.50		
Cd^{2+}	$ML/M.L$	3.47		
	$ML_2/M.L^2$	6.26		

Bibliography:

H^+, Zn^{2+}	70MSM		Ni^{2+}	73Sb
Eu^{3+}	71MSP		Cd^{2+}	71MSa

$C_{10}H_7O_8NS_2$		1-Nitroso-2-naphthol-3,6-disulfonic acid (nitroso-R acid)				H_3L
Metal ion	Equilibrium	Log K 25°, 0.1	Log K 25°, 0.5	Log K 25°, 0	ΔH 25°, 0	ΔS 25°, 0
H^+	HL/H.L	6.88	6.60	7.52 ±0.01	-3.9	21
			6.55[c]	6.64[d]		
Y^{3+}	$ML/M.L$	4.48	3.59	6.24		
	$ML_2/M.L^2$	7.83	6.83			
	$ML_3/M.L^3$	11.29	10.43			
La^{3+}	$ML/M.L$	4.37	3.49	6.19		
	$ML_2/M.L^2$	7.83	6.67			
	$ML_3/M.L^3$	11.24	10.19			
Ce^{3+}	$ML/M.L$	4.42				
Nd^{3+}	$ML/M.L$	5.01				
Sm^{3+}	$ML/M.L$	5.15				

[c] 25°, 1.0;   [d] 25°, 2.0

Nitroso-R acid (continued)

Metal ion	Equilibrium	Log K 25°, 0.1	Log K 25°, 0.5	Log K 25°, 0	ΔH 25°, 0	ΔS 25°, 0
Eu^{3+}	$ML/M.L$	4.87		6.76		
	$ML_2/M.L^2$	7.68		9.84		
Gd^{3+}	$ML/M.L$	4.92				
Dy^{3+}	$ML/M.L$	4.73				
Ho^{3+}	$ML/M.L$	4.70				
Er^{3+}	$ML/M.L$	4.65				
Yb^{3+}	$ML/M.L$	4.74				
Mn^{2+}	$ML/M.L$	2.69		3.73		
Co^{2+}	$ML/M.L$	5.40				
Ni^{2+}	$ML/M.L$	6.9		8.3		
	$ML_2/M.L^2$	12.5		13.4		
	$ML_3/M.L^3$	17.3				
Cu^{2+}	$ML/M.L$	7.7		9.9		
	$ML_2/M.L^2$	15.0		15.6		
VO^{2+}	$ML/M.L$	6.71		7.96		
Zn^{2+}	$ML/M.L$	4.46		5.73		
	$ML_2/M.L^2$	7.10		7.63		
Cd^{2+}	$ML/M.L$	3.42		4.65		
	$ML_2/M.L^2$	6.00		6.59		
Pb^{2+}	$ML/M.L$	4.64		6.07		
	$ML_2/M.L^2$	7.37		8.34		

Bibliography:

H^+ 60M,70BTL

Y^{3+},La^{3+} 67M

$Ce^{3+}-Sm^{3+},Gd^{3+}-Yb^{3+}$ 68M

Eu^{3+} 71MSP

$Mn^{2+},Ni^{2+}-Cd^{2+}$ 61M

Co^{2+} 72BTL

VO^{2+} 75ML

Pb^{2+} 66Nb

Other references: 63BG,64MSD,65MSG,66MSP,
69MSg,72GD,72L,72MD,72SMD,73MD,73MDa,
73SBa,73SBb,74AN,75BSP,75Lb

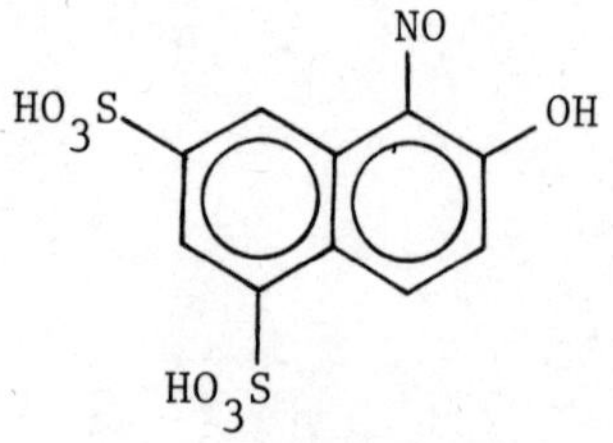

$C_{10}H_7O_8NS_2$		1-Nitroso-2-naphthol-5,7-disulfonic acid			H_3L
Metal ion	Equilibrium	Log K 25°, 0.1	Log K 25°, 0.5	Log K 25°, 0	
H^+	$HL/H.L$	6.77	6.50 6.49[c]	7.40	

[c] 25°, 1.0

E. NAPHTHOLS

1-Nitroso-2-naphthol-5,7-disulfonic acid (continued)

Metal ion	Equilibrium	Log K 25°, 0.1	Log K 25°, 0.5	Log K 25°, 0
Ni^{2+}	ML/M.L	6.54	6.01 5.80[c]	7.40
	$ML_2/M.L^2$	12.08	11.60 11.59[c]	12.78
	$ML_3/M.L^3$	16.58	16.70 16.68[c]	15.3

[c] 25°, 1.0

Bibliography: 74SR

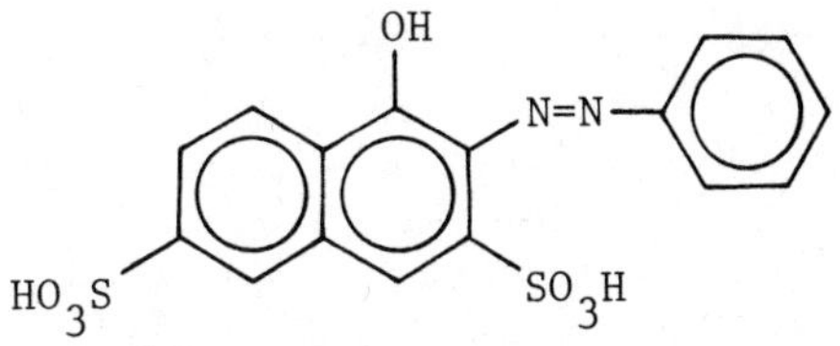

$C_{16}H_{12}O_7N_2S_2$ 1-Hydroxy-2-(phenylazo)naphthalene-3,6-disulfonic acid H_3L

Metal ion	Equilibrium	Log K 20°, 0.1
H^+	HL/H.L	10.66
Cu^{2+}	ML/M.L	8.84
	$ML_2/M.L^2$	17.55

Bibliography: 64SM

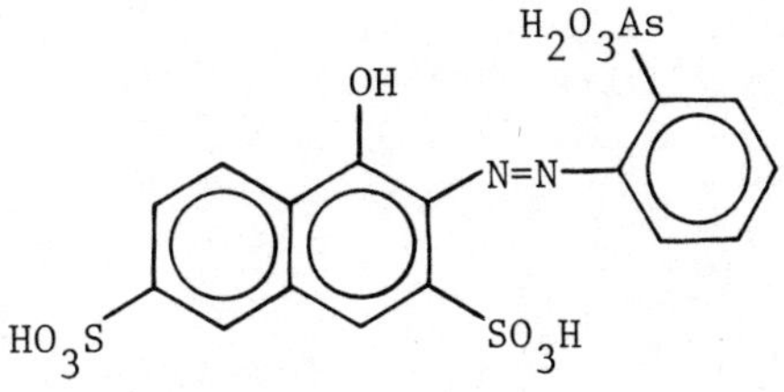

$C_{16}H_{13}O_{10}N_2AsS_2$ 2-(2-Arsonophenylazo)-1-hydroxynaphthalene-3,6-disulfonic acid H_5L

Metal ion	Equilibrium	Log K 25°, 0.1
H^+	HL/H.L	11.40
	$H_2L/HL.H$	8.05
	$H_3L/H_2L.H$	3.09
Mg^{2+}	ML/M.L	5.35
Ca^{2+}	ML/M.L	3.50
Sr^{2+}	ML/M.L	2.0
Ba^{2+}	ML/M.L	1.8

Bibliography: 71KTa

$C_{16}H_{13}O_{10}N_2ZS_2$ 1-(2-Substitutedphenylazo)-2-hydroxynaphthalene-3,6-disulfonic acid H_5L

Z =	Metal ion	Equilibrium	Log K 25°, 0.1
2-Phosphono	H^+	HL/H.L	11.10
		H_2L/HL.H	6.49
	Mg^{2+}	ML/M.L	4.83
	Ca^{2+}	ML/M.L	3.80
		MHL/M.HL	2.98
	Sr^{2+}	ML/M.L	3.06
		MHL/M.HL	2.97
2-Arsono	H^+	HL/H.L	10.80
(Thorin)		H_2L/HL.H	7.86
		H_3L/H_2L.H	3.44
	Mg^{2+}	ML/M.L	5.20
	Ca^{2+}	ML/M.L	4.17
	Sr^{2+}	ML/M.L	2.87

Bibliography: 71KMa,71KTa

Other references: 63SD,63SDa,63SDf,64SD,64SDa,64SM,66S,66Sa,66SD,67S,68GS,71EI,74ND

$C_{16}H_{12}O_5N_2S$ 1-(2-Hydroxy-5-sulfophenylazo)-2-naphthol (Solochrome Violet R) H_3L

Metal ion	Equilibrium	Log K 25°, 0.1	Log K 25°, 0	ΔH 25°, 0	ΔS 25°, 0
H^+	HL/H.L	12.78	13.40 ±0.01	−10.8	25
	H_2L/HL.H	6.86	7.23 ±0.03	−3.4	22
	H_4L_2/$(H_2L)^2$		3.29	−9.2	−16
Mg^{2+}	ML/M.L		8.6		
	ML_2/M.L^2		13.6		

Solochrome Violet R (continued)

Metal ion	Equilibrium	Log K 25°, 0.1	Log K 25°, 0	ΔH 25°, 0	ΔS 25°, 0
Ca^{2+}	$ML/M.L$		6.6		
	$ML_2/M.L^2$		10		
Ni^{2+}	$ML/M.L$		15.9		
	$ML_2/M.L^2$		26.3		
	$ML/MOHL.H$		10.6		
Cu^{2+}	$ML/M.L$		21.8		
	$ML/MOHL.H$		10.8		
Cr^{3+}	$ML_2/ML.L$		17.25[r]	(-6)[s]	(60)[r]
	$ML/MOHL.H$		6.88	(-8)[t]	(5)
	$MOHL/M(OH)_2L.H$		9.82	(-10)[t]	(15)
	$M(OH)_2L/M(OH)_3L.H$		12.12		
Zn^{2+}	$ML/M.L$		13.5		
	$ML_2/M.L^2$		20.9		
	$ML/MOHL.H$		9.6 ±0.2		
Cd^{2+}	$ML/MOHL.H$		10.5		
Pb^{2+}	$ML/M.L$		12.5		
	$ML/MOHL.H$		10.5		
Al^{3+}	$ML/M.L$		18.4		
	$ML_2/M.L^2$		31.6		
	$ML/MOHL.H$		6.4		

[r] 75°, 0; [s] 75-100°, 0; [t] 25-40°, 0

Bibliography:

H^+ 61CR,68CR Cr^{3+} 62CR

Mg^{2+}-Cu^{2+},Zn^{2+}-Al^{3+} 62CRa,63CE   Other references: 56BE,69FB

$C_{20}H_{13}O_7N_3S$		2-Hydroxy-1-(1-hydroxy-2-naphthylazo)-6-nitronaphthalene-4-sulfonic acid (Eriochrome Black T)		H_3L

Metal ion	Equilibrium	Log K 25°, 0.1	Log K 20°, 0.3	Log K 25°, 0
H^+	$HL/H.L$	11.39[h] 11.55	11.31	11.95
	$H_2L/HL.H$	6.80[h] 6.9	6.80	(5.81)

[h] 20°, 0.1

Eriochrome Black T (continued)

Metal ion	Equilibrium	Log K 25°, 0.1	Log K 20°, 0.3	Log K 25°, 0
Co^{2+}	ML/M.L		20.0	
Cu^{2+}	ML/M.L		21.38	
Zn^{2+}	ML/M.L		12.31	
Cd^{2+}	ML/M.L		12.74	
Pb^{2+}	ML/M.L		13.19	

Bibliography:

H^+ 48SB,59DL,63CRM,68CR,68KS Cd^{2+},Pb^{2+} 68KS

Co^{2+} 67K   Other references: 63R,71Ka,73MP

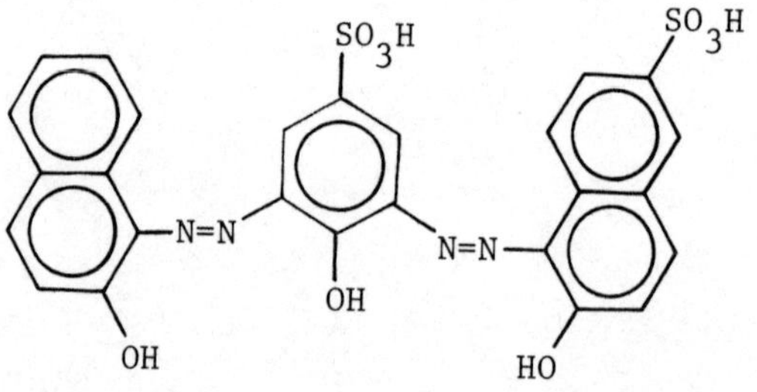

Cu^{2+},Zn^{2+} 67KE

$C_{26}H_{18}O_9N_4S_2$ 1-[3-(2-Hydroxy-1-naphthylazo)-2-hydroxy-5-sulfophenylazo]-2-hydroxynaphthalene-6-sulfonic acid (alizarin acid black SN) H_5L

Metal ion	Equilibrium	Log K 25°, 0.1
H^+	$H_3L/H_2L.H$	5.79
Ca^{2+}	$MHL.H/M.H_2L$	-6.3
	$M_2L.H^2/M^2.H_2L$	-13.9
	$M_2L_2.H^4/M^2.(H_2L)^2$	-30.1

Bibliography: 62RA

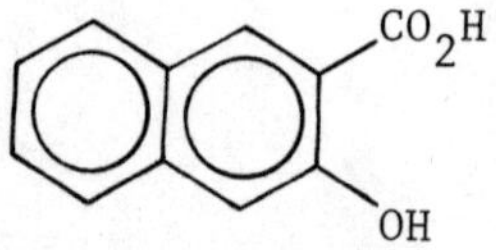

$C_{11}H_8O_3$ 3-Hydroxy-2-naphthoic acid H_2L

Metal ion	Equilibrium	Log K 25°, 0.1	Log K 25°, 0.5	Log K 25°, 0
H^+	HL/H.L	12.48	12.37 12.39[c]	12.84
	$H_2L/HL.H$	2.54	2.41 2.41[c]	2.75 ±0.04

[c] 25°, 1.0

3-Hydroxy-2-naphthoic acid (continued)

Metal ion	Equilibrium	Log K 25°, 0.1	Log K 25°, 0.5	Log K 25°, 0
Be^{2+}	ML/M.L	11.74	11.42 11.37[c]	12.51
Cu^{2+}	ML/M.L	9.49	9.20 9.22[c]	10.28
	$ML_2/M.L^2$			19.8
Al^{3+}	ML/M.L	12.69	12.94 13.08[c]	13.38

[c] 25°, 1.0

Bibliography:

H^+ 57BDH,66Ma Cu^{2+} 66Ma

Be^{2+},Al^{3+} 66Mc   Other reference: 69GSa

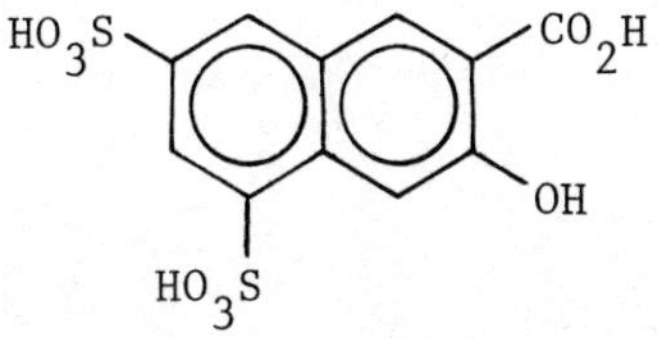

$C_{11}H_8O_9S_2$		3-Hydroxy-5,7-disulfo-2-naphthoic acid		H_4L

Metal ion	Equilibrium	Log K 25°, 0.1	Log K 25°, 1.0	Log K 25°, 0
H^+	HL/H.L	11.28 10.91[b]	10.81	12.03
	$H_2L/HL.H$	2.37 2.14[b]	2.18	2.98
Cu^{2+}	ML/M.L	8.74 8.13[b]	8.03	10.29
	$ML_2/M.L^2$	14.13 14.11[b]	14.08	

[b] 25°, 0.5

Bibliography: 75L

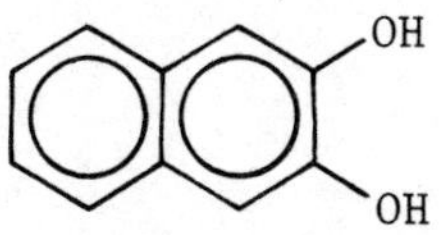

$C_{10}H_8O_2$		2,3-Dihydroxynaphthalene	H_2L

Metal ion	Equilibrium	Log K 25°, 0.1	Log K 20°, 0.1
H^+	HL/H.L		12.5
	$H_2L/HL.H$	8.55	8.68

2,3-Dihydroxynaphthalene (continued)

Metal ion	Equilibrium	Log K 25°, 0.1	Log K 20°, 0.1
UO_2^{2+}	$ML/M.L$		15.0
	$ML_2/M.L^2$		25.8
	$MHL/M.HL$		6.45
	$MHL_2/ML.HL$		4.75
Mo(VI)	$MO_2L_2/MO_4.(H_2L)^2$		6.3 +0.1
W(VI)	$MO_2L_2/MO_4.(H_2L)^2$		7.9
B(III)	$M(OH)_2L.H/M(OH)_3.H_2L$		-4.13
Ge(IV)	$ML_3.H^2/M(OH)_4.(H_2L)^3$		1.5

Bibliography:

H^+	67Ba,70A	B(III)	68HBa,70A
UO_2^{2+}	67Ba	Ge(IV)	73SB
Mo(VI)	71SB	Other references:	72JM,72MJ,73PAT,75VH
W(VI)	73B		

$C_{10}H_8O_5S$　　　　　2,3-Dihydroxynaphthalene-6-sulfonic acid　　　　　H_3L

Metal ion	Equilibrium	Log K 25°, 0.1	Log K 20°, 0.1
H^+	$HL/H.L$	12.0 ±0.3	12.16
	$H_2L/HL.H$	8.09 ±0.04	8.21 ±0.03
UO_2^{2+}	$ML/M.L$		15.6
	$ML_2/M.L^2$		26.2
	$MHL/M.HL$		6.2
	$MHL_2/ML.HL$		4.2
Cu^{2+}	$ML/M.L$	13.5 -0.1	
	$ML_2/M.L^2$	24.0 ±0.2	
Fe^{3+}	$ML/M.L$		19.9
	$ML_2/M.L^2$		34.4
	$ML_3/M.L^3$		44.2
Ti(IV)	$MOL_2/MO.L^2$		38.1
	$MOL_3/MO.L^3$		54.7
	$ML_3/MO.H^2.L^3$		56.5
	$MO(HL)_2/MO.(HL)^2$		15.7
Mo(VI)	$MO_2L_2/MO_4.(H_2L)^2$		6.3
W(VI)	$MO_2L_2/MO_4.(H_2L)^2$		7.6
B(III)	$M(OH)_2L.H/M(OH)_3.H_2L$	-3.9 ±0.2	-3.98
Ge(IV)	$ML_3.H^2/M(OH)_4.(H_2L)^3$	2.0	2.6

E. NAPHTHOLS

2,3-Dihydroxynaphthalene-6-sulfonic acid (continued)

Bibliography:

H^+	52HS,65ONa,68HBa,71AW,73SHG	Mo(VI)	71SB
UO_2^{2+}	65BS,65SSb,66BR	W(VI)	73B
Cu^{2+}	65ONa,73SHG	B(III)	68HBa,70NT,71AW
Fe^{3+}	52HS	Ge(IV)	67PB,73SB
Ti(IV)	63Sd	Other references:	73SO,75VH

$C_{10}H_8O_8S_2$	1,8-Dihydroxynaphthalene-3,6-disulfonic acid (chromotropic acid)				H_4L
Metal ion	Equilibrium	Log K 25°, 0.1	Log K 25°, 0.5	ΔH 25°, 0.1	ΔS 25°, 0.1
H^+	$H_2L/HL.H$	5.35 ±0.05	5.13	-3.3	13
	$H_3L/H_2L.H$	(0.73)[h]			
	$H_4L/H_3L.H$	(0.61)[h]			
Be^{2+}	$ML.H/M.HL$	0.74[h]			
	$ML_2.H^2/M.(HL)^2$	-3.01[h]			
	$MHL/M.HL$	2.9[h]			
UO_2^{2+}	$ML.H/M.HL$	1.0[h]			
	$ML_2.H^2/M.(HL)^2$	-3.1[h]			
	$MHL/M.HL$	4.0[h]			
	$MHL_2/ML.HL$	1.5[h]			
Cu^{2+}	$ML.H/M.HL$	-2.13 ±0.04	-2.51		
	$ML_2.H^2/M.(HL)^2$	-8.02	-8.17		
	$ML/MOHL.H$	8.3			
Ti(IV)	$MOL_2.H_3^2/MO.(HL)_3^2$	9.3[h]			
	$MOL_2.H^3/MO.(HL)^3$	9.6[h]			
	$ML_3.H/MO.(HL)^3$	13.7[h]			
	$MO(HL)_2/MOL_2.H^2$	4.4[h]			
VO^{2+}	$ML/MOHL.H$	6.0			
Mo(VI)	$MO_2L_2/MO_4.(H_2L)^2$	4.1[h]			
	$HMO_3L.H/H_2MO_4.H_2L$	-0.8[h]			
	$MO_2L_2.H/HMO_3L.H_2L$	-2.8[h]			
Nb(V)	$ML_3/MO_2.H.(HL)_3^3$	17.9[h]			
	$MOL_2/MO_2.(HL)^2$	11.3[1]			

[h] 20°, 0.1;   [1] 20°, 3.0

Chromotropic acid (continued)

Metal ion	Equilibrium	Log K 25°, 0.1	Log K 25°, 0.5	ΔH 25°, 0.1	ΔS 25°, 0.1
B(III)	$M(OH)_2L.H/M(OH)_3.H_2L$	-1.55[h]			
	$ML_2.H/M(OH)_3.(H_2L)^2$	-2.4 [h]			
Al^{3+}	$ML.H/M.HL$	1.5 [h]			
	$ML_2.H^2/M.(HL)^2$	-1.3 [h]			

[h] 20°, 0.1

Bibliography:

H^+ 63SM,66LM,67BZ,68WM,69CMa,69B,70KC, 74RS,75La

Be^{2+} 67BZ

UO_2^{2+} 65BS,65SSK,66BR

Cu^{2+} 66LM,68WM,69CMa,75La

Ti(IV)   63Sb

Vo^{2+} 59CG

Mo(VI)   70BG

Nb(V)   64SH

B(III)   67BHa

Al^{3+} 69HBa

Other references: 51HS,57BP,57J,58Sa,59S,59Sa, 59SH,60B,63RM,65BQ,65DM,65DMa,66CS,66MC, 67AM,67LA,68AS,68BDa,68BN,68TK,69MD, 69NB,72KE,74ZBD,75Ba,75L

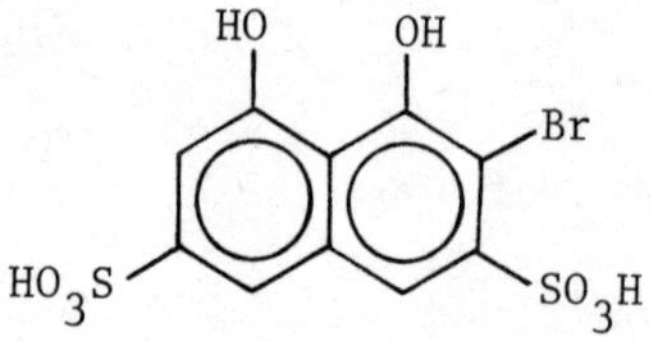

$C_{10}H_7O_8BrS_2$ 2-Bromochromotropic acid H_4L

Metal ion	Equilibrium	Log K 20°, 0.1
H^+	$H_2L/HL.H$	3.97
B(III)	$M(OH)_2L.H/M(OH)_3.H_2L$	-1.70

Bibliography: 69BBH

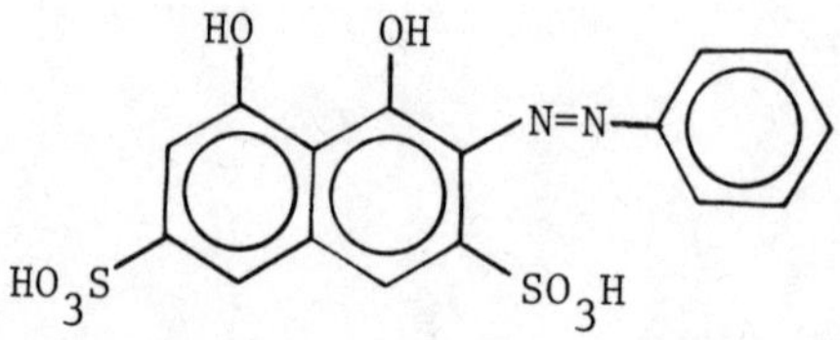

$C_{16}H_{12}O_8N_2S_2$ 2-(Phenylazo)chromotropic acid (chromotrope 2R) H_4L

Metal ion	Equilibrium	Log K 25°, 0.1	Log K 20°, 0.1
H^+	$H_2L/HL.H$	9.17 ±0.02	9.29
Mg^{2+}	$MHL/M.HL$	3.64	

E. NAPHTHOLS

Chromotrope 2R (continued)

Metal ion	Equilibrium	Log K 25°, 0.1	Log K 20°, 0.1
Ca^{2+}	MHL/M.HL	2.70	
Sr^{2+}	MHL/M.HL	2.08	-0.3
Ba^{2+}	MHL/M.HL	1.84	-0.8
Ni^{2+}	ML.H/M.HL		-2.65
Cu^{2+}	ML.H/M.HL		2.59
Fe^{3+}	ML.H/M.HL		7.77
Al^{3+}	ML.H/M.HL		3.77

Bibliography:

H^+ 64SM,68NM,71KMb

$Mg^{2+}-Ba^{2+}$ 68NM,71KMb

$Ni^{2+}-Al^{3+}$ 64SM

Other references: 63M,67PM,71DS,72DSB,73DS

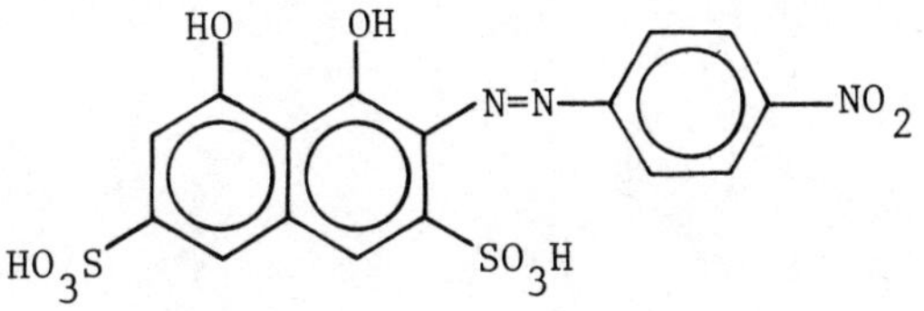

| $C_{16}H_{11}O_{10}N_3S_2$ | 2-(4-Nitrophenylazo)chromotropic acid (chromotrope 2B) | | H_4L |

Metal ion	Equilibrium	Log K 25°, 0.1
H^+	H_2L/HL.H	8.59
Cu^{2+}	ML.H/M.HL	1.5
	MHL/M.HL	6.8

Bibliography: 74RS

Other references: 61BDb,63SDe,63SSb,64MD,67S,67TMK

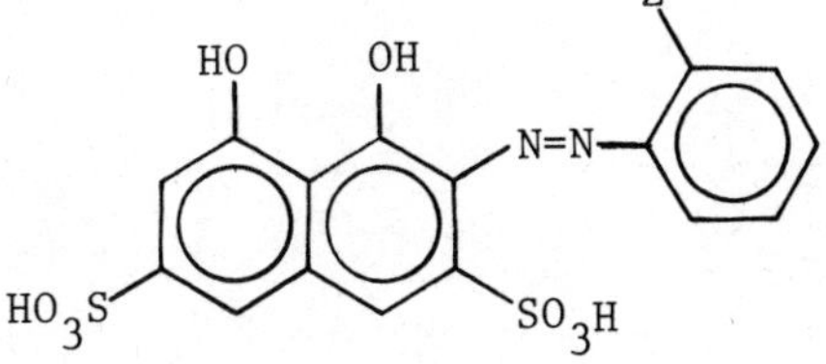

| $C_mH_nO_pN_2S_2$ | 2-(2-Substitutedphenylazo)chromotropic acid | | | H_4L |

Z =	Metal ion	Equilibrium	Log K 25°, 0.1
2-Methyl	H^+	H_2L/HL.H	9.60
$(C_{17}H_{14}O_8N_2S_2)$	Mg^{2+}	MHL/M.HL	3.47
	Ca^{2+}	MHL/M.HL	2.56

2-(2-Substitutedphenylazo)chromotropic acid (continued)

Z =	Metal ion	Equilibrium	Log K 25°, 0.1
2-Methoxy	H^+	$H_2L/HL.H$	9.92
($C_{17}H_{14}O_9N_2S_2$)	Mg^{2+}	MHL/M.HL	3.95
	Ca^{2+}	MHL/M.HL	3.25
	Sr^{2+}	MHL/M.HL	2.40
	Ba^{2+}	MHL/M.HL	2.08
2-Acetyl	H^+	$H_2L/HL.H$	9.65 +0.01
($C_{18}H_{14}O_9N_2S_2$)	Mg^{2+}	MHL/M.HL	3.66
	Ca^{2+}	MHL/M.HL	2.95
	Sr^{2+}	MHL/M.HL	2.35
	Cu^{2+}	MHL/M.HL	8.8

Bibliography:

H^+-Sr^{2+} 71KMb Cu^{2+} 71NM

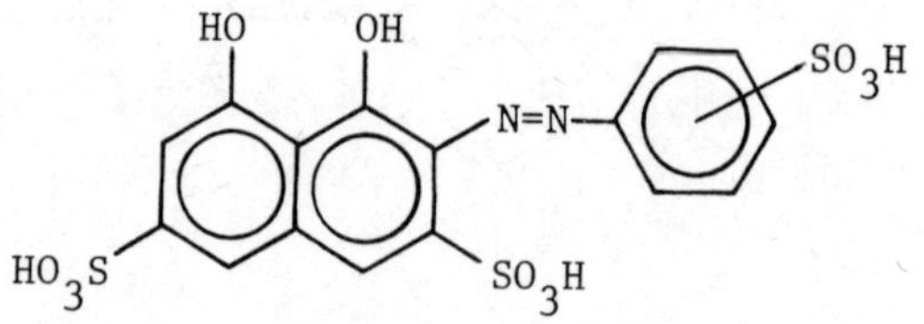

$C_{16}H_{12}O_{11}N_2S_3$ 2-(Sulfophenylazo)chromotropic acid H_5L

Isomer	Metal ion	Equilibrium	Log K 25°, 0.1
2-Sulfo	H^+	$H_2L/HL.H$	9.35
	Mg^{2+}	MHL/M.HL	3.58
	Ca^{2+}	MHL/M.HL	2.91
	Sr^{2+}	MHL/M.HL	2.58
	Ba^{2+}	MHL/M.HL	(2.66)
3-Sulfo	H^+	$H_2L/HL.H$	8.85
	Ba^{2+}	MHL/M.HL	1.44
4-Sulfo	H^+	$H_2L/HL.H$	8.90
	Ba^{2+}	MHL/M.HL	1.58

Bibliography: 68NM

Other reference: 63Ma

E. NAPHTHOLS

	C16H13O11N2ZS2		2-(Substitutedphenylazo)chromotropic acid		H6L
Z =	Metal ion	Equilibrium	Log K 25°, 0.1		
2-Phosphono	H^+	$H_2L/HL.H$	10.12		
		$H_3L/H_2L.H$	6.43		
	Mg^{2+}	MHL/M.HL	5.71		
	Ca^{2+}	MHL/M.HL	4.95		
	Sr^{2+}	MHL/M.HL	4.34		
	Ba^{2+}	MHL/M.HL	4.15		
3-Phosphono	H^+	$H_2L/HL.H$	9.46		
		$H_3L/H_2L.H$	6.78		
	Mg^{2+}	MHL/M.HL	4.34		
	Ca^{2+}	MHL/M.HL	3.48		
	Sr^{2+}	MHL/M.HL	2.72		
4-Phosphono	H^+	$H_2L/HL.H$	9.42		
		$H_3L/H_2L.H$	6.88		
	Mg^{2+}	MHL/M.HL	3.93		
	Ca^{2+}	MHL/M.HL	3.04		
	Sr^{2+}	MHL/M.HL	1.8		
2-Arsono (Neo-Thorin) (Arsenazo I)	H^+	$H_2L/HL.H$	10.07 ±0.09		
		$H_3L/H_2L.H$	7.61 ±0.04		
		$H_4L/H_3L.H$	2.95 ±0.03		
	Mg^{2+}	MHL/M.HL	5.57 +0.01		
	Ca^{2+}	MHL/M.HL	5.51 ±0.06		
	Sr^{2+}	MHL/M.HL	4.40 ±0.01		
	Ba^{2+}	MHL/M.HL	4.19 ±0.04		

Bibliography: 68NM,71KM,71KT

Other references: 60KP,61KPa,65Ba,69B,70NM,72PM

$C_nH_mO_pN_2S_2$		2-(2-Substitutedphenylazo)chromotropic acid		H_5L
HO_2C-R =	Metal ion	Equilibrium	Log K 25°, 0.1	
2-Carboxy	H^+	$H_2L/HL.H$	9.97 ±0.03	
		$H_3L/H_2L.H$	3.65 ±0.06	
$(C_{17}H_{12}O_{10}N_2S_2)$	Mg^{2+}	$MHL/M.HL$	4.54 ±0.01	
	Ca^{2+}	$MHL/M.HL$	3.39 ±0.02	
	Sr^{2+}	$MHL/M.HL$	2.82 −0.01	
	Ba^{2+}	$MHL/M.HL$	2.81 ±0.00	
2-Carboxymethyl	H^+	$H_2L/HL.H$	9.64	
		$H_3L/H_2L.H$	3.77	
$(C_{18}H_{14}O_{10}N_2S_2)$	Mg^{2+}	$MHL/M.HL$	4.00	
	Ca^{2+}	$MHL/M.HL$	3.50	
	Sr^{2+}	$MHL/M.HL$	2.75	
	Ba^{2+}	$MHL/M.HL$	2.43	
2-Oxalo	H^+	$H_2L/HL.H$	9.96	
		$H_3L/H_2L.H$	3.54	
$(C_{18}H_{12}O_{11}N_2S_2)$	Mg^{2+}	$MHL/M.HL$	4.55	
	Ca^{2+}	$MHL/M.HL$	3.41	
	Sr^{2+}	$MHL/M.HL$	2.88	
	Ba^{2+}	$MHL/M.HL$	2.73	
2-Carboxymethoxy	H^+	$H_2L/HL.H$	9.83	
		$H_3L/H_2L.H$	2.84	
$(C_{18}H_{14}O_{11}N_2S_2)$	Mg^{2+}	$MHL/M.HL$	4.31	
	Ca^{2+}	$MHL/M.HL$	5.13	
	Sr^{2+}	$MHL/M.HL$	3.65	
	Ba^{2+}	$MHL/M.HL$	3.00	
2-Carboxy-(hydroxy)methyl	H^+	$H_2L/HL.H$	9.39	
		$H_3L/H_2L.H$	3.03	
$(C_{18}H_{14}O_{11}N_2S_2)$	Mg^{2+}	$MHL/M.HL$	3.96	
	Ca^{2+}	$MHL/M.HL$	3.92	
	Sr^{2+}	$MHL/M.HL$	3.40	
	Ba^{2+}	$MHL/M.HL$	3.12	

Bibliography: 68NM,71KMb

Other references: 65TMS,67Md,67TMH,70TM

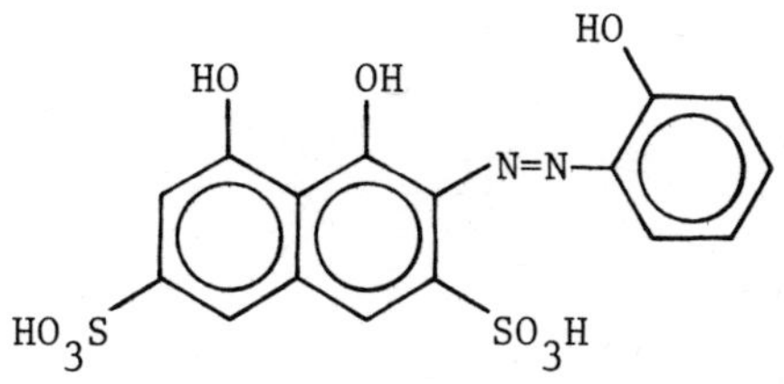

$C_{16}H_{12}O_9N_2S_2$ 2-(2-Hydroxyphenylazo)chromotropic acid H_5L

Metal ion	Equilibrium	Log K 25°, 0.1
H^+	$H_2L/HL.H$	10.60
	$H_3L/H_2L.H$	7.60
Mg^{2+}	MHL/M.HL	6.15
Ca^{2+}	MHL/M.HL	5.01
Sr^{2+}	MHL/M.HL	3.49
Ba^{2+}	MHL/M.HL	2.73

Bibliography: 67NMT,68NM

Other reference: 67TMN

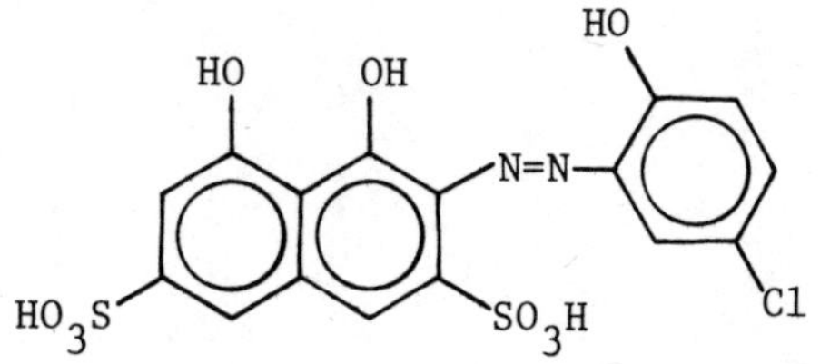

$C_{16}H_{11}O_9N_2ClS_2$ 2-(5-Chloro-2-hydroxyphenylazo)chromotropic acid H_5L

Metal ion	Equilibrium	Log K 25°, 0.1
H^+	$H_2L/HL.H$	10.35
	$H_3L/H_2L.H$	7.56
Mg^{2+}	MHL/M.HL	6.22
Ca^{2+}	MHL/M.HL	5.22

Bibliography: 67NMT

$C_{14}H_8O_7S$ <u>1,2-Dihydroxyanthraquinone-3-sulfonic acid</u> (Alizarin Red S) H_3L

Metal ion	Equilibrium	Log K 20°, 0.1	Log K 20°, 0
H^+	HL/H.L	10.9 +0.1	11.36
	$H_2L/HL.H$	5.77 ±0.03	6.17
	$H_3L/H_2L.H$		0.97
Be^{2+}	ML/M.L	10.96	
B(III)	$M(OH)_2HL.H/M(OH)_3.H_3L$	-3.4	

Bibliography:

H^+	67Ba,70AM,73KE
Be^{2+}	67BZ
B(III)	68HB

Other references: 57MD,59DB,61BD,61BDa,62BD,
63SDa,63SDd,64S,65Sd,66SM,67S,67ZF,
68NA,70DY,71VS,72GB,72GBa,72GDa,73GB

$$O$$
$$\parallel$$
$$CH_3CCH_3$$

C_3H_6O	Propanone (acetone)	L

Metal ion	Equilibrium	Log K 25°, 1.0
Ag^+	ML/M.L	−0.85

Bibliography: 38WL   Other reference: 63FP

$$O$$
$$\parallel$$
$$CH_2=CHCCH_3$$

C_4H_6O	But-1-en-3-one (methylvinylketone)	L

Metal ion	Equilibrium	Log K 25°, 1.0
Ag^+	ML/M.L	−0.38

Bibliography: 68FK

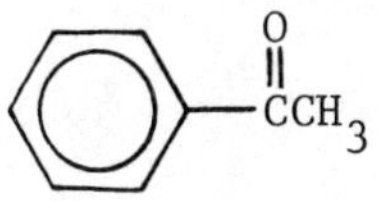

$$O$$
$$\parallel$$
$$-CCH_3$$

C_8H_8O	Acetylbenzene (acetophenone)	L

Metal ion	Equilibrium	Log K 25°, 1.0
Ag^+	ML/M.L	−0.27

Bibliography: 50AK

$$CH_3CH=CHCHO$$

C_4H_6O	But-2-enal (crotonaldehyde)	L

Metal ion	Equilibrium	Log K 25°, 1.0
Ag^+	ML/M.L	−0.72

Bibliography: 38WL

$$O \qquad O$$
$$\parallel \qquad \parallel$$
$$HCCH_2CH$$

$C_3H_4O_2$	Propanedial (malondialdehyde)	HL

Metal ion	Equilibrium	Log K 25°, 1.0
H^+	HL/H.L	4.46
Ni^{2+}	ML/M.L	2.07
Cu^{2+}	ML/M.L	3.57

Bibliography: 720

$$\underset{CH_3CCH_2CCH_3}{\overset{\displaystyle O \quad\ \ O}{\overset{\displaystyle \|\ \quad\ \|}{}}}$$

$C_5H_8O_2$		Pentane-2,4-dione (acetylacetone)				HL
Metal ion	Equilibrium	Log K 25°, 0.1	Log K 25°, 1.0	Log K 25°, 0	ΔH 25°, 0	ΔS 25°, 0
H^+	HL/H.L	8.80 ±0.06	8.80 ±0.09	8.99 ±0.06	-3.3^a±0.1	30^a
Be^{2+}	$ML/M.L$		7.55	7.90 ±0.06	$(-2)^r$	(30)
	$ML_2/M.L^2$		14.35	14.59 ±0.09	$(-9)^r$	(35)
	ML/MOHL.H			6.4		
	$MOHL/M(OH)_2L.H$			9.8		
Mg^{2+}	$ML/M.L$			3.65	$(-2)^r$	(10)
	$ML_2/M.L^2$			6.25	$(-6)^r$	(15)
Sc^{3+}	$ML/M.L$			8.0^s		
	$ML_2/M.L^2$			15.2^s		
Y^{3+}	$ML/M.L$	5.89 ±0.01		6.4^s	-0.9^a	24^a
	$ML_2/M.L^2$	10.85^p		11.1^s		
	$ML_3/M.L^3$	14.1^p		$(13.9)^s$		
La^{3+}	$ML/M.L$	4.94 ±0.02		5.1^s	-0.1^a	22^a
	$ML_2/M.L^2$	8.41^p		9.0^s		
	$ML_3/M.L^3$	10.9^p		11.9^s		
Ce^{3+}	$ML/M.L$	5.15 ±0.06		5.3	0.0^a	24^a
	$ML_2/M.L^2$	8.4^p		9.3		
	$ML_3/M.L^3$	11.3^p		12.7		
Pr^{3+}	$ML/M.L$	5.35 ±0.08		5.4^s	-0.2^a	24^a
	$ML_2/M.L^2$	9.2^p		9.5^s		
	$ML_3/M.L^3$	12.4^p		12.5^s		
Nd^{3+}	$ML/M.L$	5.36 ±0.06		5.6^s	-0.3^a	24^a
	$ML_2/M.L^2$	9.4^p		9.9^s		
	$ML_3/M.L^3$	12.6^p		13.1^s		
Sm^{3+}	$ML/M.L$	5.67 ±0.07		5.9^s	-0.8^a	24^a
	$ML_2/M.L^2$	10.01^p		10.4^s		
	$ML_3/M.L^3$	12.9^p		13.6^s		
Eu^{3+}	$ML/M.L$	5.94 ±0.06		6.1^s	-0.9^a	24^a
	$ML_2/M.L^2$	10.37^p		10.7^s		
	$ML_3/M.L^3$	13.7^p		14.1^s		
Gd^{3+}	$ML/M.L$	5.90 ±0.01			-1.0^a	24^a
	$ML_2/M.L^2$	10.38^p				
	$ML_3/M.L^3$	13.8^p				
Tb^{3+}	$ML/M.L$	6.02 +0.01			-1.1^a	24^a
	$ML_2/M.L^2$	10.63^p				
	$ML_3/M.L^3$	14.0^p				
Dy^{3+}	$ML/M.L$	6.06 ±0.02			-1.1^a	24^a
	$ML_2/M.L^2$	10.70^p				
	$ML_3/M.L^3$	14.0^p				

[a] 25°, 0.1; [p] 30°, 0.1; [r] 10-40°, 0; [s] 30°, 0

Acetylacetone (continued)

Metal ion	Equilibrium	Log K 25°, 0.1	Log K 25°, 1.0	Log K 25°, 0	ΔH 25°, 0	ΔS 25°, 0
Ho^{3+}	$ML/M \cdot L$	6.07 −0.01			-1.1^a	24^a
	$ML_2/M \cdot L^2$	10.73^P				
	$ML_3/M \cdot L^3$	14.1^P				
Er^{3+}	$ML/M \cdot L$	6.08 ±0.08			-1.1^a	24^a
	$ML_2/M \cdot L^2$	10.67^P				
	$ML_3/M \cdot L^3$	14.1^P				
Tm^{3+}	$ML/M \cdot L$	6.14 ±0.04			-1.0^a	25^a
	$ML_2/M \cdot L^2$	10.85^P				
	$ML_3/M \cdot L^3$	14.3^P				
Yb^{3+}	$ML/M \cdot L$	6.18 +0.01			-1.0^a	25^a
	$ML_2/M \cdot L^2$	11.04^P				
	$ML_3/M \cdot L^3$	14.6^P				
Lu^{3+}	$ML/M \cdot L$	6.15 ±0.05			-0.9^a	25^a
	$ML_2/M \cdot L^2$	11.00^P				
	$ML_3/M \cdot L^3$	14.6^P				
Th^{4+}	$ML/M \cdot L$	7.7		8.8^s		
	$ML_2/M \cdot L^2$	14.9		16.2^s		
	$ML_3/M \cdot L^3$	20.8		22.5^s		
	$ML_4/M \cdot L^4$	25.8		26.7^s		
U^{4+}	$ML/M \cdot L$	8.6				
	$ML_2/M \cdot L^2$	17.0				
	$ML_3/M \cdot L^3$	23.4				
	$ML_4/M \cdot L^4$	29.5				
Np^{4+}	$ML/M \cdot L$		8.58			
	$ML_2/M \cdot L^2$		17.23			
	$ML_3/M \cdot L^3$		23.94			
	$ML_4/M \cdot L^4$		30.22			
Pu^{4+}	$ML/M \cdot L$	10.5				
	$ML_2/M \cdot L^2$	19.7				
	$ML_3/M \cdot L^3$	28.1				
	$ML_4/M \cdot L^4$	34.1				
UO_2^{2+}	$ML/M \cdot L$	6.8		7.7		
	$ML_2/M \cdot L^2$	13.1		14.1		
	$MHL_2/ML_2 \cdot H$	4.4				
	$MHL_3/ML_2 \cdot HL$	1.7				
NpO_2^+	$ML/M \cdot L$	4.08	4.08		$(-6)^t$	$(-2)^a$
	$ML_2/M \cdot L^2$	7.00	7.09		$(-9)^t$	$(0)^a$
V^{2+}	$ML/M \cdot L$		5.38			
	$ML_2/M \cdot L^2$		10.19			
	$ML_3/M \cdot L^3$		14.70			
Cr^{2+}	$ML/M \cdot L$		5.96			
	$ML_2/M \cdot L^2$		11.70			
Mn^{2+}	$ML/M \cdot L$	4.07		4.21	-1.5^a	14^a
	$ML_2/M \cdot L^2$			7.30	$(-5)^r$	(20)
Fe^{2+}	$ML/M \cdot L$			5.07^s		
	$ML_2/M \cdot L^2$			8.67^s		

---

^{a} 25°, 0.1; ^{P} 30°, 0.1; ^{r} 10-40°, 0; ^{s} 30°, 0; ^{t} 18-32°, 0.1

Acetylacetone (continued)

Metal ion	Equilibrium	Log K 25°, 0.1	Log K 25°, 1.0	Log K 25°, 0	ΔH 25°, 0	ΔS 25°, 0
Co^{2+}	$ML/M.L$			5.40	(-1)[r]	(20)
	$ML_2/M.L^2$			9.54	(-6)[r]	(25)
Ni^{2+}	$ML/M.L$	5.72		6.00	-3.4[a]	15[a]
	$ML_2/M.L^2$	9.66		10.60	-7.6[a]	19[a]
Cu^{2+}	$ML/M.L$	8.16	8.22	8.25	-4.8[a]	21[a]
	$ML_2/M.L^2$	14.76	14.81	15.05	-10.1[a]	34[a]
Ti^{3+}	$ML/M.L$		10.43			
	$ML_2/M.L^2$		18.82			
	$ML_3^2/M.L^3$		24.9			
Mn^{3+}	$ML_3/ML_2.L$	3.86[u]				
Fe^{3+}	$ML/M.L$			9.8[s]		
	$ML_2/M.L^2$			18.8[s]		
	$ML_3^2/M.L^3$			26.2[s]		
Zr^{4+}	$ML/M.L$		11.25[j]			
VO^{2+}	$ML/M.L$	8.40		8.68	(-2)[v]	(30)
	$ML_2/M.L^2$	15.40		15.84	(-9)[v]	(40)
Pd^{2+}	$ML/M.L$			16.4	(-20)[w]	(10)
	$ML_2/M.L^2$			27.5	(-40)[x]	(-10)
Zn^{2+}	$ML/M.L$	4.68		5.06	-1.5[a]	16[a]
	$ML_2/M.L^2$	7.92		9.00	-3.4[a]	24[a]
Cd^{2+}	$ML/M.L$			3.83	(-2)[r]	(10)
	$ML_2/M.L^2$		6.12[y]	6.65	(-7)[r]	(10)
Hg^{2+}	$ML_2/M.L^2$		21.5[z]			
$(CH_3)_2Sn^{2+}$	$ML/M.L$	6.6				
Pb^{2+}	$ML_2/M.L^2$		6.32[y]			
Al^{3+}	$ML/M.L$			8.6[s]		
	$ML_2/M.L^2$			16.5[s]		
	$ML_3^2/M.L^3$			22.3[s]		
Ga^{3+}	$ML/M.L$			9.4[s]		
	$ML_2/M.L^2$			17.8[s]		
	$ML_3^2/M.L^3$			23.7[s]		
In^{3+}	$ML/M.L$			8.0[s]		
	$ML_2/M.L^2$			15.1[s]		

---

[a] 25°, 0.1; [j] 19°, 1.0; [r] 10-40°, 0; [s] 30°, 0; [u] 25°, 0.2; [v] 25-40°, 0; [w] 20-40°, 0; [x] 30-40°, 0; [y] 30°, 0.7; [z] 30°, 0.5

Bibliography:

H^+ 50R,51C,52V,53R,54IH,55IFb,61LR,61PA, Pu^{4+} 56R
         63GA,63YT,65CM,65SMb,68GF,71DC
                                                   NpO_2^+ 72GK
Be^{2+} 55IF,55IFb,63GA
                                                   UO_2^{2+} 55IF,55IFb,55R
Mg^{2+},Co^{2+} 55IF,55IFb
                                                   V^{2+} 65S
Sc^{3+}-Lu^{3+} 54IH,55IFa,55IFb,60GF,71DC
                                                   Cr^{2+} 65SMb
Th^{4+} 53R,55IFa
                                                   Mn^{2+} 55IF,55IFb,68GF
U^{4+} 56RR
                                                   Fe^{2+},Hg^{2+} 55IF
Np^{4+} 70LS

Acetylacetone (continued)

Ni^{2+}, Zn^{2+}	54IH,55IFb,68GF	Cd^{2+}	55IF,55IFb,62SSS
Cu^{2+}	55IF,55IFb,68GF,69AA	$(CH_3)_2Sn^{2+}$	63TY,63YT
Ti^{3+}	67VD	Pb^{2+}	62SSS
Mn^{3+}	51C		

$Fe^{3+}, Al^{3+}-In^{3+}$ 55IFa

Zr^{4+} 66KSG

VO^{2+} 56TB

Pd^{2+} 57DB

Other references: 50Ra,53B,58PZ,59RS,60R,60S,
63PB,64P,64SAa,64YC,65IA,66CB,66KZ,
66SKa,68AB,68DKa,68GD,68RSc,69BF,69CV,
69HI,69L,70L,71FN,71KO,71SI,72SMS,
73SHI,74HBP,74L,75LB

$$\underset{CF_3}{\overset{O}{\parallel}}\underset{C}{}\underset{CH_2}{}\underset{C}{\overset{O}{\parallel}}CF_3$$

$C_5H_2O_2F_6$ 1,1,1,5,5,5-Hexafluoropentane-2,4-dione (hexafluoroacetylacetone)   HL

Metal ion	Equilibrium	Log K 20°, 0.1
H^+	HL/H.L	4.35
NpO_2^+	ML/M.L	1.9
Cu^{2+}	ML/M.L	2.70

Bibliography:

H^+, Cu^{2+}	51V	Other references: 70G,71SI,73SHI,74SMN
NpO_2^+	72GK	

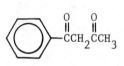

$C_{10}H_{10}O_2$ 1-Phenylbutane-1,3-dione (benzoylacetone)   HL

Metal ion	Equilibrium	Log K 25°, 0.1	Log K 25°, 1.0
H^+	HL/H.L	(8.89)	(8.24)
Ce^{3+}	ML/M.L	(6.19)[h]	
	$ML_2/M.L^2$	(12.10)[h]	
	$ML_3/M.L^3$	(16.99)[h]	
Eu^{3+}	ML/M.L	(6.89)	
	$ML_2/M.L^2$	(13.37)	
	$ML_3/M.L^3$	(19.60)	
Zr^{4+}	ML/M.L		(12.71)
	$ML_2/M.L^2$		(24.57)
	$ML_3/M.L^3$		(35.91)
	$ML_4/M.L^4$		(47.00)

[h] 20°, 0.1

Benzoylacetone (continued)

Metal ion	Equilibrium	Log K 25°, 0.1	Log K 25°, 1.0
Hf^{4+}	$ML_4/M(OH)_2 \cdot H^2 \cdot L^4$		(41.8)
VO^{2+}	$ML/M \cdot L$	(10.49)	
	$ML_2/M \cdot L^2$	(20.54)	
$VOOH^{2+}$	$ML_2/M \cdot L^2$	(21.3) ±0.3	

Bibliography:

H^+ 59ZM,61PM VO^{2+} 70MKP

Ce^{3+} 69EV $VOOH^{2+}$ 72KMP,74NM

Eu^{3+} 68BB   Other references: 60S,62EP,66KZ,68RSc,69BF,

Zr^{4+} 61PM   70G,72GK,73SHI,74SMN

Hf^{4+} 75Lc

$$Z_1-\overset{\overset{O}{\|}}{C}CH_2\overset{\overset{O}{\|}}{C}-Z_2$$

$C_nH_mO_pF_qS_r$ Substitutedalkane-1,3-dione   HL

Z =	Metal ion	Equilibrium	Log K 25°, 0.1	Log K 25°, 1.0
$Z_1=C_6H_5, Z_2=CF_3$ (1-phenyl-4,4,4-tri-fluorobutane-1,3-dione) ($C_{10}H_7O_2F_3$)	H^+	HL/H.L	6.30	6.22
	NpO_2^+	$ML/M \cdot L$	4.11	
		$ML_2/M \cdot L^2$	7.86	
$Z_1=C_4H_3O, Z_2=CH_3$ [1-(2-furanyl)-propane-1,3-dione] ($C_8H_8O_3$)	H^+	HL/H.L	8.64	
	NpO_2^+	$ML/M \cdot L$	4.40	
		$ML_2/M \cdot L^2$	7.85	
$Z_1=C_4H_3O, Z_2=CF_3$ [1-(2-furanyl)-4,4,4-trifluoro-butane-1,3-dione] ($C_8H_5O_3F_3$)	H^+	HL/H.L	5.87	
	NpO_2^+	$ML/M \cdot L$	2.23	
		$ML_2/M \cdot L^2$	4.6	
$Z_1=Z_2=C_4H_3O$ [1,3-di(2-furanyl)propane-1,3-dione) ($C_{11}H_8O_4$)	H^+	HL/H.L	8.15	
	NpO_2^+	$ML/M \cdot L$	4.03	
		$ML_2/M \cdot L^2$	7.1	
$Z_1=C_4H_3S, Z_2=CH_3$ [1-(2-thienyl)-butane-1,3-dione] ($C_8H_8O_2S$)	H^+	HL/H.L	8.36	
	NpO_2^+	$ML/M \cdot L$	4.23	
		$ML_2/M \cdot L^2$	7.41	

Substitutedalkane-1,3-dione (continued)

Bibliography:

H^+ 67Sa,72GK   Other references: 69KS,70G,73SHI,74SMN

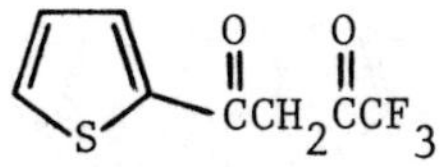

NpO_2^+ 72GK

| $C_8H_5O_2SF_3$ | 1-(2-Thienyl)-4,4,4-trifluorobutane-1,3-dione | HL |
| | (thenoyltrifluoroacetone) | |

Metal ion	Equilibrium	Log K 25°, 1.0
H^+	HL/H.L	6.53
Co^{2+}	ML/M.L	3.68
Ni^{2+}	ML/M.L	4.45
Cu^{2+}	ML/M.L	5.68
Fe^{3+}	ML/M.L	7.18
Zr^{4+}	ML/M.L	10.98
	$ML_2/M.L^2$	21.88
	$ML_3/M.L^3$	32.24
	$ML_4/M.L^4$	42.17
Hf^{4+}	ML/M.L	10.60
	$ML_2/M.L^2$	21.44
	$ML_3/M.L^3$	31.50
	$ML_4/M.L^4$	41.52

Bibliography:

H^+ 62PA   Other references: 51Ma,72TC,54KS,59TC,60GM,
Co^{2+}-Fe^{3+} 71JF   64SM,66KF,66SKa,67Sa,68DK,68RSc,68SA,
Zr^{4+} 67MO   68ZS,69BF,69BR,70G,71AD,72GK,72KMP,
Hf^{4+} 62PA   72SMS,72SS,73SHI,74SMN

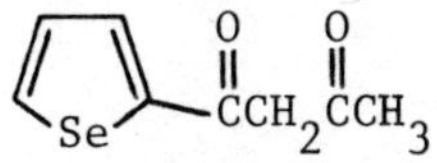

| $C_8H_8O_2Se$ | 1-(2-Selenoyl)butane-1,3-dione (selenoylacetone) | HL |

Metal ion	Equilibrium	Log K 25°, 0.1
H^+	HL/H.L	8.55
Eu^{3+}	ML/M.L	6.24
	$ML_2/M.L^2$	12.29
	$ML_3/M.L^3$	17.88
Th^{4+}	ML/M.L	10.04
	$ML_2/M.L^2$	19.04
	$ML_3/M.L^3$	27.94
	$ML_4/M.L^4$	36.78

Selenoylacetone (continued)

Bibliography:

H$^+$ 59ZM   Th^{4+} 59PZ

Eu^{3+} 68BB

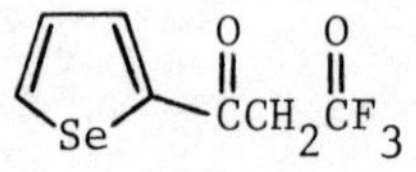

C$_8$H$_5$O$_2$F$_3$Se   <u>1-(2-Selenoyl)-4,4,4-trifluorobutane-1,3-dione</u>   HL
                                    <u>(selenoyltrifluoroacetone)</u>

Metal ion	Equilibrium	Log K 25°, 1.0
H$^+$	HL/H.L	6.32
Zr^{4+}	ML/M.L	11.35
	ML$_2$/M.L^2	22.10
	ML$_3$/M.L^3	32.25
	ML$_4$/M.L^4	41.80
Hf^{4+}	ML/M.L	10.46
	ML$_2$/M.L^2	20.74
	ML$_3$/M.L^3	30.22
	ML$_4$/M.L^4	39.70

Bibliography:

H$^+$,Hf^{4+} 62PA   Zr^{4+} 63MP

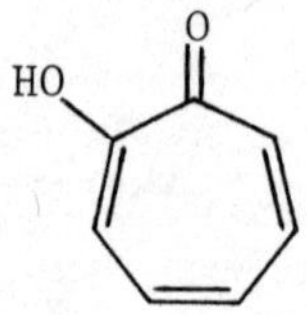

C$_7$H$_6$O$_2$ <u>1-Hydroxycyclohepta-3,5,7-trien-2-one</u> (tropolone)   HL

Metal ion	Equilibrium	Log K 25°, 0.1	Log K 25°, 0.5	Log K 25°, 0	ΔH 25°, 0	ΔS 25°, 0
H$^+$	HL/H.L	6.70 ±0.02	6.64	6.88	-2.5	23
			6.75[j]		-2.8[a]	21[a]
Be^{2+}	ML/M.L	7.40				
Mg^{2+}	ML/M.L	3.82				
Ca^{2+}	ML/M.L	3.06				
Sr^{2+}	ML/M.L	2.45				
Y^{3+}	ML/M.L	7.18 +0.3			-2.6[a]	24[a]
	ML$_2$/M.L^2	13.26				
	ML$_3$/M.L^3	18.07				
	ML$_4$/M.L^4	21.49				
La^{3+}	ML/M.L	6.19 +0.06			-2.5[a]	20[a]
	ML$_2$/M.L^2	11.12				
	ML$_3$/M.L^3	15.31				

[a] 25°, 0.1;   [j] 20°, 1.0

Tropolone (continued)

Metal ion	Equilibrium	Log K 25°, 0.1	Log K 25°, 0.5	Log K 25°, 0	ΔH 25°, 0	ΔS 25°, 0
Ce^{3+}	$ML/M.L$	6.56			-2.7[a]	21[a]
	$ML_2/M.L^2$	11.76				
	$ML_3/M.L^3$	16.12				
Pr^{3+}	$ML/M.L$	6.61			-2.8[a]	22[a]
	$ML_2/M.L^2$	11.94				
	$ML_3/M.L^3$	16.39				
Nd^{3+}	$ML/M.L$	6.77			-2.9[a]	22[a]
	$ML_2/M.L^2$	12.21				
	$ML_3/M.L^3$	16.61				
Sm^{3+}	$ML/M.L$	6.91			-2.8[a]	22[a]
	$ML_2/M.L^2$	12.59				
	$ML_3/M.L^3$	17.19				
Eu^{3+}	$ML/M.L$	7.10			-2.8[a]	23[a]
	$ML_2/M.L^2$	12.81				
	$ML_3/M.L^3$	17.62				
Gd^{3+}	$ML/M.L$	7.04			-2.7[a]	23[a]
	$ML_2/M.L^2$	12.90				
	$ML_3/M.L^3$	17.72				
	$ML_4/M.L^4$	21.02				
Tb^{3+}	$ML/M.L$	7.15			-2.7[a]	24[a]
	$ML_2/M.L^2$	13.18				
	$ML_3/M.L^3$	18.00				
	$ML_4/M.L^4$	21.38				
Dy^{3+}	$ML/M.L$	7.23			-2.8[a]	24[a]
	$ML_2/M.L^2$	13.38				
	$ML_3/M.L^3$	18.40				
	$ML_4/M.L^4$	22.12				
Ho^{3+}	$ML/M.L$	7.41			-2.8[a]	25[a]
	$ML_2/M.L^2$	13.61				
	$ML_3/M.L^3$	18.81				
	$ML_4/M.L^4$	22.58				
Er^{3+}	$ML/M.L$	7.54			-2.8[a]	25[a]
	$ML_2/M.L^2$	13.91				
	$ML_3/M.L^3$	19.15				
	$ML_4/M.L^4$	23.11				
Tm^{3+}	$ML/M.L$	7.60			-2.9[a]	25[a]
	$ML_2/M.L^2$	13.99				
	$ML_3/M.L^3$	19.39				
	$ML_4/M.L^4$	23.19				
Yb^{3+}	$ML/M.L$	7.85			-3.1[a]	26[a]
	$ML_2/M.L^2$	14.35				
	$ML_3/M.L^3$	19.83				
	$ML_4/M.L^4$	23.73				
Lu^{3+}	$ML/M.L$	7.69			-2.9[a]	26[a]
	$ML_2/M.L^2$	14.33				
	$ML_3/M.L^3$	19.77				
	$ML_4/M.L^4$	23.73				
Pu^{3+}	$ML/M.L$		7.20[j]			

[a] 25°, 0.1;   [j] 20°, 1.0

Tropolone (continued)

Metal ion	Equilibrium	Log K 25°, 0.1	Log K 25°, 0.5	Log K 25°, 0	ΔH 25°, 0	ΔS 25°, 0
Th^{4+}	$ML/M.L$	9.61				
	$ML_2/M.L^2$	18.24				
	$ML_3/M.L^3$	25.89				
	$ML_4/M.L^4$	32.56				
	$ML_5/M.L^5$	34.85				
	$ML_6/M.L^6$	36.72				
NpO_2^+	$ML/M.L$		5.45[j]			
	$ML_2/M.L^2$		9.81[j]			
UO_2^{2+}	$ML/M.L$		8.18[j]			
	$ML_2/M.L^2$		15.07[j]			
Mn^{2+}	$ML/M.L$	4.60				
Co^{2+}	$ML/M.L$	5.59				
Ni^{2+}	$ML/M.L$	5.97				
Cu^{2+}	$ML/M.L$	8.35				
Fe^{3+}	$ML/M.L$		10.50[d]			
Zn^{2+}	$ML/M.L$	5.84				
Cd^{2+}	$ML/M.L$	4.60				
Pb^{2+}	$ML/M.L$	6.64[p]				
Ge(IV)	$M(OH)_2L_2/M(OH)_4.(HL)^2$		8.03			
	$ML_3/M(OH)_4.(HL)^3.H$		13.3			

[d] 25°, 2.0; [j] 20°, 1.0; [p] 30°, 0.16

Bibliography:

H^+ 54Da,66BB,68OH,73MBP,75KO

$Be^{2+}-Sr^{2+},Cd^{2+}$ 70HO

$Y^{3+}-Lu^{3+}$ 69CM,70CM,70HO

$Pu^{3+},NpO_2^+,UO_2^{2+}$ 73MBP

Th^{4+} 55D,60R

$Mn^{2+}-Ni^{2+},Zn^{2+}$ 68OW

Cu^{2+} 68OH

Fe^{3+} 62OU

Pb^{2+} 71RG

Ge(IV) 66BB

Other reference: 52JS

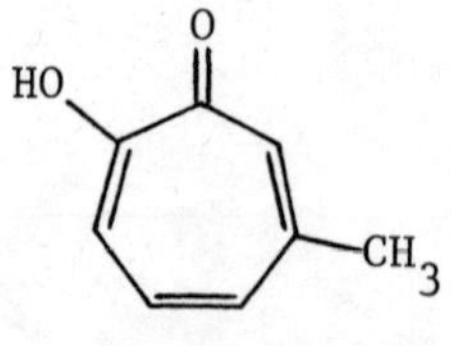

4-Methyltropolone

$C_8H_8O_2$ HL

Metal ion	Equilibrium	Log K 30°, 0.16
H^+	$HL/H.L$	6.95
Pb^{2+}	$ML/M.L$	6.9

Bibliography: 71RG

$C_{10}H_{12}O_2$ 5-(1-Methylethyl)tropolone   HL

Metal ion	Equilibrium	Log K 25°, 0.1	Log K 25°, 2.0
H^+	HL/H.L	7.10	6.86
Fe^{3+}	ML/M.L		10.64

Bibliography:

H^+ 62Da,62Ua Fe^{3+} 62Ua

$C_{10}H_{12}O_2$ 4-(1-Methylethyl)tropolone   HL

Metal ion	Equilibrium	Log K 25°, 0.1	Log K 25°, 2.0
H^+	HL/H.L	7.04	6.72
Ca^{2+}	ML/M.L	2.77	
	$ML_2/M.L^2$	4.54	
Sr^{2+}	ML/M.L	2.70	
	$ML_2/M.L^2$	4.40	
Ba^{2+}	ML/M.L	1.87	
	$ML_2/M.L^2$	2.74	
UO_2^{2+}	ML/M.L	9.5	
	$ML_2/M.L^2$	18.0	
Co^{2+}	ML/M.L	5.8	
	$ML_2/M.L^2$	10.8	
Ni^{2+}	ML/M.L	5.90	
	$ML_2/M.L^2$	11.10	
	$ML_3/M.L^3$	14.80	
Cu^{2+}	ML/M.L	9.55	
	$ML_2/M.L^2$	18.30	
Fe^{3+}	ML/M.L		11.55
Zn^{2+}	ML/M.L	6.18	
	$ML_2/M.L^2$	11.56	
	$ML_3/M.L^3$	15.00	
Cd^{2+}	ML/M.L	5.27	
	$ML_2/M.L^2$	9.94	

4-Isopropyltropolone (continued)

Bibliography:

H^+ 62Da,64OYS Fe^{3+} 64OYS

$Ca^{2+}-Cu^{2+},Zn^{2+},Cd^{2+}$ 62Da   Other references: 67Sa,71MMB

$C_nH_mO_pN_qX_r$		Substitutedtropolone			HL
Z =	Metal ion	Equilibrium	Log K 25°, 0.1	Log K 25°, 2.0	
5-Chloro	H^+	HL/H.L		5.62	
$(C_7H_5O_2Cl)$	Fe^{3+}	ML/M.L		9.92	
3-Bromo	H^+	HL/H.L		5.14	
$(C_7H_5O_2Br)$	Fe^{3+}	ML/M.L		9.25	
4-Bromo	H^+	HL/H.L		5.72	
$(C_7H_5O_2Br)$	Fe^{3+}	ML/M.L		10.04	
5-Bromo	H^+	HL/H.L		5.54	
$(C_7H_5O_2Br)$	Fe^{3+}	ML/M.L		9.74	
5-Nitro	H^+	HL/H.L	3.21	2.64	
$(C_7H_5O_4N)$	Cu^{2+}	ML/M.L	5.53		
	Fe^{3+}	ML/M.L		6.57	
3-Cyano	H^+	HL/H.L		3.41	
$(C_8H_5O_2N)$	Fe^{3+}	ML/M.L		7.28	
5-Cyano	H^+	HL/H.L		3.72	
$(C_8H_5O_2N)$	Fe^{3+}	ML/M.L		7.53	
4-Acetyl	H^+	HL/H.L		5.54	
$(C_9H_8O_3)$	Fe^{3+}	ML/M.L		9.60	
4-Amino	H^+	HL/H.L		7.47	
$(C_7H_7O_2N)$	Fe^{3+}	ML/M.L		12.58	

Bibliography:

H^+ 62OUa,64OY,64OYS,65OY,68OH Fe^{3+} 62OUa,64OY,64OYS,65OY

Cu^{2+} 68OH

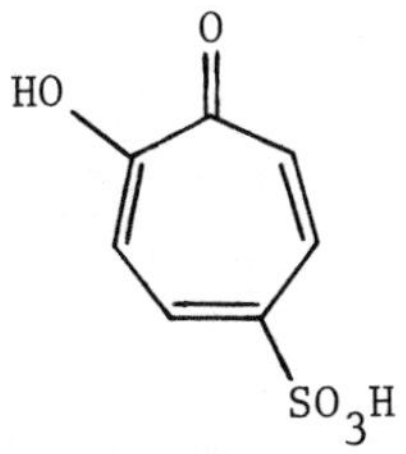

C. HYDROXYCARBONYL LIGANDS

Tropolone-5-sulfonic acid

$C_7H_6O_5S$

H_2L

Metal ion	Equilibrium	Log K 25°, 0.1	Log K 25°, 2.0
H^+	HL/H.L	4.92	4.68
Th^{4+}	$ML/M.L$		7.95
	$ML_2/M.L^2$		14.09
UO_2^{2+}	$ML/M.L$	6.86	
	$ML_2/M.L^2$	11.52	
Cu^{2+}	ML/M.L	6.90	
Fe^{3+}	$ML/M.L$		8.71
	$ML_2/M.L^2$		16.14
	$ML_3/M.L^3$		21.86

Bibliography:

H^+	62OU,68OH		Cu^{2+}	68OH
Th^{4+}	63OU		Fe^{3+}	62OU
UO_2^{2+}	63OY			

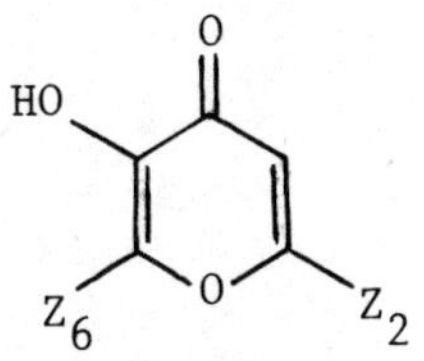

$C_nH_mO_pI_u$

Substituted-3-hydroxy-4-pyrone

HL

Z =	Metal ion	Equilibrium	Log K 25°, 0.5
$Z_2=Z_6=H$	H^+	HL/H.L	7.69
(pyromeconic	Zn^{2+}	$ML/M.L$	5.03
acid)		$ML_2/M.L^2$	9.18
$(C_5H_4O_3)$	Ge(IV)	$M(OH)_2L_2/M(OH)_4.(HL)^2$	2.86
$Z_2=CH_3,Z_6=H$	H^+	HL/H.L	7.98
(allomaltol)	Zn^{2+}	$ML/M.L$	5.28
$(C_6H_6O_3)$		$ML_2/M.L^2$	9.57
	Ge(IV)	$M(OH)_2L_2/M(OH)_4.(HL)^2$	3.43
$Z_2=CH_2I,Z_6=H$	H^+	HL/H.L	7.50
(iodokojic	Zn^{2+}	ML/M.L	4.92
acid)	Ge(IV)	$M(OH)_2L_2/M(OH)_4.(HL)^2$	2.49
$(C_6H_5O_3I)$			

Substituted-3-hydroxy-4-pyrone (continued)

Z =	Metal ion	Equilibrium	Log K 25°, 0.5	
$Z_2=CO_2H, Z_6=H$	H^+	HL/H.L	7.29	H_2L
(comenic acid)	Zn^{2+}	ML/M.L	4.86	
$(C_6H_4O_5)$		$ML_2/M.L^2$	8.76	
	Ge(IV)	$M(OH)_2L_2/M(OH)_4 \cdot (HL)^2$	2.25	
$Z_2=Z_6=CO_2H$	H^+	HL/H.L	9.35	H_3L
(meconic acid)	Zn^{2+}	ML/M.L	7.25	
$(C_7H_4O_7)$				

Bibliography: 67CB

Other references: 61SL,65BS,66SK

$C_6H_6O_3$		3-Hydroxy-2-methyl-4-pyrone (maltol)			HL
Metal ion	Equilibrium	Log K 25°, 0.1	Log K 25°, 0.5	Log K 20°, 2.0	
H^+	HL/H.L	8.50	8.40 ±0.05	8.66	
		8.51^h	8.49^j		
Y^{3+}	ML/M.L	6.70^P			
	$ML_2/M.L^2$	12.09^P			
	$ML_3/M.L^3$	15.97^P			
La^{3+}	ML/M.L	5.76^P			
	$ML_2/M.L^2$	10.17^P			
	$ML_3/M.L^3$	13.34^P			
Pr^{3+}	ML/M.L	6.13^P			
	$ML_2/M.L^2$	10.96^P			
	$ML_3/M.L^3$	14.51^P			
Nd^{3+}	ML/M.L	6.22^P			
	$ML_2/M.L^2$	11.14^P			
	$ML_3/M.L^3$	14.68^P			
Sm^{3+}	ML/M.L	6.51^P			
	$ML_2/M.L^2$	11.70^P			
	$ML_3/M.L^3$	15.43^P			
Eu^{3+}	ML/M.L	6.72^P			
	$ML_2/M.L^2$	12.03^P			
	$ML_3/M.L^3$	15.83^P			
Gd^{3+}	ML/M.L	6.58^P			
	$ML_2/M.L^2$	11.91^P			
	$ML_3/M.L^3$	15.86^P			

[h] 20°, 0.1;   [j] 20°, 1.0;   [P] 30°, 0.1

Maltol (continued)

Metal ion	Equilibrium	Log K 25°, 0.1	Log K 25°, 0.5	Log K 20°, 2.0
Dy^{3+}	$ML/M.L$	6.88^P		
	$ML_2/M.L^2$	12.49^P		
	$ML_3/M.L^3$	16.61^P		
Er^{3+}	$ML/M.L$	6.98^P		
	$ML_2/M.L^2$	12.66^P		
	$ML_3/M.L^3$	16.87^P		
Yb^{3+}	$ML/M.L$	7.06^P		
	$ML_2/M.L^2$	12.89^P		
	$ML_3/M.L^3$	17.28^P		
UO_2^{2+}	$ML/M.L$	8.3 −0.1		
	$ML_2/M.L^2$	14.9 ±0.1		
	$ML_3/M.L^3$	18.3 ±0.1		
Ni^{2+}	$ML/M.L$			5.48
	$ML_2/M.L^2$			9.80
	$ML_3/M.L^3$			12.5
Cu^{2+}	$ML/M.L$		7.69	
	$ML_2/M.L^2$		13.73	
Fe^{3+}	$ML/M.L$	11.5^h		
	$ML_2/M.L^2$	21.4^h		
	$ML_3/M.L^3$	29.7^h		
Hf^{4+}	$ML/M.L$		13.16^j	
	$ML_2/M.L^2$		24.48^j	
Zn^{2+}	$ML/M.L$		5.53	
	$ML_2/M.L^2$		10.20	
Al^{3+}	$ML/M.L$	7.7		
	$ML_2/M.L^2$	15.2		
	$ML_3/M.L^3$	21.8		
Ge(IV)	$M(OH)_2L_2/M(OH)_4 \cdot (HL)^2$		3.90	

---

[h] 20°, 0.1;   [j] 20°, 1.0;   [P] 30°, 0.1

Bibliography:

H^+	66BB,68CH,68SH,72HS,73CDB,75MR	Fe^{3+}	68SH
$Y^{3+}-Yb^{3+}$	70DS	Hf^{4+}	72HS
UO_2^{2+}	68CH,69CB	Zn^{2+}	67CB
Ni^{2+}	75MR	Al^{3+}	69CB
Cu^{2+}	73CDB	Ge(IV)	66BB

$C_6H_5O_3Cl$ 2-Chloromethyl-5-hydroxy-4-pyrone (<u>chlorokojic acid</u>)   HL

Metal ion	Equilibrium	Log K 30°, 0.1	Log K 25°, 0.5
H^+	HL/H.L	7.43	7.40
Y^{3+}	ML/M.L	6.00	
	$ML_2/M.L^2$	11.42	
La^{3+}	ML/M.L	5.28	
	$ML_2/M.L^2$	9.76	
Pr^{3+}	ML/M.L	5.70	
	$ML_2/M.L^2$	10.52	
Nd^{3+}	ML/M.L	5.73	
	$ML_2/M.L^2$	10.65	
Sm^{3+}	ML/M.L	5.83	
	$ML_2/M.L^2$	10.91	
Eu^{3+}	ML/M.L	5.98	
	$ML_2/M.L^2$	11.19	
Gd^{3+}	ML/M.L	5.98	
	$ML_2/M.L^2$	11.12	
Dy^{3+}	ML/M.L	6.13	
	$ML_2/M.L^2$	11.51	
Er^{3+}	ML/M.L	6.19	
	$ML_2/M.L^2$	11.63	
Yb^{3+}	ML/M.L	6.28	
	$ML_2/M.L^2$	11.97	
Zn^{2+}	ML/M.L		4.88
$Ge(IV)$	$M(OH)_2L_2/M(OH)_4.(HL)^2$		2.33

Bibliography:

H^+ 67CB,72DSS Zn^{2+},Ge(IV)   67CB

Y^{3+}-Yb^{3+} 72DSS

$C_6H_6O_4$		5-Hydroxy-2-hydroxymethyl-4-pyrone (kojic acid)				HL
Metal ion	Equilibrium	Log K 25°, 0.1	Log K 25°, 1.0	Log K 25°, 0	ΔH 25°, 0.1	ΔS 25°, 0.1
H^+	HL/H.L	7.66 ±0.05	7.67 -0.01	7.88	-3.5	23
		7.62[b]±0.04	7.88[d]		(-5)[r]	
Mg^{2+}	ML/M.L	2.92[h]				
		3.0				
	$ML_2/M.L^2$	5.11				
Ca^{2+}	ML/M.L	2.5[h]				
Y^{3+}	ML/M.L	6.18[p]	5.84[d]		-1.0[d]	23[d]
	$ML_2/M.L^2$	11.37[p]				
	$ML_3/M.L^3$	15.52[p]				
La^{3+}	ML/M.L	5.38 ±0.03			-0.6[d]	22[d]
			5.11[d]		-1.4[d]	19[d]
	$ML_2/M.L^2$	9.56[p]				
	$ML_3/M.L^3$	12.87[p]				
Ce^{3+}	ML/M.L		5.28[d]		-1.5[d]	19[d]
Pr^{3+}	ML/M.L	5.73 ±0.05			-0.9[d]	23[d]
			5.41[d]		-1.4[d]	20[d]
	$ML_2/M.L^2$	10.54[p]				
	$ML_3/M.L^3$	14.43[p]				
Nd^{3+}	ML/M.L	5.78 ±0.03			-0.6[d]	24[d]
			5.42[d]		-1.4[d]	20[d]
	$ML_2/M.L^2$	10.63[p]				
	$ML_3/M.L^3$	14.66[p]				
Sm^{3+}	ML/M.L	6.02 ±0.02			-0.5[d]	26[d]
			5.66[d]		-1.2[d]	22[d]
	$ML_2/M.L^2$	11.05[p]				
	$ML_3/M.L^3$	14.99[p]				
Eu^{3+}	ML/M.L	6.14 ±0.02			-0.3[d]	27[d]
			5.72[d]		-1.2[d]	22[d]
	$ML_2/M.L^2$	11.25[p]				
	$ML_3/M.L^3$	15.26[p]				
Gd^{3+}	ML/M.L	6.09 +0.01			-0.3[d]	27[d]
			5.72[d]		-1.1[d]	22[d]
	$ML_2/M.L^2$	11.21[p]				
	$ML_3/M.L^3$	15.26[p]				
Tb^{3+}	ML/M.L	6.25			-0.3[d]	28[d]
			5.88[d]		-1.1[d]	23[d]
Dy^{3+}	ML/M.L	6.37 ±0.03			-0.4[d]	28[d]
			5.97[d]		-1.1[d]	23[d]
	$ML_2/M.L^2$	11.86[p]				
	$ML_3/M.L^3$	16.21[p]				
Ho^{3+}	ML/M.L	6.35			-0.5[d]	28[d]
			5.94[d]		-1.1[d]	23[d]
Er^{3+}	ML/M.L	6.35 ±0.04			-0.2[d]	29[d]
			6.00[d]		-1.0[d]	24[d]
	$ML_2/M.L^2$	11.77[p]				
	$ML_3/M.L^3$	16.17[p]				
Tm^{3+}	ML/M.L	6.46			-0.3[d]	29[d]
			6.01[d]		-1.0[d]	24[d]

b 25°, 0.5;   d 25°, 2.0;   h 20°, 0.1;   p 30°, 0.1;   r 20-40°, 2.0

Kojic acid (continued)

Metal ion	Equilibrium	Log K 25°, 0.1	Log K 25°, 1.0	Log K 25°, 0	ΔH 25°, 0.1	ΔS 25°, 0.1
Yb^{3+}	ML/M.L	6.53 +0.01				
			6.09[d]		-0.6 / -0.8[d]	28 / 25[d]
	$ML_2/M.L^2$	12.23[p]				
	$ML_3/M.L^3$	17.05[p]				
Lu^{3+}	ML/M.L	6.50				
			6.00[d]		-0.2 / -0.5[d]	29 / 26[d]
UO_2^{2+}	ML/M.L	7.1[h] ±0.1				
	$ML_2/M.L^2$	12.6[h] ±0.1				
	$ML_3/M.L^3$	16.1[h] -0.1				
Mn^{2+}	ML/M.L	3.95	3.66[d]		(-1)[r]	(13)[d]
	$ML_2/M.L^2$	6.78	6.65[d]			
	$ML_3/M.L^3$		8.5[d]			
Co^{2+}	ML/M.L		4.55[d]		(-3)[r]	(11)[d]
	$ML_2/M.L^2$		8.26[d]		(-5)[r]	(21)[d]
	$ML_3/M.L^3$		10.70[d]		(-9)[r]	(20)[d]
Ni^{2+}	ML/M.L	4.9[h]	4.86[d]		(-2)[r]	(14)[d]
	$ML_2/M.L^2$	8.7[h]	8.81[d]		(-5)[r]	(25)[d]
	$ML_3/M.L^3$		11.62[d]		(-9)[r]	(24)[d]
Cu^{2+}	ML/M.L	6.6[h]	6.6[d]			
	$ML_2/M.L^2$	11.8[h]	11.7[d]			
Fe^{3+}	ML/M.L	9.2[h]				
	$ML_2/M.L^2$	17.2[h]		17.5[o]		
	$ML_3/M.L^3$	24.4[h]		24.6[o]		
	ML/MOHL.H	3.16				
	$MOHL/M(OH)_2L.H$	4.24				
	$(ML)^2/M_2(OH)_2L_2.H^2$	3.92				
	$(ML)^2/M_2(OH)_4L_2.H^4$	9.92				
Hf^{4+}	ML/M.L		12.04[j]			
	$ML_2/M.L^2$		22.59[j]			
Zn^{2+}	ML/M.L	4.9[h]	5.03[d]			
		4.98[b]				
	$ML_2/M.L^2$	9.1[h]	9.34[d]			
		8.95[b]				
	$ML_3/M.L^3$		12.4[d]			
Cd^{2+}	ML/M.L	4.6[h]				
Al^{3+}	ML/M.L	7.7[h]				
	$ML_2/M.L^2$	14.2[h]				
	$ML_3/M.L^3$	19.5[h]				
Ge(IV)	$M(OH)_2L_2/M(OH)_4.(HL)^2$	2.81[b]				

---

[b] 25°, 0.5;   [d] 25°, 2.0;   [h] 20°, 0.1;   [j] 20°, 1.0;   [o] 20°, 0;   [p] 30°, 0.1;   [r] 20-40°, 2.0

Bibliography:

H^+ 59OKa,61MA,62M,67CB,71SC,72DSS,72HS, UO_2^{2+} 65BS,66BR,66SK
          75GH

Mg^{2+} 59OKa,62M Mn^{2+} 62M,75GH

Ca^{2+},Cd^{2+},Al^{3+} 59OKa Co^{2+} 75GH

$Y^{3+}-Lu^{3+}$ 71SC,72DSS Ni^{2+},Cu^{2+} 59OKa,75GH

Kojic acid (continued)

Fe^{3+}	59OKa,61SL,62Ma	Ge(IV)	67CB
Hf^{4+}	72HS	Other references:	64YC,67H
Zn^{2+}	59OKa,67CB,75GH		

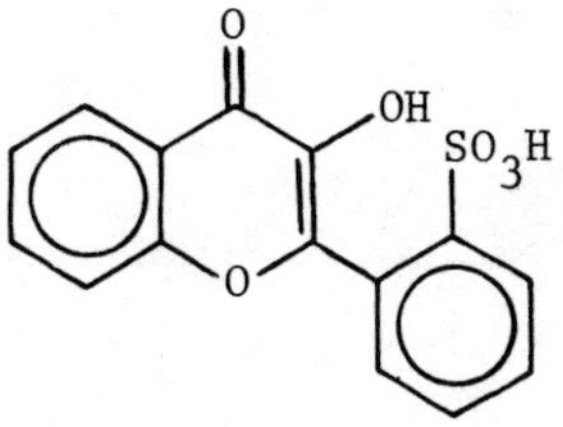

$C_{15}H_{10}O_6S$	3-Hydroxy-2-(2-sulfophenyl)-1-benzopyran-4-one (flavonol-2'-sulfonic acid)			H_2L

Metal ion	Equilibrium	Log K 25°, 0.1	Log K 25°, 0.5	Log K 25°, 1.0
H^+	$HL/H.L$	8.92	8.51	8.26 8.30[d]
Th^{4+}	$ML/M.L$	10.53	10.28	
	$ML_2/M.L^2$		18.06	
Zr^{4+}	$ML/M.L$			13.65[d]
	$ML_2/M.L^2$			24.40[d]
Tl^{3+}	$ML/M.L$	9.2		
	$ML_2/M.L^2$	16.4		
Bi^{3+}	$ML/M.L$		12.3	
	$ML_2/M.L^2$		20.4	

[d] 25°, 2.0

Bibliography:

H^+	64OYa,64OUA,67YT	Tl^{3+}	67YT
Th^{4+}	64OYA	Bi^{3+}	68YN
Zr^{4+}	64OYa		

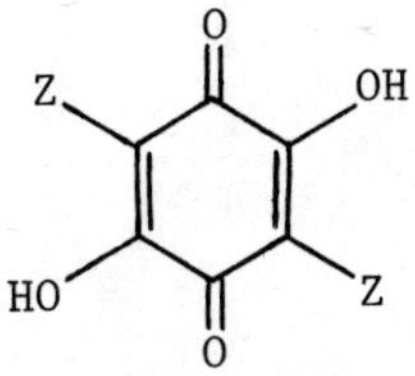

$C_nH_mO_pN_rS_t$	3,6-Disubstituted-2,5-dihydroxy-1,4-benzoquinone	H_2L

Z =	Metal ion	Equilibrium	Log K 25°, 0.5
H	H^+	$HL/H.L$	5.22
$(C_6H_4O_4)$		$H_2L/HL.H$	2.81
	Ge(IV)	$M(OH)_2L_2/M(OH)_4.(HL)^2$	8.08 ±0.01
		$HM(OH)_2L_2/M(OH)_2L_2.H$	1.7 ±0.1
		$H_2M(OH)_2L_2/HM(OH)_2L_2.H$	0.6

3,6-Disubstituted-2,5-dihydroxy-1,4-benzoquinone (continued)

Z =	Metal ion	Equilibrium	Log K 25°, 0.5
Diphenoxy	H^+	HL/H.L	3.00
$(C_{18}H_{12}O_6)$		$H_2L/HL.H$	2.0
	Ge(IV)	$M(OH)_2L_2/M(OH)_4 \cdot (HL)^2$	8.8
Dinitro	H^+	HL/H.L	−0.5
$(C_6H_2O_8N_2)$	Ge(IV)	$M(OH)_2L_2/M(OH)_4 \cdot (HL)^2$	4.9
Disulfo	H^+	HL/H.L	3.81
$(C_6H_4O_{10}S_2)$		$H_2L/HL.H$	0.8
	Ge(IV)	$M(OH)_2L_2/M(OH)_4 \cdot (HL)^2$	6.35

Bibliography: 67BB,72KLa

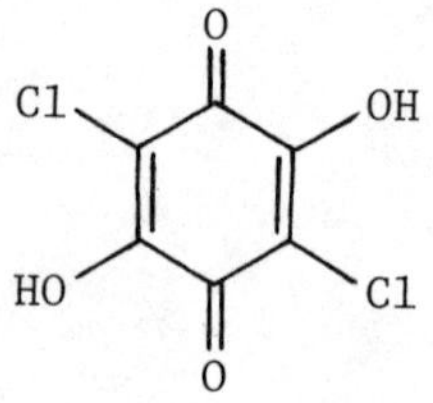

$C_6H_2O_4Cl_2$ 2,5-Dichloro-3,6-dihydroxy-1,4-benzoquinone (chloranilic acid) H_2L

Metal ion	Equilibrium	Log K 25°, 0.15	Log K 25°, 0.5
H^+	HL/H.L	2.60	2.37
	$H_2L/HL.H$	0.69	0.66
Ni^{2+}	ML/M.L	4.3	
Fe^{3+}	ML/M.L	5.8	
	$ML_2/M.L^2$	9.6	
Zr^{4+}	ML/M.L		$(9.27)^d$
	$ML_2/M.L^2$		$(16.56)^d$
Hf^{4+}	$ML.H^2/M.H_2L$		$(3.73)^e$
	$ML_3.H^6/M.(H_2L)^3$		$(11.63)^e$
Ge(IV)	$M(OH)_2L_2/M(OH)_4 \cdot (HL)^2$		6.57
	$HM(OH)_2L_2/M(OH)_2L_2.H$		0.8

---

^{d} 25°, 2.0 $HClO_4$; ^{e} 25°, 3.0 $HClO_4$

Bibliography:

H^+	64BB,67CA	Ge(IV)	64BB
Ni^{2+},Fe^{3+}	67CA		
Zr^{4+}	51TV,52TV	Other references:	62NF,64LSA,70KLa,71KL, 73CSM,73MB
Hf^{4+}	67VV		

$C_4H_2O_4$		3,4-Dihydroxy-3-cyclobutene-1,2-dione (squaric acid)				H_2L
Metal ion	Equilibrium	Log K 25°, 0.1	Log K 25°, 1.0	Log K 25°, 0	ΔH 25°, 0.1	ΔS 25°, 0.1
H^+	HL/H.L	3.10 2.78[b]	2.8 3.19[e]	3.48 ±0.00	(1)[s]	(20)
	H_2L/HL.H		0.40 0.96[e]	0.55 ±0.05		
Ac^{3+}	ML/M.L $ML_2/M.L^2$		1.85 2.74			
Am^{3+}	ML/M.L $ML_2/M.L^2$		2.17 3.10			
Cm^{3+}	ML/M.L $ML_2/M.L^2$		2.34 3.46			
Cf^{3+}	ML/M.L $ML_2/M.L^2$		2.48 4.18			
Th^{4+}	ML/M.L $ML_2/M.L^2$		4.08 7.32			
UO_2^{2+}	ML/M.L	3.08[b]				
Mn^{2+}	ML/M.L	1.51[b]				
Co^{2+}	ML/M.L	1.61[b]				
Ni^{2+}	ML/M.L M_2L/ML.M	1.49[b]	1.29[e] 0.74[e]			
Cu^{2+}	ML/M.L	2.20[b]				
Fe^{3+}	ML/M.L	4.61[b]				
VO^{2+}	ML/M.L		2.47[e]			
Al^{3+}	ML/M.L	2.83[r]				

[b] 25°, 0.5;   [e] 25°, 3.0;   [r] 25°, 0.3;   [s] 0-50°, 0.1

Bibliography:

H^+ 67IW,68Mb,69TW,70SH,71G,71SH,72AV Ni^{2+} 69TW,73AV

Ac^{3+}-Th^{4+} 72CSK VO^{2+} 74AV

UO_2^{2+},Mn^{2+},Co^{2+},Cu^{2+},Fe^{3+},Al^{3+} 69TW

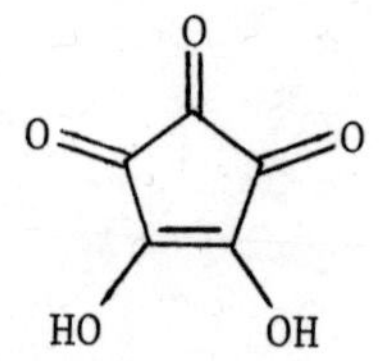

$C_5H_2O_5$ 4,5-Dihydroxy-4-cyclopentene-1,2,3-trione (croconic acid) H_2L

Metal ion	Equilibrium	Log K 25°, 0.1	Log K 25°, 1.0	Log K 25°, 0	ΔH 25°, 0	ΔS 25°, 0
H^+	HL/H.L	1.8	1.51[d]	2.23	3.0	20
	H_2L/HL.H	0.6	0.32[d]	0.76	-3.9	-10
Ca^{2+}	ML/M.L	1.29[r]				
	M.L/ML(s)	-4.05[r]				
Sr^{2+}	ML/M.L	1.21[r]				
	M.L/ML(s)	-5.08[r]				
Ba^{2+}	ML/M.L	1.55[r]	1.24			
	M.L/ML(s)	-8.28[r]	-7.82			

[d] 25°, 2.0;   [r] 25°, 0.3

Bibliography:

H^+ 62CD,75SGY

Ca^{2+}-Ba^{2+} 65CD

Other reference: 70AKa

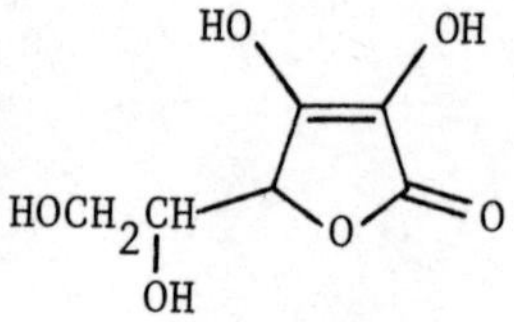

$C_6H_8O_6$ L-Ascorbic acid H_2L

Metal ion	Equilibrium	Log K 25°, 0.1	Log K 25°, 1.0	Log K 25°, 0
H^+	HL/H.L	11.34	11.35[e]-0.01	
	H_2L/HL.H	4.03 ±0.01	4.37[e]±0.01	
Ca^{2+}	MHL/M.HL	0.2[s]	0.03[e]	1.05
	ML/M.L		1.4[e]	
	$M_2L/M^2.L$		1.85[e]	
	$M_3L_3/M_3^3.L^3$		8.74[e]	
	$M_3L_4/M^3.L^4$		10.50[e]	
	$M_4OHL_3.H/M_4^4.L^3$		-2.06[e]	
	$M_4OHL_4.H/M^4.L^4$		0.5[e]	
Sr^{2+}	MHL/M.HL	(0.3)[s]		

[e] 25°, 3.0;   [s] 25°, 0.16

L-Ascorbic acid (continued)

Metal ion	Equilibrium	Log K 25°, 0.1	Log K 25°, 1.0	Log K 25°, 0
UO_2^{2+}	$MHL/M.HL$	2.35		3.04
	$M(HL)_2/M.(HL)^2$	3.32		
	$M(OH)_2HL/M.HL.(OH)^2$	19.4		
	$M(OH)(HL)_2/M.(HL)^2.OH$	12.2		
Mn^{2+}	$MHL/M.HL$			1.1
Fe^{2+}	$MHL/M.HL$		0.21^e	
	$ML/M.L$		1.99^e	
Co^{2+}	$MHL/M.HL$			1.4
Ni^{2+}	$MHL/M.HL$			1.1
Cu^{2+}	$MHL/M.HL$	1.57^t		
$Ti(IV)$	$MO(HL)_2/MO.(HL)^2$	24.8^h		
	$MO(H_2L)/MO.H_2L$	3.1^h		
	$MO(H_2L)_2/MO.(H_2L)^2$	6.25^h		
	$M(HL)_3/MO.H^2.(HL)^3$	9.3^h		
VO_2^+	$MH_2L/M.H_2L$		2.69	
Ag^+	$ML/M.L$	3.66		
Zn^{2+}	$MHL/M.HL$			1.0
Cd^{2+}	$MHL/M.HL$		0.42^e	1.3
	$M_2L.H/M^2.HL$		-5.52^e	
	$M_2H_3L_2/M^2.H.(HL)^2$		4.67^e	
	$M_3L_3.H^3/M^3.(HL)^3$		-13.65^e	
	$M_3OHL_3.H^4/M^3.(HL)^4$		-21.14^e	
	$M_4L_4.H^4/M^4.(HL)^4$		-17.16^e	
	$M_5HL_6.H^5/M^5.(HL)^6$		-20.4^e	
	$M_5L_6.H^6/M^5.(HL)^6$		-26.57^e	
	$M_5OHL_4.H^5/M^5.(HL)^4$		-23.4^e	
Pb^{2+}	$MHL/M.HL$	1.77		1.8
Al^{3+}	$MHL/M.HL$	1.89		
	$M(HL)_2/M.(HL)^2$	3.55		3.7
$As(III)$	$M(OH)_2HL/M(OH)_2.HL$	7.2		

^{e} 25°, 3.0; ^{h} 20°, 0.1; ^{t} 0°, 0.1

Bibliography:

H^+ 66VS,67TM,67WU,71W,73UW,74UW

Ca^{2+} 52SL,65VS,73UW

Sr^{2+} 52SL

UO_2^{2+} 65VS,69SH

Mn^{2+},Co^{2+},Ni^{2+},Zn^{2+} 65VS

Fe^{2+} 74UW

Cu^{2+} 64SM

$Ti(IV)$ 63S

VO_2^+ 73KT

Ag^+ 64NM

Cd^{2+} 65VS,71W

Pb^{2+},Al^{3+} 65VS,66VS

$As(III)$ 72ET

Other references: 61SMa,62SM,69TM,71EP, 71SKI,73FJ,74CC,74CP,74I,75Y

$$HOCH_2CH_2OH$$

		Ethane-1,2-diol (ethylene glycol)		L
$C_2H_6O_2$				
Metal ion	Equilibrium	Log K 25°, 0.1	Log K 25°, 1.0	
Y^{3}	$MH_{-1}L.H/M.L$	-6.95		
La^{3+}	$MH_{-1}L.H/M.L$	-8.50		
Ce^{3+}	$MH_{-1}L.H/M.L$	(-8.00)		
Pr^{3+}	$MH_{-1}L.H/M.L$	-7.90		
Nd^{3+}	$MH_{-1}L.H/M.L$	-7.80		
Sm^{3+}	$MH_{-1}L.H/M.L$	-7.50		
Eu^{3+}	$MH_{-1}L.H/M.L$	-7.30		
Gd^{3+}	$MH_{-1}L.H/M.L$	-7.25		
Tb^{3+}	$MH_{-1}L.H/M.L$	-7.20		
Dy^{3+}	$MH_{-1}L.L/M.L$	-7.15		
Ho^{3+}	$MH_{-1}L.H/M.L$	-7.05		
Er^{3+}	$MH_{-1}L.H/M.L$	-6.95		
Tm^{3+}	$MH_{-1}L.H/M.L$	-6.80		
Yb^{3+}	$MH_{-1}L.H/M.L$	-6.70		
Lu^{3+}	$MH_{-1}L.H/M.L$	-6.50		
Pb(II)	$MOH(H_{-2}L)/M(OH)_3.L$		0.30	
B(III)	$M(H_{-2}L)_2.H/M(OH)_3.L^2$	-8.53		
Ge(IV)	$HMO_2(H_{-2}L).H/M(OH)_4.L$	-8.41		
	$HMO(H_{-2}L)_2.H/M(OH)_4.L^2$	-8.95		

Bibliography:

Y^{3+}-Lu^{3+}	72MC	Ge(IV)	59A
Pb(II)	68Va	Other references:	57RL,67CBa
B(III)	67NE		

$$\overset{\displaystyle OH}{\underset{\displaystyle CH_3CHCH_2OH}{|}}$$

		Propane-1,2-diol (propylene glycol)		L
$C_3H_8O_2$				
Metal ion	Equilibrium	Log K 25°, 0.1	Log K 25°, 1.0	
Y^{3+}	$MH_{-1}L.H/M.L$	-6.95		
La^{3+}	$MH_{-1}L.H/M.L$	(-8.65)		
Ce^{3+}	$MH_{-1}L.H/M.L$	-8.05		
Pr^{3+}	$MH_{-1}L.H/M.L$	(-7.90)		
Nd^{3+}	$MH_{-1}L.H/M.L$	-7.70		

Propylene glycol (continued)

Metal ion	Equilibrium	Log K 25°, 0.1	Log K 25°, 1.0
Sm^{3+}	$MH_{-1}L.H/M.L$	(-7.55)	
Eu^{3+}	$MH_{-1}L.H/M.L$	-7.20	
Gd^{3+}	$MH_{-1}L.H/M.L$	-7.15	
Tb^{3+}	$MH_{-1}L.H/M.L$	-7.15	
Dy^{3+}	$MH_{-1}L.H/M.L$	-7.10	
Ho^{3+}	$MH_{-1}L.H/M.L$	-6.95	
Er^{3+}	$MH_{-1}L.H/M.L$	-6.85	
Tm^{3+}	$MH_{-1}L.H/M.L$	(-5.85)	
Yb^{3+}	$MH_{-1}L.H/M.L$	-6.55	
Lu^{3+}	$MH_{-1}L.H/M.L$	-6.40	
Pb(II)	$MOH(H_{-2}L)/M(OH)_3.L$		0.30
Ge(IV)	$HMO_2(H_{-2}L).H/M(OH)_4.L$	-8.30	
	$HMO(H_{-2}L)_2.H/M(OH)_4.L^2$	-8.52	

Bibliography:

$Y^{3+}-Lu^{3+}$	72MC	Ge(IV)	59A
Pb(II)	68Va	Other references:	57RL,67CBa

$$\overset{\displaystyle OH}{\underset{\textstyle |}{CH_3CH_2CHCH_2OH}}$$

$C_4H_{10}O_2$		Butane-1,2-diol (butylene glycol)*		L

Metal ion	Equilibrium	Log K 25°, 0.1
Ge(IV)	$HMO_2(H_{-2}L).H/M(OH)_4.L$	-7.94
	$HMO(H_{-2}L)_2.H/M(OH)_4.L^2$	-8.19

*Isomer not stated but presumed to be 1,2-.

Bibliography: 59A

$$\overset{\displaystyle OH}{\underset{\textstyle |}{CH_3OCH_2CHCH_2OH}}$$

$C_4H_{10}O_3$		3-Methoxypropane-1,2-diol		L

Metal ion	Equilibrium	Log K 25°, 0.1
Ge(IV)	$HMO_2(H_{-2}L).H/M(OH)_4.L$	-7.74
	$HMO(H_{-2}L)_2.H/M(OH)_4.L^2$	-8.00

Bibliography: 59A
Other reference: 57RL

$$HOCH_2CH_2CH_2OH$$

$C_3H_8O_2$ Propane-1,3-diol   L

Metal ion	Equilibrium	Log K 25°, 1.0
Pb(II)	$MOH(H_{-2}L)/M(OH)_3 \cdot L$	-0.20

Bibliography: 68Va   Other reference: 67CBa

$$\overset{\displaystyle OH}{\underset{\displaystyle |}{HOCH_2CHCH_2OH}}$$

$C_3H_8O_3$ Propane-1,2,3-triol (glycerol)   L

Metal ion	Equilibrium	Log K 25°, 0.1	Log K 25°, 1.0
Y^{3+}	$MH_{-1}L \cdot H/M \cdot L$	-6.80 ±0.05	
La^{3+}	$MH_{-1}L \cdot H/M \cdot L$	-8.30 ±0.05	
Ce^{3+}	$MH_{-1}L \cdot H/M \cdot L$	-7.95	
Pr^{3+}	$MH_{-1}L \cdot H/M \cdot L$	-7.70 ±0.05	
Nd^{3+}	$MH_{-1}L \cdot H/M \cdot L$	-7.60 ±0.00	
Sm^{3+}	$MH_{-1}L \cdot H/M \cdot L$	-7.25 ±0.00	
Eu^{3+}	$MH_{-1}L \cdot H/M \cdot L$	-7.15 ±0.00	
Gd^{3+}	$MH_{-1}L \cdot H/M \cdot L$	-7.15 ±0.05	
Tb^{3+}	$MH_{-1}L \cdot H/M \cdot L$	-6.95 ±0.00	
Dy^{3+}	$MH_{-1}L \cdot H/M \cdot L$	-6.90 -0.05	
Ho^{3+}	$MH_{-1}L \cdot H/M \cdot L$	-6.85 +0.05	
Er^{3+}	$MH_{-1}L \cdot H/M \cdot L$	-6.75 ±0.00	
Tm^{3+}	$MH_{-1}L \cdot H/M \cdot L$	-6.6 ±0.1	
Yb^{3+}	$MH_{-1}L \cdot H/M \cdot L$	-6.45 +0.05	
Lu^{3+}	$MH_{-1}L \cdot H/M \cdot L$	-6.40 ±0.10	
Pb(II)	$MOH(H_{-2}L)/M(OH)_3 \cdot L$		1.2
B(III)	$M(OH)_2(H_{-2}L) \cdot H/M(OH)_3 \cdot L$	-7.54	
	$M(H_{-2}L)_2 \cdot H/M(OH)_3 \cdot L^2$	-7.17 ±0.03	
Ge(IV)	$HMO_2(H_{-2}L) \cdot H/M(OH)_4 \cdot L$	-7.32	
	$HMO(H_{-2}L)_2 \cdot H/M(OH)_4 \cdot L^2$	-6.64	
Te(VI)	$H_3MO_4(H_{-2}L) \cdot H/M(OH)_6 \cdot L$	-5.63	

Bibliography:

Y^{3+}-Lu^{3+} 70PK,72MC   Ge(IV)   57Aa
Pb(II)   67VL   Te(VI)   56Ae
B(III)   56Af,67NE   Other references: 56Ad,57RL,63NF,67CBa,67RBa,68V

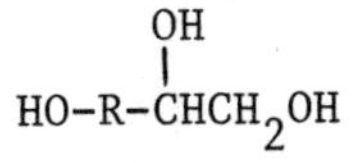

$$OH$$
$$|$$
$$HO-R-CHCH_2OH$$

$C_nH_mO_3$

<u>Alkane-triol</u>   L

R =	Metal ion	Equilibrium	Log K 25°, 1.0
CH_2CH_2 $(C_4H_{10}O_3)$ (butane-1,2,4-triol)	Pb(II)	$MOH(H_{-2}L)/M(OH)_3 \cdot L$	0.45
$CH_2CH_2CH_2$ $(C_5H_{12}O_3)$ (pentane-1,2,5-triol)	Pb(II)	$MOH(H_{-2}L)/M(OH)_3 \cdot L$	0.40

Bibliography: 68Va

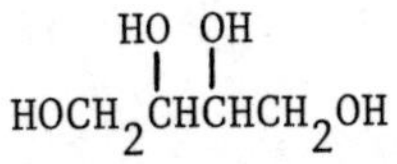

$$HO\ OH$$
$$|\ \ |$$
$$HOCH_2CHCHCH_2OH$$

$C_4H_{10}O_4$

<u>meso-Butane-1,2,3,4-tetraol</u> (<u>meso-erythritol</u>)   L

Metal ion	Equilibrium	Log K 25°, 1.0
Pb(II)	$MOH(H_{-2}L)/M(OH)_3 \cdot L$	1.93

Bibliography: 68Vb

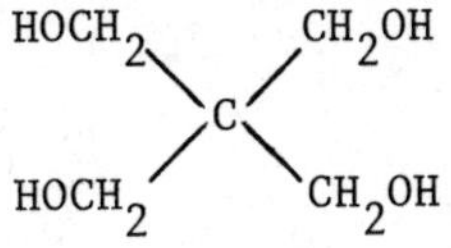

$$HOCH_2 \qquad CH_2OH$$
$$\diagdown \ C \ \diagup$$
$$\diagup \qquad \diagdown$$
$$HOCH_2 \qquad CH_2OH$$

$C_5H_{12}O_4$

<u>Tetrakis(hydroxymethyl)methane</u> (<u>pentaerythritol</u>)   L

Metal ion	Equilibrium	Log K 25°, 0.1
B(III)	$M(OH)_2(H_{-2}L) \cdot H/M(OH)_3 \cdot L$	-6.40
	$M(H_{-2}L)_2 \cdot H/M(OH)_3 \cdot L^3_2$	-5.45
As(III)	$M(OH)_2(H_{-2}L) \cdot H/M(OH)_3 \cdot L$	-8.2
Te(VI)	$H_3MO_4(H_{-2}L) \cdot H/M(OH)_6 \cdot L$	-6.8

Bibliography: 60AR

Other reference: 57RL

$$\begin{array}{cc} \text{HO} & \text{OH} \\ | & | \\ \text{HOCH}_2\text{CHCHCHCHCH}_2\text{OH} \\ | & | \\ \text{HO} & \text{OH} \end{array}$$

$C_6H_{14}O_6$ D(-)-Mannitol   L

Metal ion	Equilibrium	Log K 25°, 0.1	Log K 25°, 1.0
Mo(VI)	$H_2(MO_4)_2L/H^2 \cdot (MO_4)^2 \cdot L$		17.48^e
	$H_3(MO_4)_2L/H_2(MO_4)_2L \cdot H$		4.07^e
Pb(II)	$MOH(H_{-2}L)/M(OH)_3 \cdot L$		2.78
B(III)	$M(OH)_2(H_{-2}L) \cdot H/M(OH)_3 \cdot L$	-5.99	-6.01^e
	$M(H_{-2}L)_2 \cdot H/M(OH)_3 \cdot L^2$	-3.91 -0.3	-4.10^e
	$HM(OH)_2(H_{-2}L)/M(OH)_2(H_{-2}L) \cdot H$		5.87^e
	$HM(OH)_2(H_{-2}L)_2/M(OH)_2(H_{-2}L)_2 \cdot H$	2.70	
	$M_2(OH)_4(H_{-4}L) \cdot H^2/[M(OH)_3]^2 \cdot L$	-13.53	-13.61^e
	$M_2(OH)_2(H_{-2}L)(H_{-4}L) \cdot H^2/[M(OH)_3]^2 \cdot L^2$		-10.76^e
As(III)	$M(OH)_2(H_{-2}L) \cdot H/M(OH)_3 \cdot L$	-8.06 -0.2	
	$HM(OH)_2(H_{-2}L)/M(OH)_2(H_{-2}L) \cdot H$	8.44	
Ge(IV)	$HMO_2(H_{-2}L) \cdot H/M(OH)_4 \cdot L$		-6.43^b
	$HMO(H_{-2}L)_2 \cdot H/M(OH)_4 \cdot L^2$	-4.05	-3.95^b
	$H_2M_2O_3(H_{-2}L)(H_{-4}L) \cdot H^2/[M(OH)_4]^2 \cdot L^2$		-10.62^b
Te(VI)	$H_3MO_4(H_{-2}L) \cdot H/M(OH)_6 \cdot L$	-4.21	

^{b} 25°, 0.5; ^{e} 25°, 3.0

Bibliography:

Mo(VI)   72P   Ge(IV)   57Aa,73PA

Pb(II)   68Vb   Te(VI)   56Ac

B(III)   49RC,54Ab,57Ab,68AP,73K,73PAa   Other references: 41S,63NF,67CBa,67NE,71M

As(III)   59ARa,70AT,75VH

$$\begin{array}{cc} \text{HO} & \text{OH} \\ | & | \\ \text{HOCH}_2\text{CHCHCHCHCH}_2\text{OH} \\ | & | \\ \text{HO} & \text{OH} \end{array}$$

$C_6H_{14}O_6$ D-Galactitol (D-dulcitol)   L

Metal ion	Equilibrium	Log K 25°, 0.1	Log K 25°, 1.0
Pb(II)	$ML/M(OH)_3 \cdot L$		2.90
B(III)	$M(H_{-2}L)_2 \cdot H/M(OH)_3 \cdot L^2$	-3.87	
Ge(IV)	$HMO(H_{-2}L)_2 \cdot H/M(OH)_4 \cdot L^2$	-3.87	

Bibliography:

Pb(II)   68Vb   B(III),Ge(IV)   59AR

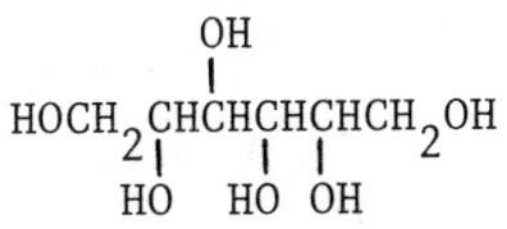

$$\underset{\underset{\text{HO}}{|}}{\text{HOCH}_2}\text{CH}\underset{\underset{\text{HO}}{|}}{\text{CH}}\underset{\underset{\text{OH}}{|}}{\text{CH}}\text{CH}_2\text{OH}$$

with OH above the second carbon:

HOCH$_2$CH(OH)CHCHCH$_2$OH with HO, HO, OH below.

D(−)-Sorbitol   L

$C_6H_{14}O_6$

Metal ion	Equilibrium	Log K 25°, 0.1	Log K 25°, 1.0
Pb(II)	$MOH(H_{-2}L)/M(OH)_3 \cdot L$		3.42
B(III)	$M(H_{-2}L)_2 \cdot H/M(OH)_3 \cdot L^2$	−3.45 −0.5	
Ge(IV)	$HMO(H_{-2}L)_2 \cdot H/M(OH)_4 \cdot L^2$	−3.49	

Bibliography:

Pb(II)   68Vb   Ge(IV)   59AR

B(III)   59AR,67NE

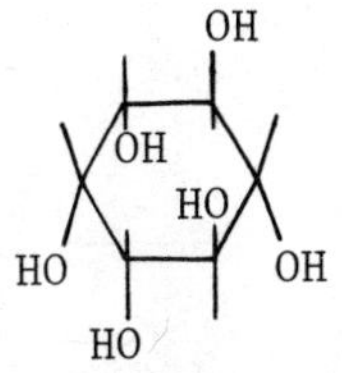

meso-Inositol   L

$C_6H_{12}O_6$

Metal ion	Equilibrium	Log K 25°, 0.1
B(III)	$M(OH)_2(H_{-1}L) \cdot H/M(OH)_3 \cdot L$	−7.46
Ge(IV)	$HMO(H_{-2}L)_2 \cdot H/M(OH)_4 \cdot L^2$	−6.44
Te(VI)	$HMO_4(H_{-2}L) \cdot H/M(OH)_6 \cdot L$	−5.61
	$HMO_2(H_{-2}L)_2 \cdot H/M(OH)_6 \cdot L^2$	−5.53

Bibliography: 67FA

Other reference: 75AHa

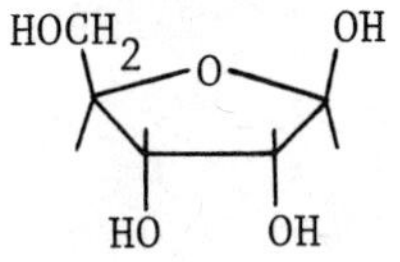

D(−)-Ribose   L

$C_5H_{10}O_5$

Metal ion	Equilibrium	Log K 25°, 0.1	Log K 25°, 0	ΔH 25°, 0	ΔS 25°, 0
H$^+$	$L/H_{-1}L \cdot H$		12.11	−8.6	27
B(III)	$M(H_{-2}L)_2 \cdot H/M(OH)_3 \cdot L^2$	−1.90			
As(III)	$M(OH)_2(H_{-2}L) \cdot H/M(OH)_3 \cdot L$	−5.50			
Ge(IV)	$HMO(H_{-2}L)_2 \cdot H/M(OH)_4 \cdot L^2$	−2.46			

D(-)-Ribose (continued)

Metal ion	Equilibrium	Log K 25°, 0.1	Log K 25°, 0	ΔH 25°, 0	ΔS 25°, 0
Te(VI)	$H_3MO_4(H_{-2}L).H/M(OH)_6.L$	-3.02			
	$HMO_2(H_{-2}L)_2.H/M(OH)_6.L^2$	-1.49			

Bibliography:

H^+ 70CRa   Other references: 66IR,67Wa

B(III)-Te(VI)   73AH

$C_5H_{10}O_5$		D(-)-Lyxose			L
Metal ion	Equilibrium	Log K 25°, 0.1	Log K 25°, 0	ΔH 25°, 0	ΔS 25°, 0
H^+	$L/H_{-1}L.H$		12.11	-8.0	29
B(III)	$M(OH)_2(H_{-2}L).H/M(OH)_3.L$	-4.28			
Ge(IV)	$HM(OH)_2(H_{-2}L).H/M(OH)_4.L$	-3.94			

Bibliography:

H^+ 70CRa

B(III),Ge(IV)   73AH

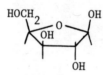

$C_5H_{10}O_5$		D(-)-Xylose			L
Metal ion	Equilibrium	Log K 25°, 0.1	Log K 25°, 0	ΔH 25°, 0	ΔS 25°, 0
H^+	$L/H_{-1}L.H$		12.15	-9.0	25
B(III)	$M(H_{-2}L)_2.H/M(OH)_3.L^2$	-5.09			
As(III)	$M(OH)_2(H_{-2}L).H/M(OH)_3.L$	-8.39			
Ge(IV)	$HMO(H_{-2}L)_2.H/M(OH)_4.L^2$	-5.20			
Te(VI)	$HMO_4(H_{-2}L).H /M(OH)_6.L$	-3.96			

Bibliography:

H^+ 70CRa   Te(VI)   73AH

B(III),Ge(IV)   59AT   Other reference: 66IR

As(III)   60AT

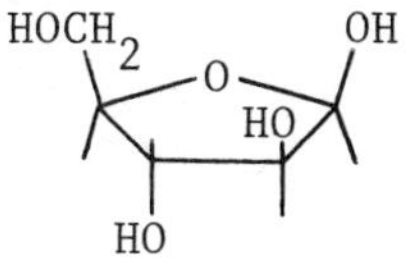

$C_5H_{10}O_5$ D(-)-Arabinose   L

Metal ion	Equilibrium	Log K 25°, 0.1	Log K 25°, 0	ΔH 25°, 0	ΔS 25°, 0
H^+	$L/H_{-1}L.H$		12.34	-9.6	24
B(III)	$M(H_{-2}L)_2.H/M(OH)_3.L^2$	-5.82			
As(III)	$M(OH)_2(H_{-2}L).H/M(OH)_3.L$	-7.85			
Ge(IV)	$HMO(H_{-2}L)_2.H/M(OH)_4.L^2$	-5.07			
Te(VI)	$H_3MO_4(H_{-2}L).H/M(OH)_6.L$	-5.08			

Bibliography:

H^+ 70CRa   Te(VI)   73AH

B(III),Ge(IV)   59AT   Other reference: 66IR

As(III)   60AT

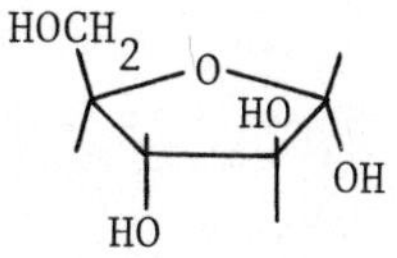

$C_5H_{10}O_5$ L(+)-Arabinose   L

Metal ion	Equilibrium	Log K 25°, 0.1
B(III)	$M(H_{-2}L)_2.H/M(OH)_3.L^2$	-5.55
As(III)	$M(OH)_2(H_{-2}L).H/M(OH)_3.L$	-7.89
Ge(IV)	$HMO(H_{-2}L)_2.H/M(OH)_4.L^2$	-4.96
Te(VI)	$H_3MO_4(H_{-2}L).H/M(OH)_6.L$	-3.69

Bibliography:

B(III),Ge(IV)   59AT   Te(VI)   73AH

As(III)   60AT   Other reference: 59RL

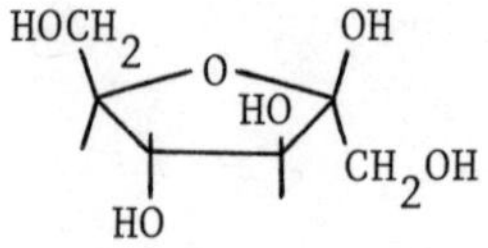

HOCH$_2$ O   OH
         HO
HO   CH$_2$OH

$C_6H_{12}O_6$ D(−)-Fructose   L

Metal ion	Equilibrium	Log K 25°, 0.1	Log K 25°, 0	ΔH 25°, 0	ΔS 25°, 0
H$^+$	L/H$_{-1}$L.H		12.03	−9.4	24
B(III)	M(OH)$_2$(H$_{-2}$L).H/M(OH)$_3$.L	−5.68			
	M(H$_{-2}$L)$_2$.H/M(OH)$_3$.L$_2^3$	−4.12 ±0.06			
As(III)	M(OH)$_2$(H$_{-2}$L).H/M(OH)$_3$.L	−8.39			
Ge(IV)	HMO(H$_{-2}$L)$_2$.H/M(OH)$_4$.L$_2$	−3.10			
Te(VI)	H$_3$MO$_4$(H$_{-2}$L).H/M(OH)$_6$.L	−5.48			

Bibliography:

H$^+$ 70CRa   Ge(IV)   57Aa

B(III)   58A,67NEa,68AP   Te(VI)   57A

As(III)   68APa   Other references: 41S,57RL,60AT,63NF,66IR,70C

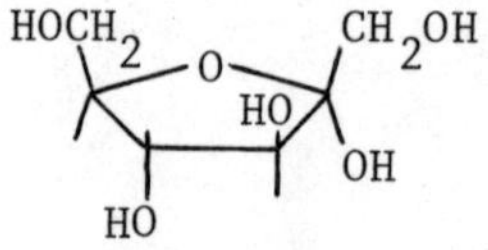

HOCH$_2$ O   CH$_2$OH
         HO
HO   OH

$C_6H_{12}O_6$ L(−)-Sorbose   L

Metal ion	Equilibrium	Log K 25°, 0.1
B(III)	M(H$_{-2}$L)$_2$.H/M(OH)$_3$.L$_2$	−3.30
As(III)	M(OH)$_2$(H$_{-2}$L).H/M(OH)$_3$.L	−8.05
Ge(IV)	HMO(H$_{-2}$L)$_2$.H/M(OH)$_4$.L$_2$	−3.24

Bibliography:

B(III),Ge(IV)   59AT   As(III)   60AT

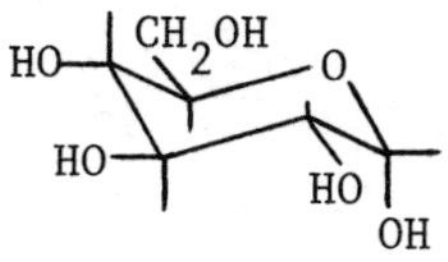

$C_6H_{12}O_6$ α-D(+)-Glucose   L

Metal ion	Equilibrium	Log K 25°, 0.1	Log K 25°, 0	ΔH 25°, 0	ΔS 25°, 0
H^+	$L/H_{-1}L.H$		12.28	-8.8	27
B(III)	$M(OH)_2(H_{-2}L).H/M(OH)_3.L$	-7.08			
	$M(H_{-2}L)_2.H/M(OH)_3.L^2$	-6.47 +0.2			
Ge(IV)	$HMO(H_{-2}L)_2.H/M(OH)_4.L^2$	-5.12			

Bibliography:

H^+ 70CRa

B(III)   58A,67NEa,68AP

Ge(IV)   59A

Other references: 57RL,63NF,66IR,67CBa,70C

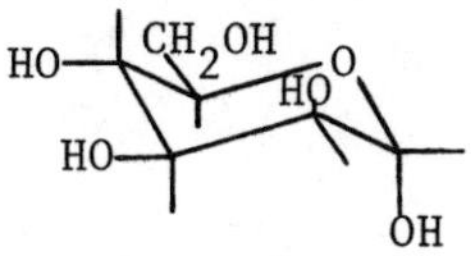

$C_6H_{12}O_6$ α-D(+)-Mannose   L

Metal ion	Equilibrium	Log K 25°, 0.1	Log K 25°, 0	ΔH 25°, 0	ΔS 25°, 0
H^+	$L/H_{-1}L.H$		12.08	-7.9	29
B(III)	$M(H_{-2}L)_2.H/M(OH)_3.L^2$	-4.58			
As(III)	$M(OH)_2(H_{-2}L).H/M(OH)_3.L$	-6.91			
	$M(H_{-2}L)_2.H/M(OH)_3.L^2$	-6.16			
Ge(IV)	$HMO(H_{-2}L)_2.H/M(OH)_4.L^2$	-4.45			

Bibliography:

H^+ 66IR,70CRa

B(III)   58A

As(III)   59ARa

Ge(IV)   59A

Other references: 57RL,70C

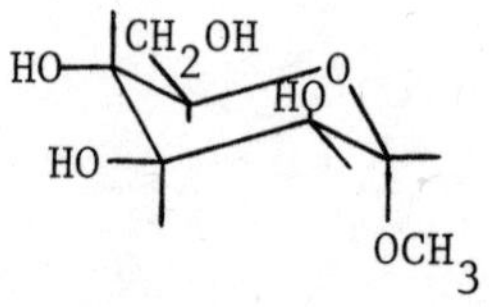

$C_7H_{14}O_6$		Methyl α-D-mannopyranoside	L
Metal ion	Equilibrium	Log K 30°, 0.1	
B(III)	$M(OH)_2(H_{-2}L)/M(OH)_4 \cdot L$	1.7	
	$M(H_{-2}L)_2/M(OH)_4 \cdot L^2$	2.8	

Bibliography: 64MG

Other reference: 65LA

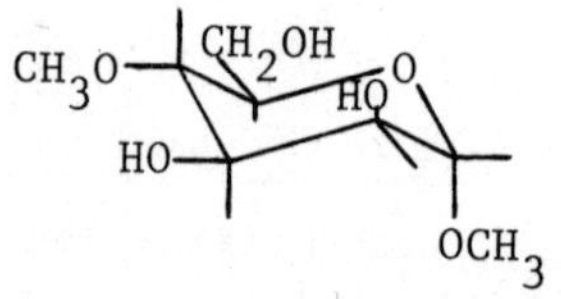

$C_8H_{16}O_6$		Methyl-4-O-methyl-α-D-mannopyranoside	L
Metal ion	Equilibrium	Log K 30°, 0.1	
B(III)	$M(OH)_2(H_{-2}L)/M(OH)_4 \cdot L$	1.5	
	$M(H_{-2}L)_2/M(OH)_4 \cdot L^2$	3.5	

Bibliography: 64MG

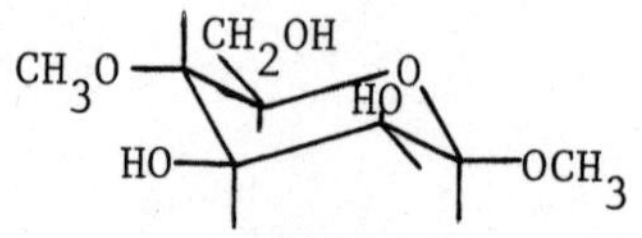

$C_8H_{16}O_6$		Methyl-4-O-methyl-β-D-mannopyranoside	L
Metal ion	Equilibrium	Log K 30°, 0.1	
B(III)	$M(OH)_2(H_{-2}L)/M(OH)_4 \cdot L$	0.3	
	$M(H_{-2}L)_2/M(OH)_4 \cdot L^2$	2.6	

Bibliography: 64MG

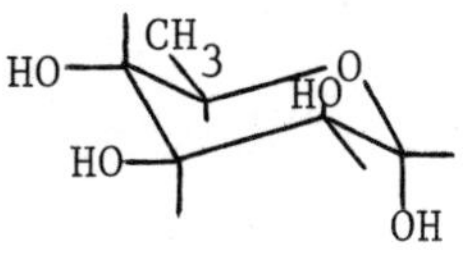

$C_6H_{12}O_5$ α-6-Deoxy-L(+)-mannose (L(+)-rhamnose)   L

Metal ion	Equilibrium	Log K 25°, 0.1
B(III)	$M(H_{-2}L)_2 \cdot H/M(OH)_3 \cdot L^2$	-6.49
As(III)	$M(OH)_2(H_{-2}L) \cdot H/M(OH)_3 \cdot L$	-8.45
Ge(IV)	$HMO(H_{-2}L)_2 \cdot H/M(OH)_4 \cdot L^2$	-5.34

Bibliography:

B(III),Ge(IV)   59AT   As(III)   60AT

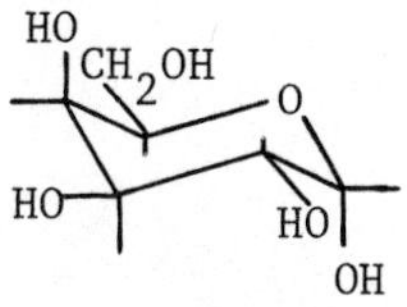

$C_6H_{12}O_6$ α-D(+)-Galactose   L

Metal ion	Equilibrium	Log K 25°, 0.1	Log K 25°, 0	ΔH 25°, 0	ΔS 25°, 0
H^+	$L/H_{-1}L \cdot H$		12.35	-9.7	24
B(III)	$M(OH)_2(H_{-2}L) \cdot H/M(OH)_3 \cdot L$	-6.79			
	$M(H_{-2}L)_2 \cdot H/M(OH)_3 \cdot L^2$	-6.71			
Ge(IV)	$HMO(H_{-2}L)_2 \cdot H/M(OH)_4 \cdot L^2$	-4.45			

Bibliography:

H^+ 70CRa   Ge(IV)   59A

B(III)   58A   Other references: 57RL,63NF,66IR,70C

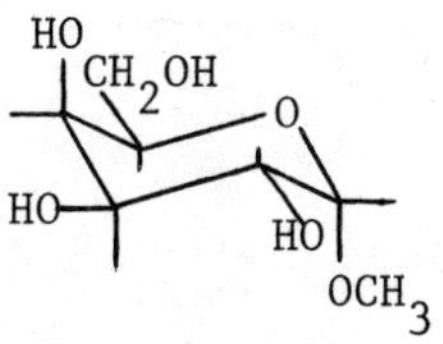

$C_7H_{14}O_6$ Methyl α-D-galactopyranoside   L

Metal ion	Equilibrium	Log K 30°, 0.1
B(III)	$M(OH)_2(H_{-2}L)/M(OH)_4 \cdot L$	2.0
	$M(H_{-2}L)_2/M(OH)_4 \cdot L^2$	2.6

Bibliography: 64MG

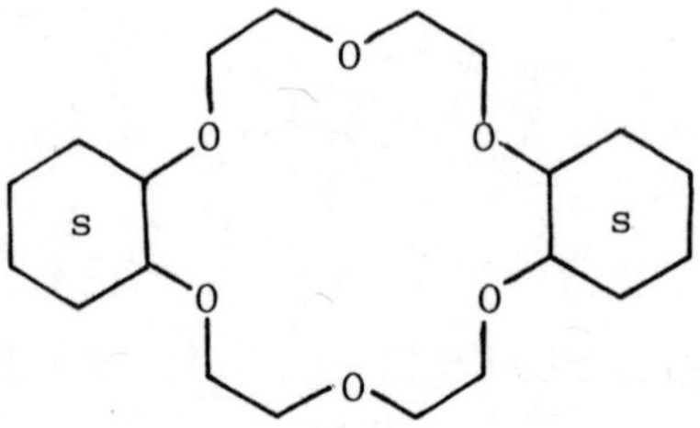

$C_{20}H_{36}O_6$ 　　　2,5,8,15,18,21-Hexaoxatricyclo$[20,4,0,0^{9,14}]$hexacosane 　　　　　L
　　　　　　　　　　　(dicyclohexyl-18-crown-6-ether)

Isomer	Metal ion	Equilibrium	Log K 25°, 0	ΔH 25°, ∿0	ΔS 25°, 0
A^z	K^+	ML/M.L	2.02	-3.9	-4
	Rb^+	ML/M.L	1.52	-3.3	-4
	Cs^+	ML/M.L	0.96	-2.4	-4
	NH_4^+	ML/M.L	1.33	-2.2	-1
	Sr^{2+}	ML/M.L	3.24	-3.7	2
	Ba^{2+}	ML/M.L	3.57	-4.9	0
B^z	K^+	ML/M.L	1.63	-5.1	-10
	Rb^+	ML/M.L	0.87	-4.0	-9
	NH_4^+	ML/M.L	0.80	-3.4	-8
	Sr^{2+}	ML/M.L	2.64	-3.2	1
	Ba^{2+}	ML/M.L	3.27	-6.2	-6
	Ag^+	ML/M.L	1.59	-2.1	0

z Stereochemistry of ligand not known; however because of the importance of the crown ethers, this one is included as an example even though not fully characterized.

Bibliography: 71IN

Other reference: 71F

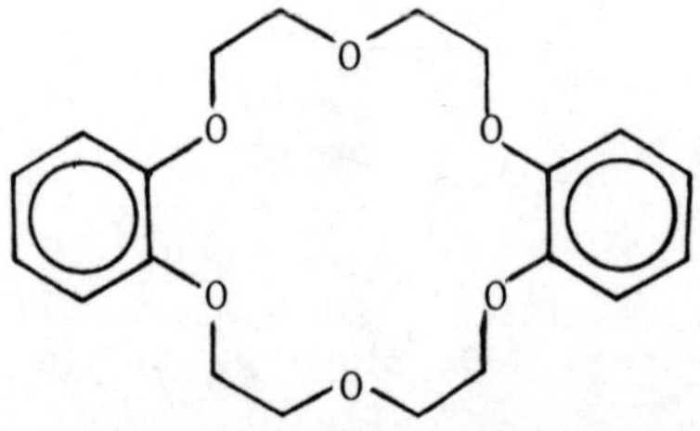

$C_{20}H_{24}O_6$ 　　　2,3,11,12-Dibenzo-1,4,7,10,13,16-hexaoxacyclooctadeca-2,11-diene 　　　L
　　　　　　　　　　　(dibenzo-18-crown-6-ether)

Metal ion	Equilibrium	Log K 25°, 0
Na^+	ML/M.L	1.16
K^+	ML/M.L	1.67

Dibenzo-18-crown-6-ether (continued)

Metal ion	Equilibrium	Log K 25°, 0
Rb^+	ML/M.L	1.08
Cs^+	ML/M.L	0.83
NH_4^+	ML/M.L	0.3
Sr^{2+}	ML/M.L	1.0
Ba^{2+}	ML/M.L	1.9
Ag^+	ML/M.L	1.41
Tl^+	ML/M.L	1.51
Pb^{2+}	ML/M.L	1.89

Bibliography: 75SN

$$HOCH_2CH_2SH$$

C_2H_6OS 2-Mercaptoethanol   HL

Metal ion	Equilibrium	Log K 25°, 0.1	Log K 25°, 0.5	Log K 25°, 0	ΔH 25°, 0	ΔS 25°, 0
H^+	HL/H.L	9.40	9.34	9.72	-6.21	23.7
		$9.48^h \pm 0.04$			-6.5^h	21^h
Ni^{2+}	$M(ML_2)_n/M^{n+1}.L^{2n}$		-2.3 +n13.03			
Ag^+	ML/M.L	13.2				
		13.0^h				
	$ML_2/M.L^2$	17.9				
		17.9^h				
	$M_2L/ML.M$	6				
	M.L/ML(s)	-19.7				
CH_3Hg^+	ML/M.L	15.87			-19.8^h	6^h
		16.12^h				
	$M_2L/ML.M$	6.27^h				
Zn^{2+}	$M_2L_3/M_3^2.L^3$		18.32			
	$M_3L_6/M_4^3.L^6$		38.52			
	$M_4L_9/M^4.L^9$		57.80			
	$M_5L_{12}/M^5.L^{12}$		77.20			
	$M_6L_{15}/M^6.L^{15}$		95.92			
Pb^{2+}	$M_2L/M^2.L$		8.94			
	$M_2L_2/M_2^2.L^2$		15.77			
	$M_2L_3/M_2^2.L^3$		22.03			
	$M_3L_4/M_3^2.L^4$		32.65			
	$M_3L_5/M^3.L^5$		38.50			
In^{3+}	ML/M.L	9.1^h				
	$ML_2/M.L^2$	17.2^h				
	$ML_3/M.L^3$	24.1^h				
	$ML_4/M.L^4$	29.9^h				
Sb(III)	$M(H_{-1}L)_2.H/M(OH)_3.(HL)^2$	-1.73^h				
	$MHL_2/ML_2.H$	7.98^h				
Ge(IV)	$MOH(H_{-1}L)_2.H/H_2MO_3.(HL)^2$	-4.21				

---

^{h} 20°, 0.1

Bibliography:

H^+ 64IN,65SS,70AM,71DV,71TS Pb^{2+} 74DT

Ni^{2+} 72DV In^{3+} 72TS

Ag^+ 58SG,65SS,71TS   Sb(III)   70AM

CH_3Hg^+ 65SS   Ge(IV) 63AT

Zn^{2+} 71DV   Other references: 61AM,61KP,62AT

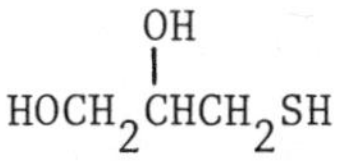

$$OH$$
$$|$$
$$HOCH_2CHCH_2SH$$

$C_3H_8O_2S$		3-Mercaptopropane-1,2-diol	HL

Metal ion	Equilibrium	Log K 25°, 0.1	Log K 25°, 0.5
H^+	HL/H.L		9.28
Ni^{2+}	$M(ML_2)_n/M^{n+1}.L^{2n}$		-2.0 +n13.25
Zn^{2+}	$M_2L_3/M^2.L^3$		18.00
	$M_3L_6/M^4.L^6$		37.85
	$M_4L_9/M^4.L^9$		56.75
	$M_5L_{12}/M^5.L^{12}$		75.75
	$M_6L_{15}/M^6.L^{15}$		93.84
Pb^{2+}	$ML/M.L$		6.63
	$ML_2/M.L^2$		12.50
	$ML_3/M.L^3$		15.90
	$M_2L/M^2.L$		7.87
	$M_3L_4/M^3.L^4$		32.42
	$M_3L_5/M^3.L^5$		38.09
B(III)	$M(OH)_2(H_{-1}L).H/M(OH)_3.HL$		
		-7.8	
	$M(H_{-1}L)_2.H/M(OH)_3.(HL)^2$		
		-6.1	

Bibliography:

H^+,Zn^{2+}	71DV	B(III)	64AT
Ni^{2+}	72DV	Other references:	62AT,72NP
Pb^{2+}	74DT		

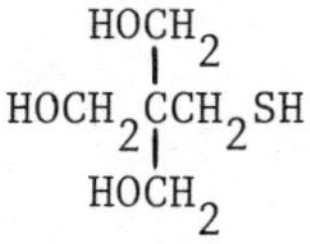

$$HOCH_2$$
$$|$$
$$HOCH_2CCH_2SH$$
$$|$$
$$HOCH_2$$

$C_5H_{12}O_3S$		2,2-Bis(hydroxymethyl)-3-mercaptopropanol (monothiopentaerythritol)	HL

Metal ion	Equilibrium	Log K 20°, 0.1
H^+	HL/H.L	9.89
Ag^+	$ML/M.L$	13
	$M_2L/M.L^2$	19.0
	$M_{10}L_9/M^{10}.L^9$	175
	M.L/ML(s)	-19.1

Bibliography: 71TS

$$HSCH_2CH_2SH$$

| $C_2H_6S_2$ | | | Ethane-1,2-dithiol | H_2L |

Metal ion	Equilibrium	Log K 30°, 0.1
H^+	HL/H.L	10.43
	H_2L/HL.H	8.85
Ni^{2+}	ML_2/M.L^2	25.6
	M_2L_3/M^2.L^3	47.4
	M.L/ML(s)	−20.7

Bibliography: 60LA

Other reference: 62AT

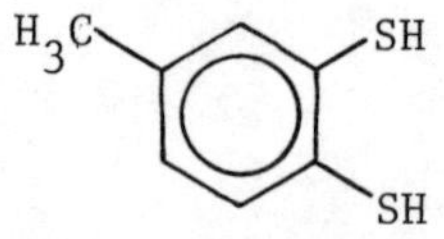

| $C_7H_8S_2$ | | | 4-Methylbenzene-1,2-dithiol (toluene-3,4-dithiol) | H_2L |

Metal ion	Equilibrium	Log K 25°, 0.1
H^+	HL/H.L	11.0
	H_2L/HL.H	5.34
Zn^{2+}	ML_2/M.L^2	25.7

Bibliography: 69HF

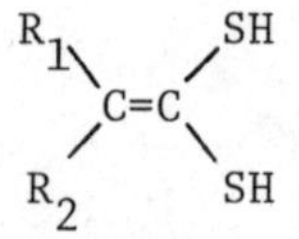

| $C_nH_mO_pN_rS_2$ | | | Substitutedethene-1,1-dithiol | H_2L |

R =	Metal ion	Equilibrium	Log K 25°, 0.2
$R_1=R_2=CN$ (2,2-dicyano) ($C_4H_2N_2S_2$)	Ni^{2+}	ML_2/M.L^2	12.2
$R_1=CH,R_2=C_6H_5$ (2-cyano-2- phenyl) ($C_9H_7NS_2$)	Ni^{2+}	ML_2/M.L^2	20.3
$R_1=NO_2,R_2=H$ (2-nitro) ($C_2H_3O_2NS_2$)	Ni^{2+}	ML_2/M.L^2	14.4

Bibliography: 70PS

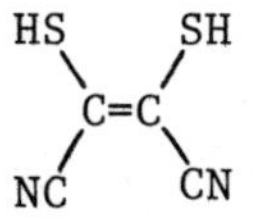

$$\begin{array}{ccc} \text{HS} & & \text{SH} \\ \diagdown & & \diagup \\ & \text{C=C} & \\ \diagup & & \diagdown \\ \text{NC} & & \text{CN} \end{array}$$

$C_4H_2N_2S_2$ <u>Dicyanoethene-1,2-dithiol</u> (<u>maleonitriledithiol</u>) H_2L

Metal ion	Equilibrium	Log K 25°, 0.1
Zn^{2+}	$ML_2/M.L^2$	17.6

Bibliography: 73PD

Other reference: 70PS

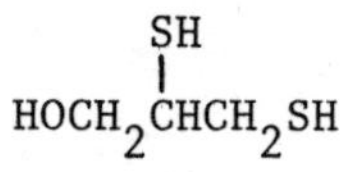

$$\begin{array}{c} \text{SH} \\ | \\ \text{HOCH}_2\text{CHCH}_2\text{SH} \end{array}$$

$C_3H_8OS_2$ <u>2,3-Dimercaptopropanol</u> (<u>BAL</u>) H_2L

Metal ion	Equilibrium	Log K 25°, 0.1	Log K 30°, 0.1
H^+	$HL/H.L$	10.68	10.61
	$H_2L/HL.H$	8.58	8.58
Mn^{2+}	$ML/M.L$		5.23
	$ML_2/M.L^2$		10.43
Fe^{2+}	$ML_2/M.L^2$		15.8
	$M_2L_3/M^2.L^3$		28
	$M.L/ML(s)$		−12.1
Ni^{2+}	$ML_2/M.L^2$		22.8
	$M_2OHL_3/M^2.OH.L^3$		45.6
Fe^{3+}	$MOHL/M.OH.L$	30.7	
Zn^{2+}	$ML/M.L$		13.5
	$ML_2/M.L^2$		23.3
	$M_2L_3/M^2.L^3$		40.4
	$M.L/ML(s)$		−17.6

Bibliography:

H^+	59L,60LM,62LJ	Ni^{2+}	59L
Mn^{2+},Zn^{2+}	61LT	Fe^{3+}	60LM
Fe^{2+}	62LJ		

Other reference: 62AT

$$\underset{\text{HSC-CSH}}{\overset{\text{O O}}{\underset{\|\quad\|}{}}}$$

C$_2$H$_2$O$_2$S$_2$ Dithiooxalic acid   H$_2$L

Metal ion	Equilibrium	Log K 25°, 0.2
Ni^{2+}	ML$_2$/M.L^2	16.1
Zn^{2+}	ML$_2$/M.L^2	10.6
Cd^{2+}	ML$_2$/M.L^2	12.7

Bibliography:

Ni^{2+} 70PS   Other reference: 68PM

Zn^{2+},Cd^{2+} 73PD

C$_2$H$_2$N$_2$S$_2$ N-Cyanodithiocarbimic acid   H$_2$L

Metal ion	Equilibrium	Log K 25°, 0.2
Ni^{2+}	ML$_2$/M.L^2	13.7

Bibliography: 70PS

$$R_1 \diagdown \qquad \diagup R_3$$
$$C=C$$
$$R_2 \diagup \qquad \diagdown R_4$$

$C_n H_m$ R =	Metal ion	Alkylethene (alkylethylene) Equilibrium	Log K 25°, 1.0	Log K 25°, 0	ΔH 25°, 0	ΔS 25°, 0 L
$R_1=R_2=R_3=R_4=H$ (ethene)(C_2H_4)	Ag^+	ML/M.L ML/MOHL.H M_2L/ML.M	1.95 ±0.02 8.7 -0.8		-7.4[c]	-16[c]
	Hg^{2+}	MOHL.H/M.L	6.54			
$R_1=CH_3,R_2=R_3=R_4=H$ (propene)(C_3H_6)	Cu^+	ML/M.L		5.02	-10.1	-11
	Ag^+	ML/M.L M_2L/ML.M	1.94 -1.0			
$R_1=CH_3CH_2,$ $R_2=R_3=R_4=H$ (but-1-ene) (C_4H_8)	Ag^+	ML/M.L		2.08		
$R_1=R_2=CH_3,R_3=R_4=H$ (2-methylpropene) (C_4H_8)	Cu^+	ML/M.L		4.49	-7.1	-3
	Ag^+	ML/M.L M_2L/ML.M	1.77[a] 1.73 -0.9	1.85	-5.1[a]	-9[a]
$R_1=R_3=CH_3,R_2=R_4=H$ (cis-but-2-ene) (C_4H_8)	Ag^+	ML/M.L		1.79		
$R_1=R_4=CH_3,R_2=R_3=H$ (trans-but-2-ene) (C_4H_8)	Ag^+	ML/M.L		1.39		
$R_1=CH_3CH_2,R_3=CH_3,$ $R_2=R_4=H$ (cis-pent-2-ene)(C_5H_{10})	Ag^+	ML/M.L	2.05[w]			
$R_1=CH_3CH_2,R_4=CH_3,$ $R_2=R_3=H$ (trans-pent-2-ene)(C_5H_{10})	Ag^+	ML/M.L	1.79[w]			
$R_1=R_2=R_3=CH_3,$ $R_4=H$ (2-methylbut-2-ene)(C_5H_{10})	Ag^+	ML/M.L	1.12 1.37[w]		-8[r]	-20[c]
$R_1=CH_3(CH_2)_3,$ $R_2=R_3=R_4=H$ (hex-1-ene)(C_6H_{12})	Ag^+	ML/M.L	2.93			

[a] 25°, 0.1;   [c] 25°, 1.0;   [r] 0-25°, 1.0;   [w] $AgNO_3$ used as background electrolyte.

Alkylethylene (continued)

Bibliography:

Cu^+ 70OT Hg^{2+} 53BP

Ag^+ 38WL,43LB,43LM,52HT,52TL,59Ba,70PTB   Other references: 70KN,71PT,72PTB

C_6H_{10}		Cyclohexene			L
Metal ion	Equilibrium	Log K 25°, 1.0		ΔH 25°, 1.0	ΔS 25°, 1.0
Ag^+	ML/M.L	1.90 1.96^w		-6^r	-10
Hg^{2+}	ML/M.L ML/MOHL.H	4.34 -0.36			

^{r} 0-25°, 1.0; ^{w} $AgNO_3$ used as background electrolyte.

Bibliography:

Ag^+ 38WL,43LB

Hg^{2+} 39LH   Other references: 56TS,59TO

XCH=CHX

$C_2H_2X_2$			1,2-Dihalogenoethene	L
X =	Metal ion	Equilibrium	Log K 25°, 1.0	
cis-1,2-Dichloro	Ag^+	ML/M.L	-0.70 ± 0.10	
trans-1,2-Dichloro	Ag^+	ML/M.L	-0.46 ± 0.06	
cis-1,2-Dibromo	Ag^+	ML/M.L	-0.17	
trans-1,2-Dibromo	Ag^+	ML/M.L	-0.25	
cis-1,2-Diiodo	Ag^+	ML/M.L $M_2L/ML.L$	1.25 0.03	
trans-1,2-Diiodo	Ag^+	ML/M.L $M_2L/ML.L$	0.74 0.34	

Bibliography: 51AK,69OF

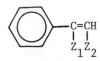

L

$C_n H_m O_p N_q X_r$ Z =	Metal ion	Equilibrium	Log K 25°, 1.0	ΔH 25°, 1.0	ΔS 25°, 1.0
H (styrene) (C_8H_8)	Ag^+	ML/M.L M_2L/ML.M	1.27 ±0.01 0.0 ±0.2	-2.1	-1
3-Methyl (C_9H_{10})	Ag^+	ML/M.L M_2L/ML.M	1.34 -0.2		
4-Methyl (C_9H_{10})	Ag^+	ML/M.L M_2L/ML.M	1.38 0.2	-1.9	0
3-Chloro (C_8H_7Cl)	Ag^+	ML/M.L M_2L/ML.M	0.88 -0.3		
4-Chloro (C_8H_7Cl)	Ag^+	ML/M.L M_2L/ML.M	1.03 0.1	-2.2	-3
3-Nitro ($C_8H_7O_2N$)	Ag^+	ML/M.L M_2L/ML.M	0.74 -0.6		
2-Methoxy ($C_9H_{10}O$)	Ag^+	ML/M.L M_2L/ML.M	1.37 0.0		
3-Methoxy ($C_9H_{10}O$)	Ag^+	ML/M.L M_2L/ML.M	1.11 0.1		
4-Methoxy ($C_9H_{10}O$)	Ag^+	ML/M.L M_2L/ML.M	1.38 -0.3		

Bibliography: 50AKa,65FO

L

$C_n H_m X_r$ Z =	Metal ion	Equilibrium	Log K 25°, 1.0
Z_1=phenyl,Z_2=H (1,1-diphenyl-ethene) ($C_{14}H_{12}$)	Ag^+	ML/M.L M_2L/ML.M	0.36 0.1
Z_2=phenyl,Z_1=H (trans-1,2-diphenylethene) (trans-stilbene) ($C_{14}H_{12}$)	Ag^+	ML/M.L M_2L/ML.M	0.80 -0.2

Substitutedphenylethene (continued)

Z =	Metal ion	Equilibrium	Log K 25°, 1.0
Z_2=methyl,Z_1=H (1-phenylpro-pene)(C_9H_{10})	Ag^+	ML/M.L M_2L/ML.M	0.59 −0.3
Z_2=bromo,Z_1=H (1-phenyl-2-bromoethene) (C_8H_7Br)	Ag^+	ML/M.L M_2L/ML.M	−0.01 0.15

Bibliography: 50AKa,65FO

$$CH_3CH_2OCH=CH_2$$

C_4H_8O Ethoxyethene (ethylvinyl ether)   L

Metal ion	Equilibrium	Log K 25°, 1.0
Ag^+	ML/M.L M_2L/ML.M	1.11 −0.2

Bibliography: 68FK

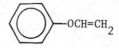

C_8H_8O Phenoxyethene (phenylvinyl ether)   L

Metal ion	Equilibrium	Log K 25°, 1.0
Ag^+	ML/M.L M_2L/ML.M	0.72 −0.1

Bibliography: 68FK

$$CH_3CH_2OCH=CHCl$$

C_4H_7OCl 1-Chloro-2-ethoxyethene   L

Isomer	Metal ion	Equilibrium	Log K 25°, 1.0
cis-	Ag^+	ML/M.L	−0.49
trans-	Ag^+	ML/M.L	−0.23

Bibliography: 69OF

$$\overset{\text{O}}{\overset{\|}{\text{CH}_3\text{OCCH=CH}_2}}$$

$C_4H_6O_2$	Propenoic acid methyl ester (methyl acrylate)	L

Metal ion	Equilibrium	Log K 25°, 1.0
Ag^+	ML/M.L	-0.43
	M_2L/ML.M	-0.6

Bibliography: 68FK

$$\overset{\text{O}}{\overset{\|}{\text{CH}_3\text{COCH=CH}_2}}$$

$C_4H_6O_2$	Acetic acid vinyl ester (vinyl acetate)	L

Metal ion	Equilibrium	Log K 25°, 1.0
Ag^+	ML/M.L	0.01
	M_2L/ML.M	-0.4

Bibliography: 68FK

Other reference: 49Ab

$$HOCH_2CH=CH_2$$

C_3H_6O		Propen-3-ol (allyl alcohol)			L
Metal ion	Equilibrium	Log K 25°, 0.1	Log K 25°, 1.0	Log K 25°, 2.0	
Cu^+	ML/M.L	4.72 ±0.0			
Ag^+	ML/M.L	1.15 ±0.01	1.08	1.36	
	$ML_2/M.L^2$	0.60	1.56	1.12	

Bibliography:

Cu^+ 49KAb,66Mf Ag^+ 38WL,49KAb,67HV,74SM

$$HOCH_2CH=CHCH_3$$

C_4H_8O		trans-But-2-en-1-ol (crotyl alcohol)			L
Metal ion	Equilibrium	Log K 25°, 0.1	Log K 25°, 1.0	Log K 25°, 2.0	
Cu^+	ML/M.L	4.00			
Ag^+	ML/M.L	0.59	0.71	0.90	
	$ML_2/M.L^2$			0.66	

Bibliography:

Cu^+ 49KAb Ag^+ 38WL,49KAb,67HV

$$\begin{array}{cc} CH_3 & CH_3 \\ | & | \\ HOCHCH_2 & C=CH_2 \end{array}$$

$C_6H_{12}O$		DL-2-Methylpent-1-en-4-ol	L
Metal ion	Equilibrium	Log K 25°, 0.1	
Cu^+	ML/M.L	4.20	

Bibliography: 49KAb

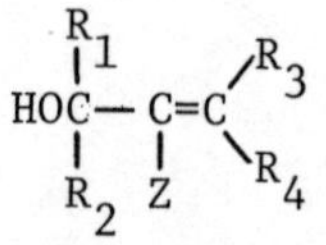

$$\begin{array}{c} R_1 \\ | \\ HOC-C=C \\ | \\ R_2 \quad Z \end{array} \begin{array}{c} R_3 \\ \\ R_4 \end{array}$$

$C_nH_mOCl_q$			Substitutedpropen-3-ol	L
R,Z =	Metal ion	Equilibrium	Log K 25°, 0.1	
$R_1=CH_3$,	Cu^+	ML/M.L	4.52	
$R_2=R_3=R_4=Z=H$	Ag^+	ML/M.L	1.15	
(DL-but-1-en-3-ol)				
(C_4H_8O)				

Substitutedpropen-3-ol (continued)

R, Z =	Metal ion	Equilibrium	Log K 25°, 0.1
$R_1=CH_3CH_2$,	Cu^+	ML/M.L	4.59
$R_2=R_3=R_4=Z=H$	Ag^+	ML/M.L	1.16
(DL-pent-1-en-3-ol)			
($C_5H_{10}O$)			
$R_1=R_2=CH_3$,	Cu^+	ML/M.L	4.40
$R_3=R_4=Z=H$			
(3-methylbut-1-en-3-ol)($C_5H_{10}O$)			
$Z=CH_3$,	Cu^+	ML/M.L	3.96
$R_1=R_2=R_3=R_4=H$	Ag^+	ML/M.L	1.04
(2-methylpropen-3-ol)(C_4H_8O)			
$Z=Cl$,	Cu^+	ML/M.L	2.66
$R_1=R_2=R_3=R_4=H$			
(2-chloropropen-3-ol)(C_3H_5OCl)			
$R_1=Z=CH_3$,	Cu^+	ML/M.L	3.60
$R_2=R_3=R_4=H$			
(DL-2-methyl-but-1-en-3-ol) ($C_5H_{10}O$)			
$Z=R_3=CH_3$,	Cu^+	ML/M.L	(3.54)[y]
$R_1=R_2=R_4=H$	Ag^+	ML/M.L	(0.71)[y]
(2-methylbut-2-en-1-ol) ($C_5H_{10}O$)			
$R_3=R_4=CH_3$,	Cu^+	ML/M.L	4.04
$R_1=R_2=Z=H$	Ag^+	ML/M.L	0.3
(3-methylbut-2-en-1-ol) ($C_5H_{10}O$)			

[y] Cis- or trans- isomer not stated.

Bibliography: 49KAb

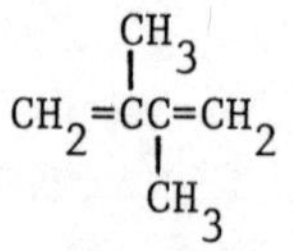

$$\begin{array}{c} CH_3 \\ | \\ CH_2=CC=CH_2 \\ | \\ CH_3 \end{array}$$

C_6H_{10}		2,3-Dimethylbut-1,3-diene	L
Metal		Log K	
ion	Equilibrium	25°, 1.0	
Ag^+	ML/M.L	1.35	
	$M_2L/ML.M$	0.0	

Bibliography: 38WL

$$CH_2=CHCH_2CH_2CH=CH_2$$

C_6H_{10}		Hex-1,5-diene	L
Metal		Log K	
ion	Equilibrium	25°, 1.0	
Ag^+	ML/M.L	3.27	
	$M_2L/ML.M$	0.4	

Bibliography: 38WL

$$R_1C\equiv CR_2$$

C_nH_m			Alkylethyne (alkylacetylene)		L
		Metal		Log K	Log K
R =		ion	Equilibrium	25°, 1.0	25°, 0
$R_1=R_2=H$		Ag^+	ML/M.L		1.63
(ethyne)(C_2H_2)					
$R_1=R_2=CH_3CH_2$		Ag^+	ML/M.L	1.26 ±0.02	
(3-hexyne)(C_6H_{10})			$M_2L/ML.M$	−0.5 ±0.1	
$R_1=(CH_3)_2CH$,		Ag^+	ML/M.L	1.31	
$R_2=CH_3CH_2$			$M_2L/ML.M$	−0.5	
(2-methyl-3-hexyne)(C_7H_{12})					
$R_1=(CH_3)_3C$,		Ag^+	ML/M.L	1.28	
$R_2=CH_3CH_2$			$M_2L/ML.M$	−0.5	
(2,2-dimethyl-3-hexyne)(C_8H_{14})					
$R_1=(CH_3)_3C$,		Ag^+	ML/M.L	1.37	
$R_2=(CH_3)_2CH$			$M_2L/ML.M$	−0.5	
(2,2,5-trimethyl-3-hexyne)(C_9H_{16})					

Alkylacetylene (continued)

R =	Metal ion	Equilibrium	Log K 25°, 1.0	Log K 25°, 0
$R_1=R_2=(CH_3)_3C$	Ag^+	ML/M.L	1.11	
(2,2,5,5-tetra-		$M_2L/ML.M$	-0.3	
methyl-3-hexyne)				
$(C_{10}H_{18})$				
$R_1=CH_3$,	Ag^+	ML/M.L	1.08	
$R_2=CH_3(CH_2)_3$		$M_2L/ML.M$	-0.6	
(2-heptyne)				
(C_7H_{12})				

Bibliography: 56DL,57HC,62TF

$$CH_3C{\equiv}N$$

		Cyanomethane (acetonitrile)		L

C_2H_3N

Metal ion	Equilibrium	Log K 25°, 0.1
Ag^+	$ML/M.L$	0.42
	$ML_2/M.L^2$	0.78

Bibliography:

Ag^+ 74SM

Other references: 24P, 63HS,64YK,67MI

$C_n H_m$ Alkylbenzene   L

R =	Metal ion	Equilibrium	Log K 25°, 1.0
$R_1=R_2=R_3=R_4=$ $R_5=R_6=H$ (benzene)(C_6H_6)	Ag^+	ML/M.L M$_2$L/ML.M	0.38 -0.7
$R_1=CH_3, R_2=R_3=$ $R_4=R_5=R_6=H$ (methylbenzene) (toluene)(C_7H_8)	Ag^+	ML/M.L M$_2$L/ML.M	0.44 ±0.03 0.76[g] -0.5 ±0.1 -0.1 [g]
$R_1=R_2=CH_3$, $R_3=R_4=R_5=R_6=H$ (1,2-dimethyl- benzene)(o-xylene) (C_8H_{10})	Ag^+	ML/M.L M$_2$L/ML.M	0.46 -0.5
$R_1=R_3=CH_3$, $R_2=R_4=R_5=R_6=H$ (1,3-dimethyl- benzene)(m-xylene) (C_8H_{10})	Ag^+	ML/M.L M$_2$L/ML.M	0.48 -0.5
$R_1=R_4=CH_3$, $R_2=R_3=R_5=R_6=H$ (1,4-dimethyl- benzene)(p-xylene) (C_8H_{10})	Ag^+	ML/M.L M$_2$L/ML.M	0.42 -0.6
$R_1=R_3=R_5=CH_3$, $R_2=R_4=R_6=H$ (1,3,5-trimethyl- benzene)(mesitylene) (C_9H_{12})	Ag^+	ML/M.L	0.26
$R_1=CH_3CH_2$, $R_2=R_3=R_4=R_5=R_6=H$ (ethylbenzene) (C_8H_{10})	Ag^+	ML/M.L M$_2$L/ML.M	0.43 -0.8
$R_1=CH_3CH(CH_3)$, $R_2=R_3=R_4=R_5=R_6=H$ (1-methylethyl- benzene)(C_9H_{12})	Ag^+	ML/M.L M$_2$L/ML.M	0.45 -1.0

[g] 25°, 5.0

Alkylbenzene (continued)

R =	Metal ion	Equilibrium	Log K 25°, 1.0
$R_1=CH_3C(CH_3)_2$,	Ag^+	ML/M.L	0.36
$R_2=R_3=R_4=R_5=R_6=H$		$M_2L/ML.M$	-0.7
(2-methyl-2-propylbenzene)			
$(C_{10}H_{14})$			
$R_1=CH_3(CH_2)_2$,	Ag^+	ML/M.L	0.46
$R_2=R_3=R_4=R_5=R_6=H$			
(propylbenzene)			
(C_9H_{12})			
$R_1=CH_3CH_2CH(CH_3)$,	Ag^+	ML/M.L	0.38
$R_2=R_3=R_4=R_5=R_6=H$		$M_2L/ML.M$	-0.9
(1-methyl-propylbenzene)			
$(C_{10}H_{14})$			
$R_1=CH_3CH_2C(CH_3)_2$,	Ag^+	ML/M.L	0.38
$R_2=R_3=R_4=R_5=R_6=H$		$M_2L/ML.M$	-0.6
(1,1-dimethyl-propylbenzene)			
$(C_{11}H_{16})$			
$R_1=CH_3(CH_2)_3$,	Ag^+	ML/M.L	0.48
$R_2=R_3=R_4=R_5=R_6=H$			
(butylbenzene)			
$(C_{10}H_{14})$			

Bibliography: 49AKa,50AKa,52KA

Other reference: 64WD

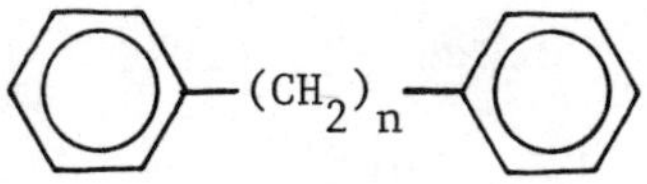

C_nH_m			Diphenylalkane	L
n =	Metal ion	Equilibrium	Log K 25°, 1.0	
0	Ag^+	ML/M.L	0.60	
(biphenyl)		$M_2L/ML.M$	0.0	
$(C_{12}H_{10})$				
1	Ag^+	ML/M.L	0.54	
(diphenyl-methane)		$M_2L/ML.M$	0.0	
$(C_{13}H_{12})$				
2	Ag^+	ML/M.L	0.90	
(dibenzyl)		$M_2L/ML.M$	-0.1	
$(C_{14}H_{14})$				

Bibliography: 49AKa,50AKa

C_10_H_8_   Benzobenzene (naphthalene)   L

Metal
 ion   Equilibrium

Ag^+^   ML.M.L   0.49
               M_2_L/ML.M   0.0

Bibliography: 49AKa

C_14_H_10_   Dibenzo[a,c]benzene (phenanthrene)   L

Metal
 ion   Equilibrium

Ag^+^   ML/M.L   0.56
               M_2_L/ML.M   0.3

Bibliography: 49AKa

C_6_H_4_I_2_   1,2-Diiodobenzene   L

Metal
 ion   Equilibrium

Ag^+^   ML/M.L   1.22
               M_2_L/ML.L   0.2

Bibliography: 51AK

C_6_H_4_I_2_   1,3-Diiodobenzene   L

Metal
 ion   Equilibrium

Ag^+^   ML/M.L   0.76
               M_2_L/ML.M   0.46

Bibliography: 51AK

$C_{10}H_8$

Benzobenzene (naphthalene)   L

Metal ion	Equilibrium	Log K 25°, 1.0
Ag^+	ML.M.L	0.49
	M_2L/ML.M	0.0

Bibliography: 49AKa

$C_{14}H_{10}$

Dibenzo[a,c]benzene (phenanthrene)   L

Metal ion	Equilibrium	Log K 25°, 1.0
Ag^+	ML/M.L	0.56
	M_2L/ML.M	0.3

Bibliography: 49AKa

$C_6H_4I_2$

1,2-Diiodobenzene   L

Metal ion	Equilibrium	Log K 25°, 1.0
Ag^+	ML/M.L	1.22
	M_2L/ML.L	0.2

Bibliography: 51AK

$C_6H_4I_2$

1,3-Diiodobenzene   L

Metal ion	Equilibrium	Log K 25°, 1.0
Ag^+	ML/M.L	0.76
	M_2L/ML.M	0.46

Bibliography: 51AK

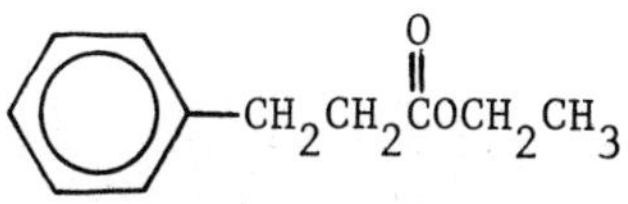

$C_{11}H_{14}O_2$ 3-Phenylpropanoic acid ethyl ester (ethyl hydrocinnamate)   L

Metal ion	Equilibrium	Log K 25°, 1.0
Ag^+	ML/M.L	0.20
	$M_2L/ML.M$	-0.5

Bibliography: 50AKa

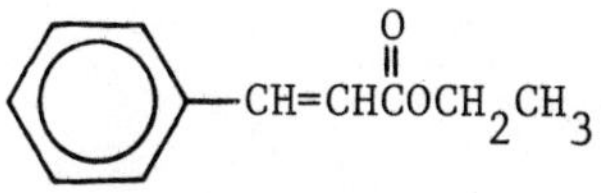

$C_{11}H_{12}O_2$ 3-Phenylpropenoic acid ethyl ester (ethyl cinnamate)   L

Metal ion	Equilibrium	Log K 25°, 1.0
Ag^+	ML/M.L	0.00
	$M_2L/ML.M$	-0.5

Bibliography: 50AKa

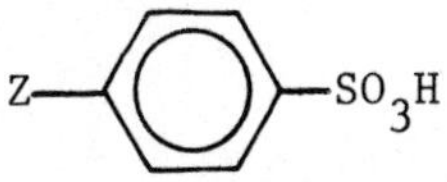

$C_nH_mO_rXS_sSe_t$ 4-(Substituted)benzenesulfonic acid   HL

Z =	Metal ion	Equilibrium	Log K 25°, 0.1	Log K 25°, 1.0
H $(C_6H_6O_3S)$	Ag^+	ML/M.L		-0.04
4-(Methoxy) $(C_7H_8O_4S)$	Ag^+	ML/M.L $ML_2/M.L^2$		-0.12 -0.2
4-(Ethylthio) $(C_8H_{10}O_3S_2)$	Ag^+	ML/M.L $ML_2/M.L^2$ $ML_3/M.L^3$ $ML_4/M.L^4$	2.62 4.30 5.7	2.59^r 4.28^r 5.51^r 6.55^r
4-(Phenylthio) $(C_{12}H_{10}O_3S_2)$	Ag^+	ML/M.L $ML_2/M.L^2$ $ML_3/M.L^3$ $ML_4/M.L^4$	1.67 3.01 4.12 5.72	1.67^r 2.75^r 3.95^r 5.88^r
	Cd^{2+}	ML/M.L $ML_3/M.L^3$		0.67^r 3.04^r

[r] 25°, 0.2

4-(Substituted)benzenesulfonic acid (continued)

Z =	Metal ion	Equilibrium	Log K 25°, 0.1	Log K 25°, 1.0
4-(4-Sulfo-	Ag^+	$ML/M.L$		1.40^r
phenylthio)		$ML_2/M.L^2$		2.3^r
$(C_{12}H_{10}O_6S_3)$		$ML_3/M.L^3$		4.0^r
4-(Phenylseleno)	Ag^+	$ML/M.L$	2.63	
$(C_{12}H_{10}O_3SSe)$		$ML_2/M.L^2$	4.89	
		$ML_3/M.L^3$	6.43	
		$ML_4/M.L^4$	8.66	

^{r} 25°, 0.2

Bibliography:

Ag^+ 48AC Cd^{2+} 58ACb

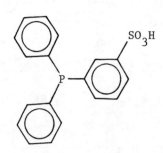

$C_6H_{15}OP$		2-Hydroxyethyldiethylphosphine		L
Metal ion	Equilibrium	Log K 22°, 1.0	ΔH 22°, 1.0	ΔS 22°, 1.0
Ag^+	$ML/M.L$	11.71	-19.3	-11
Hg^{2+}	$ML_2/M.L^2$	36.93	-52.8	-8

Bibliography: 67Mf

$C_{18}H_{15}O_3SP$		3-(Diphenylphosphino)benzenesulfonic acid (3-sulfotriphenylphosphine)		HL
Metal ion	Equilibrium	Log K 25°, 0.1	Log K 25°, 1.0	
H^+	$HL/H.L$		0.63	
Cu^+	$ML/M.L$		5.76	
	$ML_2/M.L^2$		11.21	
	$ML_3/M.L^3$		17.12	
	$ML_4/M.L^4$		19.92	
Ag^+	$ML/M.L$	8.15		
	$ML_2/M.L^2$	14.10		
	$ML_3/M.L^3$	19.50		

3-Sulfotriphenylphosphine (continued)

Metal ion	Equilibrium	Log K 25°, 0.1	Log K 25°, 1.0
Pd^{2+}	$ML/M.L$		10.2
	$ML_2/M.L^2$		20.0
	$ML_3/M.L^3$		26.3
	$ML_4/M.L^4$		31.2
Cd^{2+}	$ML/M.L$	0.9	
	$ML_2/M.L^2$	3.40	
Hg^{2+}	$ML/M.L$		14.46
	$ML_2/M.L^2$		24.72
	$ML_3/M.L^3$		29.76
	$ML_4/M.L^4$		32.4
Bi^{3+}	$ML/M.L$		3.7
	$ML_6/M.L^6$		21.8

Bibliography:

H^+, Bi^{3+} 62WB Pd^{2+} 72CB

Cu^+ 68GB Cd^{2+} 58ACb

Ag^+ 58ACa Hg^{2+} 62SB,68GB

$C_{18}H_{15}O_9S_3As$ Tris-(3-sulfophenyl)arsine H_3L

Metal ion	Equilibrium	Log K 25°, 0.2
Ag^+	$ML/M.L$	5.36

Bibliography: 58AC

$CHCl_3$

CHCl$_3$	Trichloromethane (chloroform)	L

Metal ion	Equilibrium	Log K 25°, 1.0
Ag$^+$	ML/M.L	−0.70

Bibliography: 51AK

CH_2Br_2

CH$_2$Br$_2$	Dibromomethane (methylene bromide)	L

Metal ion	Equilibrium	Log K 25°, 1.0
Ag$^+$	ML/M.L	−0.13

Bibliography: 51AK

CH_2I_2

CH$_2$I$_2$	Diiodomethane (methylene iodide)	L

Metal ion	Equilibrium	Log K 25°, 1.0
Ag$^+$	ML/M.L	1.31

Bibliography: 51AK

$$\overset{\overset{\displaystyle O}{\|}}{CH_3CNHOH}$$

$C_2H_5O_2N$ <u>Acetohydroxamic acid</u>   HL

Metal ion	Equilibrium	Log K 20°, 0.1
H^+	HL/H.L	9.36
Ca^{2+}	ML/M.L	2.4
La^{3+}	$ML/M.L$	5.16
	$ML_2/M.L_2$	9.33
	$ML_3/M.L_3$	11.88
Ce^{3+}	$ML/M.L$	5.45
	$ML_2/M.L_2$	9.79
	$ML_3/M.L_3$	12.8
Sm^{3+}	$ML/M.L$	5.96
	$ML_2/M.L_2$	10.73
	$ML_3/M.L_3$	14.41
Gd^{3+}	$ML/M.L$	6.10
	$ML_2/M.L_2$	10.86
	$ML_3/M.L_3$	13.93
Dy^{3+}	$ML/M.L$	6.52
	$ML_2/M.L_2$	11.91
	$ML_3/M.L_3$	15.95
Yb^{3+}	$ML/M.L$	6.61
	$ML_2/M.L_2$	12.20
	$ML_3/M.L_3$	16.49
Mn^{2+}	$ML/M.L$	4.0
	$ML_2/M.L_2$	6.9
Fe^{2+}	$ML/M.L$	4.8
	$ML_2/M.L_2$	8.5
Co^{2+}	$ML/M.L$	5.1
	$ML_2/M.L_2$	8.9
Ni^{2+}	$ML/M.L$	5.3
	$ML_2/M.L_2$	9.3
Cu^{2+}	ML/M.L	7.9
Fe^{3+}	$ML/M.L$	11.42
	$ML_2/M.L_2$	21.10
	$ML_3/M.L_3$	28.33
Zn^{2+}	$ML/M.L$	5.4
	$ML_2/M.L_2$	9.6
Cd^{2+}	$ML/M.L$	4.5
	$ML_2/M.L_2$	7.8
Pb^{2+}	$ML/M.L$	6.7
	$ML_2/M.L_2$	10.7
Al^{3+}	$ML/M.L$	7.95
	$ML_2/M.L_2$	15.29
	$ML_3/M.L_3$	21.47

Bibliography:

H^+ 63AL,63SS Fe^{3+} 63SS

$Ca^{2+}-Cu^{2+},Zn^{2+}-Al^{3+}$ 63AL

$$\text{Benzene ring} - \overset{\overset{\displaystyle O}{\|}}{C}NHOH$$

$C_7H_7O_2N$ Benzohydroxamic acid   HL

Metal ion	Equilibrium	Log K 20°, 0.1
H^+	HL/H.L	8.79
Fe^{3+}	ML/M.L	11.06
	ML_2/M.L$_2$	20.43
	ML_3/M.L$_3$	27.8

Bibliography: 63SS

Other references: 65BG,66BBB,66MR,69DS,70SC,71ST,73BPC,74AS

$$\text{ring} - \overset{\overset{\displaystyle O}{\|}}{C} - \overset{\overset{\displaystyle OH}{|}}{N} - \text{ring}$$

$C_{13}H_{11}O_2N$ N-Phenylbenzohydroxamic acid   HL

Metal ion	Equilibrium	Log K 25°, 0.1	Log K 20°, 1.0
H^+	HL/H.L	8.15	
Th^{4+}	ML/M.L	10.4	
	ML_2/M.L$_2$	20.3	
	ML_3/M.L$_3$	29.3	
	ML_4/M.L$_4$	37.4	
In^{3+}	ML/M.L	9.2 +0.7	
	ML_2/M.L$_2$	18.4 -0.1	
	ML_3/M.L$_3$	26.3	

Bibliography:

H^+ 56D In^{3+} 68SA,69ZS

Th^{4+} 56D,60R   Other references: 65AK,69DS,69SMK,70FR,70LSF,
                                                                              72ST

$$\text{ring} - CH=CHC - \overset{\overset{\displaystyle O\ OH}{\|\ |}}{}N - \text{ring}$$

$C_{15}H_{13}O_2N$ N,3-Diphenylpropenohydroxamic acid   HL
                          (N-cinnamoyl-N-phenylhydroxylamine)

Metal ion	Equilibrium	Log K 20°, 0.1	Log K 20°, 1.0
H^+	HL/H.L	8.80	8.81
Th^{4+}	ML/M.L	12.8	
	ML_2/M.L$_2$	24.7	
	ML_3/M.L$_3$	35.7	
Ti^{4+}	ML/M.L		13.3
	ML_2/M.L$_2$		26.4
	ML_3/M.L$_3$		39.3
	ML_4/M.L$_4$		52.0

N-Cinnamoyl-N-phenylhydroxylamine (continued)

Metal ion	Equilibrium	Log K 20°, 0.1	Log K 20°, 1.0
In^{3+}	$ML/M.L$	11.8	
	$ML_2/M.L^2$	22.0	

Bibliography:

H^+	67ZS,70LSF	In^{3+}	69ZS
Th^{4+}	67ZSO	Other references:	70FR,72ST
Ti^{4+}	70LSF		

$$CH_2CH_2CH_2CH_2\overset{\overset{O}{\|}}{N}CCH_2CH_2\overset{\overset{O}{\|}}{C}NHCH_2CH_2CH_2CH_2$$

$C_{25}H_{48}O_8N_6$ Desferriferrioxamin B H_3L

Metal ion	Equilibrium	Log K 20°, 0.1
H^+	$H_2L/HL.H$	9.70
	$H_3L/H_2L.H$	9.03
	$H_4L/H_3L.H$	8.39
Mg^{2+}	$MHL/M.HL$	4.30
Ca^{2+}	$MHL/M.HL$	2.64
Sr^{2+}	$MHL/M.HL$	2.20
La^{3+}	$MHL/M.HL$	10.89
	$MH_2L/MHL.H$	6.42
	$MH_3L/MH_2L.H$	6.02
Yb^{3+}	$MHL/M.HL$	16.0
	$MH_2L/MHL.H$	4.7
	$MH_3L/MH_2L.H$	4.4
Fe^{2+}	$MH_2L/M.H_2L$	7.2
	$MH_3L/MH_2L.H$	5.6
Co^{2+}	$MHL/M.HL$	10.31
	$MH_2L/MHL.H$	6.75
	$MH_3L/MH_2L.H$	5.85
Ni^{2+}	$MHL/M.HL$	10.90
	$MH_2L/MHL.H$	6.50
	$MH_3L/MH_2L.H$	5.69
Cu^{2+}	$MHL/M.HL$	14.12
	$MH_2L/MHL.H$	9.08
	$MH_3L/MH_2L.H$	3.13
Fe^{3+}	$MHL/M.HL$	30.60
	$MH_2L/MHL.H$	0.94
Zn^{2+}	$MHL/M.HL$	10.07
	$MH_2L/MHL.H$	6.50
	$MH_3L/MH_2L.H$	5.61

Desferriferrioxamin B (continued)

Metal ion	Equilibrium	Log K 20°, 0.1
Cd^{2+}	MHL/M.HL	7.88
	$MH_2L/MHL.H$	7.40
	$MH_3L/MH_2L.H$	6.77

Bibliography:

H^+, Fe^{3+} 63SS $Mg^{2+}-Cu^{2+}, Zn^{2+}, Cd^{2+}$ 63AL

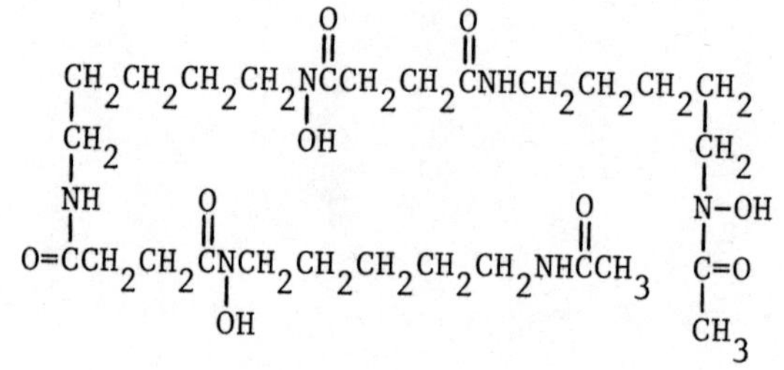

$C_{27}H_{50}O_9N_6$

N-Acetyldesferriferrioxamin B H_3L

Metal ion	Equilibrium	Log K 20°, 0.1
H^+	HL/H.L	9.69
	$H_2L/HL.H$	9.24
	$H_3L/H_2L.H$	8.50
Fe^{3+}	ML/M.L	30.76
	MHL/ML.H	0.5

Bibliography: 63ALa

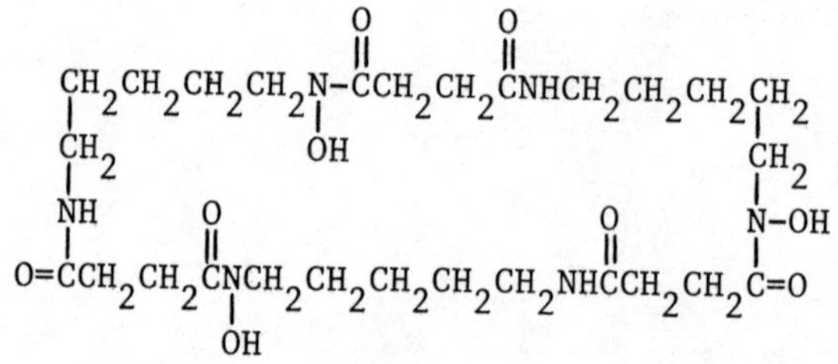

$C_{27}H_{48}O_9N_6$

Desferriferrioxamin E (Nocardamin) H_3L

Metal ion	Equilibrium	Log K 20°, 0.1
H^+	HL/H.L	9.89
	$H_2L/HL.H$	9.42
	$H_3L/H_2L.H$	8.63
Co^{2+}	ML/M.L	11.88
	MHL/ML.H	6.43
	$MH_2L/MHL.H$	5.76
Ni^{2+}	ML/M.L	12.24
	MHL/ML.H	6.22
	$MH_2L/MHL.H$	5.79

Nocardamin (continued)

Metal ion	Equilibrium	Log K 20°, 0.1
Cu^{2+}	ML/M.L	13.69
	MHL/ML.H	8.95
	$MH_2L/MHL.H$	4.01
Fe^{3+}	ML/M.L	32.49
Zn^{2+}	ML/M.L	12.07
	MHL/ML.H	6.35
	$MH_2L/MHL.H$	5.76
Cd^{2+}	ML/M.L	8.83
	MHL/ML.H	7.25
	$MH_2L/MHL.H$	(7.83)

Bibliography: 63AL

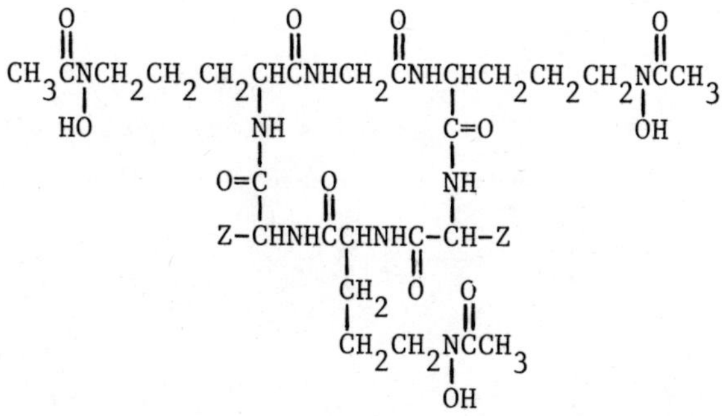

$$CH_3\overset{O}{\overset{||}{C}}NCH_2CH_2CH_2\overset{NH}{\underset{|}{C}}HCNHCH_2\overset{O}{\overset{||}{C}}NH\overset{C=O}{\underset{|}{C}}HCHCH_2CH_2CH_2\overset{O}{\overset{||}{N}}CCH_3$$

Polyamidopolyhydroxamic acid

$C_nH_mO_pN_9$

H_3L

Z =	Metal ion	Equilibrium	Log K 20°, 0.1
Z=H	H^+	HL/H.L	9.83
(Desferri-		$H_2L/HL.H$	9.00
ferrichrome)		$H_3L/H_2L.H$	8.11
$(C_{27}H_{45}O_{12}N_9)$	Fe^{3+}	ML/M.L	29.07
		MHL/ML.H	1.49
Z=CH_2OH	H^+	HL/H.L	10.01
(Desferri-		$H_2L/HL.H$	9.02
ferrichrysin)		$H_3L/H_2L.H$	8.17
$(C_{29}H_{49}O_{14}N_9)$	Fe^{3+}	ML/M.L	29.96

Bibliography: 63ALa

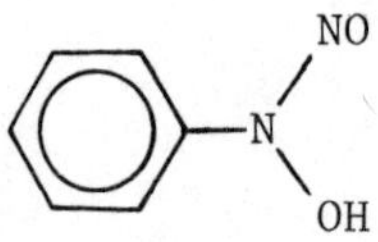

$C_6H_6O_2N_2$ N-Nitrosophenylhydroxylamine (Cupferron)   HL

Metal ion	Equilibrium	Log K 25°, 0.1
H^+	HL/H.L	4.16
La^{3+}	$ML_3/M.L^3$	12.9
Sm^{3+}	$ML_3/M.L^3$	14.3
Th^{4+}	$ML/M.L$	7.4
	$ML_2/M.L^2$	14.3
	$ML_3/M.L^3$	20.9
	$ML_4/M.L^4$	27.0

Bibliography:

H^+ 54D

La^{3+}, Sm^{3+} 54D,60R

Th^{4+} 53D,60R

Other references: 61KK,69LS

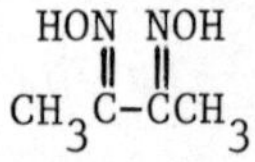

$C_4H_8O_2N_2$ Butane-2,3-dione dioxime (dimethylglyoxime)   HL

Metal ion	Equilibrium	Log K 25°, 0.1	Log K 20°, 1.0	Log K 25°, 0	ΔH 25°, 0	ΔS 25°, 0
H^+	$L/H_{-1}L.H$	11.9		12.0		
	HL/H.L	10.45 ±0.03		10.66 ±0.0	$(-2)^r$	(40)
Ni^{2+}	$ML_2/M.L^2$	17.24 +0.2		17.84		
	$ML_2/ML_2(s)$	-6.03		-6.00		
Cu^{2+}	$ML_2/M.L^2$	19.24				
	$ML_2/MOHL_2.H$	10.60				
Pd^{2+}	$ML_2/M.L^2$	34.1	34.3			
	$MOHL_2/ML_2.OH$	5.50	5.6			

^{r} 25-40°, 0

Bibliography:

H^+ 52BC,54CS,58BB,63SAP,64D

Ni^{2+} 54CS,59DK,64SA

Cu^{2+} 61DH

Pd^{2+} 63BD,66BS

Other references: 51BM,62HR,63BA,64AS,64BR, 67BK,72UC

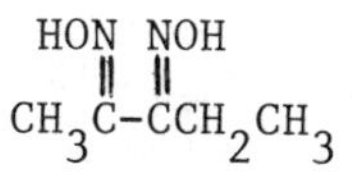

$$HON \quad NOH$$
$$\| \quad \|$$
$$CH_3C-CCH_2CH_3$$

$C_5H_{10}O_2N_2$ Pentane-2,3-dione dioxime (ethylmethylglyoxime)   HL

Metal ion	Equilibrium	Log K 25°, 0.1
H^+	$L/H_{-1}L.H$	12.02
	$HL/H.L$	10.51 −0.1
Ni^{2+}	$ML/M.L$	8.37
	$ML_2/M.L^2$	17.79
	$ML_2/ML_2(s,\alpha)$	−5.56
	$ML_2/ML_2(s,\beta)$	−4.92
Cu^{2+}	$ML/M.L$	9.03
	$ML_2/M.L^2$	19.67

Bibliography:

H^+ 63BA,68E Cu^{2+} 72E

Ni^{2+} 69E

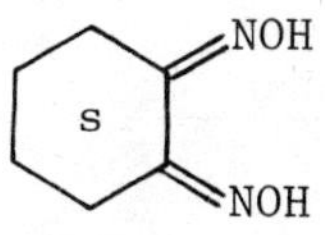

$C_6H_{10}O_2N_2$ Cyclohexane-1,2-dione dioxime (nioxime)   HL

Metal ion	Equilibrium	Log K 25°, 0.1	Log K 20°, 1.0	Log K 25°, 0	ΔH 25°, 0	ΔS 25°, 0
H^+	$L/H_{-1}L.H$	12.1		12.5		
	$HL/H.L$	10.57 ±0.02		10.7	(−3)[r]	(40)
Ni^{2+}	$ML/M.L$	10.8				
	$ML_2/M.L^2$	21.5				
	$ML_2/M L_2(s)$		−6.48			
Fe^{3+}	$ML/M.L$		11.07[b]			
	$ML_2/M.L^2$		21.74[b]			
	$ML_3/M.L^3$		31.99[b]			
Pd^{2+}	$MOHL_2/ML_2.OH$		3.9			

[b] 25°, 0.5;   [r] 25-40°, 0

Bibliography:

H^+ 49VB,52BC,58BB,63BA,63SAP Pd^{2+} 67ST

Ni^{2+} 64SA,69AI   Other references: 62HR,64D,66PS,69MV

Fe^{3+} 64ASa

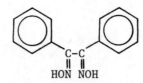

$C_7H_{12}O_2N_2$		Cyclohepta	ne-1,2-dione dio	xime (heptoxim	e)	HL
Metal ion	Equilibrium	Log K 25°, 0.1	Log K 20°, 1.0	Log K 25°, 0	ΔH 25°, 0	ΔS 25°, 0
H^+	$L/H_{-1}L.H$	12.1	12.1	12.4		
	$HL/H.L$	10.54 +0.2	10.56	10.8	(-5)[r]	(30)
Ni^{2+}	$ML/M.L$	11.0				
	$ML_2/M.L^2$	21.9				
	$ML_2/ML_2(s)$	-5.85[h]				
Pd^{2+}	$ML_3/M.L^3$		43.0			
	$MOHL_2/ML_2.OH$		3.4			

[h] 20°, 0.1; [r] 25-40°, 0

Bibliography:

H^+ 49VB,52BC,58BB,63BA,63SAP,68AT Pd^{2+} 68AT

Ni^{2+} 64AS,64SA

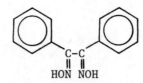

$C_{14}H_{12}O_2N_2$		1,2-Diphenylethane-1,2-dione dioxime (α-benzil dioxime)		HL
Metal ion	Equilibrium	Log K 25°, 0.1	Log K 20°, 1.0	
H^+	$L/H_{-1}L.H$	11.2	11.8	
	$HL/H.L$	10.3 ±0.3	10.15	
	$HL/HL(s)$	-4.80		
Ni^{2+}	$ML_2/M.L^2$	26.2		
Pd^{2+}	$ML_2/M.L^2$		34.6	
	$MOHL_2/ML_2.OH$		4.9	

Bibliography:

H^+ 64D,65PS,70TB Pd^{2+} 70TB

Ni^{2+} 65PS

$C_{10}H_8O_2N_2$ <u>1,2-Di(2-furanyl)ethane-1,2-dione dioxime (α-furyldioxime)</u>   HL

Metal ion	Equilibrium	Log K 25°, 0.1	Log K 20°, 1.0
H^+	$L/H_{-1}L.H$	11.1	11.4
	$HL/H.L$	9.73	9.72
Ni^{2+}	$ML/M.L$	8.2	
	$ML_2/M.L^2$	14.90	
Pd^{2+}	$ML_2/M.L^2$		(43.7)
	$MOHL_2/ML_2.OH$		3.4

Bibliography:

H^+ 49VB,63SAP,67ST Pd^{2+} 67ST

Ni^{2+} 64SA,66PS   Other reference: 66MK

$$R_2 - \overset{\overset{R_1}{|}}{\underset{\underset{R_3}{|}}{C}} - \overset{\overset{O}{\|}}{C}NH_2$$

$C_n H_m O_p N$ Substituted acetic acid amide   L

R =	Metal ion	Equilibrium	Log K 25°, 0.1
$R_1 = R_2 = R_3 = H$ (C_2H_5ON) (acetamide)	Hg^{2+}	$MH_{-1}L.H/M.L$	-7.05^s
		$M(H_{-1}L)_2.H^2/M.L^2$	-14.61^s
		$MOH(H_{-1}L)/M(H_{-1}L).OH$	5.79^s
$R_1 = R_2 = CH_3, R_3 = OH$ ($C_4H_9O_2N$) (2-methyl-2-hydroxy-propanoic acid amide)	Hg^{2+}	$MH_{-1}L.H/M.L$	-6.57^s
		$M(H_{-1}L)_2.H^2/M.L^2$	-12.53^s
		$MOH(H_{-1}L)/M(H_{-1}L).OH$	5.80^s
$R_1 = H, R_2 = C_6H_5, R_3 = OH$ ($C_8H_9O_2N$) (DL-phenylhydroxy-acetic acid amide)	Hg^{2+}	$MH_{-1}L.H/M.L$	-6.29^s
		$M(H_{-1}L)_2.H^2/M.L^2$	$(-12.94)^s$
		$MOH(H_{-1}L)/M(H_{-1}L).OH$	6.04^s
$R_1 = CH_3, R_2 = C_6H_5, R_3 = OH$ ($C_9H_{11}O_2N$) (DL-2-phenyl-2-hydroxy-propanoic acid amide)	Hg^{2+}	$MH_{-1}L.H/M.L$	-6.20^s
		$M(H_{-1}L)_2.H^2/M.L^2$	$(-12.55)^s$
		$MOH(H_{-1}L)/M(H_{-1}L).OH$	5.82^s

^{s} Not corrected for Cl^-.

Bibliography: 70GS

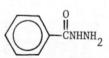

$C_7H_8ON_2$ Benzoic acid hydrazide (benzoylhydrazine)   HL

Metal ion	Equilibrium	Log K 25°, 1.0
H^+	$HL/H.L$	12.39
	$H_2L/HL.H$	3.13
Cu^{2+}	$MHL/M.HL$	3.99
	$M(HL)_2/M.(HL)^2$	6.16
	$MHL/ML.H$	5.35
	$M(HL)_2/MHL_2.H$	4.33

Bibliography:

H^+ 63NT

Cu^{2+} 63NK

Other references: 56Ab,74FSD,75PSb

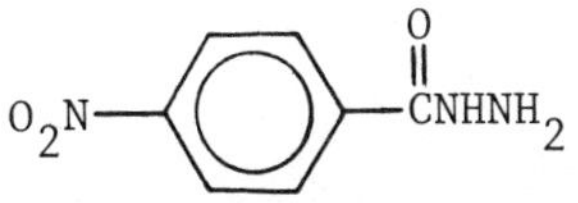

$C_7H_7O_3N_3$ 4-Nitrobenzoic acid hydrazide (4-nitrobenzoylhydrazine)   HL

Metal ion	Equilibrium	Log K 25°, 1.0
H^+	HL/H.L	11.14
	$H_2L/HL.H$	2.76
Cu^{2+}	MHL/M.HL	3.31
	$M(HL)_2/M.(HL)^2$	5.55
	MHL/ML.H	4.58
	$M(HL)_2/MHL_2.H$	3.44

Bibliography:

H^+ 63NT Cu^{2+} 63NK   Other reference: 75PSb

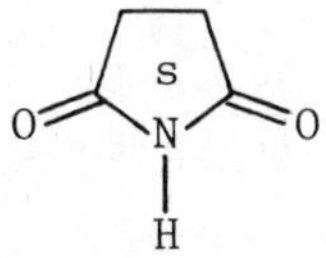

$C_4H_5O_2N$ Pyrrolidine-2,5-dione (succinimide)   HL

Metal ion	Equilibrium	Log K 25°, 0.1
H^+	HL/H.L	9.38
Ag^+	$ML/M.L$	4.45
	$ML_2/M.L^2$	9.54

Bibliography: 65CO

Other references: 61BH,65JK

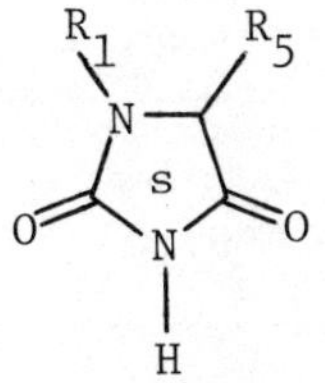

$C_nH_mO_2N_2$ Substituted-1,3-diazolidine-2,4-dione (substituted hydantoin)   HL

R =	Metal ion	Equilibrium	Log K 25°, 0.1
$R_1=R_5=H$	H^+	HL/H.L	8.93
$(C_3H_4O_2N_2)$	Ag^+	$ML/M.L$	4.29
		$ML_2/M.L^2$	9.20

Substituted hydantoin (continued)

R =	Metal ion	Equilibrium	Log K 25°, 0.1
$R_1=CH_3, R_5=H$	H^+	HL/H.L	9.09
$(C_4H_6O_2N_2)$	Ag^+	ML/M.L	4.37
		$ML_2/M.L^2$	9.34
$R_1=H, R_5=CH_3$	H^+	HL/H.L	9.07
$(C_4H_6O_2N_2)$	Ag^+	ML/M.L	4.34
		$ML_2/M.L^2$	9.27

Bibliography: 65CO

Other reference: 61BH

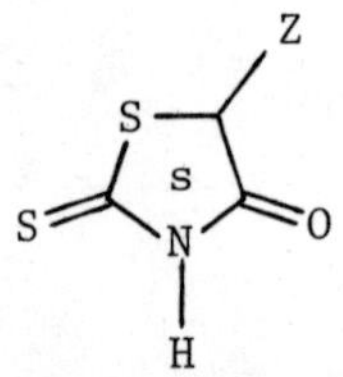

$C_n H_m O_p N_r S_t$ 5-Substituted-2-thioxo-1,3-thiazolidine-4-one   HL
                                              (substituted rhodanine)

Z =	Metal ion	Equilibrium	Log K 20°, 0.1
H	H^+	HL/H.L	5.18
$(C_3H_3ONS_2)$	Ag^+	ML/M.L	5.47
		$ML_2/M.L^2$	9.68
5-(3-Phenyl-allylidene) (5-cinnamyl-idene)	H^+	HL/H.L	7.68
	Ag^+	ML/M.L	9.08
$(C_{12}H_9ONS_2)$			
5-Benzylidene	H^+	HL/H.L	7.58
$(C_{10}H_7ONS_2)$	Ag^+	ML/M.L	8.35
		$ML_2/M.L^2$	15.85
5-(4-Methoxy-benzylidene)	H^+	HL/H.L	7.76
$(C_{11}H_9O_2NS_2)$	Ag^+	ML/M.L	8.80
5-(4-Dimethyl-amino-benzyl-idene)	H^+	HL/H.L	8.20
	Cu^{2+}	ML/M.L	6.08
$(C_{12}H_{12}ON_2S_2)$	Ag^+	ML/M.L	9.15
5-(2-Furylidene)	H^+	HL/H.L	6.37
$(C_7H_5O_2NS_2)$	Ag^+	ML/M.L	7.19

Bibliography: 65NK

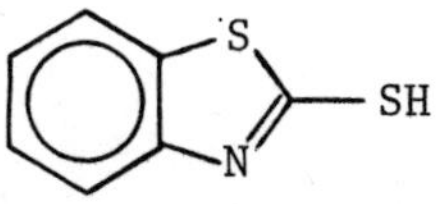

$C_7H_5NS_2$ 2-Mercaptobenzo-1,3-thiazole   HL

Metal ion	Equilibrium	Log K 20°, 0.1
H^+	$HL/H.L$	6.93
Zn^{2+}	$ML/M.L$	3.25
	$ML_2/M.L^2$	5.74

Bibliography: 68NL

CH_4ON_2 Carbamide (urea)   L

Metal ion	Equilibrium	Log K 30°, 0.1	Log K 25°, 0
H^+	$L/H_{-1}L.H$		14.3
	$HL/H.L$	0.11^h	0.1
Hg^{2+}	$ML/M.L$	2.1	

^{h} 21°, 0.1

Bibliography:
H^+ 43BG,72WH   Other references: 65SK,70KL,74GF
Hg^{2+} 69GL

CH_4N_2S Thiocarbamide (thiourea)   L

Metal ion	Equilibrium	Log K 25°, 0.1	Log K 25°, 1.0	Log K 25°, 0	ΔH 25°, 0.5	ΔS 25°, 0.5
Cu_2^{2+}	$ML_4/M.L^4$	15.4^b				
Ag^+	$ML/M.L$	$7.11^b \pm 0.07$			-2.6	24
	$ML_2/M.L^2$	$10.61^b -0.01$			-21.9	-25
	$ML_3/M.L^3$	$12.73^b \pm 0.02$			-23.1	-19
	$ML_4/M.L^4$	$13.57^b \pm 0.08$			-32.0	-45
Pd^{2+}	$ML_4/M.L^4$	30.1^b				
Zn^{2+}	$ML/M.L$		0.5^d			
	$ML_2/M.L^2$		0.8^d			
	$ML_3/M.L^3$		0.9^d			

^{b} 25°, 0.5; ^{d} 25°, 2.0

Thiourea (continued)

Metal ion	Equilibrium	Log K 25°, 0.1	Log K 25°, 1.0	Log K 25°, 0	ΔH 25°, 0.5	ΔS 25°, 0.5
Cd^{2+}	ML/M.L	1.33 ±0.1	1.82	1.5		
		1.77^d	1.9^d ±0.3	2.12^e		
	$ML_2/M.L^2$	2.18 ±0.05	2.64	2.2		
		2.57^b	2.9^d ±0.2	3.65^e		
	$ML_3/M.L^3$	2.7 ±0.0	3.0	2.6		
		2.9^b	3.5^d ±0.2	4.0^e		
	$ML_4/M.L^4$	3.3 ±0.1	3.8	3.1		
		3.6^b	4.4^d -0.1	5.5^e		
Hg^{2+}	$ML_2/M.L^2$	21.3^p	22.1		-33^p	-13^p
	$ML_3/M.L^3$	24.2^p	24.7		-35^p	-6^p
	$ML_4/M.L^4$	25.8^p	26.5		-46^p	-35^p
Pb^{2+}	ML/M.L	0.17	0.63	0.09		
		0.25^b	1.13^d	1.68^e		
	$ML_2/M.L^2$	0.86	1.37	0.83		
		1.08^b	1.88^d	2.54^e		
	$ML_3/M.L^3$	1.4	1.8	1.3		
		1.5^b	2.2^d	2.9^e		
	$ML_4/M.L^4$	1.5	2.0	1.5		
		1.7^b	2.4^d	3.1^e		
	$ML_5/M.L^5$	1.4	2.0	1.5		
		1.8^b	2.3^d	2.9^e		
	$ML_6/M.L^6$	1.8	2.0	1.7		
		1.9^b	2.2^d	3.3^e		
Bi^{3+}	ML/M.L			2.1		

---

^{b} 25°, 0.5; ^{d} 25°, 2.0; ^{e} 25°, 3.0; ^{p} 30°, 0.1

Bibliography:

Cu_2^{2+} 50OL,73EB

Ag^+ 69BL,73BD

Pd^{2+} 65FK

Zn^{2+} 66SLK

Cd^{2+} 66SL,69GL,71BL,74FF

Hg^{2+} 56T,69GL

Pb^{2+} 74FF

Bi^{3+} 68VG

Other references: 24P,49JH,52YV,55F,58LR, 60TK,63MT,64MTb,65TS,66SB,67NP,67VG, 69FD,71TM,74KL,75OM

$$\underset{R_1NHCNHR_2}{\overset{\overset{\textstyle S}{\|}}{}}$$

$C_nH_mN_2S$			N,N'-Alkylthiourea			L
R =	Metal ion	Equilibrium	Log K 25°, 0.1	Log K 25°, 0.5	ΔH 25°, 0.5	ΔS 25°, 0.5
$R_1=CH_3,R_2=H$	Ag^+	ML/M.L		6.70	-1.6	25
(N-methylthio-		$ML_2/M.L^2$		10.61	-22.4	-26
urea)($C_2H_6N_2S$)		$ML_3/M.L^3$		13.04	-22.8	-17
		$ML_4/M.L^4$		14.06	-32.4	-44

N,N'-Alkylthiourea (continued)

R =	Metal ion	Equilibrium	Log K 25°, 0.1	Log K 25°, 0.5	ΔH 25°, 0.5	ΔS 25°, 0.5
	Cd^{2+}	$ML/M.L$	1.42			
		$ML_2/M.L^2$	2.40			
		$ML_3/M.L^3$	2.9			
		$ML_4/M.L^4$	4.1			
$R_1=CH_3CH_2, R_2=H$	Cd^{2+}	$ML/M.L$	1.46			
(N-ethylthio-		$ML_2/M.L^2$	2.18			
urea)($C_3H_8N_2S$)		$ML_3/M.L^3$	3.5			
		$ML_4/M.L^4$	4.4			
$R_1=R_2=CH_3$	Ag^+	$ML/M.L$		6.08	0	28
(N,N'-dimethyl-		$ML_2/M.L^2$		10.15	-21.8	-27
thiourea)		$ML_3/M.L^3$		12.74	-22.2	-16
($C_3H_8N_2S$)		$ML_4/M.L^4$		13.85	-31.6	-42
$R_1=R_2=CH_3CH_2$	Ag^+	$ML/M.L$		6.00	-1.3	23
(N,N'-diethyl-		$ML_2/M.L^2$		10.30	-21.5	-25
thiourea)		$ML_3/M.L^3$		13.33	-21.8	-12
($C_5H_{12}N_2S$)		$ML_4/M.L^4$		14.15	-30.8	-39

Bibliography:

Ag^+ 73EB Cd^{2+} 75FF

$$\overset{S}{\underset{}{\overset{\|}{Z-NHCNH_2}}}$$

N-Substitutedthiourea   L

$C_nH_mN_2S$

Z =	Metal ion	Equilibrium	Log K 25°, 0.5	Log K 25°, 1.0	ΔH 25°, 1.0	ΔS 25°, 1.0
N-Allyl	Cd^{2+}	$ML/M.L$	1.46			
($C_4H_7N_2S$)		$ML_2/M.L^2$	(2.15)			
		$ML_3/M.L^3$	(3.41)			
	Hg^{2+}	$ML_4/M.L^4$		27.1	(-46)[r]	(-30)
	Pb^{2+}	$ML_2/M.L^2$	1.45			
N-Phenyl	Hg^{2+}	$ML_4/M.L^4$		25.0[s]		
($C_7H_8N_2S$)						

[r] 20-35°, 1.0;   [s] 40°, 0.08

Bibliography:

Cd^{2+}, Pb^{2+} 74RG Hg^{2+} 56T

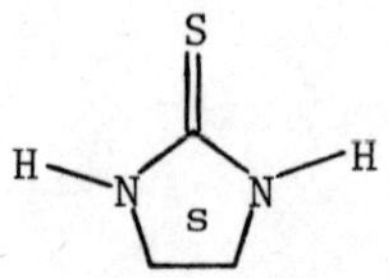

$$\underset{\underset{H-NN-H}{\underset{\diagdown/}{S}}}{\overset{S}{\|}}$$

C$_3$H$_6$N$_2$S		N,N'-Ethylenethiourea		L
Metal ion	Equilibrium	Log K 25°, 1.0		
Cd^{2+}	ML/M.L	1.31		
	ML$_2$/M.L^2	2.1		
	ML$_3$/M.L^3	2.7		
	ML$_4$/M.L^4	3.4		

Bibliography: 63CR

Other reference: 75OM

$$\underset{H_2NCNH_2}{\overset{\overset{Se}{\|}}{}}$$

CH$_4$N$_2$Se		Selenocarbamide (selenourea)		L	
Metal ion	Equilibrium	Log K 30°, 0.1		ΔH 30°, 0.1	ΔS 30°, 0.1
H$^+$	HL/H.L	0.58		0	3
Cd^{2+}	ML/M.L	(0.9)			
	ML$_2$/M.L^2	3.7			
Hg^{2+}	ML$_2$/M.L^2	24.0		-30.5	9
	ML$_3$/M.L^3	30.2		-37.4	15
	ML$_4$/M.L^4	32.9		-54.1	-28

Bibliography: 69GL

$$\underset{H_2NCNHNH_2}{\overset{\overset{O}{\|}}{}}$$

CH$_5$ON$_3$		Semicarbazide		L	
Metal ion	Equilibrium	Log K 30°, 0.1		ΔH 30°, 0.1	ΔS 30°, 0.1
H$^+$	HL/H.L	3.53		-6.1	-4
Cu^{2+}	ML/M.L	4.00		-5.8	-1
	ML$_2$/M.L^2	6.94		-12.0	-8
Ag$^+$	ML/M.L	1.95		-2.2	2
	ML$_2$/M.L^2	2.70		-12.5	-29
Zn^{2+}	ML/M.L	2.3			
	ML$_2$/M.L^2	3.7			
Cd^{2+}	ML/M.L	1.26		-2.8	-4
	ML$_2$/M.L^2	2.79		-5.0	-4
Hg^{2+}	ML$_2$/M.L^2	11.6		-16	0
	ML$_3$/M.L^3	15.2		-18	10

Semicarbazide (continued)

Metal ion	Equilibrium	Log K 30°, 0.1	ΔH 30°, 0.1	ΔS 30°, 0.1
Pb^{2+}	$ML/M.L$	2.11	-3.5	-2
	$ML_2/M.L^2$	2.86	-6.9	-10

Bibliography:

H^+, Zn^{2+}, Hg^{2+} 69GL Ag^+, Cd^{2+}, Pb^{2+} 73AG

Cu^{2+} 71AG   Other reference: 60TN

$$\overset{\text{S}}{\overset{\|}{\text{H}_2\text{NCNHNH}_2}}$$

CH₅N₃S		Thiosemicarbazide				L
Metal ion	Equilibrium	Log K 30°, 0.1	Log K 25°, 1.0	Log K 25°, 2.0	ΔH 30°, 0.1	ΔS 30°, 0.1
H^+	$H_{-1}L/H_{-2}L.H$	12.6[a]	12.5	12.6		
	$L/H_{-1}L.H$	11.9[a]	12.0	12.1		
	$HL/H.L$	1.6 ±0.1	1.8	2.0	-4.5	-8
Cu^{2+}	$ML/M.L$	6.11			-9.8	-4
	$ML_2/M.L^2$	11.59			-17.8	-6
Ag^+	$ML_3/M.L^3$		12.9[b]		(-20)[r]	(-8)[b]
Zn^{2+}	$ML_2/M.L^2$	2.8				
Cd^{2+}	$ML/M.L$	2.28	2.57		-4.3	-4
	$ML_2/M.L^2$	4.40	4.70		-8.9	-9
	$ML_3/M.L^3$		5.86			
Hg^{2+}	$ML_2/M.L^2$	22.4			-35.3	-14
	$ML_3/M.L^3$	24.8			-39.9	-18
	$ML_4/M.L^4$		25.8[b]		(-40)[r]	(-16)[b]
Pb^{2+}	$ML/M.L$	2.89			-4.2	-1

[a] 25°, 0.1;   [b] 25°, 0.8;   [r] 20-50°, 0.8

Bibliography:

H^+ 68BLD, 69GL Cd^{2+} 63CR, 73AG

Cu^{2+} 71AG Hg^{2+} 60TK, 69GL

Ag^+ 60TK Pb^{2+} 73AG

Zn^{2+} 69GL   Other reference: 60TN

$$\overset{\text{Se}}{\overset{\|}{\text{H}_2\text{NCNHNH}_2}}$$

CH₅N₃Se		Selenosemicarbazide			L
Metal ion	Equilibrium	Log K 30°, 0.1		ΔH 30°, 0.1	ΔS 30°, 0.1
H^+	$HL/H.L$	0.80		-4.5	-11
Cu^{2+}	$ML/M.L$	5.54		-11.0	-11
	$ML_2/M.L^2$	10.83		-24.7	-32

Selenosemicarbazide (continued)

Metal ion	Equilibrium	Log K 30°, 0.1	ΔH 30°, 0.1	ΔS 30°, 0.1
Cd^{2+}	$ML/M.L$	2.92	-3.1	3
	$ML_2/M.L^2$	5.04	-11.4	-14
Hg^{2+}	$ML_2/M.L^2$	26.9		
	$ML_3/M.L^3$	30.4	-44	-7
	$ML_4/M.L^4$	32.4	-61	-53
Pb^{2+}	$ML/M.L$	2.28	-4.8	-6

Bibliography:

H^+, Hg^{2+}	69GL	Cd^{2+}, Pb^{2+}	73AG
Cu^{2+}	71AG		

$$\overset{O}{\overset{\|}{H_2NNHCNHNH_2}}$$

CH_6ON_4	Carbohydrazide		L

Metal ion	Equilibrium	Log K 20°, 0.1
H^+	$HL/H.L$	4.14
	$H_2L/HL.H$	2.10
Co^{2+}	$ML/M.L$	2.83
	$ML_2/M.L^2$	5.38
Ni^{2+}	$ML/M.L$	3.44
	$ML_2/M.L^2$	6.62
	$ML_3/M.L^3$	8.64
Cu^{2+}	$ML/M.L$	4.92
	$ML_2/M.L^2$	8.97
Zn^{2+}	$ML/M.L$	2.77
Cd^{2+}	$ML/M.L$	2.37

Bibliography: 64COV

Other reference: 66KSD

$$\overset{S}{\overset{\|}{H_2NNHCNHNH_2}}$$

CH_6N_4S	Thiocarbohydrazide			L

Metal ion	Equilibrium	Log K 25°, 0.1	Log K 25°, 0.5	Log K 25°, 2.0
H^+	$H_{-1}L/H_{-2}L.H$	12.4	12.3	12.4
	$L/H_{-1}L.H$	11.3	11.3	11.3
	$HL/H.L$	3.13	3.16	3.39
	$H_2L/HL.H$	0.78	0.81	0.99
Co^{2+}	$ML/M.L$		2.97	
	$ML_2/M.L^2$		5.66	

XV. AMIDES

Thiocarbohydrazide (continued)

Metal ion	Equilibrium	Log K 25°, 0.5
Ni^{2+}	$ML/M.L$	4.40
	$ML_2/M.L^2$	8.10
	$ML_3/M.L^3$	11.21
Zn^{2+}	$ML/M.L$	2.53

Bibliography:

H^+ 68BLD $Co^{2+}-Zn^{2+}$ 69BD

$$\overset{\displaystyle SCH_3}{\underset{}{|}}$$
$$H_2NNHC=NNH_2$$

$C_2H_8N_4S$	S-Methylisothiocarbohydrazide				HL

Metal ion	Equilibrium	Log K 25°, 0.1	Log K 25°, 0.5	Log K 25°, 0	ΔH 25°, 0.1	ΔS 25°, 0.1
H^+	$HL/H.L$	7.56	7.71	7.33	$(-10)^r$	(8)
	$H_2L/HL.H$	1.3	1.4	0.9		
Mn^{2+}	$ML/M.L$	2.02				
Co^{2+}	$ML/M.L$	4.81				
	$ML_2/M.L^2$	9.25				
Ni^{2+}	$ML/M.L$	6.16				
	$ML_2/M.L^2$	11.79				
Zn^{2+}	$ML/M.L$	4.27				
	$ML_2/M.L^2$	7.62				
Cd^{2+}	$ML/M.L$	3.55				
	$ML_2/M.L^2$	5.97				

^{r} 5-35°, 0.1

Bibliography: 72BM

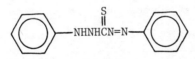

$C_{13}H_{12}N_4S$	1,5-Diphenyl-1,2,4,5-tetraazapent-1-en-3-thione (diphenylthiocarbazone) (dithizone)		HL

Metal ion	Equilibrium	Log K 25°, 0.1	Log K 25°, 1.0
H^+	$HL/H.L$	4.45 ±0.01	
	$HL/HL(s)$	-6.61	
Cu^{2+}	$ML_2/M.L^2$	22.3	
	$M.L^2/ML_2(s)$	-30.7	
Zn^{2+}	$ML/M.L$		7.75
	$ML_2/M.L^2$		15.05

Dithizone (continued)

Metal ion	Equilibrium	Log K 25°, 0.1	Log K 25°, 1.0
Pb^{2+}	$ML_2/M \cdot L^2$	15.2	
	$M \cdot L^2/ML_2(s)$	-23.7	

Bibliography:

H^+	52DH,52IB,53GS	Pb^{2+}	64MS
Cu^{2+}	53GS	Other references:	56Ba,60DT,73BSa,74KM
Zn^{2+}	62HF		

$$\underset{H_2NC-CNH_2}{\overset{\overset{\displaystyle S}{\|}\ \overset{\displaystyle O}{\|}}{}}$$

				HL
$C_2H_4ON_2S$		Monothiooxamide		

Metal ion	Equilibrium	Log K 25°, 0.1	Log K 25°, 1.0	Log K 25°, 0
H^+	$HL/H \cdot L$	11.31 \newline 11.21[b]	11.20	11.53
Ni^{2+}	$ML/M \cdot L$ \newline $ML_2/M \cdot L^2$	(4.80) \newline (12.77)		
Pd^{2+}	$ML/M \cdot L$ \newline $ML_2/M \cdot L^2$		6.81[d] \newline 13.56[d]	

[b] 25°, 0.5;   [d] 25°, 2.0

Bibliography:

H^+	70VHE	Ni^{2+}, Pd^{2+}	70VH

$$\underset{HOCH_2CH_2NHC-CNHCH_2CH_2OH}{\overset{\overset{\displaystyle S}{\|}\ \overset{\displaystyle O}{\|}}{}}$$

				HL
$C_6H_{12}O_3N_2S$		N,N'-Bis(2-hydroxyethyl)monothiooxamide		

Metal ion	Equilibrium	Log K 25°, 0.1	Log K 25°, 1.0	Log K 25°, 0
H^+	$HL/H \cdot L$	11.15 \newline 11.06[b]	11.09	11.38
Ni^{2+}	$ML/M \cdot L$ \newline $ML_2/M \cdot L^2$	5.38 \newline 10.76		
Pd^{2+}	$ML/M \cdot L$ \newline $ML_2/M \cdot L^2$		6.37[d] \newline 12.34[d]	

[b] 25°, 0.5;   [d] 25°, 2.0

Bibliography:

H^+	70VHE	Ni^{2+}, Pd^{2+}	70VH

$$\underset{\underset{\displaystyle HOCH_2CH_2CH_2NHC-CNHCH_2CH_2CH_2OH}{\overset{\displaystyle S \quad O}{\overset{\|}{} \;\; \overset{\|}{}}}{}$$

C$_8$H$_{16}$O$_3$N$_2$S	N,N'-Bis(3-hydroxypropyl)monothiooxamide			HL
Metal ion	Equilibrium	Log K 25°, 0.1	Log K 25°, 1.0	Log K 25°, 0
H$^+$	HL/H.L	11.56 11.47[b]	11.50	11.79
Ni^{2+}	ML/M.L	5.98		
	ML$_2$/M.L^2	11.82		
Pd^{2+}	ML/M.L		6.49[d]	
	ML$_2$/M.L^2		12.42[d]	

[b] 25°, 0.5;   [d] 25°, 2.0

Bibliography:

H$^+$ 70VHE   Ni^{2+},Pd^{2+} 70VH

$$\underset{\underset{\displaystyle HOCH_2CH_2CH_2CH_2NHC-CNHCH_2CH_2CH_2CH_2OH}{\overset{\displaystyle S \quad O}{\overset{\|}{} \;\; \overset{\|}{}}}{}$$

C$_{10}$H$_{20}$O$_3$N$_2$S	N,N'-Bis(4-hydroxybutyl)monothiooxamide			HL
Metal ion	Equilibrium	Log K 25°, 0.1	Log K 25°, 1.0	Log K 25°, 0
H$^+$	HL/H.L	11.80 11.70[b]	11.70	12.02
Ni^{2+}	ML/M.L	(5.04)		
	ML$_2$/M.L^2	(12.24)		
Pd^{2+}	ML/M.L		(6.39)[d]	
	ML$_2$/M.L^2		(13.25)[d]	

[b] 25°, 0.5;   [d] 25°, 2.0

Bibliography:

H$^+$ 70VHE   Ni^{2+},Pd^{2+} 70VH

$$\underset{\underset{\displaystyle HOCH_2CH_2CH_2CH_2CH_2NHC-CNHCH_2CH_2CH_2CH_2CH_2OH}{\overset{\displaystyle S \quad O}{\overset{\|}{} \;\; \overset{\|}{}}}{}$$

C$_{12}$H$_{24}$O$_3$N$_2$S	N,N'-Bis(5-hydroxypentyl)monothiooxamide			HL
Metal ion	Equilibrium	Log K 25°, 0.1	Log K 25°, 1.0	Log K 25°, 0
H$^+$	HL/H.L	11.93 11.80[b]	11.83	12.18
Ni^{2+}	ML/M.L	6.45		
	ML$_2$/M.L^2	12.65		

[b] 25°, 0.5;

N,N'Bis(5-hydroxypentyl)monothiooxamide (continued)

Metal ion	Equilibrium	Log K 25°, 0.1	Log K 25°, 1.0	Log K 25°, 0
Pd^{2+}	$ML/M.L$		6.55^d	
	$ML_2/M.L^2$		12.54^d	

------

^{d} 25°, 2.0

Bibliography:

H^+	70VHE		Ni^{2+},Pd^{2+}	70VH

$$\underset{HO_3SCH_2CH_2NHC-CNHCH_2CH_2SO_3H}{\overset{\displaystyle S\ \ S}{\overset{\displaystyle \parallel\ \parallel}{}}}$$

| $C_6H_{12}O_6N_2S_4$ | N,N'-Bis(2-sulfoethyl)dithiooxamide | | | H_3L |

Metal ion	Equilibrium	Log K 25°, 0.1	Log K 25°, 1.0	Log K 25°, 0
H^+	$HL/H.L$	11.10	10.50	11.73
	$H_2L/HL.H$		2.55	
Pd^{2+}	$MCl_2H_2L/MCl_4.H_2L$		8.98^g	
	$M(H_2L)_2/MCl_4.(H_2L)^2$		15.95^g	
	$M(H_2L)_2/MH_3L_2.H$		-0.39	
	$MH_3L_2/M(HL)_2.H$		1.16	

------

^{g} 25°, 5.0

Bibliography:

H^+	68GHa		Pd^{2+}	68GHE

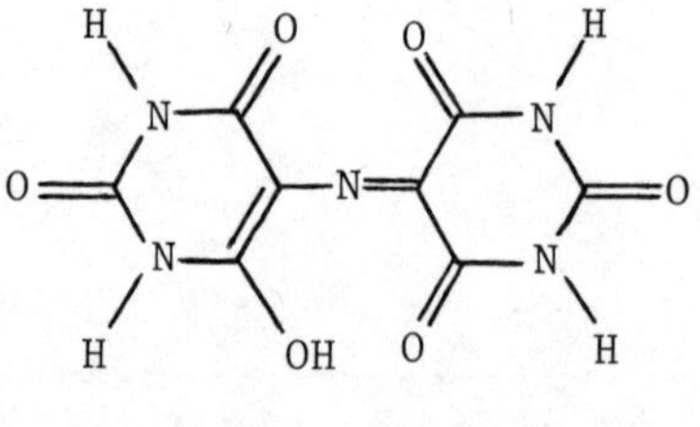

| $C_8H_5O_6N_5$ | N-(4-Hydroxy-2,6-dioxo-1,3-diazin-5-yl)-5-imino-perhydro-1,3-diazine-2,4,6-trione (purpuric acid) (murexide=L.NH$_3$) | | H_3L |

Metal ion	Equilibrium	Log K 20°, 0.1
H^+	$HL/H.L$	10.9
	$H_2L/HL.H$	9.2
Mg^{2+}	$MH_2L/M.H_2L$	2.2
Ca^{2+}	$MH_2L/M.H_2L$	2.68^r
	$MH_2L/MHL.H$	8.2
	$MHL/ML.H$	9.5
Ni^{2+}	$MH_2L/M.H_2L$	4.6

------

^{r} 25°, 0.2

Purpuric acid (continued)

Metal ion	Equilibrium	Log K 20°, 0.1
Cu^{2+}	$MH_2L/M.H_2L$	5
Zn^{2+}	$MH_2L/M.H_2L$	3.1
Cd^{2+}	$MH_2L/M.H_2L$	4.2

Bibliography:

H^+ 49SG

Mg^{2+},Zn^{2+},Cd^{2+} 49SGa

Ca^{2+} 49SGa,61Na,64SM

Ni^{2+},Cu^{2+} 64SM

Other references: 60Ca,65G,75KIF

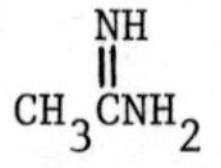

$$\overset{\displaystyle NH}{\underset{\displaystyle CH_3\overset{\displaystyle \|}{C}NH_2}{}}$$

$C_2H_6N_2$		Acetamidine	L

Metal ion	Equilibrium	Log K 25°, 0.1
Hg^{2+}	ML.H/M.HL	-6.24^s
	$ML_2.H^2/M.L^2$	-13.0^s
	MOHL/ML.OH	6.09^s

_____

^{s} Not corrected for Cl^-

Bibliography: 70GS

_____

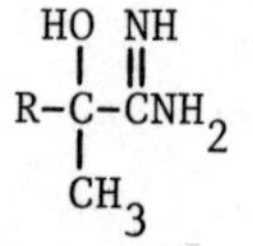

$$\overset{\displaystyle HO\ NH}{\underset{\displaystyle CH_3}{\underset{\displaystyle |}{R-\overset{\displaystyle |}{\underset{\displaystyle |}{C}}-\overset{\displaystyle \|}{C}NH_2}}}$$

$C_nH_mON_2$			2-Methyl-2-hydroxy-alkylamidine	L

R =	Metal ion	Equilibrium	Log K 25°, 0.1
CH_3	H^+	HL/H.L	11.24
$(C_4H_{10}ON_2)$	Ag^+	$M(H_{-1}L).H/M.L$	-8.93
(2-methyl-2-		$M(H_{-1}L).H^2/M.L^2$	-17.35
hydroxy-		$MOH(H_{-1}L)_2/M(H_{-1}L)_2.OH$	
propamidine)			3.56
	Hg^{2+}	$M(H_{-1}L).H/M.L$	-7.96^s
		$M(H_{-1}L)_2.H^2/M.L^2$	-16.37^s
		$MOH(H_{-1}L)/M(H_{-1}L).OH$	
			5.97^s

CH_3CH_2	H^+	HL/H.L	11.49
$(C_5H_{12}ON_2)$	Ag^+	$M(H_{-1}L).H/M.L$	-9.0
(DL-2-methyl-2-		$M(H_{-1}L)_2.H^2/M.L^2$	(-17.16)
hydroxy-		$MOH(H_{-1}L)_2/M(H_{-1}L)_2.OH$	
butamidine)			3.67
	Hg^{2+}	$M(H_{-1}L).H/M.L$	-7.88^s
		$M(H_{-1}L)_2.H^2/M.L^2$	$(-16.08)^s$
		$MOH(H_{-1}L)/M(H_{-1}L).OH$	
			5.87^s

_____

^{s} Not corrected for Cl^-

Bibliography:

H^+,Hg^{2+} 70GS Ag^+ 70GSa

$$\text{HO} \quad \text{NH}$$
$$\text{C}_6\text{H}_5\text{-CHCNH}_2$$

$C_8H_{10}ON_2$	DL-Phenylhydroxyacetamidine (mandelamidine)		HL

Metal ion	Equilibrium	Log K 25°, 0.1
H^+	HL/H.L	12.41
	$H_2L/HL.H$	10.71
Ni^{2+}	ML/M.L	7.38
	$ML_2/M.L^2$	(14.40)
Cu^{2+}	ML/M.L	12.50
	$ML_2/M.L^2$	(23.80)
Ag^+	ML/M.L	4.40
	$ML_2/M.L^2$	(9.32)
	$MOHL_2/ML_2.OH$	5.13
Cd^{2+}	ML/M.L	2.71[s]
	MOHL/ML.OH	5.09[s]
	$M(OH)_2L/ML.(OH)^2$	11.0[s]
Hg^{2+}	ML/M.L	4.86[s]
	$ML_2/M.L^2$	(9.82)[s]
	MOHL/ML.OH	5.5[s]

[s] Not corrected for Cl^-.

Bibliography:

H^+	62GJ	Ag^+,Cd^{2+}	70GSa
Ni^{2+}	63GJa	Hg^{2+}	70GS
Cu^{2+}	63GJ		

$$\text{HO} \quad \text{NH}$$
$$\text{C}_6\text{H}_5\text{-C-CNH}_2$$
$$\text{CH}_3$$

$C_9H_{12}ON_2$	DL-2-Phenyl-2-hydroxypropamidine (atrolactamidine)		HL

Metal ion	Equilibrium	Log K 25°, 0.1
H^+	HL/H.L	12.61
	$H_2L/HL.H$	10.85
Ni^{2+}	ML/M.L	7.87
	$ML_2/M.L^2$	(15.40)
Cu^{2+}	ML/M.L	12.73
	$ML_2/M.L^2$	(24.30)
Ag^+	ML/M.L	4.16
	$ML_2/M.L^2$	(8.86)
	$MOHL_2/ML_2.OH$	4.01

Atrolactamidine (continued)

Metal ion	Equilibrium	Log K 25°, 0.1
Cd^{2+}	ML/M.L	2.90^s
	MOHL/ML.OH	4.99^s
Hg^{2+}	ML/M.L	5.00^s
	$ML_2/M.L^2$	$(9.93)^s$
	MOHL/ML.OH	5.76^s

---

^{s} Not corrected for Cl^-.

Bibliography:

H^+	62GJ	Ag^+, Cd^{2+}	70GSa
Ni^{2+}	63GJa	Hg^{2+}	70GS
Cu^{2+}	63GJ		

$$\text{HO} \quad \text{NH}$$
$$C6H5-\overset{|}{\underset{|}{C}}-\overset{\|}{C}NH_2$$
$$CH_2CH_3$$

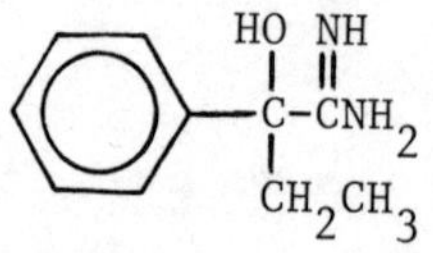

$C_{10}H_{14}ON_2$	DL-2-Phenyl-2-hydroxybutamidine	HL

Metal ion	Equilibrium	Log K 25°, 0.1
H^+	HL/H.L	12.75
	$H_2L/HL.H$	10.95
Ni^{2+}	ML/M.L	8.06
	$ML_2/M.L^2$	(15.80)
Cu^{2+}	ML/M.L	12.86
	$ML_2/M.L^2$	(24.56)

Bibliography:

H^+	62GJ	Cu^{2+}	63GJ
Ni^{2+}	63GJa		

$$\text{NH}$$
$$H_2N\overset{\|}{C}NH_2$$

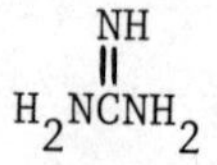

CH_5N_3	Guanidine	L

Metal ion	Equilibrium	Log K 27°, 1.0
H^+	HL/H.L	13.54
Hg^{2+}	$ML_2/M.L^2$	24.96

Bibliography: 64WDa

$$\begin{matrix} NH & NH \\ \parallel & \parallel \\ H_2NCNHCNH_2 \end{matrix}$$

$C_2H_7N_5$ Biguanide   L

Metal ion	Equilibrium	Log K 20°, 0.1
H^+	$H_2L/HL.H$	3.07
Ni^{2+}	$ML_2.H^2/M.(HL)^2$	-10.2
Cu^{2+}	$ML.H/M.HL$	-1.39
	$ML_2.H^2/M.(HL)^2$	-4.81

Bibliography: 63SA

Other references: 50DG,52BG,56SR,59RR

PROTONATION VALUES FOR OTHER LIGANDS:

A. Carboxylic acids $R-CO_2H$

Ligand	Equilibrium	Log K 25°, 0	ΔH 25°, 0	ΔS 25°, 0	Bibliography
2-Methylbutanoic acid ($C_5H_{10}O_2$), HL	HL/H.L	4.83 4.63[a] 4.58[b] 4.61[c]	1.24	26.1	32L,70CSS
Hexanoic acid (caproic acid)* ($C_6H_{12}O_2$), HL	HL/H.L	4.857 4.63[a]±0.03 4.61[b] 4.64[c]	0.67±0.03	24.5	31LAa,48CD,52EL, 70CSS,75IP Other reference: 69Sc
2-Methylpentanoic acid ($C_6H_{12}O_2$), HL	HL/H.L	4.81 4.61[a] 4.56[b] 4.60[c]	1.28	26.2	32L,70CSS
3-Methylpentanoic acid ($C_6H_{12}O_2$), HL	HL/H.L	4.85 4.65[a] 4.60[b] 4.64[c]	1.12	25.9	32L,70CSS
4-Methylpentanoic acid (isohexanoic acid) ($C_6H_{12}O_2$), HL	HL/H.L	4.845 4.65[a] 4.61[b] 4.63[c]	0.62±0.06	24.3	31LAa,52EL,70CSS
2,2-Dimethylbutanoic acid ($C_6H_{12}O_2$), HL	HL/H.L	5.04 4.85[a] 4.81[c]	0.62	25.2	32L,70CSS
2-Ethylbutanoic acid ($C_6H_{12}O_2$), HL	HL/H.L	4.736 4.54[a] 4.50[b] 4.52[c]	2.00±0.03	28.4	32L,52EL,70CSS
Heptanoic acid ($C_7H_{14}O_2$), HL	HL/H.L	4.86 +0.03 4.66[a] 4.61[b] 4.65[c]	0.61	24.3	31LAa,38Db,70CSS
2,2-Dimethylpentanoic acid ($C_7H_{14}O_2$), HL	HL/H.L	4.97	0.99	26.1	70CSS
2-Ethylpentanoic acid ($C_7H_{14}O_2$), HL	HL/H.L	4.17[n] 4.52[h] 4.46[i] 4.49[j]			32L

[a] 25°, 0.1;   [b] 25°, 0.5;   [c] 25°, 1.0;   [h] 18°, 0.1;   [i] 18°, 0.5;   [j] 18°, 1.0;   [n] 18°, 0;

* metal constants were also reported but are not included in the compilation of selected constants.

Carboxylic acids (continued)

Ligand	Equilibrium	Log K 25°, 0	ΔH 25°, 0	ΔS 25°, 0	Bibliography
4,4-Dimethylpentanoic acid ($C_7H_{14}O_2$), HL	HL/H.L	4.79^n 4.59^h 4.57^j			32L
Octanoic acid ($C_8H_{16}O_2$), HL	HL/H.L	4.89	0.62	24.5	38Db,70CSS
Cyclopropanecarboxylic acid ($C_4H_6O_2$), HL	HL/H.L	4.827 4.63^a	0.01	22.1	53KMa,72CSS
Cyclobutanecarboxylic acid ($C_5H_8O_2$), HL	HL/H.L	4.785 4.59^a	0.68	24.2	53KMa,72CSS
Cyclopentanecarboxylic acid ($C_6H_{10}O_2$), HL	HL/H.L	4.905 4.79^a	0.52	24.2	53KMa,72CSS
Cyclohexanecarboxylic acid ($C_7H_{12}O_2$), HL	HL/H.L	4.899 4.70^a	0.37	23.7	53KEM,72CSS
Cyclohexylacetic acid ($C_8H_{14}O_2$), HL	HL/H.L	4.51	1.39	25.3	72CSS
4-Methylphenylacetic acid ($C_9H_{10}O_2$), HL	HL/H.L	4.370			37DL
4-Ethylphenylacetic acid ($C_{10}H_{12}O_2$), HL	HL/H.L	4.372			37BD
4-(1-Methylethyl)phenylacetic acid ($C_{11}H_{14}O_2$), HL	HL/H.L	4.392			37BD
4-(2-Methyl-2-propyl)phenyl-acetic acid ($C_{12}H_{16}O_2$), HL	HL/H.L	4.317			37BD
Diphenylacetic acid ($C_{14}H_{12}O_2$), HL	HL/H.L	3.939			34DW
2-Phenylpropanoic acid* ($C_9H_{10}O_2$), HL	HL/H.L	4.35^n			33La Other reference: 72SSb
3-Phenylpropanoic acid ($C_9H_{10}O_2$), HL	HL/H.L	4.662±0.002 4.48^h 4.42^i 4.45^j			31LAb,37DL,61KPb
4-Phenylbutanoic acid ($C_{10}H_{12}O_2$), HL	HL/H.L	4.757			37DL

$$R-C=C-R-CO_2H$$

Ligand	Equilibrium	Log K 25°, 0	ΔH 25°, 0	ΔS 25°, 0	Bibliography
Propenoic acid (acrylic acid) ($C_3H_4O_2$), HL	HL/H.L	4.258±0.003	0.45	20.9	37DL,37GJ,53KMa, 69GW
2-Methylpropenoic acid (α-methylacrylic acid) ($C_4H_6O_2$), HL	HL/H.L	4.65^n 4.45^h 4.43^j			32La
cis-3-Phenylpropenoic acid (cis-cinnamic acid) ($C_9H_8O_2$), HL	HL/H.L	3.879			37DL
trans-3-Phenylpropenoic acid (trans-cinnamic acid) ($C_9H_8O_2$), HL	HL/H.L	4.438 4.27^a			37DL,53H

[a] 25°, 0.1; [h] 18°, 0.1; [i] 18°, 0.5; [j] 18°, 1.0; [n] 18°, 0; * metal constants were also reported but are not included in the compilation of selected constants.

Carboxylic acids (continued)

Ligand	Equilibrium	Log K 25°, 0	ΔH 25°, 0	ΔS 25°, 0	Bibliography
cis-But-2-enoic acid (isocrotonic acid) ($C_4H_6O_2$), HL	HL/H.L	4.41^n 4.20^h 4.15^i 4.17^j			32La
trans-2-Methylbut-2-enoic acid (angelic acid) ($C_5H_8O_2$), HL	HL/H.L	4.29^n 4.09^h 4.04^i 4.06^j			32La
trans-Pent-2-enoic acid ($C_5H_8O_2$), HL	HL/H.L	4.695			33IL
Pent-3-enoic acid ($C_5H_8O_2$), HL	HL/H.L	4.507			33IL
Pent-4-enoic acid (allylacetic acid) ($C_5H_8O_2$), HL	HL/H.L	4.677±0.002			33IL,37DL Other reference: 32La
cis-3-Methylpent-2-enoic acid ($C_6H_{10}O_2$), HL	HL/H.L	5.149			33IL
trans-3-Methylpent-2-enoic acid ($C_6H_{10}O_2$), HL	HL/H.L	5.131			33IL
4-Methylpent-2-enoic acid ($C_6H_{10}O_2$), HL	HL/H.L	4.701			33IL
4-Methylpent-3-enoic acid ($C_6H_{10}O_2$), HL	HL/H.L	4.600			33IL
5-Phenylpent-2,4-dienoic acid (cinnamylideneacetic acid) ($C_{11}H_{10}O_2$), HL	HL/H.L	4.426			37DL
trans-Hex-2-enoic acid ($C_6H_{10}O_2$), HL	HL/H.L	4.703			33IL
Hex-3-enoic acid ($C_6H_{10}O_2$), HL	HL/H.L	4.516			33IL
Hex-4-enoic acid ($C_6H_{10}O_2$), HL	HL/H.L	4.719			33IL
5-Methylhex-4-enoic acid ($C_7H_{12}O_2$), HL	HL/H.L	4.799			33IL

$$R-C{\equiv}C-CO_2H$$

Ligand	Equilibrium	Log K 25°, 0	ΔH 25°, 0	ΔS 25°, 0	Bibliography
Propynoic acid ($C_3H_2O_2$), HL	HL/H.L	1.89	2.49	17.1	69GW
But-2-ynoic acid (tetrolic acid) ($C_4H_4O_2$), HL	HL/H.L	2.653	0.12	12.4	37GJ,69GW Other reference: 32La

Ligand	Equilibrium	Log K 25°, 0	ΔH 25°, 0	ΔS 25°, 0	Bibliography
2-Methylbenzoic acid ($C_8H_8O_2$), HL	HL/H.L	3.903±0.005	1.45±0.05	22.8	37DLa,39EW,59ZPL

---

^{h} 18°, 0.1; ^{i} 18°, 0.5; ^{j} 18°, 1.0; ^{n} 18°, 0

Carboxylic acids (continued)

Ligand	Equilibrium	Log K 25°, 0	ΔH 25°, 0	ΔS 25°, 0	Bibliography
3-Methylbenzoic acid ($C_8H_8O_2$), HL	HL/H.L	4.273±0.001	-0.07±0.00	19.3	36DL,39EW,51BB, 59ZPL,67WG,74MK
4-Methylbenzoic acid ($C_8H_8O_2$), HL	HL/H.L	4.370±0.003	-0.27±0.03	19.1	36DL,39EW,51BB, 59ZAP,67WG
2-Ethylbenzoic acid ($C_9H_{10}O_2$), HL	HL/H.L	3.79			54DH
4-Ethylbenzoic acid ($C_9H_{10}O_2$), HL	HL/H.L	4.353			37BD
2-(1-Methylethyl)benzoic acid ($C_{10}H_{12}O_2$), HL	HL/H.L	3.64			54DH
4-(1-Methylethyl)benzoic acid ($C_{10}H_{12}O_2$), HL	HL/H.L	4.355			37BD
2-(2-Methyl-2-propyl)benzoic acid ($C_{11}H_{14}O_2$), HL	HL/H.L	3.54			54DH
3-(2-Methyl-2-propyl)benzoic acid ($C_{11}H_{14}O_2$), HL	HL/H.L	4.204	-2.47	10.9	70WB
4-(2-Methyl-2-propyl)benzoic acid ($C_{11}H_{14}O_2$), HL	HL/H.L	4.382	-2.90	10.4	70WB
2-Phenylbenzoic acid ($C_{13}H_{10}O_2$), HL	HL/H.L	3.560			37DLa
2,3-Dimethylbenzoic acid ($C_9H_{10}O_2$), HL	HL/H.L	3.716	2.01	23.7	67WG
2,4-Dimethylbenzoic acid ($C_9H_{10}O_2$), HL	HL/H.L	4.219	2.32	27.1	67WG
2,5-Dimethylbenzoic acid ($C_9H_{10}O_2$), HL	HL/H.L	4.001	2.72	27.4	67WG
2,6-Dimethylbenzoic acid ($C_9H_{10}O_2$), HL	HL/H.L	3.354	4.28	29.7	67WG
3,4-Dimethylbenzoic acid ($C_9H_{10}O_2$), HL	HL/H.L	4.41			54DH
3,5-Dimethylbenzoic acid ($C_9H_{10}O_2$), HL	HL/H.L	4.298	0.23	20.7	67WG
2,4,6-Trimethylbenzoic acid ($C_{10}H_{12}O_2$), HL	HL/H.L	3.446	4.37	30.4	67WG
2,3,5,6-Tetramethylbenzoic acid ($C_{11}H_{14}O_2$), HL	HL/H.L	3.416	3.50	27.3	67WG
4-Allylbenzoic acid ($C_{10}H_{10}O_2$), HL	HL/H.L	4.34			64BA
2,6-Dimethyl-4-(2-methyl-2-propyl)benzoic acid ($C_{12}H_{18}O_2$), HL	HL/H.L	3.44			54DH

Ligand	Equilibrium	Log K 25°, 0	ΔH 25°, 0	ΔS 25°, 0	Bibliography
1-Naphthoic acid ($C_{11}H_8O_2$), HL	HL/H.L	3.70			54DH
2-Naphthoic acid ($C_{11}H_8O_2$), HL	HL/H.L	4.16			54DH

Carboxylic acids (continued) $X-R-CO_2H$

Ligand	Equilibrium	Log K 25°, 0	ΔH 25°, 0	ΔS 25°, 0	Bibliography
Fluoroacetic acid ($C_2H_3O_2F$), HL	HL/H.L	2.586	1.39	16.5	55IP, Other references: 66A, 68Me, 74CH
2-Chloropropanoic acid ($C_3H_5O_2Cl$), HL	HL/H.L	2.90 2.71[a] 2.66[b] 2.68[c]	1.5	18	33L, 67CIH Other references: 66A, 74Ja
2,2-Dichloropropanoic acid ($C_3H_4O_2Cl_2$), HL	HL/H.L	2.06	0.4	11	66A, 68Me
2,3-Dichloropropanoic acid ($C_3H_4O_2Cl_2$), HL	HL/H.L	2.85	0.2	14	66A, 68Me
2-Bromopropanoic acid ($C_3H_5O_2Br$), HL	HL/H.L	2.99 −0.01 2.81[a] −0.01 2.75[b] 2.77[c]	1.3 ±0.0	18	33L, 66A, 67CIH, 69SMM
3-Bromopropanoic acid ($C_3H_5O_2Br$), HL	HL/H.L	3.99 −0.01 3.79[a] 3.75[b] 3.78[c]	0.17	18.9	33L, 68CO, 69SMM Other references: 66A, 67CIH
2,3-Dibromopropanoic acid ($C_3H_4O_2Br_2$), HL	HL/H.L	2.33	(1.9)	(17)	66A, 68Me
2-Iodopropanoic acid ($C_3H_5O_2I$), HL	HL/H.L	3.11[n] 2.92[h] 2.86[i] 2.89[j]			33L
2-Chlorobutanoic acid ($C_4H_7O_2Cl$), HL	HL/H.L	2.92			68Me
3-Chlorobutanoic acid ($C_4H_7O_2Cl$), HL	HL/H.L	4.17			68Me
3-Bromobutanoic acid ($C_4H_7O_2Br$), HL	HL/H.L	4.01			68Me
2-Chloro-2-methylpropanoic acid ($C_4H_7O_2Cl$), HL	HL/H.L	2.98[n] 2.78[h] 2.72[i] 2.73[j]			33L
2-Chlorophenylacetic acid ($C_8H_7O_2Cl$), HL	HL/H.L	4.066			34DWa
3-Chlorophenylacetic acid ($C_8H_7O_2Cl$), HL	HL/H.L	4.140			34DWa
4-Chlorophenylacetic acid ($C_8H_7O_2Cl$), HL	HL/H.L	4.190			34DW
2-Bromophenylacetic acid ($C_8H_7O_2Br$), HL	HL/H.L	4.054			34DWa
4-Bromophenylacetic acid ($C_8H_7O_2Br$), HL	HL/H.L	4.188			34DW
2-Iodophenylacetic acid ($C_8H_7O_2I$), HL	HL/H.L	4.038			36DL

[a] 25°, 0.1; [b] 25°, 0.5; [c] 25°, 1.0; [h] 18°, 0.1; [i] 18°, 0.5; [j] 18°, 1.0; [n] 18°, 0

Carboxylic acids (continued)

Ligand	Equilibrium	Log K 25°, 0	ΔH 25°, 0	ΔS 25°, 0	Bibliography
3-Iodophenylacetic acid ($C_8H_7O_2I$), HL	HL/H.L	4.159			36DL
4-Iodophenylacetic acid ($C_8H_7O_2I$), HL	HL/H.L	4.178			34DW

$$X-R-C=C-R-CO_2H$$

Ligand	Equilibrium	Log K 25°, 0	ΔH 25°, 0	ΔS 25°, 0	Bibliography
trans-2-Bromo-3-phenylpropenoic acid ($C_9H_7O_2Br$), HL	HL/H.L	2.01			39EW
trans-3-(2-Chlorophenyl)propenoic acid ($C_9H_7O_2Cl$), HL	HL/H.L	4.234			37DLa
trans-3-(4-Chlorophenyl)propenoic acid ($C_9H_7O_2Cl$), HL	HL/H.L	4.413			37DLa

$$X-C\equiv C-CO_2H$$

Ligand	Equilibrium	Log K 25°, 0	ΔH 25°, 0	ΔS 25°, 0	Bibliography
Chloropropynoic acid (C_3HO_2Cl), HL	HL/H.L	1.85	1.70	14.1	69GW
Bromopropynoic acid (C_3HO_2Br), HL	HL/H.L	1.86	1.72	14.4	69GW

Ligand	Equilibrium	Log K 25°, 0	ΔH 25°, 0	ΔS 25°, 0	Bibliography
2-Fluorobenzoic acid ($C_7H_5O_2F$), HL	HL/H.L	3.267			36DL
3-Fluorobenzoic acid ($C_7H_5O_2F$), HL	HL/H.L	3.865	-0.22	17.0	36DL,74MK
4-Fluorobenzoic acid ($C_7H_5O_2F$), HL	HL/H.L	4.14			35DW
2-Chlorobenzoic acid ($C_7H_5O_2Cl$), HL	HL/H.L	2.922±0.02	2.47	21.7	34SM,35DW,39EW
3-Chlorobenzoic acid ($C_7H_5O_2Cl$), HL	HL/H.L	3.825±0.002	-0.20±0.03	16.9	34SM,51BB,67WG, 72BF
4-Chlorobenzoic acid ($C_7H_5O_2Cl$), HL	HL/H.L	3.986±0.001	-0.17±0.07	17.7	34SM,51BB,67WG, 72BF
2-Bromobenzoic acid ($C_7H_5O_2Br$), HL	HL/H.L	2.85			35DW
3-Bromobenzoic acid ($C_7H_5O_2Br$), HL	HL/H.L	3.811±0.002	-0.18±0.02	16.9	36DL,51BB,67WG, 72BF
4-Bromobenzoic acid ($C_7H_5O_2Br$), HL	HL/H.L	4.002±0.000	-0.15±0.04	17.8	51BB,67WG,72BF
2-Iodobenzoic acid ($C_7H_5O_2I$), HL	HL/H.L	2.863±0.001	3.25	24.0	36DL,39EW
3-Iodobenzoic acid ($C_7H_5O_2I$), HL	HL/H.L	3.854±0.003	-0.19±0.00	17.0	36DL,39EW,51BB
4-Iodobenzoic acid ($C_7H_5O_2I$), HL	HL/H.L	3.996	-0.08	18.0	72BF

$$Z-R-CO_2H$$

Ligand	Equilibrium	Log K 25°, 0	ΔH 25°, 0	ΔS 25°, 0	Bibliography
2-Methyl-2-nitropropanoic acid ($C_4H_7O_4N$), HL	HL/H.L	1.93^n 1.70^h	$(4)^s$	(22)	34P

^{h} 18°, 0.1; ^{n} 18°, 0; ^{s} 10-18°, 0

Carboxylic acids (continued)

Ligand	Equilibrium	Log K 25°, 0	ΔH 25°, 0	ΔS 25°, 0	Bibliography
2-Nitrophenylacetic acid ($C_8H_7O_4N$), HL	HL/H.L	4.004			37DLa
3-Nitrophenylacetic acid ($C_8H_7O_4N$), HL	HL/H.L	3.967			34DWa
4-Nitrophenylacetic acid ($C_8H_7O_4N$), HL	HL/H.L	3.851			34DW
2,4-Dinitrophenylacetic acid ($C_8H_6O_6N_2$), HL	HL/H.L	3.502			34DWa
2-Cyano-2-methylpropanoic acid ($C_5H_7O_2N$), HL	HL/H.L	2.422	1.87	17.4	70IM
2-Cyano-3-methylbutanoic acid ($C_6H_9O_2N$), HL	HL/H.L	2.401	2.18	18.3	70IM
2-Cyano-3-methyl-2-(1-methyl-ethyl)butanoic acid ($C_9H_{15}O_2N$), HL	HL/H.L	2.556	3.40	23.1	65IM
Sulfoacetic acid ($C_2H_4O_5S$), HL	HL/H.L	3.71[h]			75VB

Ligand	Equilibrium	Log K 25°, 0	ΔH 25°, 0	ΔS 25°, 0	Bibliography
3,5-Dinitrobenzoic acid ($C_7H_4O_6N_2$), HL	HL/H.L	2.68[a]			53H
3-Cyanobenzoic acid ($C_8H_5O_2N$), HL	HL/H.L	3.596+0.001	−0.18	15.8	51BB,67WG
4-Cyanobenzoic acid ($C_8H_5O_2N$), HL	HL/H.L	3.550+0.001	−0.04+0.01	16.1	51BB,67WG
2-Carboxy-6-(methoxyformyl)-phenol 1-dihydrogenphosphate ($C_9H_9O_8P$), H_3L	HL/H.L H$_2$L/HL.H	7.63[v] 3.38[v]			64MM

HO-R-CO$_2$H

Ligand	Equilibrium	Log K 25°, 0	ΔH 25°, 0	ΔS 25°, 0	Bibliography
1-Hydroxycycloheptanecarboxylic acid ($C_8H_{14}O_3$), HL	HL/H.L	4.210 4.03[a] 3.96[b] 3.98[c] 4.17[d]			68PF
sym-Dimethylcitrate ($C_8H_{12}O_7$), HL	HL/H.L	3.21 3.02[a,y]			75PC
3-Hydroxy-2-methylbutanoic acid ($C_5H_{10}O_3$), HL	HL/H.L	4.65[n]			33La
3-Hydroxy-2,2-dimethylpropanoic acid ($C_5H_{10}O_3$), HL	HL/H.L	4.87[n]			33La
3-Phenyl-3-hydroxypropanoic acid ($C_9H_{10}O_3$), HL	HL/H.L	4.40[n]			33La
cis-2-Hydroxycyclohexane-carboxylic acid ($C_7H_{12}O_3$), HL	HL/H.L	4.796[a] 4.60[b] 4.53			53KM

[a] 25°, 0.1; [b] 25°, 0.5; [c] 25°, 1.0; [d] 25°, 2.0; [h] 20°, 0.1; [n] 18°, 0; [v] 35°, 0.1; [y] corrected for background electrolyte.

Carboxylic acids (continued)

Ligand	Equilibrium	Log K 25°, 0	ΔH 25°, 0	ΔS 25°, 0	Bibliography
trans-2-Hydroxycyclohexane-carboxylic acid ($C_7H_{12}O_3$), HL	HL/H.L	4.692 4.49[a]			53KM
Tris(hydroxymethyl)acetic acid ($C_5H_{10}O_5$), HL	HL/H.L	4.460	-1.54	15.3	71GSB
as-Dimethylcitrate ($C_8H_{12}O_7$), HL	HL/H.L	3.99 3.78[a,y]			75PC
4-Hydroxypentanoic acid ($C_5H_{10}O_3$), HL	HL/H.L	4.69[n] 4.49[h] 4.46[i] 4.49[j]			31LAb
4-Hydroxy-4-methylpentanoic acid ($C_6H_{12}O_3$), HL	HL/H.L	4.69[n]			33La
cis-3-Hydroxycyclohexane-carboxylic acid ($C_7H_{12}O_3$), HL	HL/H.L	4.602 4.41[a]			53KM
trans-3-Hydroxycyclohexane-carboxylic acid ($C_7H_{12}O_3$), HL	HL/H.L	4.815 4.53[a]			53KM
cis-4-Hydroxycyclohexane-carboxylic acid ($C_7H_{12}O_3$), HL	HL/H.L	4.836 4.64[a]			53KM
trans-4-Hydroxycyclohexane-carboxylic acid ($C_7H_{12}O_3$), HL	HL/H.L	4.678 4.48[a]			53KM
6-Hydroxy-3,7-dimethyloctanoic acid ($C_{10}H_{20}O_3$), HL	HL/H.L	5.08[n]			33La

$$\underset{R-C-R'-CO_2H}{\overset{\overset{\textstyle O}{\|}}{}}$$

Ligand	Equilibrium	Log K 25°, 0	ΔH 25°, 0	ΔS 25°, 0	Bibliography
2-Oxobutanoic acid ($C_4H_6O_3$), HL	HL/H.L	2.50	-2.82 -2.93[t]	2.0	67OW
Oxaloacetic acid ethyl ester ($C_6H_8O_5$), HL	HL/H.L	2.74			58GH
3-Oxobutanoic acid (acetylacetic acid ($C_4H_6O_3$), HL	HL/H.L	3.65[n]			33La
4-Oxopentanoic acid (levulinic acid)* ($C_5H_8O_3$), HL	HL/H.L	4.64[n] 4.44[h] 4.38[i] 4.41[j]			31LAb Other reference: 70GP
5-Oxohexanoic acid (4-acetyl-butanoic acid) ($C_6H_{10}O_3$), HL	HL/H.L	4.66[n]			33La

$$\underset{R-O-C-R'-CO_2H}{\overset{\overset{\textstyle O}{\|}}{}}$$

Ligand	Equilibrium	Log K 25°, 0	ΔH 25°, 0	ΔS 25°, 0	Bibliography
cis-Butenedioic acid ethyl ester (ethylmaleate) ($C_6H_8O_4$), HL	HL/H.L	3.077			60DL
trans-Butenedioic acid ethyl ester (ethylfumarate) ($C_6H_8O_4$), HL	HL/H.L	3.396			60DL

---

[a] 25°, 0.1;   [h] 18°, 0.1;   [i] 18°, 0.5;   [j] 18°, 1.0;   [n] 18°, 0;   [t] 25°, 0.05;

[y] corrected for background electrolyte;   * metal constants were also reported but are not included in the compilation of selected constants.

Carboxylic acids (continued)

$$\underset{R-CNH-R'-CO_2H}{\overset{\displaystyle O}{\overset{\displaystyle \|}{}}}$$

Ligand	Equilibrium	Log K 25°, 0	ΔH 25°, 0	ΔS 25°, 0	Bibliography
N-Acetylalanine ($C_5H_9O_3N$), HL	HL/H.L	3.715	0.63	19.1	56KK
N-Acetyl-2-aminobutanoic acid ($C_6H_{11}O_3N$), HL	HL/H.L	3.716	0.77	19.6	56KK
N-Acetyltyrosine ($C_{11}H_{13}O_4N$), H_2L	HL/H.L	9.88[r]			63KN
N-Acetylcysteine ($C_5H_9O_3NS$), H_2L	HL/H.L	9.38[b]			65FCW, Other reference: 75IM
N-Acetylpenicillamine ($C_7H_{13}O_3NS$), H_2L	HL/H.L	9.86[b]			65FCW
		10.33[h]			65D
	$H_2L/HL.H$	3.33[h]			
N-Acetyl-2-amino-3-mercapto-3-methylpentanoic acid ($C_8H_{15}O_3NS$), H_2L	HL/H.L	10.26[b]			65FCW
2-(Acetylamino)-3-(2-aminoethyl-thio)propanoic acid ($C_7H_{13}O_3NS$), HL	HL/H.L	9.39[h]			68HL
N-Acetyl-3-aminopropanoic acid ($C_5H_9O_3N$), HL	HL/H.L	4.445	-0.26	19.5	56KK
N-Propanoylglycine ($C_5H_9O_3N$), HL	HL/H.L	3.718	0.14	17.5	56KK
N-Carbamoylglycine ($C_3H_6O_3N_2$), HL	HL/H.L	3.876	-0.29	16.8	56Ka
N-Carbamoylalanine ($C_4H_8O_3N_2$), HL	HL/H.L	3.892	0.23	18.6	56Ka
N-Carbamoyl-2-aminobutanoic acid ($C_5H_{10}O_3N_2$), HL	HL/H.L	3.886	0.49	19.4	56Ka
N-Carbamoyl-2-amino-2-methyl-propanoic acid ($C_5H_{10}O_3N_2$), HL	HL/H.L	4.463	-0.22	19.7	56Ka
N-Carbamoyl-3-aminopropanoic acid ($C_4H_8O_3N_2$), HL	HL/H.L	4.487	-0.19	19.9	56Ka
N-Carbamoyl-4-aminobutanoic acid ($C_5H_{10}O_3N_2$), HL	HL/H.L	4.683	0.12	21.8	56Ka

$$R-O-R'-CO_2H$$

Ligand	Equilibrium	Log K 25°, 0	ΔH 25°, 0	ΔS 25°, 0	Bibliography
2-Methylphenoxyacetic acid ($C_9H_{10}O_3$), HL	HL/H.L	3.227			43HB
3-Methylphenoxyacetic acid ($C_9H_{10}O_3$), HL	HL/H.L	3.203			43HB
4-Methylphenoxyacetic acid ($C_9H_{10}O_3$), HL	HL/H.L	3.215			43HB
2-Fluorophenoxyacetic acid ($C_8H_7O_3F$), HL	HL/H.L	3.085			43HB
3-Fluorophenoxyacetic acid ($C_8H_7O_3F$), HL	HL/H.L	3.082			43HB
4-Fluorophenoxyacetic acid ($C_8H_7O_3F$), HL	HL/H.L	3.130			43HB
2-Chlorophenoxyacetic acid ($C_8H_7O_3Cl$), HL	HL/H.L	3.051			43HB

[b] 30°, 0.3;   [h] 20°, 0.1;   [r] 25°, 0.16

Carboxylic acids (continued)

Ligand	Equilibrium	Log K 25°, 0	Bibliography
3-Chlorophenoxyacetic acid $(C_8H_7O_3Cl)$, HL	HL/H.L	3.070	43HB
4-Chlorophenoxyacetic acid $(C_8H_7O_3Cl)$, HL	HL/H.L	3.103	43HB
2-Bromophenoxyacetic acid $(C_8H_7O_3Br)$, HL	HL/H.L	3.123	43HB
3-Bromophenoxyacetic acid $(C_8H_7O_3Br)$, HL	HL/H.L	3.095	43HB
4-Bromophenoxyacetic acid $(C_8H_7O_3Br)$, HL	HL/H.L	3.133	43HB
2-Iodophenoxyacetic acid $(C_8H_7O_3I)$, HL	HL/H.L	3.173	43HB
3-Iodophenoxyacetic acid $(C_8H_7O_3I)$, HL	HL/H.L	3.128	43HB
4-Iodophenoxyacetic acid $(C_8H_7O_3I)$, HL	HL/H.L	3.159	43HB
2-Nitrophenoxyacetic acid $(C_8H_7O_5N)$, HL	HL/H.L	2.896	43HB
3-Nitrophenoxyacetic acid $(C_8H_7O_5N)$, HL	HL/H.L	2.951	43HB
4-Nitrophenoxyacetic acid $(C_8H_7O_5N)$, HL	HL/H.L	2.891	43HB
2,6-Dimethylphenoxyacetic acid $(C_{10}H_{12}O_3)$, HL	HL/H.L	3.356	43HB
4-Chloro-3-nitrophenoxyacetic acid $(C_8H_6O_5NCl)$, HL	HL/H.L	2.959	43HB
2-Cyanophenoxyacetic acid $(C_9H_7O_3N)$, HL	HL/H.L	2.975	43HB
3-Cyanophenoxyacetic acid $(C_9H_7O_3N)$, HL	HL/H.L	3.034	43HB
4-Cyanophenoxyacetic acid $(C_9H_7O_3N)$, HL	HL/H.L	2.932	43HB
2-Methoxyphenoxyacetic acid $(C_9H_{10}O_4)$, HL	HL/H.L	3.231	43HB
3-Methoxyphenoxyacetic acid $(C_9H_{10}O_4)$, HL	HL/H.L	3.142	43HB
4-Methoxyphenoxyacetic acid $(C_9H_{10}O_4)$, HL	HL/H.L	3.213	43HB
4-Methoxyphenylacetic acid $(C_9H_{10}O_3)$, HL	HL/H.L	4.362	34DWa
3,4-Dimethoxyphenylacetic acid $(C_{10}H_{12}O_4)$, HL	HL/H.L	4.334	34DWa

Carboxylic acids (continued)

$$R-O-C_6H_4-CO_2H$$

Ligand	Equilibrium	Log K 25°, 0	ΔH 25°, 0	ΔS 25°, 0	Bibliography
3-Methoxybenzoic acid ($C_8H_8O_3$), HL	HL/H.L	4.090±0.003	-0.05±0.01	18.6	36DL,59ZPL,72BF, 74MK
4-Methoxybenzoic acid (p-anisic acid ($C_8H_8O_3$), HL	HL/H.L	4.475±0.004	-0.57±0.3	18.6	34DWa,39EW,59ZPL, 72BF
2-Phenoxybenzoic acid ($C_{13}H_{10}O_3$), HL	HL/H.L	3.527			37DLa
3-Phenoxybenzoic acid ($C_{13}H_{10}O_3$), HL	HL/H.L	3.951			37DLa
4-Phenoxybenzoic acid ($C_{13}H_{10}O_3$), HL	HL/H.L	4.523			37DLa
3-Methoxy-2-naphthoic acid ($C_{12}H_{10}O_3$), HL	HL/H.L	3.82			57BDH

$$HS-R-CO_2H$$

Ligand	Equilibrium	Log K 25°, 0	ΔH 25°, 0	ΔS 25°, 0	Bibliography
2-Mercapto-2-methylpropanoic acid ($C_4H_8O_2S$), H_2L	H_2L/HL.H	3.90			28L
DL-3-Mercapto-4-phenylbutanoic acid ($C_{10}H_{12}O_2S$), H_2L	H_2L/HL.H	4.46[n]			39A

$$R-S-R'-CO_2H$$

Ligand	Equilibrium	Log K 25°, 0	ΔH 25°, 0	ΔS 25°, 0	Bibliography
(4-Methylthiophenylthio)acetic acid ($C_9H_{10}O_2S_2$), HL	HL/H.L	3.41[h]			68PRS
(2-Phenylethylthio)acetic acid ($C_{10}H_{12}O_2S$), HL	HL/H.L	3.80[n]			39A
Thiodiacetic acid monoethyl ester ($C_6H_{10}O_4S$), HL	HL/H.L	3.66			66S
2'-Methylthiodiacetic acid 1-monoethylester ($C_7H_{12}O_4S$), HL	HL/H.L	3.71			66S
2-Methylthiodiacetic acid 1-monoethylester ($C_7H_{12}O_4S$), HL	HL/H.L	3.77			66S

$$R-Se-R'-CO_2H$$

Ligand	Equilibrium	Log K 25°, 0	ΔH 25°, 0	ΔS 25°, 0	Bibliography
(4-Methylthiophenylseleno)acetic acid ($C_9H_{10}O_2SSe$), HL	HL/H.L	3.72[h]			68PRS

$$R_3Si-R'-CO_2H$$

Ligand	Equilibrium	Log K 25°, 0	ΔH 25°, 0	ΔS 25°, 0	Bibliography
3-(Trimethylsilyl)benzoic acid ($C_{10}H_{14}O_2Si$), HL	HL/H.L	4.089	-2.32	10.9	70WB
4-(Trimethylsilyl)benzoic acid ($C_{10}H_{14}O_2Si$), HL	HL/H.L	4.198	-3.06	8.9	70WB

$$\overset{NH}{\underset{CH_3}{H_2N\overset{\|}{C}\underset{|}{N}CH_2CO_2H}}$$

Ligand	Equilibrium	Log K 25°, 0	ΔH 25°, 0	ΔS 25°, 0	Bibliography
N-Amidino-N-methylaminoacetic acid (creatine)($C_4H_9O_2N_3$), HL	HL/H.L	2.631	-4	-10	63DGa, Other reference: 54S

[h] 20°, 0.1;   [n] 18°, 0

Carboxylic acids (continued

$$HO_2CCCO_2H$$ with R above and R' below

Ligand	Equilibrium	Log K 25°, 0	Bibliography
Propylpropanedioic acid (propylmalonic acid)[*] $(C_6H_{10}O_4)$, H_2L	HL/H.L H_2L/HL.H	5.85 3.00	31GI Other references: 30R,31IR
Ethylmethylmalonic acid $(C_6H_{10}O_4)$, H_2L	HL/H.L H_2L/HL.H	6.41 2.86	31GI
Ethylpropylmalonic acid $(C_8H_{14}O_4)$, H_2L	HL/H.L H_2L/HL.H	7.43 2.15	31GI
Ethyl-2-propylmalonic acid $(C_8H_{14}O_4)$, H_2L	HL/H.L H_2L/HL.H	7.99[a] 1.92[a]	65ME
Ethylbutylmalonic acid $(C_9H_{16}O_4)$, H_2L	HL/H.L H_2L/HL.H	7.14[a] 2.04[a]	65ME
Ethyl-3-methylbutylmalonic acid $(C_{10}H_{18}O_4)$, H_2L	HL/H.L H_2L/HL.H	7.20[a] 2.04[a]	65ME
Ethylphenylmalonic acid $(C_{11}H_{12}O_4)$, H_2L	HL/H.L H_2L/HL.H	7.01[a] 1.8[a]	65ME
Di-2-propylmalonic acid $(C_9H_{16}O_4)$, H_2L	HL/H.L H_2L/HL.H	8.49[a] 2.07[a]	65ME
Diheptylmalonic acid $(C_{17}H_{32}O_4)$, H_2L	HL/H.L	7.34[a]	65ME
Cyclopentane-1,1-diacetic acid $(C_9H_{14}O_4)$, H_2L	HL/H.L H_2L/HL.H	6.59 3.78	31GI
Cyclohexane-1,1-diacetic acid $(C_{10}H_{16}O_4)$, H_2L	HL/H.L H_2L/HL.H	7.48 3.00	31GI
Cycloheptane-1,1-diacetic acid $(C_{11}H_{18}O_4)$, H_2L	HL/H.L H_2L/HL.H	6.56 3.96	31GI

$$HO_2C-R-CO_2H$$

Ligand	Equilibrium	Log K 25°, 0	Bibliography
Tetramethylbutanedioic acid (tetramethylsuccinic acid) $(C_8H_{14}O_4)$, H_2L	HL/H.L H_2L/HL.H	7.28 3.50	31GI
3-Methylpentanedioic acid (3-methylglutaric acid) $(C_6H_{10}O_4)$, H_2L	HL/H.L H_2L/HL.H	6.23 4.25	31GI
3-Ethylglutaric acid $(C_7H_{12}O_4)$, H_2L	HL/H.L H_2L/HL.H	6.33 4.29	31GI
3-Propylglutaric acid $(C_8H_{14}O_4)$, H_2L	HL/H.L H_2L/HL.H	6.39 4.31	31GI
3,3-Dimethylglutaric acid $(C_7H_{12}O_4)$, H_2L	HL/H.L H_2L/HL.H	6.29 3.70	31GI
3-Ethyl-3-methylglutaric acid $(C_8H_{14}O_4)$, H_2L	HL/H.L H_2L/HL.H	6.70 3.62	31GI
trans-Pentenedioic acid (trans-glutaconic acid) $(C_5H_6O_4)$, H_2L	HL/H.L H_2L/HL.H	5.077 3.769±0.002	37GJ

[a] 25°, 0.1;   [*] metal constants were reported but are not included in the compilation of selected constants

Carboxylic acids (continued)

Ligand	Equilibrium	Log K 25°, 0	ΔH 25°, 0	ΔS 25°, 0	Bibliography
Dodecaborane-10-1,12-dicarboxylic acid ($C_2H_{12}O_4B_{12}$), H_2L	HL/H.L $H_2L/HL.H$	10.24 9.07	-2.30 -2.15	39.1 34.4	66HP

| 2,6-Dicarboxyphenol 1-dihydrogen phosphate ($C_8H_7O_8P$), H_4L | HL/H.L
$H_2L/HL.H$
$H_3L/H_2L.H$ | 8.35[v]
4.82[v]
2.98[v] | | | 64MM |

| as-Monomethylcitrate ($C_7H_{10}O_7$), H_2L | HL/H.L

$H_2L/HL.H$ | 5.05
4.66[a,y]
3.0
2.7 [a,y] | | | 75PC |
| sym-Monomethylcitrate ($C_7H_{10}O_7$), H_2L | HL/H.L

$H_2L/HL.H$ | 4.94
4.63[a,y]
3.63
3.39[a,y] | | | 75PC |

| 2-Acetoxybenzene-1,3-dicarboxylic acid (2-acetoxyisophthalic acid) ($C_{10}H_8O_6$), H_2L | HL/H.L
$H_2L/HL.H$ | 3.80[a]
2.85[a] | | | 64MM |

$$HO_2C-R-O-R'-CO_2H$$

DL-2-Methyloxydiacetic acid ($C_5H_8O_5$), H_2L	HL/H.L $H_2L/HL.H$	4.58 3.03			66S
2,2-Dimethyloxydiacetic acid ($C_6H_{10}O_5$), H_2L	HL/H.L $H_2L/HL.H$	4.87 3.44			66S
DL-2-Phenyloxydiacetic acid ($C_{10}H_{10}O_5$), H_2L	HL/H.L $H_2L/HL.H$	4.40 2.87			66S

$$HO_2C-R-S-R'-CO_2H$$

DL-2-Methylthiodiacetic acid ($C_5H_8O_4S$), H_2L	HL/H.L $H_2L/HL.H$	4.46 3.38			66S
2,2-Dimethylthiodiacetic acid ($C_6H_{10}O_4S$), H_2L	HL/H.L $H_2L/HL.H$	4.30 3.74			66S
2,2,2',2'-Tetramethylthiodiacetic acid ($C_8H_{14}O_4S$), H_2L	HL/H.L $H_2L/HL.H$	5.19 3.83			66S

[a] 25°, 0.1;   [y] corrected for background electrolyte;   [v] 35°, 0.1

Carboxylic acids (continued)

Ligand	Equilibrium	Log K 25°, 0	Bibliography
DL-2-Phenylthiodiacetic acid ($C_{10}H_{10}O_4S$), H_2L	HL/H.L	4.65	66S
	H_2L/HL.H	2.90	
4,4'-Thiodibutanoic acid ($C_8H_{14}O_4S$), H_2L	HL/H.L	5.26^n	39A
		4.89^h	
		4.79^i	
		4.80^j	
		4.94^k	
		5.12^l	
	H_2L/HL.H	4.35^n	
		4.20^h	
		4.13^i	
		4.17^j	
		4.31^k	
		4.50^l	

$$HO_2CCH_2S-R-SCH_2CO_2H$$

Ligand	Equilibrium	Log K 25°, 0	Bibliography
(tetramethylenedithio)diacetic acid ($C_8H_{14}O_4S_2$), H_2L	HL/H.L	4.42^n	39A, Other reference: 44L
		4.06^h	
		3.95^i	
		3.94^j	
		4.06^k	
		4.21^l	
	H_2L/HL.H	3.46^n	
		3.30^h	
		3.22^i	
		3.25^j	
		3.38^k	
		3.57^l	
(Pentamethylenedithio)diacetic acid ($C_9H_{16}O_4S_2$), H_2L	HL/H.L	4.41^n	39A
		4.05^h	
		3.93^i	
		3.93^j	
		4.05^k	
		4.22^l	
	H_2L/HL.H	3.49^n	
		3.30^h	
		3.25^i	
		3.28^j	
		3.42^k	
		3.60^l	
Benzylidenebis(thioacetic acid) ($C_{11}H_{12}O_4S_2$), H_2L	HL/H.L	3.98^r	61JW
	H_2L/HL.H	3.09^r	

$$HO_2C-R-SO_2-R-CO_2H$$

Ligand	Equilibrium	Log K 25°, 0	Bibliography
Sulfonyldiacetic acid ($C_4H_6O_6S$), H_2L	HL/H.L	2.70^d	72AN
	H_2L/HL.H	1.82^d	

$$HO_2C-R-Te-R-CO_2H$$

Ligand	Equilibrium	Log K 25°, 0	Bibliography
Tellurodiacetic acid ($C_4H_6O_4Te$), H_2L	HL/H.L	4.77^a	75LP
	H_2L/HL.H	3.26^a	

---

[a] 25°, 0.1;   [d] 25°, 2.0;   [h] 18°, 0.1;   [i] 18°, 0.5;   [j] 18°, 1.0;   [k] 18°, 2.0;   [l] 18°, 3.0;   [n] 18°, 0;   [r] 20°, 0.2

Carboxylic acids (continued)

Ligand	Equilibrium	Log K 25°, 0	ΔH 25°, 0	ΔS 25°, 0	Bibliography
As-(3-Trifluoromethylphenyl)-arsinodiacetic acid $(C_{11}H_{10}O_4AsF_3)$, H_2L	$HL/H.L$	4.85^h			69PR
	$H_2L/HL.H$	3.47^h			
As-(4-Carboxyphenyl)arsino-diacetic acid $(C_{11}H_{11}O_6As)$, H_3L	$HL/H.L$	5.16^h			69PR
	$H_2L/HL.H$	4.03^h			
	$H_3L/H_2L.H$	3.35^h			
As-(3-Methylthiophenyl)-arsinodiacetic acid $(C_{11}H_{13}O_4AsS)$, H_2L	$HL/H.L$	5.02^a			71FP
	$H_2L/HL.H$	3.71^a			

Ligand	Equilibrium	Log K 25°, 0	ΔH 25°, 0	ΔS 25°, 0	Bibliography
Benzene-1,3,5-tricarboxylic acid $(C_9H_6O_6)$, H_3L	$HL/H.L$	5.18 −0.1 4.47^h	1.17	27.7	41A,72PTR
	$H_2L/HL.H$	4.10 −0.2 3.70^h	0.49	20.4	
	$H_3L/H_2L.H$	3.12 +0.06 3.02^h	−0.87	11.4	
(Carboxymethoxy)butanedioic acid $(C_6H_8O_7)$, H_3L	$HL/H.L$	5.00^a			74RD
	$H_2L/HL.H$	3.77^a			
	$H_3L/H_2L.H$	2.52^a			

Ligand	Equilibrium	Log K 25°, 0	ΔH 25°, 0	ΔS 25°, 0	Bibliography
Butane-1,2,3,4-tetracarboxylic acid $(C_8H_{10}O_8)$, H_4L	$HL/H.L$	7.16	−1.2	37	72PTR
	$H_2L/HL.H$	5.85	−0.5	28	
	$H_3L/H_2L.H$	4.58	−0.3	25	
	$H_4L/H_3L.H$	3.43	−0.2	15	
Benzene-1,2,4,5-tetracarboxylic acid (pyromellitic acid) $(C_{10}H_6O_8)^*$, H_4L	$HL/H.L$	6.23	1.60	33.9	72PTR
	$H_2L/HL.H$	4.92	0.79	25.2	Other reference:
	$H_3L/H_2L.H$	3.12	1.57	19.6	72NS
	$H_4L/H_3L.H$	1.70	3.11	18.2	
1,2-Phenylenebis(arsenodiacetic acid) $(C_{14}H_{16}O_8As_2)$, H_4L	$HL/H.L$	6.18^a			71FP
	$H_2L/HL.H$	5.07^a			
	$H_3L/H_2L.H$	4.50^a			
	$H_4L/H_3L.H$	3.94^a			

---

^{a} 25°, 0.1; ^{h} 18°, 0.1;   * metal constants were also reported but are not included in the compilation of selected constants.

Carboxylic acids (continued)

Ligand	Equilibrium	Log K 25°, 0	ΔH 25°, 0	ΔS 25°, 0	Bibliography
Benzenehexacarboxylic acid	HL/H.L	7.49	-0.27	35.5	62PTR
(mellitic acid)	$H_2L/HL.H$	6.32	-0.05	28.7	
$(C_{12}H_6O_{12})$, H_6L	$H_3L/H_2L.H$	5.09	1.13	27.1	
	$H_4L/H_3L.H$	3.52	2.75	25.3	
	$H_5L/H_4L.H$	2.21	3.6	22	
	$H_6L/H_5L.H$	0.7	7	(27)	

B. Phosphorus acids

Ligand	Equilibrium	Log K		Bibliography
Benzenephosphonic acid[*]	HL/H.L	7.45 ±0.02		69MK,740
$(C_6H_7O_3P)$, H_2L		6.88[a]		Other reference: 69M
	$H_2L/HL.H$	1.83		
3,4-Dimethylbenzenephosphonic	HL/H.L	7.69		740
acid $(C_8H_{11}O_3P)$, H_2L	$H_2L/HL.H$	2.03		
3-Nitrobenzenephosphonic acid	HL/H.L	6.53		740
$(C_6H_6O_5NP)$, H_2L	$H_2L/HL.H$	1.14		
4-Nitrobenzenephosphonic acid	HL/H.L	6.50		740
$(C_6H_6O_5NP)$, H_2L	$H_2L/HL.H$	1.06		
3-Hydroxybenzenephosphonic	HL/H.L	10.03		740
acid $(C_6H_7O_4P)$, H_3L	$H_2L/HL.H$	7.24		
	$H_3L/H_2L.H$	1.78		
4-Hydroxybenzenephosphonic	HL/H.L	10.56		740
acid $(C_6H_7O_4P)$, H_3L	$H_2L/HL.H$	7.69		
	$H_3L/H_2L.H$	2.00		
2-Methoxybenzenephosphonic	HL/H.L	7.90		740
acid $(C_7H_9O_4P)$, H_2L	$H_2L/HL.H$	2.17		
4-Methoxybenzenephosphonic	HL/H.L	7.65		740
acid $(C_7H_9O_4P)$, H_2L	$H_2L/HL.H$	2.10		
4-Methylbenzenephosphonic	HL/H.L	7.65		740
acid $(C_7H_9O_3P)$, H_2L	$H_2L/HL.H$	1.96		

---

[a] 25°, 0.1; [*] metal constants were also reported but are not included in the compilation of selected constants.

Phosphorus acids (continued) $H_2O_3P-R-PO_3H_2$

Ligand	Equilibrium	Log K 25°, 0	Bibliography
Hexamethylenediphosphonic acid ($C_6H_{12}O_6P_2$), H_4L	HL/H.L	8.56 8.34[a,x] 8.27[c,x]	62IM Other references: 67KL,68DM
	H_2L/HL.H	7.73 7.65[a,x] 7.59[c,x]	
	H_3L/H_2L.H	3.12 3.07[a,x] 3.00[c,x]	
	H_4L/H_3L.H	1.8 1.8[a,x] 1.9[c,x]	
Decamethylenediphosphonic acid ($C_{10}H_{24}O_6P_2$), H_4L	HL/H.L	(8.94)[a,x] (8.83)[a,x] (8.73)[c,x]	62IM
	H_2L/HL.H	7.93 7.74[a,x] 7.68[c,x]	
	H_3L/H_2L.H	3.27 3.15[a,x] 3.06[c,x]	
	H_4L/H_3L.H	2.1[a,x] 2.1[c,x] 2.0	
Bromomethanediphosphonic acid ($CH_5O_6BrP_2$), H_2L	HL/H.L H_2L/HL.H H_3L/H_2L.H	10.00[t] 6.40[t] 2.10[t]	67GQ
Dichloromethanediphosphonic acid ($CH_4O_6Cl_2P_2$), H_4L	HL/H.L	9.78 9.50[a,x]	67GQ
	H_2L/HL.H	6.11 5.89[a,x]	
Hydroxymethanediphosphonic acid ($CH_6O_7P_2$), H_4L	HL/H.L	10.56 10.26[a,x]	67GQ
	H_2L/HL.H	7.05 6.78[a,x]	
	H_3L/H_2L.H	2.74 2.60[a,x]	
Oxomethanediphosphonic acid ($CH_4O_7P_2$), H_4L	HL/H.L	8.42 8.16[a,x]	67GQ
	H_2L/HL.H	5.50 5.38[a,x]	

$R-OPO_3H_2$

Ligand	Equilibrium	Log K 25°, 0	Bibliography
Phosphoric acid phenyl ester ($C_6H_7O_4P$), H_2L	HL/H.L	6.28 5.76[a]±0.02	55CF,69MK
	H_2L/HL.H	0.9[a]	
Phosphoric acid 1-naphthyl ester ($C_{10}H_9O_4P$), H_2L	HL/H.L	6.37 5.74[a]±0.00 5.43[c]	55CF,71MMa
	H_2L/HL.H	0.8[a]	

[a] 25°, 0.1; [c] 25°, 1.0; [t] 25°, 0.03; [x] $(CH_3)_4NBr$ used as background electrolyte.

Phorphorus acids (continued)

Ligand	Equilibrium	Log K 25°, 0	ΔH 25°, 0	ΔS 25°, 0	Bibliography
Phosphoric acid 2-naphthyl ester $(C_{10}H_9O_4P)$, H_2L	HL/H.L	6.19			55CF,71MMa
		$5.69^a \pm 0.04$			
		5.34^c_a			
	$H_2L/HL.H$	1.2^a			
Phosphoric acid 4-nitrophenyl ester $(C_6H_6O_6NP)$, H_2L	HL/H.L		2.6		55S
Ribose-5-dihydrogenphosphate $(C_5H_{11}O_8P)$, H_2L	$L/H_{-1}L.H$	13.05	-6.1	39	62IC,66IR
	HL/H.L	6.70	2.7	40	
Glucose-6-dihydrogenphosphate* $(C_6H_{13}O_9P)$, H_2L	$L/H_{-1}L.H$	11.71	-8.4	25	66IR,Other reference: 74AG

$$(R-O)_2PO_2H$$

Ligand	Equilibrium	Log K 25°, 0	ΔH 25°, 0	ΔS 25°, 0	Bibliography
Phosphoric acid dibutyl ester (dibutylphosphate)* $(C_8H_{19}O_4P)$, HL	HL/H.L	1.00^a			57D,Other references: 61SK,61SS, 62SK,63A,65SSa, 68KH

C. Arsonic acids

Ligand	Equilibrium	Log K 25°, 0	ΔH 25°, 0	ΔS 25°, 0	Bibliography
Benzenearsonic acid* $(C_6H_7O_3As)$, H_2L	HL/H.L	8.75 ±0.02			68NM,6800,7300 Other reference: 69M
		8.25^a			
	$H_2L/HL.H$	3.61 ±0.04			
		3.39^a			
4-Hydroxybenzenearsonic acid $(C_6H_7O_4As)$, H_3L	HL/H.L	10.24 ±0.1			6800,7300 Other reference: 73VG
	$H_2L/HL.H$	8.65 ±0.03			
	$H_3L/H_2L.H$	3.85 ±0.00			
2,4-Dihydroxy-5-phenylazo- benzenearsonic acid $(C_{12}H_{11}O_5N_2As)$, H_4L	HL/H.L	13.99			74Ob
	$H_2L/HL.H$	8.99			
	$H_3L/H_2L.H$	6.56			
2-Methoxybenzenearsonic acid $(C_7H_9O_4As)$, H_2L	HL/H.L	9.40			7300
	$H_2L/HL.H$	4.08			
4-Methoxybenzenearsonic acid $(C_7H_9O_4As)$, H_2L	HL/H.L	8.93			7300
	$H_2L/HL.H$	3.79			
2,4-Dimethoxybenzenearsonic acid $(C_8H_{11}O_5As)$, H_2L	HL/H.L	9.55			7300
	$H_2L/HL.H$	4.35			

---

[a] 25°, 0.1; [c] 25°, 1.0; * metal constants were also reported but are not included in the compilation of selected constants.

D. Phenols

Ligand	Equilibrium	Log K 25°, 0	ΔH 25°, 0	ΔS 25°, 0	Bibliography
2-Methylphenol (o-cresol) (C_7H_7O), HL	HL/H.L	10.28 ±0.01	-5.7	28	56B,57HK,62CL Other reference: 75GJ
2-Propylphenol ($C_9H_{12}O$), HL	HL/H.L	10.50	-5.8	28	63OH
3,5-Dimethylphenol (3,5-xylenol)($C_8H_{10}O$), HL	HL/H.L	9.87[c]	-7.5[c]	20[c]	73A
2,6-Dimethylphenol (2,6-xylenol)($C_8H_{10}O$), HL	HL/H.L	10.58[c]	-5.8[c]	29[c]	73A
2,3,5-Trimethylphenol ($C_9H_{12}O$), HL	HL/H.L	10.67	-5.8	29	64KO
2,4,5-Trimethylphenol ($C_9H_{12}O$), HL	HL/H.L	10.57	-6.2	27	64KO
2,4,6-Trimethylphenol ($C_9H_{12}O$), HL	HL/H.L	10.89	-5.3	32	64KO
3,4,5-Trimethylphenol ($C_9H_{12}O$), HL	HL/H.L	10.25	-5.5	28	64KO
2-Allylphenol ($C_9H_{10}O$), HL	HL/H.L	10.27	-5.8	27	63OH
4-Allylphenol (chavicol) ($C_9H_{10}O$), HL	HL/H.L	10.23			64BA
trans-4-Propenylphenol (anol) ($C_9H_{10}O$), HL	HL/H.L	9.824			62LN,64BA
2-Hydroxyazobenzene ($C_{12}H_{10}ON_2$), HL	HL/H.L	9.38			74Ob
4-Hydroxyazobenzene ($C_{12}H_{10}ON_2$), HL	HL/H.L	8.36			74Ob
2-Hydroxy-5-methylazobenzene ($C_{13}H_{12}ON_2$), HL	HL/H.L	9.68			74Ob
4-Hydroxy-3-methylazobenzene ($C_{13}H_{12}ON_2$), HL	HL/H.L	8.46			74Ob

Ligand	Equilibrium	Log K 25°, 0	ΔH 25°, 0	ΔS 25°, 0	Bibliography
3-Trifluoromethylphenol ($C_7H_5OF_3$), HL	HL/H.L	8.95	-5.2	23	72LSH
4-Trifluoromethylphenol ($C_7H_5OF_3$), HL	HL/H.L	8.68	-5.0	23	72LSH
3-Chloro-2-methylphenol (C_7H_7OCl), HL	HL/H.L	9.19[c]	-5.6[c]	23[c]	73A
4-Chloro-3-methylphenol (C_7H_7OCl), HL	HL/H.L	9.27[c]	-5.5[c]	24[c]	73A

---

[c] 25°, 1.0

Phenols (continued)

Ligand	Equilibrium	Log K 25°, 0	ΔH 25°, 0	ΔS 25°, 0	Bibliography
2,3-Dichlorophenol ($C_6H_4OCl_2$), HL	HL/H.L	7.696			64R
2,4-Dichlorophenol ($C_6H_4OCl_2$), HL	HL/H.L	7.892			64R
2,5-Dichlorophenol ($C_6H_4OCl_2$), HL	HL/H.L	7.508			64R
2,6-Dichlorophenol ($C_6H_4OCl_2$), HL	HL/H.L	6.791			64R
3,4-Dichlorophenol ($C_6H_4OCl_2$), HL	HL/H.L	8.585			64R
3,5-Dichlorophenol ($C_6H_4OCl_2$), HL	HL/H.L	8.185			64R
3-Methyl-4-nitrophenol ($C_7H_7O_3N$), HL	HL/H.L	7.16[c]	-4.6[c]	17[c]	73A
2,6-Dinitrophenol ($C_6H_4O_5N_2$), HL	HL/H.L	3.704			71KK,74KK
2,4,6-Trinitrophenol (picric acid) ($C_6H_3O_7N_3$), HL	HL/H.L	0.33 ±0.0			54BSa,63DP Other references: 71YI,72I,73I
4-Hydroxybenzoic acid methyl ester ($C_8H_8O_3$), HL	HL/H.L	8.47			45SS
4-Hydroxybenzoic acid ethyl ester ($C_9H_{10}O_3$), HL	HL/H.L	8.50			45SS
4-Hydroxybenzoic acid butyl ester ($C_{11}H_{14}O_3$), HL	HL/H.L	8.47			45SS
4-Hydroxybenzoic acid benzyl ester ($C_{14}H_{12}O_3$), HL	HL/H.L	8.41			45SS
2-Methoxyphenol ($C_7H_8O_2$), HL	HL/H.L	9.98	-5.6	27	56B,64MA
3-Methoxyphenol ($C_7H_8O_2$), HL	HL/H.L	9.65	-5.1	27	56B,64MA
4-Methoxyphenol ($C_7H_8O_2$), HL	HL/H.L	10.21	-5.5	28	61BR,64MA
2-Methoxy-4-methylphenol (creosol)($C_8H_{10}O_2$), HL	HL/H.L	10.28			62LN
3-Allyl-2-methoxyphenol ($C_{10}H_{11}O_2$), HL	HL/H.L	9.92			64BA
4-Allyl-2-methoxyphenol (eugenol) ($C_{10}H_{11}O_2$), HL	HL/H.L	10.17 ±0.02			62LN,64BA
5-Allyl-2-methoxyphenol (chavibetol)($C_{10}H_{11}O_2$), HL	HL/H.L	10.02			64BA
6-Allyl-2-methoxyphenol (o-eugenol) ($C_{10}H_{11}O_2$), HL	HL/H.L	10.38			64BA
trans-2-Methoxy-4-propenyl-phenol (isoeugenol) ($C_{10}H_{11}O_2$), HL	HL/H.L	9.89 -0.01			62LN,64BA

---

[a] 25°, 0.1;   [b] 25°, 0.5;   [c] 25°, 1.0;   [*] metal constants were also reported but are not included in the compilation of selected constants.

Phenols (continued)

Ligand	Equilibrium	Log K 25°, 0	ΔH 25°, 0	ΔS 25°, 0	Bibliography
2-Methoxy-5-propenylphenol (isochavibetol) ($C_{10}H_{11}O_2$), HL	HL/H.L	9.90			64BA
2-Methoxy-6-propenylphenol (o-isoeugenol) ($C_{10}H_{11}O_2$), HL	HL/H.L	10.20			64BA
5-Methoxy-2-nitrosophenol ($C_7H_7O_3N$), HL	HL/H.L	6.59 6.38[a]			68Ma
2-Hydroxybenzaldehyde oxime (salicylaldoxime)* ($C_7H_7O_2N$), H_2L	HL/H.L	12.11 11.69[a] 11.46[b] 11.39[c]			57L, Other references: 58L, 67Mb, 68MD, 69BM
	H_2L/HL.H	9.18 8.95[a] 8.84[b] 8.85[c]			
	H_3L/H_2L.H	1.37			
3-Hydroxyphenyltrimethyl-ammonium chloride ($C_9H_{14}ON^+$), HL^+	HL/H.L	8.06	-5.9	17	64KO
4-Hydroxyphenyltrimethyl-ammonium chloride ($C_9H_{14}ON^+$), HL^+	HL/H.L	8.35	-5.3	20	64KO

Ligand	Equilibrium	Log K 25°, 0	ΔH 25°, 0	ΔS 25°, 0	Bibliography
3-Hydroxybenzoic acid ($C_7H_6O_3$), H_2L	HL/H.L	9.96 ±0.04			29L, 39EW, 39MK, 48CD, 51BB, 57BDH Other references: 31LAb, 73SN
	H_2L/HL.H	4.080±0.003	-0.17+0.01	18.1	
4-Hydroxybenzoic acid* ($C_7H_6O_3$), H_2L	HL/H.L	9.46			44JS, 45SS, 48CD, 51BB, 72BF Other references: 52B, 55VG, 71RBS, 73SN
	H_2L/HL.H	4.58 ±0.01	-0.37±0.02	19.7	
6-Chloro-2-hydroxybenzoic acid ($C_7H_5O_3Cl$), H_2L	H_2L/HL.H	2.63			57BDH
2-Hydroxy-4-nitrobenzoic acid ($C_7H_5O_5N$), H_2L	H_2L/HL.H	2.23			57BDH
2-Hydroxy-6-nitrobenzoic acid ($C_7H_5O_5N$), H_2L	H_2L/HL.H	2.24			57BDH
3-Bromo-2-hydroxy-5-sulfo-benzoic acid ($C_7H_6O_6BrS$), H_3L	H_2L/HL.H	2.03[e]			73L
5-Amino-2-hydroxybenzoic acid*[3] ($C_7H_7O_3N$), H_2L	HL/H.L	13.74[e]			55Aa, Other references: 60DA, 61GK, 66NV, 66RS

[a] 25°, 0.1;   [b] 25°, 0.5;   [c] 25°, 1.0;   [e] 25°, 3.0;   * metal constants were also reported but are not included in the compilation of selected constants.

Phenols (continued)

Ligand	Equilibrium	Log K 25°, 0	Bibliography
2,6-Dicarboxyphenol (2-hydroxyisophthalic acid) ($C_8H_6O_5$), H_3L	$H_2L/HL.H$ $H_3L/H_2L.H$	4.53^P 2.13^P	64MM

Ligand	Equilibrium	Log K 25°, 0	Bibliography
2,4-Dihydroxyazobenzene ($C_{12}H_{10}O_2N_2$), H_2L	$HL/H.L$ $H_2L/HL.H$	12.50 6.65	74Ob
4-Bromo-1,2-dihydroxybenzene* ($C_6H_5O_2Br$), H_2L	$HL/H.L$ $H_2L/HL.H$	11.32^a 8.37^a	74NV
3-Nitro-1,2-dihydroxybenzene* ($C_6H_5O_4N$), H_2L	$HL/H.L$ $H_2L/HL.H$	11.83^a 6.49^a	70NLb, Other reference: 72HK
3,5-Dinitro-1,2-dihydroxy-benzene* ($C_6H_4O_6N_2$), H_2L	$HL/H.L$ $H_2L/HL.H$	10.0^a 3.39^a	70NLa, Other references: 72HK, 75PS
1,3-Dihydroxy-2,4,6-trinitro-benzene ($C_6H_3O_8N_3$), H_2L	$HL/H.L$ $H_2L/HL.H$	4.23 0.06	64WM
trans-3-(3,4-Dihydroxyphenyl)-propenoic acid (caffeic acid)* ($C_9H_8O_4$), H_3L	$H_2L/HL.H$ $H_3L/H_2L.H$	9.07 8.76^t 4.62 4.49^t	59T
trans-Chlorogenic acid* ($C_{16}H_{18}O_9$), H_3L	$H_2L/HL.H$ $H_3L/H_2L.H$	8.59 8.35^t 3.59 3.50^t	59T

Ligand	Equilibrium	Log K 25°, 0	Bibliography
3,3'-Dimethyl-4'-hydroxy-fuchsone-2''-sulfonic acid (Cresol Red)($C_{21}H_{18}O_5S$), H_3L	$HL/H.L$ $H_2L/HL.H$	7.95^a 1.00^a	67MY, Other reference: 75NF

Ligand	Equilibrium	Log K 25°, 0	Bibliography
1-Naphthol ($C_{10}H_8O$), HL	$HL/H.L$	9.34 9.14^a 9.06^b 9.05^c	66M

---

[a] 25°, 0.1; [b] 25°, 0.5; [c] 25°, 1.0; [P] 30°, 0.1; [t] 25°, 0.05; * metal constants were also reported but are not included in the compilation of selected constants.

Phenols (continued)

Ligand	Equilibrium	Log K 25°, 0	Bibliography
1-Naphthol-2-sulfonic acid* $(C_{10}H_8O_4S)$, HL	HL/H.L	9.59 9.11^a 8.78^b 8.64^c	72MPM Other references: 64M,69BM
1-Naphthol-3-sulfonic acid $(C_{10}H_8O_4S)$, HL	HL/H.L	8.80 8.43^a 8.27^b 8.23^c	72MPM
1-Naphthol-4-sulfonic acid $(C_{10}H_8O_4S)$, HL	HL/H.L	8.44 8.04^a 7.87^b 7.85^c	72MPM Other reference: 62Mb
1-Naphthol-5-sulfonic acid $(C_{10}H_8O_4S)$, HL	HL/H.L	9.19 8.81^a 8.68^b 8.68^c	64M,72MPM
1-Naphthol-8-sulfonic acid $(C_{10}H_8O_4S)$, HL	HL/H.L	13.12 12.65^a 12.33^b 12.19^c	75PM
1-Naphthol-3,6-disulfonic acid $(C_{10}H_8O_7S_2)$, HL	HL/H.L	8.65 8.06^a 7.81^b 7.77^c	72MPM Other reference: 64M
1-Naphthol-3,8-disulfonic acid $(C_{10}H_8O_7S_2)$, HL	HL/H.L	12.19 11.49^a 11.05^b 10.88^c	75PM
2-Nitro-1-naphthol $(C_{10}H_7O_3N)$, HL	HL/H.L	5.89 5.71^a 5.66^b 5.66^c	66M
2-Naphthol $(C_{10}H_8O)$, HL	HL/H.L	9.51 9.31^a 9.24^b 9.25^c	66M
2-Naphthol-6-sulfonic acid $(C_{10}H_8O_4S)$, HL	HL/H.L	9.20 8.86^a 8.77^b 8.79^c	72MPA Other reference: 62Mb
2-Naphthol-7-sulfonic acid $(C_{10}H_8O_4S)$, HL	HL/H.L	9.37 9.00^a 8.90^b 8.93^c	72MPA
2-Naphthol-3,6-disulfonic acid* $(C_{10}H_8O_7S_2)$, HL	HL/H.L	9.79 9.16^a 8.86^b 8.75^c	72MPA Other references: 62B,62Mb,68BD

^{a} 25°, 0.1; ^{b} 25°, 0.5; ^{c} 25°, 1.0; * metal constants were also reported but are not included in the compilation of selected constants.

Phenols (continued)

Ligand	Equilibrium	Log K 25°, 0	ΔH 25°, 0	ΔS 25°, 0	Bibliography
2-Naphthol-6,8-disulfonic acid ($C_{10}H_8O_7S_2$), HL	HL/H.L	9.15 8.59[a] 8.37[b] 8.33[c]			62MPA Other reference: 64M
1-Nitro-2-naphthol ($C_{10}H_7O_3N$), HL	HL/H.L	5.93 5.75[a] 5.68[b] 5.66[c]			66M
2-Hydroxy-1-(1-hydroxy-2-naphthylazo)naphthalene-4-sulfonic acid (Eriochrome Black B)($C_{20}H_{14}O_5N_2S$), H_3L	HL/H.L H$_2$L/HL.H	12.81 6.49			68CR Other reference: 48SB,73MP
2-Hydroxy-1-(2-hydroxy-1-naphthylazo)naphthalene-4-sulfonic acid (Eriochrome Black R)($C_{20}H_{14}O_5N_2S$), H_3L	HL/H.L H$_2$L/HL.H	13.80 7.31			68CR Other references: 48SB,57HR,68MN, 71AA
2-Hydroxy-1-(2-hydroxy-1-naphthylazo)-5-nitro-naphthalene-4-sulfonic acid (Eriochrome Black A) ($C_{20}H_{13}O_7N_3S$), H_3L	HL/H.L H$_2$L/HL.H	13.10 6.22			68CR Other reference: 48SB
1-Hydroxy-4-sulfo-2-naphthoic acid ($C_{11}H_8O_6S$), H_3L	HL/H.L H$_2$L/HL.H	12.51 11.87[a] 2.97 2.57[a]			72MA
1,8-Dihydroxynaphthalene* ($C_{10}H_8O_2$), H_2L	H$_2$L/HL.H	6.36			68BN

Ligand	Equilibrium	Log K 25°, 0	ΔH 25°, 0	ΔS 25°, 0	Bibliography
2-(2-Nitrophenylazo)chromotropic acid* ($C_{16}H_{11}O_{10}N_3S_2$), H_4L	H$_2$L/HL.H	8.85			67TMK
2-(3-Nitrophenylazo)chromotropic acid* ($C_{16}H_{11}O_{10}N_3S_2$), H_4L	H$_2$L/HL.H	8.49			67TMK
2-(4,5,6-Trihydroxy-3-oxo-3-H-9-xanthenyl)-benzene-1-sulfonic acid (Pyrogallol Red) ($C_{19}H_{12}O_8S$), H_4L	H$_3$L/H$_2$L.H	6.5 [h]			70AM

E. Carbonyl ligands

$$R-\overset{O}{\overset{\|}{C}}CH_2\overset{O}{\overset{\|}{C}}-R'$$

Ligand	Equilibrium	Log K 25°, 0	ΔH 25°, 0	ΔS 25°, 0	Bibliography
Hexane-2,4-dione ($C_6H_{10}O_2$), HL	HL/H.L	9.38	(-5)[s]	(25)	65CM,68CC
3-Methylpentane-2,4-dione ($C_6H_{10}O_2$), HL	HL/H.L	10.85	(-5)[t]	(30)	61LR

[a] 25°, 0.1; [b] 25°, 0.5; [c] 25°, 1.0; [h] 20°, 0.1; [s] 5-45°, 0; [t] 10-38°, 0; * metal constants were also reported but are not included in the compilation of selected constants.

Carbonyl ligands (continued)

Ligand	Equilibrium	Log K 25°, 0	ΔH 25°, 0	ΔS 25°, 0	Bibliography
Heptane-2,4-dione ($C_7H_{12}O_2$), HL	HL/H.L	9.29 ±0.06	(-4)[s]	(30)	65CM,68CC,75VSK
Heptane-3,5-dione ($C_7H_{12}O_2$), HL	HL/H.L	10.04			75VSK
3-Methylhexane-2,4-dione ($C_7H_{12}O_2$), HL	HL/H.L	11.20			75VSK
5-Methylhexane-2,4-dione ($C_7H_{12}O_2$), HL	HL/H.L	9.42 ±0.02	(-3)[s]	(35)	61LR,65CM,68CC, 75VSK
3-Ethylpentane-2,4-dione ($C_7H_{12}O_2$), HL	HL/H.L	11.34			61LR
6-Methylheptane-2,4-dione ($C_8H_{14}O_2$), HL	HL/H.L	9.20	(-2)[s]	(35)	65CM,68CC
5,5-Dimethylhexane-2,4-dione (pivaloylacetone) ($C_8H_{14}O_2$), HL	HL/H.L	10.00	(-4)[t]	(30)	61LR
5,5-Dimethylcyclohexane-1,3-dione (dimedone) ($C_8H_{12}O_2$), HL	HL/H.L	5.23[u]			52V,Other reference: 75DS
6,6-Dimethylheptane-2,4-dione ($C_9H_{16}O_2$), HL	HL/H.L	9.26			65CM
2-Methylnonane-4,6-dione ($C_{10}H_{18}O_2$), HL	HL/H.L	9.55			65CM
2,7-Dimethyloctane-3,5-dione ($C_{10}H_{18}O_2$), HL	HL/H.L	9.66			65CM
2,6,6-Trimethylheptane-3,5-dione ($C_{10}H_{18}O_2$), HL	HL/H.L	10.72			65CM
2,8-Dimethylnonane-4,6-dione ($C_{11}H_{20}O_2$), HL	HL/H.L	9.48			65CM
1-(2-Naphthyl)-4,4,4-trifluoro-butane-1,3-dione (2-naphthyoyltrifluoro-acetone)* ($C_{14}H_9O_2F_3$), HL	HL/H.L	6.28[a]			67Sa, Other reference: 69KS

Ligand	Equilibrium	Log K 25°, 0	ΔH 25°, 0	ΔS 25°, 0	Bibliography
Trichlorohydroxy-1,4-benzo-quinone ($C_6HO_3Cl_3$), HL	HL/H.L	1.05			66BB
Tetrahydroxyquinone ($C_6H_4O_6$), H_4L	HL/H.L H_2L/HL.H	6.65[b] 4.55[b]			72MF
2-Hydroxy-1,4-naphthoquinone* (lawsone) ($C_{10}H_6O_3$), HL	HL/H.L	4.00			66BB,Other references: 59ZP,66SP

---

[a] 25°, 0.1;   [b] 25°, 0.3;   [s] 5-45°, 0;   [t] 10-38°, 0;   [u] 30°, 0;   * metal constants were also reported but are not included in the compilation of selected constants.

Carbonyl ligands (continued)

Ligand	Equilibrium	Log K 25°, 0	ΔH 25°, 0	ΔS 25°, 0	Bibliography
2-Carboxycyclohepta-2,4,6-trien-1-one (2-carboxy-tropone) ($C_8H_6O_3$), HL	HL/H.L	3.73			59TOK
3-Carboxy-1-hydroxycyclohepta-3,5,7-trien-2-one (3-carboxytropolone) ($C_8H_6O_4$), H_2L	HL/H.L H_2L/HL.H	7.98 3.20			59TOK
4-Carboxytropolone ($C_8H_6O_4$), H_2L	HL/H.L H_2L/HL.H	7.03 3.42			59TOK
5-Carboxytropolone ($C_8H_6O_4$), H_2L	HL/H.L H_2L/HL.H	6.41 4.05			59TOK
3-Carboxy-4-methyltropolone ($C_9H_8O_4$), H_2L	HL/H.L H_2L/HL.H	8.29 3.15			59TOK, Other references: 65DS, 66GDS, 66SD, 67GD, 67GDa, 71RG
3-Carboxymethyltropolone ($C_9H_8O_4$), H_2L	HL/H.L H_2L/HL.H	8.77 4.52			59TOK
3,4-Dicarboxytropolone ($C_9H_6O_6$), H_3L	HL/H.L H_2L/HL.H H_3L/H_2L.H	8.47 6.26 3.24			59TOK
3-Carboxy-4-carboxymethyl-tropolone ($C_{10}H_8O_6$), H_3L	HL/H.L H_2L/HL.H H_3L/H_2L.H	8.08 5.90 3.75			59TOK

Ligand	Equilibrium	Log K 25°, 0	ΔH 25°, 0	ΔS 25°, 0	Bibliography
5,6-Dihydroxy-5-cyclohexene-1,2,3,4-tetrone (rhodizonic acid ($C_6H_2O_6$), H_2L	HL/H.L H_2L/HL.H	3.58[e] 3.45[e]			72AV

F. Alcohols   R-OH

Ligand	Equilibrium	Log K 25°, 0	ΔH 25°, 0	ΔS 25°, 0	Bibliography
1,1,1,3,3,3-Hexafluoropropane-2-ol ($C_3H_2OF_6$), HL	HL/H.L	9.42	-6.4	22	71WH
Hexafluoropropane-2,2-diol ($C_3H_2O_2F_6$), HL	HL/H.L	6.65	-6.0	10	71WH

---

[e] 25°, 3.0

Alcohols (continued)

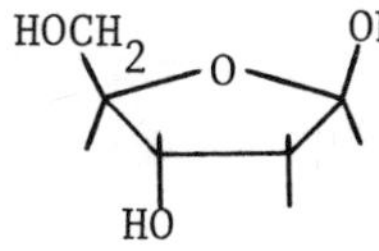

Ligand	Equilibrium	Log K 25°, 0	ΔH 25°, 0	ΔS 25°, 0	Bibliography
2-Deoxyribose ($C_5H_{10}O_4$), L	$L/H_{-1}L.H$	12.61	-8.2	30	66IR,70CRa

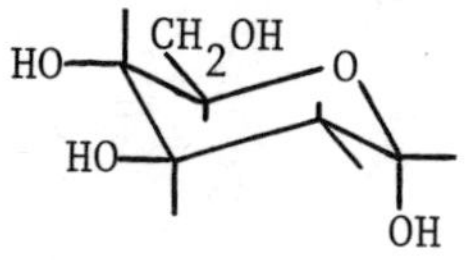

2-Deoxyglucose ($C_6H_{12}O_5$), L	$L/H_{-1}L.H$	12.52	-8.2	30	66IR,70CRa

G. Thiols   R-SH

Ethanethiol (C_2H_6S), HL	HL/H.L	10.61	-6.4	27	64IN
Propane-2-thiol (C_3H_8S), HL	HL/H.L	10.86	-5.4	32	64IN
2-Methylpropane-2-thiol ($C_4H_{10}S$), HL	HL/H.L	11.22	-5.3	34	64IN
N-Acetyl-2-aminoethanethiol (C_4H_9ONS), HL	HL/H.L	9.92	-6.3	24	64IN

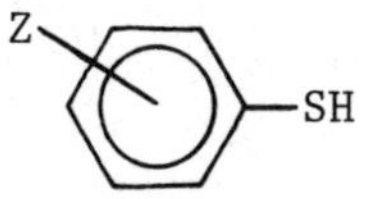

Benzenethiol (thiophenol) (C_6H_6S), HL	HL/H.L	6.615 6.46[a] 6.43[c]	-4.0	16.9	71JS,73DF,74LP
2-Methylbenzenethiol (C_7H_8S), HL	HL/H.L	6.995			74DF
3-Methylbenzenethiol (C_7H_8S), HL	HL/H.L	6.660			73DF
4-Methylbenzenethiol (C_7H_8S), HL	HL/H.L	6.820			73DF
2,6-Dimethylbenzenethiol ($C_8H_{10}S$), HL	HL/H.L	7.377			74DF
2-Chlorobenzenethiol (C_6H_5ClS), HL	HL/H.L	5.675			74DF
3-Chlorobenzenethiol (C_6H_5ClS), HL	HL/H.L	5.780			73DF
4-Chlorobenzenethiol (C_6H_5ClS), HL	HL/H.L	6.135			73DF
2-Bromobenzenethiol (C_6H_5BrS), HL	HL/H.L	5.659			74DF

[a] 25°, 0.1;   [c] 25°, 1.0

Thiols (continued)

Ligand	Equilibrium	Log K 25°, 0	Bibliography
4-Bromobenzenethiol (C_6H_5BrS), HL	HL/H.L	6.020	73DF
2-Nitrobenzenethiol ($C_6H_5O_2NS$), HL	HL/H.L	5.453	74DF
3-Nitrobenzenethiol ($C_6H_5O_2NS$), HL	HL/H.L	5.241	73DF
4-Nitrobenzenethiol ($C_6H_5O_2NS$), HL	HL/H.L	4.715 4.50[c]	71JS, 73DF
2-Methoxybenzenethiol (C_7H_8OS), HL	HL/H.L	6.890	74DF
3-Methoxybenzenethiol (C_7H_8OS), HL	HL/H.L	6.385	73DF
4-Methoxybenzenethiol (C_7H_8OS), HL	HL/H.L	6.775	73DF
2-Ethoxybenzenethiol ($C_8H_{10}OS$), HL	HL/H.L	7.063	74DF
2-Methylthiobenzenethiol ($C_7H_8S_2$), HL	HL/H.L	5.754	74DF
4-Acetylbenzenethiol (C_8H_8OS), HL	HL/H.L	5.330	73DF
Pentafluorobenzenethiol (C_6HF_5S), HL	HL/H.L	2.68[c]	71JS

H. Thioacids

$$\overset{\text{S}}{\overset{\|}{R\text{-}C}}\text{-OH}$$

Thioacetic acid (C_2H_4OS), HL	HL/H.L	3.20[c]	71JS

$$\overset{\text{S}}{\overset{\|}{R\text{-}C}}\text{-SH}$$

N,N-Diethyldithiocarbamic acid* ($C_5H_{11}NS_2$), HL	HL/H.L	3.42 3.31[a] 3.22[b] 3.16[c]	70AJ Other references: 56Z, 69BH, 69JA, 73SSC

I. Sulfonic acids

$$R\text{-}SO_3H$$

Methylsulfonic acid (CH_4O_3S), HL	HL/H.L	1.92	74CT
Ethylsulfonic acid ($C_2H_6O_3S$), HL	HL/H.L	1.68	74CT
Propylsulfonic acid ($C_3H_8O_3S$), HL	HL/H.L	1.53	74CT

[a] 25°, 0.1; [b] 25°, 0.5; [c] 25°, 1.0; * metal constants were also reported but are not included in the compilation of selected constants.

J. Hydroxamic acids

$$\overset{\displaystyle O}{\overset{\|}{R-CNHOH}}$$
$$\underset{\displaystyle Z}{|}$$

Ligand	Equilibrium	Log K 25°, 0	ΔH 25°, 0	ΔS 25°, 0	Bibliography
N-Phenylphenylacethydroxamic acid ($C_{14}H_{13}O_2N$), HL	HL/H.L	7.87[h]			66ZO

K. Oximes

$$\underset{\displaystyle R}{\overset{\displaystyle R-C=NOH}{|}}$$

Ligand	Equilibrium	Log K 25°, 0	ΔH 25°, 0	ΔS 25°, 0	Bibliography
Butane-2,3-dione monoxime (diacetylmonoxime) ($C_4H_7O_2N$), HL	HL/H.L	9.51 9.28[a]			68MS
1-Phenylpropane-1,2-dione 2-oxime ($C_9H_9O_2N$), HL	HL/H.L	9.31 9.10[a]			68MS
Hexane-3,4-dione dioxime* (diethylglyoxime) ($C_6H_{12}O_2N_2$), HL	HL/H.L	10.67[a]			63BA
3-Methylcyclohexane-1,2-dione dioxime* (3-methylnioxime) ($C_7H_{12}O_2N_2$), HL	HL/H.L	10.61[a]			63BA
4-Methylnioxime* ($C_7H_{12}O_2N_2$), HL	HL/H.L	10.54[a]			63BA
4-(1-Methylethyl)nioxime* ($C_9H_{16}O_2N_2$), HL	HL/H.L	10.53[a]			63BA
4-Carboxynioxime* ($C_7H_{10}O_4N_2$), H_2L	L/H$_{-1}$L.H HL/H.L H_2L/HL.H	12.3[a] 10.35[a] 4.75[a]			58BL, Other reference: 62BL

L. Nitrocompounds

$$R-NO_2$$

Ligand	Equilibrium	Log K 25°, 0	ΔH 25°, 0	ΔS 25°, 0	Bibliography
Nitromethane (CH_3O_2N), HL	HL/H.L	10.2	-5.9[t]	27	73MH
Nitroethane ($C_2H_5O_2N$), HL	HL/H.L	8.57	-2.4	31	73MH,74MH
1-Nitropropane ($C_3H_7O_2N$), HL	HL/H.L	9.0	-2.6[t]	32	73MH
2-Nitropropane ($C_3H_7O_2N$), HL	HL/H.L	7.7	-0.1[t]	35	73MH

M. Amides

$$\overset{\displaystyle O}{\overset{\|}{R-CNH-R'}}$$

Ligand	Equilibrium	Log K 25°, 0	ΔH 25°, 0	ΔS 25°, 0	Bibliography
N-Methylacetamide (C_3H_7ON), L	HL/H.L	0.4			72WH
N-Methylformamide (C_2H_5ON), L	HL/H.L	-0.1			72WH
2,6,8-Trihydroxypurine (uric acid) ($C_5H_4O_3N_4$), HL	HL/H.L	5.61	-5.2	8	74FS

---

[a] 25°, 0.1; [h] 20°, 0.1; [t] 25°, 0.06; * Metal constants were also reported but are not included in the compilation of selected constants.

Amides (continued)

$$\underset{R-NHC-CNH-R}{\overset{\overset{O}{\|}\ \overset{S}{\|}}{}}$$

Ligand	Equilibrium	Log K 25°, 0	Bibliography
N,N'-Bis(2-sulfoethyl)monothio-oxamide ($C_6H_{12}O_7N_2S_3$), H_2L	HL/H.L	11.97 11.32[a] 11.05[b]	68GH
N,N'-Bis(4-sulfobenzyl)mono-thiooxamide ($C_{16}H_{16}O_7N_2S_3$) H_2L	HL/H.L	11.67 11.06[a] 10.84[b]	68GH

$$\underset{R-NHC-CNH-R}{\overset{\overset{S}{\|}\ \overset{S}{\|}}{}}$$

N,N'-Bis(3-sulfobenzyl)dithio-oxamide ($C_{16}H_{16}O_6N_2S_4$), H_3L	HL/H.L	11.29 10.67[a] 10.50[b]	68HGa
	H_2L/HL.H	2.89[b]	
N,N'-Bis(4-sulfobenzyl)dithio-oxamide ($C_{16}H_{16}O_6N_2S_4$), H_3L	HL/H.L	11.27 10.68[a] 10.40[b]	68GHa
	H_2L/HL.H	2.90[b]	

N. Biguanides

$$\underset{R_2NCNHCNH_2}{\overset{\overset{HN}{\|}\ \ \overset{NH}{\|}}{}}$$

1,1-Dimethylbiguanide* ($C_4H_{11}N_5$), L	HL/H.L H_2L/HL.H	12.40[i] 3.24[i] 2.96[h]	65D, Other references: 56SR, 60Ra
1-Butylbiguanide ($C_6H_{13}N_5$), L	HL/H.L H_2L/HL.H	12.66[i] 3.57[i] 3.31[h]	65D, Other reference: 61DS
Anhydro-1,1-bis(2-hydroxy-ethyl)biguanide ($C_6H_{15}ON_6$), L	HL/H.L H_2L/HL.H	12.61[i] 2.70[i] 2.43[h]	65D, Other reference: 71KLa
1-(4-Chlorophenyl)-4-propyl-biguanide ($C_{11}H_{17}N_5Cl$), L	HL/H.L H_2L/HL.H	11.54[i] 11.43[h] 2.59[i] 2.33[h]	65D
1-(3,4-Dichlorophenyl)-4-propylbiguanide ($C_{11}H_{16}N_5Cl_2$), L	HL/H.L H_2L/HL.H	11.04[i] 10.96[h] 2.22[i] 1.97[h]	65D

[a] 25°, 0.1; [b] 25°, 0.5; [h] 20°, 0.1; [i] 20°, 0.5; * Metal constants were also reported but are not included in the compilation of selected constants.

XVIII. LIGANDS CONSIDERED BUT NOT INCLUDED IN THE COMPILATION OF SELECTED CONSTANTS

A. Carboxylic acids

Ligand	Bibliography
2-Sulfobenzoic acid ($C_7H_6O_5S$)	73DP
2-Formylbenzoic acid (salicylamide) ($C_7H_7O_2N$)	69KA
Hydroxyacetic acid dihydrogen phosphate (phosphonoacetic acid) ($C_2H_5O_5P$)	72EZb
4-Bromophenylhydroxyacetic acid ($C_8H_7O_3Br$)	70KKM
DL-2-Mercaptobutanoic acid ($C_4H_8O_2NS$)	64SM
2-Mercaptobenzoic acid ($C_7H_6O_2S$)	73KD
(Methoxyethylthio)acetic acid ($C_5H_{10}O_3S$)	71SSa,71SSb
(Hexamethylenedithio)diacetic acid ($C_{10}H_{18}O_4S_2$)	44L
3-(Benzylthio)propanoic acid ($C_{10}H_{12}O_2S$)	44L
2-(Diethylphosphinyl)propanoic acid ($C_7H_{15}O_3P$)	72EZ
1,2,2-Trimethylcyclopentane-1,3-dicarboxylic acid (camphoric acid) ($C_{10}H_{16}O_4$)	70RJ
erthro-2-Methyltartaric acid ($C_5H_8O_6$)	71HM
threo-2-Methyltartaric acid ($C_5H_8O_6$)	71HM
meso-2,3-Dimethyltartaric acid ($C_6H_{10}O_6$)	71HM
threo-2,3-Dimethyltartaric acid ($C_6H_{10}O_6$)	71HM
2,3,4-Trihydroxypentanedioic acid (trihydroxyglutaric acid) ($C_5H_8O_7$)	56FG,57FG,60RE, 61ZK,62BA,62GL, 62GLa,62GM,63Cb, 64RME,66DD,66Ga, 67CO,69PD,69VD, 70KK,75F,75PSa, 75SP
Dihydroxymalonic acid ($C_3H_4O_6$)	70PL,73LP
Saccharic acid ($C_6H_{10}O_8$)	74KKc,75F
DL-2,4-Dimercaptopentanedioic acid ($C_5H_8O_4S_2$)	73ERa
meso-2,5-Dimercaptohexanedioic acid ($C_6H_{10}O_4S_2$)	73ER

B. Phosphorus acids

Ligand	Bibliography
Methylphosphonic acid (CH_5O_3P)	67BE
Hydroxyethane-1,2-diphosphonic acid ($C_2H_8O_7P_2$)	73VN
Phosphorothioic acid $O,O^{\llcorner}$-diethylester ($C_4H_{11}O_3PS$)	61TSL
Phosphoroselenoic acid $O,O^{\llcorner}$-diethylester ($C_4H_{11}O_3PSe$)	66TS
Hydroxymethane(ethylphosphinic acid) ($C_3H_9O_4P$)	72EZc
(Methylphenylphosphinylmethyl)phenylphosphinic acid ($C_{14}H_{16}O_3P_2$)	72EZa
Phosphoric acid methylester (methylphosphate) (CH_5O_4P)	65B
Phosphoric acid diethylester (diethylphosphate) ($C_4H_{11}O_4P$)	69US,71MG
Phosphoric acid tributylester (tributylphosphate) ($C_{12}H_{27}O_4P$)	61SKa,62PB
Phosphoric acid dipentylester (diamylphosphate) ($C_{10}H_{23}O_4P$)	70SK
Phosphoric acid bis(2-ethylhexyl)ester ($C_{16}H_{35}O_4P$)	63US,69SL,72GS

C. Arsenic acids

Dimethylarsinic acid ($C_2H_7O_2As$)	04J,10H,60Ba
4-Nitrobenzenearsonic acid ($C_6H_6O_5NAs$)	69M
Naphthalene-1-arsonic acid ($C_{10}H_9O_3As$)	69M

D. Phenols

5-Hydroxy-2-methyl-3-nitrobenzenesulfonic acid ($C_7H_7O_6NS$)	71B,72B
4-Ethyl-2-(2-thiazolylazo)phenol ($C_{11}H_{11}ON_3S$)	75KS
5-Dimethylamino-2-(2-thiazolylazo)phenol ($C_{11}H_{11}ON_4S$)	75KS
4-Chloro-2-(2-thiazolylazo)phenol ($C_9H_6ON_3S$)	74KS
4-Methoxy-2-(2-thiazolylazo)phenol ($C_{10}H_9O_2N_3S$)	74KS
2-Hydroxybenzohydroxamic acid (salicylohydroxamic acid) ($C_7H_7O_3N$)	69DS,70Sc
N-(3-Hydroxyphenyl)thiourea ($C_7H_8ON_2S$)	58H
3-Acetyl-1,5-bis(2-hydroxy-5-sulfophenyl)formazan ($C_{15}H_{14}O_9N_4S_2$)	71SE
3-Cyano-1,5-bis(2-hydroxy-5-sulfophenyl)formazan ($C_{14}H_{11}O_8N_5S_2$)	71SE
1,5-Bis(2-hydroxy-5-sulfophenyl)formazan ($C_{13}H_{11}O_{10}N_5S_2$)	71SE
3-Bromo-2-hydroxybenzoic acid ($C_7H_5O_3Br$)	73JK,74J,75JK
4-Bromo-2-hydroxybenzoic acid ($C_7H_5O_3Br$)	75JK
2-Hydroxy-4-iodobenzoic acid ($C_7H_5O_3I$)	73JK,75JK
2-Hydroxy-5-iodobenzoic acid ($C_7H_5O_3I$)	74J,75JK
3,5-Dibromo-2-hydroxybenzoic acid ($C_7H_4O_3Br_2$)	73JK,74J,75JK
3,5-Dichloro-2-hydroxybenzoic acid ($C_7H_4O_3Cl_2$)	73JK,74J,75JK
2-Hydroxyphenylpropenoic acid (o-coumaric acid) ($C_9H_8O_3$)	72TP,73TP,75TB
3,4-Dihydroxyazobenzene-2'-carboxylic acid ($C_{13}H_{10}O_4N_2$)	68OY
Acetylsalicylic acid ($C_9H_8O_4$)	68GJP

Phenols (continued)

Ligand	Bibliography
3-(3,4-Dihydroxyphenyl)propanoic acid ($C_9H_{10}O_4$)	69Ha, 69HO
Gallein ($C_{20}H_{12}O_7$)	62PL,72SL
2,6-Dibromo-3,4,5-trihydroxybenzoic acid ($C_7H_4O_5Br_2$)	69AK
2-Acetyl-4-methylphenol (2-hydroxy-5-methylacetophenone) ($C_9H_{10}O_2$)	70GMb
1,4-Dihydroxybenzene (hydroquinone) ($C_6H_6O_2$)	67RB
1,2-Dihydroxy-4-methylbenzene (4-methylcatechol) ($C_7H_8O_2$)	62H,63Ha,69Ha
2,4-Dihydroxybenzoic acid amide (β-resorcylamide) ($C_7H_7O_3N$)	66D
1-(2,4-Dihydroxyphenyl)acetophenone oxime (resacetophenone oxime) ($C_8H_9O_3N$)	64BR
3-(2,4-Dihydroxy)propiophenone oxime ($C_9H_{11}O_2N$)	67GDb
4'-Chloro-3,4-dihydroxyazobenzene ($C_{12}H_9O_2N_2Cl$)	58IS
3,4-Dihydroxybenzophenone ($C_{13}H_{10}O_3$)	62Ha
4-Nitroso-5,6-dihydroxybenzene-1,3-disulfonic acid ($C_6H_5O_9NS_2$)	67BH
4-(2,4-Dihydroxyphenylazomethyl)benzenesulfonic acid ($C_{13}H_{12}O_5N_2S$)	63SDc
4-Phenyl-7,8-dihydroxycoumarin ($C_{15}H_{10}O_4$)	73KS
D-2-(3,4-Dihydroxyphenyl)-3,5,7-trihydroxy-2,3-dihydropyran (catechin) ($C_{15}H_{14}O_6$)	59T
2'-Bromo-4',5'-dihydroxyazobenzene-4-sulfonic acid ($C_{12}H_9O_5N_2BrS$)	70BA
5-Chloro-2,2',4'-trihydroxyazobenzene-3-sulfonic acid (lumogallion) ($C_{12}H_9O_6N_2ClS$)	68MN,69AN
4,4'-Bis(3,4-dihydroxyphenylazo)stilbene-2,2'-disulfonic acid (stilbazo) ($C_{26}H_{26}O_{11}N_6S_2$)	68TL
4-Nitrophenylazocatechol ($C_{12}H_9O_4N_3$)	68TL
2,2',4'-Trihydroxyazobenzene ($C_{12}H_{10}O_3N_2$)	60DE
2,2'-Dihydroxyazobenzene ($C_{12}H_{10}O_2N_2$)	60DE
1,2,4-Trihydroxybenzene ($C_6H_6O_3$)	73RD
3,6-Dichloro-1,2,4,5-tetrahydroxybenzene ($C_6H_4O_4Cl_2$)	72KL
4-Hydroxy-3-nitrosocoumarin ($C_9H_5O_4N$)	69MBJ
Hamatoxylin (oxidized) ($C_{16}H_{12}O_6$)	72PMa
2"-Chloro-4"-nitro-4'-hydroxy-3,3'-dimethylfuchsone-5,5'-dicarboxylic acid (Chromal Blue G) ($C_{22}H_{16}O_8NCl$)	69U,69Ub
2",5",6"-Trichloro-4'-hydroxy-3,3'-dimethylfuchsone-5,5'-dicarboxylic acid (Eriochrome Azurol G) ($C_{22}H_{15}O_6Cl_3$)	69Ua
4',4"-Dihydroxyfuchsone-3,3',3"-tricarboxylic acid (aurintricarboxylic acid) (aluminon) ($C_{22}H_{14}O_9$)	54SL,57MD,58MD, 58MDa,58MDb,58MDc, 59MD,64Sa,65Sa, 66MSD,67S,68BDb, 70AD,72AB
4',3"-Dihydroxy-3,3',4"-trimethylfuchsone-5,5',5"-tricarboxylic acid (Chromoxane Violet R) ($C_{25}H_{20}O_9$)	67LM
Chrome Azurol C	68TL

Phenols (continued)

Ligand	Bibliography
Bromopyrogallol Red	69SF
1-Hydroxy-2-naphthoic acid ($C_{11}H_8O_3$)	65KC,71PSa
2-Hydroxy-1-naphthoic acid ($C_{11}H_8O_3$)	72PS
1-(4-Sulfophenylazo)-2-naphthol (tropoeolin 000) (Orange II) ($C_{16}H_{12}O_4N_2S$)	52Sb,68SD,70HL, 71HK
2-Hydroxy-1-phenylazonaphthalene-5,7-disulfonic acid ($C_{16}H_{12}O_7N_2S_2$)	52Sc
2-Hydroxy-1-(1-hydroxy-8-sulfo-2-naphthylazo)naphthalene-4-sulfonic acid (Palatine Fast Blue GGNA) ($C_{20}H_{14}O_8N_2S_2$)	59RS
2-Hydroxy-1-(2-hydroxy-5-methylphenylazo)naphthalene-4-sulfonic acid (calmagite) ($C_{17}H_{14}O_5N_2S$)	60LD,71JM,73MP
1-[4-(4-Chloro-3-sulfophenylazo)-5-methyl-2-hydroxyphenylazo]-2-naphthol (Solochrome Green) ($C_{23}H_{17}O_5N_4ClS$)	73JA
1-(5-Chloro-2-hydroxy-3-sulfophenylazo)-2-hydroxynaphthalene-3,6-disulfonic acid ($C_{16}H_{11}O_{11}N_2ClS_3$)	68SY
2-Hydroxy-1-(2-hydroxy-5-nitro-3-sulfophenylazo)naphthalene-3,6-disulfonic acid ($C_{16}H_{11}O_{13}N_3S_3$)	68SDY
8-Amino-1-hydroxy-2-(2-hydroxy-3,5-dinitrophenylazo)naphthalene-3,6-disulfonic acid ($C_{16}H_{12}O_{12}N_3S_2$)	68DY
7-Amino-1-hydroxynaphthalene-3,6-disulfonic acid ($C_{10}H_9O_7NS_2$)	68BD
8-Amino-1-hydroxynaphthalene-3,6-disulfonic acid ($C_{10}H_9O_7NS_2$)	63RS
1-(2,4-Dihydroxyphenylazo)-2-hydroxynaphthalene-4-sulfonic acid ($C_{16}H_{12}O_6N_2S$)	68MN
1-Hydroxy-2-(2-hydroxyphenylazo)naphthalene-4-sulfonic acid ($C_{16}H_{12}O_5N_2S$)	69BY,74BL,74BLa
1-Hydroxy-2-(5-chloro-2-hydroxyphenylazo)naphthalene-4-sulfonic acid ($C_{16}H_{11}O_5N_2ClS$)	69BY,74BL,74BLa
1-Hydroxy-2-(2-hydroxy-5-sulfophenylazo)naphthalene-4-sulfonic acid ($C_{16}H_{12}O_8N_2S_2$)	74BL,74BLa
1-Hydroxy-2-(5-chloro-2-hydroxy-3-sulfophenylazo)naphthalene-4-sulfonic acid ($C_{16}H_{11}O_8N_2ClS_2$)	69BY,74BL,74BLa
1-Hydroxy-2-(2-hydroxy-3,5-disulfophenylazo)naphthalene-4-sulfonic acid ($C_{16}H_{12}O_{11}N_2S_3$)	69BY,74BL,74BLa
1-Hydroxy-2-(2-hydroxy-5-nitrophenylazo)naphthalene-4-sulfonic acid ($C_{16}H_{11}O_7N_3S$)	69BY,74BL,74BLa
1-Hydroxy-2-(2-hydroxy-5-nitro-3-sulfophenylazo)naphthalene-4-sulfonic acid ($C_{16}H_{11}O_{10}N_3S_2$)	69BY,74BL,74BLa
1-Hydroxy-2-(2-hydroxy-3,5-dinitrophenylazo)naphthalene-4-sulfonic acid ($C_{16}H_{10}O_9N_4S$)	69BY,74BL,74BLa
2-Hydroxy-1-(2-hydroxy-3,5-dinitrophenylazo)naphthalene-3,6-disulfonic acid (Picramine R) ($C_{16}H_{10}O_{12}N_4S_2$)	68GT

Phenols (continued)

Ligand	Bibliography
Cyclo-tris-7-(1-azo-8-hydroxynaphthalene-3,6-disulfonic acid (Calcichrome) ($C_{30}H_{18}O_{21}N_6S_6$)	65BB
2-(2-Hydroxy-3,5-dinitrophenylazo)chromotropic acid ($C_{16}H_{10}O_{13}N_4S_2$)	68GT
2-(2-Hydroxy-5-sulfophenylazo)chromotropic acid ($C_{16}H_{12}O_{12}N_2S_3$)	66BBa
2,7-Bis(2-sulfophenylazo)chromotropic acid (Sulphonazo III) ($C_{22}H_{16}O_{14}N_4S_4$)	65BV,69B
2,7-Bis(5-chloro-2-hydroxy-3-sulfophenylazo)chromotropic acid (Chlorosulphophenol S) ($C_{22}H_{14}O_{16}N_4Cl_2S_4$)	65BSb,68PI
2,7-Bis(2-phosphonophenylazo)chromotropic acid (Phosphonazo III) ($C_{22}H_{18}O_{14}N_4P_2S_2$)	67BHB
2-(4-Chloro-2-phosphonophenylazo)-7-(2-phosphonophenylazo)chromo-tropic acid ($C_{22}H_{17}O_{14}N_4ClP_2S_2$)	67BHB
2,7-Bis(4-chloro-2-phosphonophenylazo)chromotropic acid (Chlorophos-phonazo III) ($C_{22}H_{16}O_{14}N_4Cl_2P_2S_2$)	67BH,69B
2-(4-Chloro-5-methyl-2-phosphonophenylazo)-7-(4-chloro-2-phosphono-phenylazo)chromotropic acid ($C_{23}H_{18}O_{14}N_4ClP_2S_2$)	67BHB
2-(2-Arsonophenylazo)-7-(phenylazo)chromotropic acid (Monoarsenazo III) ($C_{22}H_{17}O_{11}N_4AsS_2$)	65Ba
2,7-Bis(2-arsonophenylazo)chromotropic acid (Arsenazo III) ($C_{22}H_{18}O_{14}N_4As_2S_2$)	63B,64B,64Ba,68TL, 69B,70BS,72SSP, 75M,75NM
2,7-Bis(2-arsono-6-carboxyphenylazo)chromotropic acid ($C_{24}H_{18}O_{18}N_4As_2S_2$)	65BH
2,7-Dibromochromotropic acid ($C_{10}H_6O_8Br_2S_2$)	75MD
8-Hydroxy-1,2-napthoquinone-3,6-disulfonic acid ($C_{10}H_6O_9S_2$)	75MD
2,7-Dichlorochromotropic acid ($C_{10}H_6O_8Cl_2S_2$)	69NB,70OM,73DMa
1,4-Dihydroxy-9,10-anthraquinone-2-sulfonic acid (quinizarin-2-sulfonic acid) ($C_{14}H_8O_7S$)	64JJ,69TA,70JJ, 71RB
1,3,4,6-Tetrahydroxy-2-(1-oxo-2,3,4,5-tetrahydroxyhexyl)-8-methyl-anthraquinone-5-carboxylic acid (Carminic acid) ($C_{22}H_{20}O_{13}$)	70PLS

E. Carbonyl ligands

2-Methyl-4-oxopent-2-ene ($C_6H_{10}O$)	72ZG
2-Hydroxy-2-methyl-4-oxopentane ($C_6H_{12}O_2$)	70MSa
2,2-Dimethyl-5,5,5-trifluorohexane-3,5-dione (pivaloyltrifluoroacetone) ($C_8H_{11}O_2F_3$)	73SHI,74SMN
1,3-Diphenylpropane-1,3-dione (dibenzoylmethane) ($C_{15}H_{12}O_2$)	61MS,68BB,68RSc, 70G,70VE
1,1,1,2,2,3,3-Heptafluoro-7,7-dimethyloctane-4,6-dione ($C_{10}H_{11}O_2F_7$)	70SB
1,3-Di(2-thienyl)propane-1,3-dione (dithenoylacetone) ($C_{11}H_8O_2S_2$)	70G
1,1,1-Trifluoropentane-2,4-dione (trifluoroacetylacetone) ($C_5H_5O_2F_3$)	70G,71SI,72GK, 73SHI,74SMN

Carbonyl ligands (continued)

Ligand	Bibliography
2,2,6,6-Tetramethylheptane-3,5-dione ($C_{11}H_{18}O_2$)	68SP
1-Phenyl-3-(2-selenoyl)propane-1,3-dione (selenenoylbenzoylmethane) ($C_{13}H_{10}O_2Se$)	68BB
3-Methyl-1-phenyl-4-benzoylpyrazolone ($C_{17}H_{14}O_2N_2$)	69LSa
3-Acetyl-2-hydroxy-6-methyl-4-pyrone (dehydracetic acid) ($C_8H_8O_3$)	56AR
6-Ethyl-2-hydroxy-3-propionyl-4-pyrone ($C_{10}H_{12}O_3$)	56AR

F. Alcohols

Ethanol (C_2H_6O)	63FP
Butanol ($C_4H_{10}O$)	65OS
Butan-2-ol (s-butanol) ($C_4H_{10}O$)	65OS
2-Methylpropan-2-ol (t-butanol) ($C_4H_{10}O$)	65OS
meso-Butane-2,3-diol ($C_4H_{10}O_2$)	57RL,67CBa
L-Butane-2,3-diol ($C_4H_{10}O_2$)	67CBa
2-Methylbutane-2,3-diol ($C_5H_{12}O_2$)	67CBa
cis-Cyclohexane-1,2-diol ($C_6H_{12}O_2$)	67CBa
2,2'-Oxydiethanol (diethyleneglycol) ($C_4H_{10}O$)	65OS
2,2'-Thiodiethanol (β-thiodiglycol) ($C_4H_{10}S$)	71MST
Methyl α-D-glucopyranoside ($C_7H_{14}O_6$)	65LA
Phenyl β-D-glucopyranoside ($C_{12}H_{16}O_6$)	65LA

G. Ethers

Ethoxyethane (diethyl ether) ($C_4H_{10}O$)	65PB
Bis(t-butylcyclohexyl)-18-crown-6-ether ($C_{28}H_{44}O_6$)	72ES

H. Thiols

Furylmethanethiol (furfurylmercaptan) (C_5H_6OS)	74SS
1,3-Dimercaptopropane-2-sulfonic acid ($C_3H_8O_3S_3$)	66PRa,67PRa,68PRM
2,3-Dimercaptopropane-1-sulfonic acid (unithiol) ($C_3H_8O_3S_3$)	66PRa,67PRa,68OF, 68OR,68PRM
2-(2,3-Dimercaptopropoxy)ethane-1-sulfonic acid ($C_5H_{12}O_4S_3$)	66PRa,67PRa,68PRM
2-(2,3-Dimercaptopropylthio)ethane-1-sulfonic acid ($C_5H_{12}O_3S_4$)	66PRa,67PRa,68PRM
2-(2,3-Dimercaptopropylsulfonyl)ethane-1-sulfonic acid ($C_5H_{12}O_5S_4$)	66PRa,67PRa,68PRM
Ethylenebis(3-mercaptopropanoate) ($C_8H_{14}O_4S_2$)	72SCa,72SCb

I. Thioacids

Ligand	Bibliography
Propanethioic acid (thiopropanoic acid) (C_3H_6OS)	67MS,70MN
Ethanedidithioic acid (tetrathiooxalic acid) ($C_2H_2S_4$)	57JB,58D
Propanedithioic acid (dithiomalonic acid) ($C_3H_4O_2S_2$)	58D
Dimethylaminocarbodithioic acid (dimethyldithiocarbamic acid) ($C_3H_7NS_2$)	56J,57Ja
Dipropylaminocarbodithioic acid (dipropyldithiocarbamic acid) ($C_7H_{14}NS_2$)	57Ja
Tetramethyleneaminocarbodithioic acid (pyrrolidine-N-dithiocarboxylic acid) ($C_5H_9NS_2$)	57Ja,69JA,73SSC
Pentamethyleneaminocarbodithioic acid (piperidine-N-dithiocarboxylic acid) ($C_6H_{11}NS_2$)	57Ja,73SSC
Hexamethyleneaminocarbodithioic acid ($C_7H_{13}NS_2$)	73SSC
Morpholine-N-carbodithioic acid ($C_5H_9ONS_2$)	72GKK
3-Carboxypiperidine-1-carbodithioic acid ($C_7H_{11}O_2NS_2$)	57Ja
Bis(carboxymethyl)aminocarbodithioic acid ($C_5H_7O_4NS_2$)	67HM
Ethoxycarbodithioic acid (ethyl xanthate) ($C_3H_6OS_2$)	67KHa,69NY
Ethylthiosulfonic acid ($C_2H_6O_2S_2$)	74DS

J. Unsaturated hydrocarbons

	Bibliography
3-Chloropropene (allyl chloride) (C_3H_5Cl)	65TF
Cyclopentadiene (C_5H_6)	72BS
Cyanopropane (butyronitrile) (C_4H_7N)	24P
Cyanoethene (acrylonitrile) (C_3H_3N)	68Sa
1-Chloro-2-cyanoethene (2-chloroacrylonitrile) (C_3H_2NCl)	68S
2-Cyanopropene (methacrylonitrile) (C_4H_5N)	68S
3-Cyanopropene (crotononitrile) (C_4H_5N)	68S
1,2-Dicyanoethane (succinonitrile) ($C_4H_4N_2$)	24P
1,4-Dimethoxybenzene (4-methoxyanisole) ($C_8H_{10}O_2$)	64WD
Dihexylphenylsulfonylamidophosphonate ($C_{18}H_{32}O_6NPS$)	68SK

K. Hydroxamic acids

	Bibliography
Oxalodihydroxamic acid ($C_2H_4O_4N_2$)	57MJ,70Sc
N-Phenylbutanohydroxamic acid ($C_{10}H_{13}O_2N$)	72ST
N-Phenylhexanohydroxamic acid ($C_{12}H_{17}O_2N$)	72ST
N-(4-Methylphenyl)nonanohydroxamic acid ($C_{17}H_{27}O_2N$)	72ST
N-(4-Methylphenyl)tetradecanohydroxamic acid ($C_{21}H_{35}O_2N$)	72ST
N-Phenyl-2-furohydroxamic acid ($C_{11}H_9O_3N$)	69DS,70FR,72ST
N-Phenyl-2-thienohydroxamic acid ($C_{11}H_9O_2NS$)	70FR
4-Methyl-N-phenylbenzohydroxamic acid ($C_{14}H_{13}O_2N$)	69DS

Hydroxamic acids (continued)

Ligand	Bibliography
N-(2-Methylphenyl)benzohydroxamic acid $(C_{14}H_{13}O_2N)$	69DS
N-(4-Methylphenyl)benzohydroxamic acid $(C_{14}H_{13}O_2N)$	69DS
N-Phenyl-2-trifluoromethylbenzohydroxamic acid $(C_{14}H_{10}O_2NF_3)$	70RF
N-Phenyl-3-trifluoromethylbenzohydroxamic acid $(C_{14}H_{10}O_2NF_3)$	70RF
N-Phenyl-4-trifluoromethylbenzohydroxamic acid $(C_{14}H_{10}O_2NF_3)$	70RF
5-Nitro-N-phenyl-3-trifluoromethylbenzohydroxamic acid $(C_{14}H_9O_4N_2F_3)$	70RF
4-Carboxy-N-phenylbenzohydroxamic acid ethyl ester $(C_{16}H_{15}O_4N)$	70FR
4-Methoxy-N-phenylbenzohydroxamic acid $(C_{14}H_{13}O_3N)$	69DS

L. Oximes

Formaldoxime (CH_3ON)	71BJ
Acetaldoxime (C_2H_5ON)	68SGe
5-Methylcyclohexane-1,2-dione dioxime (5-methylnioxime) $(C_7H_{12}O_2N_2)$	62HR
Phenyloximinoacetamide $(C_8H_8O_2N_2)$	73ES
Benzamidoxime $(C_7H_8ON_2)$	68Mf
O-Methylhydroxylamine (CH_5ON)	68SGe

M. Amides

N-Methylcarbamide $(C_2H_6N_2)$	72KM
N-Ethylcarbamide $(C_3H_8N_2)$	72KM
Acetic acid hydrazide (acetohydrazide) $(C_2H_6ON_2)$	74FSD
Butanedioic acid dihydrazide (succinodihydrazide) $(C_4H_{10}O_2N_4)$	66KSD,68Sb,74FSD
Hexanedioic acid dihydrazide (adipodihydrazide) $(C_6H_{14}O_2N_4)$	66KSD,74FSD
Cyanoacetic acid hydrazide $(C_3H_5ON_3)$	68Z
1-Phenyl-3-methyl-4-benzoylpyrazol-5-one $(C_{17}H_{14}O_2N_2)$	72KEM
Dimethylol thiourea $(C_3H_8O_2N_2S)$	70TK
Propanedioic acid diamide (malonamide) $(C_3H_6O_2N_2)$	69KAM
Thioacetic acid amide (C_2H_5NS)	66PL
Dithiooxamide (rubeanic acid) $(C_2H_4N_2S_2)$	52YVa
Biuret $(C_2H_5O_2N_3)$	60Ka,75SS
1,3,4-Thiadiazole-2,5-dithiol $(C_2H_2N_2S_3)$	58MC
2-Chlorobenzoic acid hydrazide (C_7H_7ONCl)	75PSb
4-Methoxybenzoic acid hydrazide $(C_8H_{10}O_2N)$	75PSb
Bis(1-phenyl-2,3-dimethyl-5-oxo-1,2-diazol-4-yl)methane (4,4'-diantipyrylmethane) $(C_{23}H_{24}O_2N_4)$	62BT,63BS,75HS
3-Furylprop-2-enethiosemicarbazone (furylacroleinthiosemicarbazone) $(C_8H_9ON_3S)$	71KLM,72KLM

Amides (continued)

Ligand	Bibliography
3-(5-Methylfuryl)prop-2-enethiosemicarbazone ($C_9H_{11}ON_3S$)	71KLM,72KLM
3-(5-Nitrofuryl)prop-2-enethiosemicarbazone ($C_8H_8O_3N_4S$)	71KLM,72KLM
3-(5-Chlorofuryl)prop-2-enethiosemicarbazone ($C_8H_8ON_3ClS$)	71KLM,72KLM
2-Ethyl-3-furylprop-2-enethiosemicarbazone ($C_{10}H_{13}ON_3S$)	71KLM,72KLM
3-Phenylprop-2-enealthiosemicarbazone (cinnamaldehydethiosemicarbazone) ($C_{10}H_{11}N_3S$)	71KLM,72KLM

N. Amidines

Dithio-1,1'-diformamidine ($C_2H_6N_4S_2$)	74AL
N-Guanylurea ($C_2H_7ON_5$)	60D
N-Guanyl-N'-methylurea ($C_3H_9ON_5$)	60D
N-Guanyl-N'-ethylurea ($C_4H_{11}ON_5$)	60D
1-Methylbiguanide ($C_3H_9N_5$)	56SR,59RR
1-Ethylbiguanide ($C_4H_{11}N_5$)	56SR
1,1-Diethylbiguanide ($C_6H_{19}N_5$)	56SR,60R
1-Phenylbiguanide ($C_8H_{12}N_5$)	52BG,56SR,59RR
1-(4-Chlorophenyl)-5-(2-propyl)biguanide ($C_{11}H_{17}N_5Cl$)	60R
1-(2-Hydroxyethyl)biguanide ($C_4H_{11}ON_5$)	60SR
1-(3-Hydroxypropyl)biguanide ($C_5H_{13}ON_5$)	60SR
1-(2-Methoxyethyl)biguanide ($C_5H_{13}ON_5$)	60SR
1-(3-Methoxypropyl)biguanide ($C_5H_{15}ON_5$)	60SR
Ethylenedibiguanide ($C_6H_{16}N_{10}$)	50SG,56SR,59RR, 69LM
1,3-Phenylenedibiguanide ($C_8H_{11}N_5$)	56SR

XIX. BIBLIOGRAPHY

Russian translations have the page of the original in parentheses.

04J   J. Johnston, Chem. Ber., 1904, 37, 3625

04K   F. Kunschert, Z. Anorg. Allg. Chem., 1904, 41, 337

05AS   R. Abegg and J.F. Spencer, Z. Anorg. Allg. Chem., 1905, 46, 406

05SA   J.F. Spencer and R. Abegg, Z. Anorg. Allg. Chem., 1905, 44, 379 (see also 64SM)

05SAa   H. Schafer and R. Abegg, Z. Anorg. Allg. Chem., 1905, 45, 293

10H   B. Holmberg, Z. Phys. Chem., 1910, 70, 153

10J   A. Jacques, Trans. Faraday Soc., 1910, 5, 225

24L   E. Larrson, Z. Anorg. Allg. Chem., 1924, 140, 292

24P   F.G. Pawelka, Z. Elektrochem., 1924, 30, 180

27D   C.W. Davies, Trans. Faraday Soc., 1927, 23, 351

27S   R. Scholder, Chem. Ber., 1927, 60, 1510

28DH   H.M. Dawson, G.V. Hall, and A. Key, J. Chem. Soc., 1928, 2844

28L   E.Larsson, Z. Anorg. Allg. Chem., 1928, 172, 375

29BU   N. Bjerrum and A. Unmack, Kgl. Danske Videnskabernes Selskab, Math.-Fys. Medd.,
          1929, 9, No. 1, 1-208

29R   H.L. Riley, J. Chem. Soc., 1929, 1307

29Ra   H.L. Riley, J. Chem. Soc., 1929, 1387

29RF   H.L. Riley and N.I. Fisher, J. Chem. Soc., 1929, 2006

30L   E. Larsson, Chem. Ber., 1930, 63B, 1347

30R   H.L. Riley, J. Chem. Soc., 1930, 1640, 1642

31BR   W.H. Banks, E.C. Righellato, and C.W. Davies, Trans. Faraday Soc., 1931, 27, 621

31GI   R. Gane and C.K. Ingold, J. Chem. Soc., 1931, 2153

31IR   D.J.G. Ives and H.L. Riley, J. Chem. Soc., 1931, 1998

31LA   E. Larsson and B. Adell, Z. Phys. Chem., 1931, A156, 352

31LAa   E. Larsson and B. Adell, Z. Phys. Chem., 1931, A156, 381

31LAb   E. Larsson and B. Adell, Z. Phys. Chem., 1931, A157, 342

31MS   D.A. MacInnes and T. Shedlovsky, J. Amer. Chem. Soc., 1931, 53, 2419

32L   E. Larsson, Z. Phys. Chem., 1932, A159, 306

32La   E. Larsson, Z. Phys. Chem., 1932, A159, 315

32MD   R.W. Money and C.W. Davies, Trans. Faraday Soc., 1932, 28, 609

33CC   A.K. Chibnall and R.A. Cannan, Biochem. J., 1933, 27, 945

33HE   H.S. Harned and R.W. Ehlers, J. Amer. Chem. Soc., 1933, 55, 652

33HEa   H.S. Harned and R.W. Ehlers, J. Amer. Chem. Soc., 1933, 55, 2379

33IL   D.J.G. Ives, R.P. Linstead, and H.L. Riley, J. Chem. Soc., 1933, 561

33L   E. Larsson, Z. Phys. Chem., 1933, A165, 53

33La   E. Larsson, Z. Phys. Chem., 1933, A166, 241

33SL   B. Saxton and T. Langer, J. Amer. Chem. Soc., 1933, 55, 3638

34BK   F.G. Brockman and M. Kilpatrick, J. Amer. Chem. Soc., 1934, 56, 1483

34DW   J.F. Dippy and F.R. Williams, J. Chem, Soc., 1934, 161

34DWa   J.F.J. Dippy and F.R. Williams, J. Chem. Soc., 1934, 1888

34FR   E. Ferrell, J.M. Ridgion, and H.L. Riley, J. Chem. Soc., 1934, 1440

34HE   H.S. Harned and N.D. Embree, J. Amer. Chem. Soc., 1934, 56, 1042

34HM   A.B. Hastings, F.C. McLean, L. Eichelberger, J.L. Hall, and E. DaCosta, J. Biol.
          Chem., 1934, 107, 351

34HS   H.S. Harned and R.D. Sutherland, J. Amer. Chem. Soc., 1934, 56, 2039

34JV   G.H. Jeffery and A.I. Vogel, J. Chem. Soc., 1934, 166

34MD   R.W. Money and C.W. Davies, J. Chem. Soc., 1934, 400

34P   K.J. Pedersen, J. Phys. Chem., 1934, 38, 559

34S   G. Sartori, Gazz. Chim. Ital., 1934, 64, 3

34SM   B. Saxton and H.F. Meier, J. Amer. Chem. Soc., 1934, 56, 1918

34W   D.D. Wright, J. Amer. Chem. Soc., 1934, 56, 314

35BB   J.H. Backer and C.C. Bolt, Rec. Trav. Chim., 1935, 54, 186

35BJ   H.T.S. Britton and M.E.D. Jarret, J. Chem. Soc., 1935, 168

35D   C.W. Davies, J. Chem. Soc., 1935, 910

35DW   J.F.J. Dippy, F.R. Williams, and R.H. Lewis, J. Chem. Soc., 1935, 343

36BJ   H.T.S. Britton and M.E.D. Jarrett, J. Chem. Soc., 1936, 1489

36CE   G.H. Cartledge and W.P. Ericks, J. Amer. Chem. Soc., 1936, 58, 2065

36DL   J.F.J. Dippy and R.H. Lewis, J. Chem. Soc., 1936, 644

36JS   I. Jones and F.G. Soper, J. Chem. Soc., 1936, 133 (see also 63EW)

36ML   J. Muus and H. Lebel, Kgl. Danske Vid. Selsk. Math.-Fys. Medd., 1936, 13, No. 19

36N   L.F.Nims, J. Amer. Chem. Soc., 1936, 58, 987

36NS   L.F. Nims and P.K. Smith, J. Biol. Chem., 1936, 113, 145

37BD   J.W. Baker, J.F.J. Dippy, and J.E. Page, J. Chem. Soc., 1937, 1774

37CV   W.J. Clayton and W.C. Vosburg, J. Amer. Chem. Soc., 1937, 59, 2414

37DL   J.F.J. Dippy and R.H. Lewis, J. Chem. Soc., 1937, 1008

37DLa   J.F.J. Dippy and R.H. Lewis, J. Chem. Soc., 1937, 1426

37GJ   W.L. German, G.H. Jeffery, and A.I. Vogel, J. Chem. Soc., 1937, 1604

37GV   W.L. German and A.I. Vogel, J. Chem. Soc., 1937, 1108

37HH   H.S. Harned and F.C. Hickey, J. Amer. Chem. Soc., 1937, 59, 1284, 2303

37MT   A.W. Martin and H.V. Tartar, J. Amer. Chem. Soc., 1937, 59, 2672

37N   A. Neuberger, Proc. Roy. Soc. (London), 1937, A158, 68

37RD   R.A. Robinson and C.W. Davies, J. Chem. Soc., 1937, 574

38BD   W.H. Banks and C.W. Davies, J. Chem. Soc., 1938, 73

38CK   R.K. Cannan and A. Kibrick, J. Amer. Chem. Soc., 1938, 60, 2314

38D   C.W. Davies, J. Chem. Soc., 1938, 271, 273

38Da   C.W. Davies, J. Chem. Soc., 1938, 277

38Db   J.F.J. Dippy, J. Chem. Soc., 1938, 1222

38MD   R.W. Money and C.W. Davies, J. Chem. Soc., 1938, 2098

38N   R. Nordbo, Skand. Archiv Physiol., 1938, 80, 341

38WL   S. Winstein and H.J. Lucas, J. Amer. Chem. Soc., 1938, 60, 836

39A   B. Adell, Z. Phys. Chem., 1939, A185, 161

39EW   D.H. Everett and W.F.K. Wynne-Jones, Trans. Faraday Soc., 1939, 35, 1380

39HF   H.S. Harned and L.O. Fallon, J. Amer. Chem. Soc., 1939, 61, 3111

39LH   H.J. Lucas, F.R. Hepner, and S. Winstein, J. Amer. Chem. Soc., 1939, 61, 3102

39P   K.J. Pedersen, Trans. Faraday Soc., 1939, 35, 277

39PG   H.N. Parton and R.C. Gibbons, Trans. Faraday Soc., 1939, 35, 542

40A   B. Adell, Z. Phys. Chem., 1940, A187, 66

40CN   G.H. Cartledge and P.N. Nichols, J. Amer. Chem. Soc., 1940, 34, 269

40EB   S.M. Edmonds and N. Birnbaum, J. Amer. Chem. Soc., 1940, 62, 2367

40HB   W.J. Hamer, J.O. Burton and S.F. Acree, J. Res. Nat. Bur. Stand., 1940, 24, 269

40SD   B. Saxton and L.S. Darken, J. Amer. Chem. Soc., 1940, 62, 846

40TD   N.E. Topp and C.W. Davies, J. Chem. Soc., 1940, 87

40VB   W.C. Vosburg and J.F. Beckman, J. Amer. Chem. Soc., 1940, 62, 1028

41A   B. Adell, Svensk Kem. Tid., 1941, 53, No. 6, 89

41S   H. Schafer, Z. Anorg. Allg. Chem., 1941, 247, 96

42K   S.S. Kety, J. Biol. Chem., 1942, 142, 181

42KP   I.M. Kolthoff, R.W. Perlich, and D. Weiblen, J. Phys. Chem., 1942, 46, 561

42MA   F.H. MacDougall and M. Allen, J. Phys. Chem., 1942, 46, 730, 738

42S   H. Schafer, Z. Anorg. Allg. Chem., 1942, 250, 82

42Sa   H. Schafer, Z. Anorg. Allg. Chem., 1942, 250, 127

43BG   J. Bell, W.A. Gillespie, and D.B. Taylor, Trans. Faraday Soc., 1943, 39, 137

43BS   R.G. Bates, G.L. Siegel and S.F. Acree, J. Res. Nat. Bur. Stand., 1943, 31, 205

43HB   N.V. Hayes and G.E.K. Branch, J. Amer. Chem. Soc., 1943, 65, 1555

43L   I. Leden, Diss., Lund, 1943

43LB   H.J. Lucas, F.W. Billmeyer, Jr., and D. Pressman, J. Amer. Chem. Soc., 1943, 65,
          230

43LM   H.J. Lucas, R.S. Moore, and D. Pressman, J. Amer. Chem. Soc., 1943, 65, 227

44L   E. Larson, The Svedberg, Uppsala, 1944, 311

45B   A.K. Babko, Zh. Obsh. Khim., 1945, 15, 745

45HP   W.J. Hamer, G.D. Pinching and S.F. Acree, J. Res. Nat. Bur. Stand., 1945, 35,
          381, 539

45P   K.J. Pedersen, Kgl. Danske Videnskab. Selskab, Mat.-Fys. Medd., 1945, 22, 12

45SS   E.E. Sager, M.R. Schooley, A.S. Carr, and S.F. Acree, J. Res. Nat. Bur. Stand.,
          1945, 35, 521

46J   N.R. Joseph, J. Biol. Chem., 1946, 164, 529

46L   J. Leden, Svenck. Kem. Tidskr., 1946, 58, 129

46PS   B.C. Purkayastha, R.N. Sen-Sarma, J. Indian Chem. Soc., 1946, 23, 31

47B   A.K. Babko, Zh. Obsh. Khim., 1947, 17, 443

47L   S. Lacroix, Bull. Soc. Chim. France, 1947, 408

47MP   F.H. MacDougall and S. Peterson, J. Phys. Chem., 1947, 51, 1346

47T   A.D. Terebaugh, AECD-2749, 1947

47TM   E.R. Tompkins and S.W. Major, J. Amer. Chem. Soc., 1947, 69, 2859

48AK   L.J. Andrews and R.M. Keefer, J. Amer. Chem. Soc., 1948, 70, 3261

48CD   T.L. Cottrell, G.W. Drake, D.L. Levi, K.J. Tully, and J.H. Wolfenden, J. Chem.
        Soc., 1948, 1016

48CW   T.L. Cottrell and J.H. Wolfenden, J. Chem. Soc., 1948, 1019

48F   S. Fronaeus, Doct. Diss., Lund, 1948

48FA   R.T. Foley and R.C. Anderson, J. Amer. Chem. Soc., 1948, 70, 1195

48LQ   O.E. Lanford and J.R. Quinan, J. Amer. Chem. Soc., 1948, 70, 2900

48PB   G.D. Pinching and R.G. Bates, J. Res. Nat. Bur. Stand., 1948, 40, 405

48SB   G. Schwarzenbach and W. Biedermann, Helv. Chim. Acta, 1948, 31, 678

48SR   J. Schubert and J.W. Richter, J. Phys. Chem., 1948, 52, 350

48SRa   J. Schubert and J.W. Richter, J. Amer. Chem. Soc., 1948, 70, 4259

48T   H. Taube, J. Amer. Chem. Soc., 1948, 70, 3928

48WB   G.W. Wheland, R.M. Brownell, and E.C. Mayo, J. Amer. Chem. Soc., 1948, 70, 2493

49A   S. Ahrland, Acta Chem. Scand., 1949, 3, 783

49AK   L.J. Andrews and R.M. Keefer, J. Amer. Chem. Soc., 1949, 71, 2379
49AKa   L.J. Andrews and R.M. Keefer, J. Amer. Chem. Soc., 1949, 71, 3644
49BP   R.G. Bates and G.D. Pinching, J. Amer. Chem. Soc., 1949, 71, 1274

49DW   C.W. Davies and P.A.H. Wyatt, Trans. Faraday Soc., 1949, 45, 770

49FA   R.T. Foley and R.C. Anderson, J. Amer. Chem. Soc., 1949, 71, 909

49JH   W. Jacnicke and K. Hauffe, Z. Naturforsch, 1949, 4a, 363

49KA   R.M. Keefer and L.J. Andrews, J. Amer. Chem. Soc., 1949, 71, 1723

49KAa   R.M. Keefer, L.J. Andrews, and R.E. Kepner, J. Amer. Chem. Soc., 1949, 71, 2381

49KAb   R.M. Keefer, L.J. Andrews, and R.E. Kepner, J. Amer. Chem. Soc., 1949, 71, 3906

49L   I. Leden, Acta Chem. Scand., 1949, 3, 1318

49La   S. Lacroix, Ann. Chim. (France), 1949, 4, 5

49Lb   J. Lambling, Bull. Soc. Chim. France, 1949, 495

49M   L. Meides, J. Amer. Chem. Soc., 1949, 71, 3269

49P   K.J. Pedersen, Acta Chem. Scand., 1949, 3, 676

49PH   J.V. Parker, C. Hirayama, and F.H. MacDougall, J. Phys. Chem., 1949, 53, 912

49RC   S.D. Ross and A.J. Catotti, J. Amer. Chem. Soc., 1949, 71, 3563

49SD   D.I. Stock and C.W. Davies, J. Chem. Soc., 1949, 1371

49SG   G. Schwarzenbach and H. Gysling, Helv. Chim. Acta, 1949, 32, 1108

49SGa   G. Schwarzenbach and H. Gysling, Helv. Chim. Acta, 1949, 32, 1314

49TA   S.E. Turner and R.C. Anderson, J. Amer. Chem. Soc., 1949, 71, 912

49VB   R.C. Voter and C.V. Banks, Anal. Chem., 1949, 21, 1320

50AK   L.J. Andrews and R.M. Keefer. J. Amer. Chem. Soc., 1950, 72, 3113

50AKa   L.J. Andrews and R.M. Keefer, J. Amer. Chem. Soc., 1950, 72, 5034

50CM   C.E. Crouthamel and D.S. Martin, J. Amer. Chem. Soc., 1950, 72, 1382

50DG   A.K. Dey, S.P. Ghosh, and P. Ray, J. Indian Chem. Soc., 1950, 27, 493

50DS   R.A. Day, Jr., and R.W. Stroughton, J. Amer. Chem. Soc., 1950, 72, 5662

50DW   C.W. Davies and G.M. Waind, J. Chem. Soc., 1950, 301

50M   L. Meites, J. Amer. Chem. Soc., 1950, 72, 180

50Ma   L. Meites, J. Amer. Chem. Soc., 1950, 72, 184

50OL   E.L. Onstott and H.A. Laitinen, J. Amer. Chem. Soc., 1950, 72, 4724

50PB   G.D. Pinching and R.G. Bates, J. Res. Nat. Bur. Stand., 1950, 45, 322, 444

50R   J. Rydberg, Svensk Kem. Tidskr., 1950, 62, 179

50Ra   J. Rydberg, Acta Chem. Scand., 1950, 4, 1503

50S   J. Schubert, Anal. Chem., 1950, 22, 1359

50SG   D. Sen, N.N. Ghosh, and P. Ray, J. Indian Chem. Soc., 1950, 27, 619

50SR   J. Schubert, E. Russell, and L. Myers, J. Biol. Chem., 1950, 185, 387

50SZ   G. Schwarzenbach and J. Zurc, Monat. Chem., 1950, 81, 202

51A   S. Ahrland, Acta Chem. Scand., 1951, 5, 199

51AK   L.J. Andrews and R.M. Keefer, J. Amer. Chem. Soc., 1951, 73, 5733

51B   R.G. Bates, J. Amer. Chem. Soc., 1951, 73, 2259

51BA   J.E. Barney, W.J. Argersinger. Jr., and C.A. Reynolds, J. Amer. Chem. Soc.,
           1951, 73, 3785

51BB   G. Briegleb and A. Bieber, Z. Elektrochem., 1951, 55, 250

51BC   R.G. Bates and A.G. Canham, J. Res. Nat. Bur. Stand., 1951, 47, 343

51BM   A.K. Babko and P.B. Mikhelson, Zh. Anal. Khim., 1951, 6, 267 (see also 64D)

51BP   C.V. Banks and J.H. Patterson, J. Amer. Chem. Soc., 1951, 73, 3062

51BW   R.P. Bell and G.M. Waind, J. Chem. Soc., 1951, 2357

51C   C.H. Cartledge, J. Amer. Chem. Soc., 1951, 73, 4416

51CM   C.E. Crouthamel and D.S. Martin, J. Amer. Chem. Soc., 1951, 73, 569

51DM   C.W. Davies and C.B. Monk, J. Chem. Soc., 1951, 2718

51F   S. Fronaeus, Acta Chem. Scand., 1951, 5, 859

51H   E.Heinz, Biochem. Z., 1951, 321, 314

51HS   J. Heller and G. Schwarzenbach, Helv. Chim. Acta, 1951, 34, 1876

51LW   M. Lloyd, V. Wycherley, and C.B. Monk, J. Chem. Soc., 1951, 1786

51M   L. Meites, J. Amer. Chem. Soc., 1951, 73, 3727

51Ma   W.H. McVey, HW-21487, 1951

51MB   H.V. Meek and C.V. Banks, J. Amer. Chem. Soc., 1951, 73, 4108

51MM   T. Meites and L. Meites, J. Amer. Chem. Soc., 1951, 73, 1161

51PJ   J.M. Peacock and J.C. James, J. Chem. Soc., 1951, 2233

51S   S. Suzuki, Nippon Kagaku Zasshi, 1951, 72, 524, 721

51Sa   S. Suzuki, Nippon Kagaku Zasshi, 1951, 72, 974; Sci. Reports, Tohoku Univ., 1952,
             4, 464

51SR   G. Schwarzenbach, P. Ruckstuhl and J. Zurc, Helv. Chim. Acta, 1951, 34, 455

51TV   B.J. Thamer and A.F. Voight, J. Amer. Chem. Soc., 1951, 73, 3197

51V   L.G. Van Uitert, Thesis, Penn. State Coll., 1951

51WS   A. Willi and G. Schwarzenbach, Helv. Chim. Acta, 1951, 34, 528

52B   C. Bertin-Batsch, Ann. Chim (France), 1952, 7, 481

52BC   C.V. Banks and A.B. Carlson, Anal. Chim. Acta, 1952, 7, 291

52BG   D. Banerjea, N.N. Ghosh, and P. Ray, J. Indian Chem. Soc., 1952, 29, 157

52CM   C.A. Coleman-Porter and C.B. Monk, J. Chem. Soc., 1952, 4363

52DA   N.K. Das, S. Aditya, and B. Prasad, J. Indian Chem. Soc., 1952, 29, 169

52DH   D.Dyrssen and B. Hok, Svensk Kem. Tidskr., 1952, 64, 80

52EL   D.H. Everett, D.A. Landsman, and B.R.W. Pinsent, Proc. Roy. Soc. (London), 1952,
             A215, 403

52EM   J.I. Evans and C.B. Monk, Trans. Faraday Soc., 1952, 48, 934

52F   S. Fronaeus, Acta Chem. Scand., 1952, 6, 1200

52HS   G. Heller and G. Schwarzenbach, Helv. Chim. Acta, 1952, 35, 812

52HT   F.R. Hepner, K.N. Trueblood, and J.H. Lucas, J. Amer. Chem. Soc., 1952, 74, 1333

52IB   H. Irving and C.F. Bell, J. Chem. Soc., 1952, 1216

52JP   A.V. Jones and H.N. Parton, Trans. Faraday Soc., 1952, 48, 8

52JS   J.C. James and J.C. Speakman, Trans. Faraday Soc., 1952, 48, 474

52KA   R.M. Keefer and L.J. Andrews, J. Amer. Chem. Soc., 1952, 74, 640

52MT   F.H. MacDougall and L.E. Topol, J. Phys. Chem., 1952, 56, 1090

52P   K.J. Pedersen, Acta Chem. Scand., 1952, 6, 243

52Pa   K.J. Pedersen, Acta Chem. Scand., 1952, 6, 285

52PD   R.W. Parry and F.W. Dubois, J. Amer. Chem. Soc., 1952, 74, 3749

52S   J. Schubert, J. Phys. Chem., 1952, 56, 113

52Sa   S. Suzuki, Nippon Kagaku Zasshi, 1952, 73, 92; Sci. Reports, Tohoku Univ., 1953,
           5, 16

52Sb   S. Suzuki, Nippon Kagaku Zasshi, 1952, 73, 317, 448

52Sc   F.A. Snavely, "Investigations of the Coordination Tendencies of o-Substituted
           Arylazo Compounds," Dept. Chem., Penn. State Coll., 1952

52SL   J. Schubert and A. Lindenbaum, J. Amer. Chem. Soc., 1952, 74, 3529

52TL   K.N. Trueblood and H.J. Lucas, J. Amer. Chem. Soc., 1952, 74, 1338

52TM   W.E. Trevelyan, P.F.E. Mann, and J.S. Harrison, Arch. Biochem. Biophys., 1952,
           39, 419

52TV   P.J. Thamer and A.F. Voigt, J. Phys. Chem., 1952, 56, 225

52V   L.G. VanUitert, Diss., Penn. State Coll., 1952

52YV   R.P. Yaffee and A.F. Voigt, J. Amer. Chem. Soc., 1952, 74, 2503

52YVa   R.P. Yaffee and A.F. Voigt, J. Amer. Chem. Soc., 1952, 74, 2941, 3163

53A   S. Ahrland, Acta Chem. Scand., 1953, 7, 485

53AP   S. Aditya and B. Prasad, J. Indian Chem. Soc., 1953, 30, 255

53B   J. Badoz-Lambling, Ann. Chim. (France), 1953, 8, 586

53BB   M. Bobtelsky and I. Bar-Gadda, Bull. Soc. Chim. France, 1953, 276, 382, 687

53BP   P. Brandt and O. Plum, Acta Chem. Scand., 1953, 7, 97

53CG   J.D. Chanley and E.M. Gindler, J. Amer. Chem. Soc., 1953, 75, 4035

53D   D.Dyrssen, Svensk Kem. Tidskr., 1953, 65, 43

53DH   C.W. Davies and B.E. Hoyle, J. Chem. Soc., 1953, 4134; ibid. 1955, 1038

53F   S. Fronaeus, Svensk Kem. Tidskr., 1953, 65, 19

53GK   D.M. Gruen and J.J. Katz, J. Amer. Chem. Soc., 1953, 75, 3772

53GS   R.W. Geiger and E.B. Sandell, Anal. Chim. Acta, 1953, 8, 197

53H   B. Hok, Svensk Kem. Tidskr., 1953, 65, 182

53JA   P.K. Jena, S. Aditya, and B. Prasad, J. Indian Chem. Soc., 1953, 30, 735

53K   M. Kilpatrick, J. Amer. Chem. Soc., 1953, 75, 584

53KE   M. Kilpatrick and R.D. Eanes, J. Amer. Chem. Soc., 1953, 75, 586

53KEa   M. Kilpatrick and R.D. Eanes, J. Amer. Chem. Soc., 1953, 75, 587

53KEM   M. Kilpatrick, R.D. Eanes, and J.G. Morse, J. Amer. Chem. Soc., 1953, 75, 588

53KM   M. Kilpatrick and J.G. Morse, J. Amer. Chem. Soc., 1953, 75, 1846

53KMa   M. Kilpatrick and J.G. Morse, J. Amer. Chem. Soc., 1953, 75, 1854

53LK   D.L. Leussing and I.M. Kolthoff, J. Amer. Chem. Soc., 1953, 75, 3904

53MA   P. Mahapatra, S. Aditya, and B. Prasad, J. Indian Chem. Soc., 1953, 30, 509

53R   J. Rydberg, Svensk Kem. Tidskr., 1953, 65, 37

53Ra   J. Rydberg, Arkiv. Kemi, 1953, 5, 413

53S   N. Sunden, Svensk Kem. Tidskr., 1953, 65, 257

53Sa   S. Suzuki, Nippon Kagaku Zasshi, 1953, 74, 531

53Sb   S. Suzuki, Nippon Kagaku Zasshi, 1953, 74, 590

53SA   S.C. Sircar, S. Aditya, and B. Prasad, J. Indian Chem. Soc., 1953, 30, 255, 633

53WW   R.C. Warner and I. Weber, J. Amer. Chem. Soc., 1953, 75, 5086

54A   A. Agren, Acta Chem. Scand., 1954, 8, 266

54Aa   A. Agren, Acta Chem. Scand., 1954, 8, 1059

54Ab   P.J. Antikainen, Ann. Acad. Sci. Fenn., Ser. A II, No. 56, 1954; Acta Chem.
          Scand., 1955, 9, 1008

54AC   J.H. Ashby, E.M. Crook, and S.P. Datta, Biochem. J., 1954, 56, 198

54BC   M. Bose and D.M. Chowdhury, J. Indian Chem. Soc., 1954, 31, 111

54BS   R.J. Bates and G. Schwarzenbach, Helv. Chim. Acta, 1954, 37, 1069

54BSa   R.G. Bates and G. Schwarzenbach, Experientia, 1954, 10, 482

54CC   H.B. Clarke, D.G. Cusworth, and S.P. Datta, Biochem. J., 1954, 58, 146

54CS   H. Christopherson and E.B. Sandell, Anal. Chim. Acta, 1954, 10, 1

54D   D. Dryssen, Svensk Kem. Tidskr., 1954, 66, 234

54Da   D. Dyrssen, Acta Chem. Scand., 1954, 8, 1394

54DH   J.F.J. Dippy, S.R.C. Hughes, and J.W. Laxton, J. Chem. Soc., 1954, 1470

54DM   P.R. Davies and C.B. Monk, Trans. Faraday Soc., 1954, 50, 128

54DMa   C.W. Davies and C.B. Monk, Trans. Faraday Soc., 1954, 50, 132

54DS   H.S. Dunsmore and J.C. Speakman, Trans. Faraday Soc., 1954, 50, 236

54EM   W.P. Evans and C.B. Monk, J. Chem. Soc., 1954, 550

54HS   R.E. Hamm, S.M. Shull, Jr., and D.M. Grant, J. Amer. Chem. Soc., 1954, 76, 2111

54IH   R.M. Izatt, C.G. Haas, Jr., B.P. Block, and W.C. Fernelius, J. Phys. Chem., 1954,
              58, 1133

54KS   T.K. Keenan and J.F. Suttle, J. Amer. Chem. Soc., 1954, 76, 2184

54MP   B.C. Mohanty and S. Pani, J. Indian Chem. Soc., 1954, 31, 593

54NP   A.C. Nanda and S. Pani, J. Indian Chem. Soc., 1954, 31, 588

54S   J. Schubert, J. Amer. Chem. Soc., 1954, 76, 3442

54Sa   S. Suzuki, Nippon Kagaku Zasshi, 1954, 75, 1088

54SK   W. Stricks, I.M. Kolthoff, and A. Heyndrick, J. Amer. Chem. Soc., 1954, 76, 1515

54SL   J. Schubert and A. Lindenbaum, J. Biol. Chem., 1954, 208, 359

54SLB   W.B. Schaap, H.A. Laitinen, and J.C. Bailar, J. Amer. Chem. Soc., 1954, 76, 5868

54WS   A.V. Willi and J.F. Stocker, Helv. Chim. Acta, 1954, 37, 1113

55A   A. Agren, Acta Chem. Scand., 1955, 9, 39

55Aa   A. Agren, Acta Chem. Scand., 1955, 9, 49

55AC   J.H. Ashby, H.B. Clarke, E.M. Crook, and S.P. Datta, Biochem. J., 1955, 59, 203

55AS   A. Agren and G. Schwarzenbach, Helv. Chim. Acta, 1955, 38, 1920

55BA   S. Bardham and S. Aditya, J. Indian Chem. Soc., 1955, 32, 102, 105, 109

55CF   J.D. Chanley and E.J. Feageson, J. Amer. Chem. Soc., 1955, 77, 4002

55D   D. Dyrssen, Acta Chem. Scand., 1955, 9, 1567

55DJ   D. Dyrssen and E. Johansson, Acta Chem. Scand., 1955, 9, 763

55F   W.S. Fyfe, J. Chem. Soc., 1955, 1032

55FT   I. Feldman, T.Y. Toribara, J.R. Havill, and W.F. Neuman, J. Amer. Chem. Soc.,
              1955, 77, 878

55GL   H.P. Gregor, L.B. Luttinger, and E.M. Loebl, J. Phys. Chem., 1955, 59, 34

55IF   R.M. Izatt, W.C. Fernelius, and B.P. Block, J. Phys. Chem., 1955, 59, 80

55IFa   R.M. Izatt, W.C. Fernelius, C.G. Haas, Jr., and B.P. Block, J. Phys. Chem., 1955,
          59, 170

55IFb   R.M. Izatt, W.C. Fernelius, and B.P. Block, J. Phys. Chem., 1955, 59, 235

55IP   D.J.G. Ives and J.H. Pryor, J. Chem. Soc., 1955, 2104

55LM   N.C. Li and R.A. Manning, J. Amer. Chem. Soc., 1955, 77, 5225

55LR   A.M. Liquori and A. Ripamonti, Gazz. Chim. Ital., 1955, 85, 578

55M   R.M. Milburn, J. Amer. Chem. Soc., 1955, 77, 2064

55MA   R.C. Mohanty and S. Aditya, J. Indian Chem. Soc., 1955, 32, 234

55MAa   R.C. Mohanty and S. Aditya, J. Indian Chem. Soc., 1955, 32, 249

55PJ   R.L. Pecsok and R.S. Juvet, Jr., J. Amer. Chem. Soc., 1955, 77, 202

55PP   G. Patra and S. Pani. J. Indian Chem. Soc., 1955, 32, 217

55PPa   G. Patra and S. Pani, J. Indian Chem. Soc., 1955, 32, 572

55PS   R.L. Pecsok and J. Sandera, J. Amer. Chem. Soc., 1955, 77, 1489

55R   J. Rydberg, Svensk Kem. Tidskr., 1955, 67, 499; Arkiv Kemi, 1955, 8, 101, 113

55RB   R.A. Robinson and A.I. Biggs, Trans. Faraday Soc., 1955, 51, 901

55S   J.M. Sturtevant, J. Amer. Chem. Soc., 1955, 77, 255

55Sa   S. Suzuki, Nippon Kagaku Zasshi, 1955, 76, 287

55SR   K.C. Samantora, D.V. Raman, and S. Pani, J. Indian Chem. Soc., 1955, 32, 165, 197

55V   L. Vareille, Bull. Soc. Chim. France, 1955, 872

55Va   L. Vareille, Bull. Soc. Chim. France, 1955, 1496

55VG   A.M. Vasiliev and V.M. Gorokhovskii, Uch. Zap. Kazansk. Gosudvest Univ., 1955,
          115, No. 3, 27; Chem. Abs., 1958, 52, 955a

55VGa   A.M. Vasiliev and V.M. Gorokhovskii, Uch. Zap. Kazansk. Gosudvest Univ., 1955,
          115, No. 3, 39

56A   A. Agren, Svensk Kem. Tidskr., 1956, 68, 181

56Aa   A. Agren, Svensk Kem. Tidskr., 1956, 68, 185

56Ab   A. Albert, Nature, 1956, 177, 525

56Ac   P.J. Antikainen, Suomen Kem., 1956, B29, 14

56Ad   P.J. Antikainen, Suomen Kem., 1956, B29, 123

56Ae   P.J. Antikainen, Suomen Kem., 1956, B29, 135

56Af   P.J. Antikainen, Suomen Kem., 1956, B29, 179

56AR   A. Albert, C.W. Reese, and A.J.H. Tomlinson, Brit. J. Exp. Path., 1956, 37, 500

56B   A.I. Biggs, <u>Trans. Faraday Soc.</u>, 1956, <u>52</u>, 35

56Ba   M. Breant, <u>Bull. Soc. Chim. France</u>, 1956, 948

56BE   R.B. Bentley and J.P. Elder, <u>J. Soc. Dyers Colourists</u>, 1956, <u>72</u>, 332

56BH   E.A. Burns and D.N. Hume, <u>J. Amer. Chem. Soc.</u>, 1956, <u>78</u>, 3958

56CD   H.B. Clarke and S.P. Datta, <u>Biochem. J.</u>, 1956, <u>64</u>, 604

56CF   J.D. Chanley and F. Feageson, <u>J. Amer. Chem. Soc.</u>, 1956, <u>78</u>, 2237

56D   D. Dyrssen, <u>Acta Chem. Scand.</u>, 1956, <u>10</u>, 353

56Da   D. Dyrssen, <u>Svensk Kem. Tidskr.</u>, 1956, <u>68</u>, 212

56DD   D. Dyrssen, M. Dyrssen, and E. Johansson, <u>Acta Chem. Scand.</u>, 1956, <u>10</u>, 106

56DL   W.S. Dorsey and H.J. Lucas, <u>J. Amer. Chem. Soc.</u>, 1956, <u>78</u>, 1665

56FG   Ya.A. Fialkov and V.V. Grigoreva, <u>J. Inorg. Chem. USSR</u>, 1956, <u>1</u>, No. 11, 74 (2504)

56FI   F.S. Feates and D.J.G. Ives, <u>J. Chem. Soc.</u>, 1956, 2798

56FS   V.V. Formin and V.V. Sinkowskii, <u>J. Inorg. Chem. USSR</u>, 1956, <u>1</u>, No. 10, 149 (2316)

56GN   E. Gelles and G.N. Nancollas, <u>Trans. Faraday Soc.</u>, 1956, <u>52</u>, 98

56GNa   E. Gelles and G.H. Nancollas, <u>Trans. Faraday Soc.</u>, 1956, <u>52</u>, 680

56GNb   E. Gelles and G.H. Nancollas, <u>J. Chem. Soc.</u>, 1956, 4847

56H   B. Hok-Bernstrom, <u>Acta Chem. Scand.</u>, 1956, <u>10</u>, 163

56Ha   B. Hok-Bernstrom, <u>Acta Chem. Scand.</u>, 1956, <u>10</u>, 174

56HE   C. Heitner and I. Eliezer, <u>Bull. Soc. Chim. France</u>, 1956, 174

56IT   M. Ishibashi, T. Tanaka, and T. Kawai, <u>Nippon Kagaku Zasshi</u>, 1956, <u>77</u>, 1613, 1603

56J   M.J. Jassen, <u>Rec. Trav. Chim.</u>, 1956, <u>75</u>, 1411

56Ka   E.J. King, <u>J. Amer. Chem. Soc.</u>, 1956, <u>78</u>, 6020

56KF   G.A. Knyazev, V.V. Fomin, and O.I. Zakhaeov-Martisissov, <u>J. Inorg. Chem. USSR</u>,
          1956, <u>1</u>, No. 2, 159 (342)

56KK   E.J. King and G.W. King, <u>J. Amer. Chem. Soc.</u>, 1956, <u>78</u>, 1089

56L   P.O. Lumme, <u>Suomen Kem.</u>, 1956, <u>B29</u>, 217

56LP   S.K. Lee, E.O. Price, and J.E. Land, <u>J. Amer. Chem. Soc.</u>, 1956, <u>78</u>, 1325

56N   G.N. Nancollas, <u>J. Chem. Soc.</u>, 1956, 744

56Na   R. Nasanen, <u>Suomen Kem.</u>, 1956, <u>B29</u>, 91

56NM   R. Nasanen and R. Markkanen, <u>Suomen Kem.</u>, 1956, <u>B29</u>, 119

56NV   R. Nasanen and J. Veivo, <u>Suomen Kem.</u>, 1956, <u>B29</u>, 213

560C   B.G. Odenheimer and G.R. Choppin, UCRL-3515, 1956

56P   V.E. Panova, J. Inorg. Chem. USSR, 1956, 1, No. 3, 75 (422)

56PJ   R.L. Pecsok and R.S. Juvet, J. Amer. Chem. Soc., 1956, 78, 3967

56R   J. Rydberg, Arkiv Kemi, 1956, 9, 109

56RC   O. Ryba, J. Cifka, M. Malat, and V. Serk, Coll. Czech. Chem. Comm., 1956, 21, 349

56RK   R.A. Robinson and A.K. Kiang, Trans. Faraday Soc., 1956, 52, 327

56RR   J. Rydberg and B. Rydberg, Arkiv Kemi, 1956, 9, 81

56S   B. Sarma, J. Sci. Ind. Research, India, 1956, 15 B, 696

56SD   J.A. Schufle and C. D'Agostine, Jr., J. Phys. Chem., 1956, 60, 1623

56SR   B.D. Sarma and P. Ray, J. Indian Chem. Soc., 1956, 33, 841

56T   V.F. Toropova, J. Inorg. Chem. USSR, 1956, 1, No. 5, 52 (930)

56TB   R. Trujillo and F. Brito, An. Fis. Quim., 1956, 52B, 407

56TS   J.G. Traynham and M.F. Sehnert, J. Amer. Chem. Soc., 1956, 78, 4024

56VP   L.I. Vinogradova and B.V. Ptitsyn, J. Inorg. Chem. USSR, 1956, 1, No. 3, 81 (427),
          87 (432)

56YA   K.B. Yatsimirskii and I.I. Alekseeva, J. Inorg. Chem. USSR, 1956, 1, No. 5,
          76 (952)

56YF   K.B. Yatsimirskii and T.I. Fedorova, J. Inorg. Chem. USSR, 1956, 1, No. 10,
          144 (2310)

56YS   M. Yasuda, K. Suzuki, and K. Yamasaki, J. Phys. Chem., 1956, 60, 1649

56Z   R.A. Zingaro, J. Amer. Chem. Soc., 1956, 78, 3568

56Za   E.K. Zolotorev, Diss., Chem. Tech. Ivanovo, 1956

57A   P.J. Antikainen, Suomen Kem., 1957, B30, 45

57Aa   P.J. Antikainen, Suomen Kem., 1957, B30, 147

57Ab   P.J. Antikainen, Suomen Kem., 1957, B30, 185

57BD   A.K. Babko and L.I. Dubovenko, J. Inorg. Chem. USSR, 1957, 2, No. 4, 149 (808)

57BDa   A.K. Babko and L.I. Dubovenko, J. Inorg. Chem. USSR, 1957, 2, No. 6, 128 (1294)

57BDH   L.G. Bray, J.F.J. Dippy, S.R.C. Hughes, and L.W. Laxton, J. Chem. Soc., 1957,
          2405

57BDM   W.D. Bale, E.W. Davies, D.B. Morgan, and C.B. Monk, Disc. Faraday Soc., 1957,
          24, 94

57BP   A.K. Babko and O.I. Popova, J. Inorg. Chem. USSR, 1957, 2, No. 1, 214 (138)

57CR   D. Cozzi and G. Raspi, Ric. Sci., 1957, 27, 2392

57D   D. Dyrssen, Acta Chem. Scand., 1957, 11, 1771

57DB   H.A. Droll, B.P. Block, and W.C. Fernelius, J. Phys. Chem., 1957, 61, 1000

57DS   N.K. Dutt and B. Sur, Z Anorg. Allg. Chem., 1957. 293, 195

57FG   Ya.A. Fialkov and V.V. Grigoreva, J. Inorg. Chem. USSR, 1957, 2, No. 2, 93 (287)

57FP   Ya.A. Fialkov and N.G. Peryshkina, J. Inorg. Chem. USSR, 1957, 2, No. 4, 59 (749)

57GL   V.M. Gorokhorskii and Ya.A. Levin, J. Inorg. Chem. USSR, 1957, 2, No. 2, 176 (343)

57GM   A.D. Gelman, N.N. Matorina, and A.I. Moskvin, Proc. Acad. Sci. USSR, Chem. Sect.,
           1957, 117, 979 (88)

57HB   H.M. Hershenson, R.J. Brooks, and M.E. Murphy, J. Amer. Chem. Soc., 1957, 79,
           2046

57HC   G.K. Helmkamp, F.L. Carter, and H.J. Lucas, J. Amer. Chem. Soc., 1957, 79, 1306

57HE   C. Heitner-Wirgin and I. Eliezer, Bull. Soc. Chim. France, 1957, 149

57HK   E.F.G. Herington and W. Kynaston, Trans. Faraday Soc., 1957, 53, 138

57HR   G.P. Hildebrand and C.N. Reilly, Anal. Chem., 1957, 29, 258

57J   O. Jantti, Suomen Kem., 1957, B30, 136

57Ja   M.J. Jassen, Rec. Trav. Chim., 1957, 76, 827

57JB   E. Jungrois and Bobtelski, Bull. Res. Council Israel, 1957, 7A, 35

57KP   I.A. Korshunov, A.P. Pochinaib, and V.M. Tikhomirova, J. Inorg. Chem. USSR, 1957,
           2, No. 1, 101 (68)

57L   P.O. Lumme, Suomen Kem., 1957, B30, 194

57La   J. Lefebvre, J. Chim. Phys., 1957, 54, 567

57Lb   J. Lefebvre, J. Chim. Phys., 1957, 54, 581

57Lc   J. Lefebvre. J. Chim. Phys., 1957, 54, 601

57LW   N.C. Li, W.M. Westfall, A. Lindenbaum, J.M. White, and J. Schubert, J. Amer.
           Chem. Soc., 1957, 79, 5864

57MC   A.E. Martell, S. Chaberek, S. Westerback, and H. Hyytiainen, J. Amer. Chem. Soc.,
           1957, 79, 3036

57MD   A.K. Mukerjee and A.K. Dey, J. Indian Chem. Soc., 1957, 34, 461

57MJ   E. Monnier and C. Jegge, Helv. Chim. Acta, 1957, 40, 513

57N   R. Nasanen, Suomen Kem., 1957, B30, 61

57Na   R. Nasanen, Acta Chem. Scand., 1957, 11, 1308

57NA   R.K. Nanda and S. Aditya, J. Indian Chem. Soc., 1957, 34, 577

57P   V.E. Panova, J. Inorg. Chem. USSR, 1957, 2, No. 2, 156 (330)

57PC   D. Peltier and M. Conti, Compt. Rend. Acad. Sci. Paris, 1957, 244, 2811

57PP   R.K. Patnaik and S. Pani, J. Indian Chem. Soc., 1957, 34, 19, 619, 673

57PS   R.L. Pecsok and J. Sandera, J. Amer. Chem. Soc., 1957, 79, 4069

57RB   R.A. Robinson and A.I. Biggs, Aust. J. Chem., 1957, 10, 128

57RL   G.L. Roy, A.L. Lafedriere, and J.O. Edwards, J. Inorg. Nucl. Chem., 1957, 4, 106

57SA   G. Schwarzenbach and G. Anderegg, Helv. Chim. Acta, 1957, 40, 1229

57SAa   G. Schwarzenbach and G. Anderegg, Helv. Chim. Acta, 1957, 40, 1773

57T   C.F. Timberlake, J. Chem. Soc., 1957, 4987

57TB   R.M. Tichane and W.E. Bennett, J. Amer. Chem. Soc., 1957, 79, 1293

57TT   R. Trujillo and F. Torres, An. Real. Soc. Espan. Fis. Quim., 1957, 53 B, 263

57V   O. Vartapetian, Ann. Chim. (France), 1957, 2, 916

57WB   F. Wold and C.E. Ballou, J. Biol. Chem., 1957, 227, 301

58A   P.J. Antikainen, Suomen Kem., 1958, B31, 255

58AC   S. Ahrland, J. Chatt, N.R. Davies, and A.A. Williams, J. Chem. Soc., 1958, 264

58ACa   S. Ahrland, J. Chatt, N.R. Davies, and A.A. Williams, J. Chem. Soc., 1958, 276

58ACb   S. Ahrland, J. Chatt, N.R. Davies, and A.A. Williams, J. Chem. Soc., 1958, 1403

58AO   A.W. Adamson, H. Ogata, J. Grossman, and R. Newbury, J. Inorg. Nucl. Chem., 1958,
            6, 319

58BB   C.V. Banks and D.W. Barnum, J. Amer. Chem. Soc., 1958, 80, 3579

58BG   S.M. Banerjee, A.K. SenGupta, S.K. Siddhanta, J. Indian Chem. Soc., 1958, 35, 269

58BL   C.V. Banks, J.P. Laplante, and J.J. Richard, J. Org. Chem., 1958, 23, 1210

58BS   S.M. Banerjee, A.K. SenGupta, and S.K. Siddhanta, J. Indian Chem. Soc., 1958,
            35, 268

58CG   R.C. Courtney, R.L. Gustafson, S. Chaberek, Jr., and A.E. Martell, J. Amer. Chem.
            Soc., 1958, 80, 2121

58CP   W.J. Canady, H.M. Papee, and K.J. Laidler, Trans. Faraday Soc., 1958, 54, 502

58D   W.A. Deskin, J. Amer. Chem. Soc., 1958, 80, 5680

58DB   N.K. Dutt and P. Bose, Z. Anorg. Allg. Chem., 1958, 295, 131

58DG   S.P. Datta and A.K. Grzybowski, Biochem. J., 1958, 69, 218

58G   D.P. Graddon, J. Inorg. Nucl. Chem., 1958, 5, 219

58GD   A.D. Gelman, L.E. Drabkina, and A.I. Moskvin, J. Inorg. Chem. USSR, 1958, 3,
            No. 7, 96 (1546); No. 8, 290 (1934)

58GH   E. Gelles and R.W. Hay, J. Chem. Soc., 1958, 3673

58GS   E. Gelles and A. Salama, J. Chem. Soc., 1958, 3683

58H   N. Hojo, Kogyo Kagaku Zasshi, 1958, 61, 1514

58HF   C. Heitner-Wirguin, D. Friedmann, J.H. Goldschmidt, and J. Shamin, Bull. Soc. Chim. France, 1958, 864

58IS   F. Iimura, M. Shima, and H. Sano, Nippon Kagaku Zasshi, 1958, 79, 1032, 1037, 1041, 1045, 1048

58KY   N.A. Kostromina and S.I. Yakubson, J. Inorg. Chem. USSR, 1958, 3, No. 11, 104 (2506)

58L   P.O. Lumme, Suomen Kem., 1958, B31, 253

58La   D.L. Leussing, J. Amer. Chem. Soc., 1958, 80, 4180

58LR   T.J. Lane, J.A. Ryan, and E.F. Britten, J. Amer. Chem. Soc., 1958, 80, 315

58MC   A.K. Majumdar and M.M. Chakrabartty, Anal. Chim. Acta, 1958, 19, 372

58MD   A.K. Mukherji and A.K. Dey, Anal. Chim. Acta, 1958, 18, 324

58MDa   A.K. Mukherji and A.K. Dey, J. Inorg. Nucl. Chem., 1958, 6, 314

58MDb   A.K. Mukherji and A.K. Dey, J. Sci. Ind. Res. India, 1958, 17B, 312

58MDc   A.K. Mukherji and A.K. Dey, J. Indian Chem. Soc., 1958, 35, 237

58MG   A.I. Moskvin and A.D. Gelman, J. Inorg. Chem. USSR, 1958, 3, No. 4, 188 (956), 198 (962)

58MS   P.K. Migal and A.Y. Sychev, J. Inorg. Chem. USSR, 1958, 3, No. 2, 104 (314)

58N   R. Nasanen, Suomen Kem., 1958, B31, 19

58Na   R. Nasanen, Suomen Kem., 1958, B31, 261

58NP   C.B. Nanda and S. Pani, J. Indian Chem. Soc., 1958, 35, 355

58P   D.D. Perrin, Nature, 1958, 182, 741

58Pa   E. Pisko, Chem. Zvesti, 1958, 12, 95

58PM   V.I. Paramonova, A.N. Mosenvich, and A.I. Subbotina, J. Inorg. Chem. USSR, 1958, 3, No. 1, 131 (88)

58PZ   V.M. Peshkova and A.P. Zozulya, Nauch. Dokl. Vys. Shk., Khim., 1958, 470

58RC   P. Romain and J.C. Colleter, Bull. Soc. Chim. France, 1958, 867

58S   A. Sonesson, Acta Chem. Scand., 1958, 12, 165, 1937

58Sa   Sommer, Z. Anal. Chem., 1958, 164, 299

58SB   S.K. Siddhanta and S.N. Banerjee, J. Indian Chem. Soc., 1958, 35, 279

58SBa   S.K. Siddhanta and S.N. Banerjee, J. Indian Chem. Soc., 1958, 35, 323, 339, 343, 349, 419, 425, 426

58SBb   S.K. Siddhanta and S.N. Banerjee, J. Indian Chem. Soc., 1958, 35, 279, 323

58SG   G. Schwarzenbach, O. Gubeli, and H. Zust, Chimia (Switz.), 1958, 12, 84

58SL   J.L. Schubert, E.L. Lind, W.M. Westfall, R. Pleger, and N.C. Li, J. Amer. Chem. Soc., 1958, 80, 4799

58T   A.S. Tikhonov, J. Inorg. Chem. USSR, 1958, 3, No. 2, 75 (296)

58TF   W.B. Truemann and L.M. Ferris, J. Amer. Chem. Soc., 1958, 80, 5050

58W   G. Watelle-Morion, Compt. Rend. Acad. Sci. Paris, 1958, 246, 3610

58Wa   J.S. Wiberg, Arch. Biochem. Biophys., 1958, 73, 337

58YF   K.B. Yatsimirskii and T.I. Fedorova, Izv. Vyssh. Ucheb. Zaved., Khim., 1958,
           No. 3, 40

59A   P.J. Antikainen, Acta Chem. Scand., 1959, 13, 312

59AK   P.J. Antikainen and A. Kauppila, Suomen Kem., 1959, B32, 141

59AM   P.J. Antikainen and P.J. Malkonen, Suomen Kem., 1959, B32, 179

59AR   P.J. Antikainen and V.M.K. Rossi, Suomen Kem., 1959, B32, 182

59ARa   P.J. Antikainen and V.M.K. Rossi, Suomen Kem., 1959, B32, 185

59AT   P.J. Antikainen and K. Tevanen, Suomen Kem., 1959, B32, 214

59B   E.H. Binns, Trans. Faraday Soc., 1959, 55, 1900

59Ba   P. Brandt, Acta Chem. Scand., 1959, 13, 1639

59BD   A.K. Babko and L.I. Dubovenko, Russ. J. Inorg. Chem., 1959, 4, 165 (372)

59BS   C.V. Banks and R.S. Singh, J. Amer. Chem. Soc., 1959, 81, 6159

59BSR   A.K. Bhattacharya, V.V. Subbanna, and G.S. Rao, J. Sci. Ind. Res., India, 1959,
           18 B, 127

59CF   G.E. Cheney, Q. Fernando, and H. Freiser, J. Phys. Chem., 1959, 63, 2055

59CG   S. Chaberek, Jr., R.L. Gustafson, R.C. Courtney, and A.E. Martell, J. Amer. Chem.
           Soc., 1959, 81, 515

59DA   B. Das and S. Aditya, J. Indian Chem. Soc., 1959, 36, 473

59DB   A.K. Dey and S.K. Banerji, Proc. Symposium on Chem. of Coordination Compounds,
           Agra, Part 3, 1959, 198

59DG   N.K. Dutt and N. Goswami, Z. Anorg. Allg. Chem., 1959, 298, 258

59DGa   N.K. Dutt and N. Goswami, Z. Anorg. Allg. Chem., 1959, 298, 265

59DK   D. Dyrssen, F. Krasovec and L.G. Sillen, Acta Chem. Scand., 1959, 13, 50
59DL   H. Diehl and F. Lindstrom, Anal. Chem., 1959, 31, 414
59EB   M. Eden and R.G. Bates, J. Res. Nat. Bur. Stand., 1959, 62, 161

59FE   Ya.A. Fialkov and V.I. Ermolenko, Russ. J. Inorg. Chem., 1959, 4, 159 (359),
           615 (1369)

59FF   M.M. Fickling, A. Fisher, B.R. Mann, J. Packer, and J. Vaughan, J. Amer. Chem.
           Soc., 1959, 81, 4226

59FH   L.P. Fernandez and L.G. Hepler, J. Amer. Chem. Soc., 1959, 81, 1783

59FHa   L.P. Fernandez and L.G. Hepler, J. Phys. Chem., 1959, 63, 110

59H   J. Halmekoski, Ann. Acad. Sci. Fenn., 1959, A II, 96

59KV   K.N. Kovalenko and L.I. Vistyak, Russ. J. Inorg. Chem., 1959, 4, 364 (801)

59L   D.L. Leussing, J. Amer. Chem. Soc., 1959, 81, 4208

59LL   N.C. Li, A. Lindenbaum, and J.M. White, J. Inorg. Nucl. Chem., 1959, 12, 122

59MD   A.K. Mukherji and A.K. Dey, Z. Phys. Chem. (Frankfurt), 1959, 210, 114

59MR   D.L. Martin and R.J.C. Rossotti, Proc. Chem. Soc., 1959, 60

59MZ   A.I. Moskvin and F.A. Zakharova, Russ, J. Inorg. Chem., 1959, 4, 975 (2151)

59N   R. Nasanen, Acta Chem. Scand., 1959, 13, 869

59Na   R. Nasanen, Suomen Kem., 1959, B32, 7

59NH   R. Nasanen and T. Heikkila, Suomen Kem., 1959, B32, 73

59OK   A. Okac and Z. Kolarik, Coll. Czech. Chem. Comm., 1959, 24, 1

59OKa   A. Okac and Z. Kolarik, Coll. Czech. Chem. Comm., 1959, 24, 266

59P   D.D. Perrin, J. Chem. Soc., 1959, 1710

59PT   B.V. Ptitsyn and E.N. Tekster, Russ. J. Inorg. Chem., 1959, 4, 1024 (2248)

59PZ   V.M. Peshkova and A.P. Zozulya, J. Anal. Chem. USSR, 1959, 14, 433 (411)

59RE   D.I. Ryabchikov, A.N. Ermakov, V.K. Belyaeva, and I.N. Marov, Russ. J. Inorg.
          Chem., 1959, 4, 818 (1814)

59RR   M.M. Ray and P. Ray, J. Indian Chem. Soc., 1959, 36, 849

59RS   J. Rydberg and J.C. Sullivan, Acta Chem. Scand., 1959, 13, 2057

59S   L. Sommer, Coll. Czech. Chem. Comm., 1959, 24, 1649

59Sa   L. Sommer, Bull. Soc. Chim. France, 1959, 862

59Sb   A. Sonesson, Acta Chem. Scand., 1959, 13, 998

59Sc   A. Sonesson, Acta Chem. Scand., 1959, 13, 1437

59SH   L. Sommer and M. Hnilickova, Bull. Soc. Chim. France, 1959, 36

59SM   V. Suk and V. Miketukova, Coll. Czech. Chem. Comm., 1959, 24, 3629

59T   C.F. Timberlake, J. Chem. Soc., 1959, 2795

59TC   R.W. Taft, Jr. and E.H. Cook, J. Amer. Chem. Soc., 1959, 81, 46

59TK   N. Tanaka and K. Kato, Bull. Chem. Soc. Japan, 1959, 32, 516

59TKa   N. Tanaka and K. Kato, Bull. Chem. Soc. Japan, 1959, 32, 1376

59TO   J.G. Traynham and J.R. Olechowski, J. Amer. Chem. Soc., 1959, 81, 571

59TOK   N. Tanaka, I.T. Oiwa, T. Kurosawa, and T. Nozoe, Bull. Chem. Soc. Japan, 1959,
          32, 92

59TT   R. Trujillo and F. Torres, An. Real. Soc. Espan. Fis. Quim., 1959, 52 B, 157

59TV   E.N. Tekster, L.I. Vinogradova, and B.V. Ptitsyn, Russ. J. Inorg. Chem., 1959,
        4, 347 (764)

59WH   C.W. Wu and K.H. Hsu, Kexue Tongbao, 1959, 330

59Z   V.L. Zolotavin, Russ. J. Inorg. Chem., 1959, 4, 1254 (2713)

59ZM   A.P. Zozulya, N.N. Mezentseva, V.M. Peshkova, and Yu.K. Yurev, J. Anal. Chem.
        USSR, 1959, 14, 15 (17)

59ZP   A.P. Zozulya and V.M. Peshkova, Russ. J. Inorg. Chem., 1959, 4, 168 (379)

59ZPL   T.W. Zawidzki, H.M. Papee, and K.J. Laidler, Trans. Faraday Soc., 1959, 55, 1743

60AR   P.J. Antikainen and V.M.K. Rossi, Suomen Kem., 1960, B33, 94

60AS   A. Adams and T.D. Smith, J. Chem. Soc., 1960, 4846

60AT   P.J. Antikainen and K. Tevanen, Suomen Kem., 1960, B33, 7

60B   L. Benisek, Coll. Czech. Chem. Comm., 1960, 25, 2688

60Ba   E. Baumgartel, Habil-Schrift, T.H. Dresden, 1960

60BC   M.T. Beck, B. Csiszar, and P. Szarvas, Nature, 1960, 188, 846

60BD   A.K. Babko and L.I. Dubovenko, Izv. Vyssh. Ucheb. Zaved., Khim., 1960, 3, 226

60BS   C.V. Banks and R.S. Singh, J. Inorg. Nucl. Chem., 1960, 15, 125

60BSa   A.K. Bhattacharya and M.C. Saxena, Curr. Sci. (India), 1960, 29, 128

60C   J.C. Colleter, Ann. Chim. (France), 1960, 5, 415

60Ca   R.K. Chaturvedi, Curr. Sci. (India), 1960, 29, 128

60CI   S.H. Cohen, R.J. Iwomota, and J. Kleinberg. J. Amer. Chem. Soc., 1960, 82, 1844

60D   R.L. Dutta, J. Indian Chem. Soc., 1960, 37, 499

60DA   R.C. Das and S. Aditya, J. Indian Chem. Soc., 1960, 37, 557

60DE   H. Diehl and J. Ellingboe, Anal. Chem., 1960, 32, 1120

60DL   G. Dahlgren, Jr., and F.A. Long, J. Amer. Chem. Soc., 1960, 82, 1303

60DP   R. Das, R.K. Pattanaik, and S. Pani, J. Indian Chem. Soc., 1960, 37, 59

60DT   J.F. Duncan and F.G. Thomas, J. Chem. Soc., 1960, 2814

60FN   I. Feldman, C.A. North, and H.B. Hunter, J. Phys. Chem., 1960, 64, 1224

60FS   W.W. Forest and J.M. Sturtevant, J. Amer. Chem. Soc., 1960, 82, 585

60GF   I. Grenthe and W.C. Fernelius, J. Amer. Chem. Soc., 1960, 82, 6258

60GS   A.A. Grinberg and L.V. Shikheeva, Russ. J. Inorg. Chem., 1960, 5, 287 (599)

60GM   G. Goldstein, O. Menis, and D.L. Manning, Anal. Chem., 1960, 32, 400

60GR   R.L. Gustafson, C. Richard, and A.E. Martell, J. Amer. Chem. Soc., 1960, 82, 1526

60HP   G.P. Haight, Jr., and V. Pargamian, Anal. Chem., 1960, 32, 642

60JP   B. Jezowska-Trzebiatowska and L. Pajdowski, Rocz. Chem., 1960, 34, 787

60K   E.J. King, J. Amer. Chem. Soc., 1960, 82, 3575

60Ka   M. Kato, Z. Phys. Chem., 1960, 23, 375

60KF   W.L. Koltun, M. Fried, and F.R.N. Gurd, J. Amer. Chem. Soc., 1960, 82, 233

60KP   A.E. Klygin and V.K. Pavlova, Russ. J. Inorg. Chem., 1960, 5, 734 (1516)

60Lb   P.O. Lumme, Suomen Kem., 1960, B33, 87

60LA   D.L. Leussing and G.S. Alberts, J. Amer. Chem. Soc., 1960, 82, 4458

60LD   F. Lindstrom and H. Diehl, Anal. Chem., 1960, 32, 1123

60LL   D.L. Leussing, R.E. Laramy, and G.S. Alberts, J. Amer. Chem. Soc., 1960, 82, 4826

60LM   D.L. Leussing and J.P. Mislan, J. Phys. Chem., 1960, 64, 1908

60LP   I.A. Lebedev, S.V. Pirozhkov, and G.N. Yakovlev, Radiokhimiya, 1960, 2, 549

60M   O. Makitie, Suomen Kem., 1960, B33, 207

60MK   A.I. Moskvin, G.V. Khalturin, and A.D. Gelman, Soviet Radiochem., 1960, 2,
         69 (141)

60MN   A. McAuley and G.H. Nancollas, Trans. Faraday Soc., 1960, 56, 1165

60N   R. Nasanen, Suomen Kem., 1960, B33, 1

60Na   R. Nasanen, Suomen Kem., 1960, B33, 111

60R   J. Rydberg, Acta Chem. Scand., 1960, 14, 157

60Ra   A.K. Ray, Z. Anorg. Allg. Chem., 1960, 305, 207

60RE   D.I. Ryabchikov, A.N. Ermakov, V.K. Belyaeva, and I.N. Marov, Russ. J. Inorg.
         Chem., 1960, 5, 505 (1051)

60S   J. Stary, Coll. Czech. Chem. Comm., 1960, 25, 86, 890

60Sa   J. Stary, Coll. Czech. Chem. Comm., 1960, 25, 2630

60Sb   A. Sonesson, Acta Chem. Scand., 1960, 14, 1495

60Sc   S.C. Srivastava, Doctor's Thesis, Allahabad Univ., India, 1960

60SR   N.R. Sengupta and P. Ray, J. Indian Chem. Soc., 1960, 37, 303

60SV   V.G. Spidsyn and O. Vortekh, Proc. Acad. Sci. USSR, 1960, 133, 859 (613)

60TK   V.F. Toropova and L.S. Kirillova, Russ. J. Inorg. Chem., 1960, 5, 276 (575)

60TKa   N. Tanaka and K. Kato, Bull. Chem. Soc. Japan, 1960, 33, 417

60TKO   N. Tanaka, M. Kamada, H. Osawa, and G. Sato, Bull. Chem. Soc. Japan, 1960, 33,
         1412

60TN   V.F. Toropova and K.V. Naimushina, Russ. J. Inorg. Chem., 1960, 5, 421 (874)

60WT   J.M. White, P. Tang, and N.C. Li, J. Inorg. Nucl. Chem., 1960, 14, 255

60YD   Y. Yamane and N. Davidson, J. Amer. Chem. Soc., 1960, 82, 2123

60YY   M. Yasuda, K. Yamsaki, and H. Ohtake, Bull. Chem. Soc. Japan, 1960, 23, 1067

61AM   J.P. Armanet and J.C. Merlin, Bull Soc. Chim. France, 1961, 440

61BB   S.D. Bhardwaj and G.V. Bakore, J. Indian Chem. Soc., 1961, 38, 967

61BD   S.K. Banerji and A.K. Dey, Z. Anorg. Allg. Chem., 1961 309, 226

61BDa   S.K. Banerji and A.K. Dey, J. Indian Chem. Soc., 1961, 38, 121

61BDb   S.K. Banerji and A.K. Dey, J. Indian Chem. Soc., 1961, 38, 139

61BH   P. Bamberg and P. Hemmerich, Helv. Chim. Acta, 1961, 44, 1001

61BR   A.I. Biggs and R.A. Robinson, J. Chem. Soc., 1961, 388, 2572

61CC   G.R. Choppin and J.A. Chopoorian, J. Inorg. Nucl. Chem., 1961, 22, 97

61CO   I. Cadariu and L. Oniciu, Studii si Cercetari Chim. (Cluj), 1961, 12, 69

61COC   E. Campi, G. Ostacoli, N. Cibrario, and G. Saini, Gazz. Chim. Ital., 1961, 91,
         361

61CR   E. Coates and B. Rigg, Trans. Faraday Soc., 1961, 57, 1088, 1637

61DA   R.C. Das and S. Aditya, J. Indian Chem. Soc., 1961, 38, 19

61DH   D. Dyrssen and M. Hennichs, Acta Chem. Scand., 1961, 15, 47

61DI   S.M. Das and S.J.G. Ives, Proc. Chem. Soc., 1961, 373

61DS   R.L. Dutta and N.R. Sengupta, J. Indian Chem. Soc., 1961, 38, 741

61GA   A.A. Grinberg and V.I. Astapovich, Russ. J. Inorg. Chem., 1961, 6, 164 (321)

61GK   A.A. Grinberg and Kh.Kh. Khakimov, Russ. J. Inorg. Chem., 1961, 6, 71 (144)

61JS   T.J. Jin, L. Sommer, and A. Okac, Spisy Prir. Fak. Univ. Purkyne, Brno, 1961,
         No. 420, 93

61JW   B.R. James and R.J.P. Williams, J. Chem. Soc., 1961, 1007

61KK   A.E. Klygin and N.S. Kolyada, Russ. J. Inorg. Chem., 1961, 6, 107 (216)

61KP   E.C. Knoblock and W.C. Purdy, J. Electroanal. Chem., 1961, 2, 493

61KPa   A.E. Klygin and V.K. Pavlova, Russ. J. Inorg. Chem., 1961, 6, 536 (1050)

61KPb   E.J. King and J.E. Prue, J. Chem. Soc., 1961, 275

61LR   L. Laloi and P. Rumpf, Bull. Soc. Chim. France, 1961, 1645

61LT   D.L. Leussing and T.N. Tischer, J. Amer. Chem. Soc., 1961, 83, 65

61M   O. Makitie, Agricultural Research Center, Helsinki, Agrogeological Publications
         No. 79, 1961

61MA   W.A.E. McBryde and G.F. Atkinson, Canad. J. Chem., 1961, 39, 510

61MM   P.G. Manning and C.B. Monk, Trans. Faraday Soc., 1961, 57, 1996

61MMZ   A.I. Moskvin, I.N. Marov, and Yu.A. Zolotov, Russ. J. Inorg. Chem., 1961, 6, 926
          (1813)

61MN   A. McAuley and G.H. Nancollas, J. Chem. Soc., 1961, 2215

61MNa   A. McAuley and G.H. Nancollas, J. Chem. Soc., 1961, 4458

61MS   V. Moucka and J. Stary, Coll. Czech. Chem. Comm., 1961, 26, 763

61N   V.S.K. Nair, Trans. Faraday Soc., 1961, 57, 1988

61Na   L.B. Nanninga, Biochim. Biophys. Acta, 1961, 54, 330

61NN   V.S.K. Nair and G.H. Nancollas, J. Chem. Soc., 1961, 4367

61NP   R. Nasanen and K. Paakkola, Suomen Kem., 1961, B34, 19

61NPG   B.P. Nikolskii, V.V. Palchevskii, and R.G. Gorbunova, Russ. J. Inorg. Chem.,
          1961, 6, 309 (606)

61OC   G. Ostacoli, E. Campi, N. Cibrario, and G. Saini, Gazz. Chim. Ital., 1961, 91,
          349

61PA   V.M. Peshkova and P. Ang, Russ. J. Inorg. Chem., 1961, 6, 1064 (2082)

61PM   V.M. Peshkova, N.V. Melchakova, and S.G. Zhemchuzin, Russ. J. Inorg. Chem., 1961,
          6, 630 (1233)

61PP   R.K. Patnaik and S. Pani, J. Indian Chem. Soc., 1961, 38, 229, 233, 364, 379,
          709, 896

61PS   R.L. Pecsok and W.P. Schaefer, J. Amer. Chem. Soc., 1961, 83, 62

61RM   A.K. Rai and R.C. Mehrotra, Z. Phys. Chem. (Frankfurt), 1961, 29, 237

61S   A. Sonesson, Acta Chem. Scand., 1961, 15, 1

61Sa   I.C. Smith, Diss., Kansas State Univ., 1961

61Sb   A. Sandell, Acta Chem. Scand., 1961, 15, 190

61SK   S.Ya. Shnaiderman and I.E. Kalinichenko, Russ. J. Inorg. Chem., 1961, 6, 941
          (1843)

61SKa   Z.A. Sheka and E.E. Kriss, Russ. J. Inorg. Chem., 1961, 6, 984 (1930)

61SL   L. Sommer and A. Losmanova, Coll. Czech. Chem. Comm., 1961, 26, 2781

61SM   K. Schlyter and D.L. Martin, Trans. Roy. Inst. Technol., Stockholm, 1961, No. 175

61SMa   A. Sobkowska and J. Minczewsky, Rocz. Chim., 1961, 35, 47

61SO   G. Saini, G. Ostacoli, E. Campi, and N. Cibrario, Gazz. Chim. Ital., 1961, 91,
          242

61SOa   G. Saini, G. Ostacoli, E. Campi, and N. Cibrario, Gazz. Chim. Ital., 1961, 91,
          904

61SP   L. Sommer and K. Pliska, Coll. Czech. Chem. Comm., 1961, 26, 2754

61SS   V.B. Shevchenko and V.S. Smelov, Russ. J. Inorg. Chem., 1961, 6, 372 (732)

61TD   E.R. Tucci, E. Doody, and N.C. Li, J. Phys. Chem., 1961, 65, 1570

61TSL   V.F. Toropova, M.K. Saikina, and N.K. Lutskaga, Russ. J. Inorg. Chem., 1961, 6,
            1066 (2086)

61W   M. Walser, J. Phys. Chem., 1961, 65, 159

61Y   M. Yasada, Z. Phys. Chem. (Frankfurt), 1961, 27, 333

61YB   K.B. Yatsimirskii and L.I. Budarin, Russ. J. Inorg. Chem., 1961, 6, 944 (1840)

61YR   K.B. Yatsimirskii and L.P. Raizman, Russ. J. Inorg. Chem., 1961, 6, 1263 (2496)

61ZK   O.E. Zvyagintsev and L.G. Khromenkov, Russ. J. Inorg. Chem., 1961, 6, 548 (1074)

61ZM   Yu.A. Zolotov, I.N. Marov, and A.I. Moskvin, Russ. J. Inorg. Chem., 1961, 6, 539
            (1055)

62AK   Z.F. Andreeva and I.V. Kolosor, Izu. Timirazevsk. Selskokhoz. Acad., 1962, 212;
            Chem. Abs., 1963, 58, 64a

62AM   R.P. Agarwal and R.C. Mehrotra, J. Inorg. Nucl. Chem., 1962, 24, 821

62AMa   H. Asai and M. Morales, Arch. Biochem. Biophys., 1962, 99, 383

62AMY   L.P. Adamovich, O.V. Morgyl-Meshkova, and B.V. Yutsis, J. Anal. Chem. USSR,
            1962, 17, 673 (678)

62AT   P.J. Antikainen and K. Tevanen, Suomen Kem., 1962, B35, 224

62B   S.K. Banerji, Z. Anorg. Allg. Chem., 1962, 315, 229

62BA   I.E. Bukolov, K.V. Astakhov, V.I. Zimin, and V.S. Tairov, Russ. J. Inorg. Chem.,
            1962, 7, 816 (1577)

62BC   J.L. Bear, G.R. Choppin, and J.V. Quagliano, J. Inorg. Nucl. Chem., 1962, 24,
            1601

62BD   S.K. Banerji and A.K. Dey, Bull. Chem. Soc. Japan, 1962, 35, 2051

62BK   H.J. deBruin, D. Kaitis, and R.B. Temple, Aust. J. Chem., 1962, 15, 457

62BL   C.V. Banks and J.P. Laplante, Anal. Chim. Acta, 1962, 27, 80

62BN   J.R. Brannan and G.H. Nancollas, Trans. Faraday Soc., 1962, 58, 354

62BT   A.K. Babko and M.M. Tananaiko, Russ. J. Inorg. Chem., 1962, 7, 286 (562)

62BTS   A.I. Busev, V.G. Tiptsova, and L.M. Sorokina, Russ. J. Inorg. Chem., 1962, 7,
            1098 (2122)

62BV   A.K. Babko and A.I. Volkova, Russ. J. Inorg. Chem., 1962, 7, 1216 (2345)

62BVG   A.K. Babko, A.I. Volkova, and T.E. Gefman, Russ. J. Inorg. Chem., 1962, 7, 145
            (284), 1121 (2167)

62C   E.R. Clark, J. Inorg. Nucl. Chem., 1962, 24, 81

62CD   B. Carlqvist and D. Dyrssen, Acta Chem. Scand., 1962, 16, 94

62CL   D.T. Chen and K.J. Laidler, Trans. Faraday Soc., 1962, 58, 480

62CM   C.A. Crutchfield, Jr., W.M. McNabb, and J.F. Hazel, J. Inorg. Nucl. Chem., 1962,
           24, 291

62CR   E. Coates and B. Rigg, Trans. Faraday Soc., 1962, 58, 88

62CRa   E. Coates and B. Rigg, Trans. Faraday Soc., 1962, 58, 2058

62CTC   M. Cefola, A.S. Tompa, A.V. Celiano, and P.S. Gentile, Inorg. Chem., 1962, 1, 290

62CTG   M. Cefola, R.C. Taylor, P.S. Gentile, and A.V. Celiano, J. Phys. Chem., 1962, 66,
           790

62D   N.K. Davidenko, Russ. J. Inorg. Chem., 1962, 7, 1412 (2709)

62Da   D. Dyrssen, Roy. Inst. Tech. Stockholm, No. 188, 1962

62DN   I.R. Desai and V.S.K. Nair, J. Chem. Soc., 1962, 2360

62EP   I.P. Efimov and V.M. Peshkova, Vestn. Mosk. Univ., Ser. II Khim., 1962, No. 3, 63

62FC   C. Furlani and E. Cervone, Ann. Chim. (Rome), 1962, 52, 564

62G   I. Grenthe, Acta Chem. Scand., 1962, 16, 1695

62GJ   R.O. Gould and R.F. Jameson, J. Chem. Soc., 1962, 296

62GL   I. Geletseanu and A.V. Lapitskii, Proc. Acad. Sci. USSR, 1962, 144, 460 (573)

62GLa   I. Geletseanu and A.V. Lapitskii, Proc. Acad. Sci. USSR, 1962, 147, 983 (372)

62GM   V.V. Grigoreva and I.M. Maister, Russ. J. Inorg. Chem., 1962, 7, 1107 (2140)

62H   J. Halmekoski, Suomen Kem., 1962, B35, 41
62Ha   J. Halmekoski, Suomen Kem., 1962, B35, 108
62Hb   J. Halmekoski. Suomen Kem., 1962, B35, 171
62HF   C.B. Honaker and H. Freiser, J. Phys. Chem., 1962, 66, 127
62HO   J. Hala and A. Okac, Coll. Czech. Chem. Comm., 1962, 27, 1697

62HR   R.A. Haines, D.E. Ryan, and G.E. Cheney, Canad. J. Chem., 1962, 40, 1149

62IC   R.M. Izatt and J.J. Christensen, J. Phys. Chem., 1962, 66, 359

62IM   R.R. Irani and K. Moedritzer, J. Phys. Chem., 1962, 66, 1349

62IN   T. Ishimori and E. Nakamura, Radiochim. Acta, 1962, 1, 6

62JS   R.H. Jones and D.I. Stock, J. Chem. Soc., 1962, 306

62JSa   T.J. Jin, L. Sommer, and A. Okac, Coll. Czech. Chem. Comm., 1962, 27, 1150

62JSb   T.J. Jin, L. Sommer, and A. Okac, Coll. Czech. Chem. Comm., 1962, 27, 1161

62JSc   T.J. Jin, L. Sommer, and A. Okac, Coll. Czech. Chem. Comm., 1962, 27, 1171

62K   N.A. Kostromina, Russ. J. Inorg. Chem., 1962, 7, 806 (1559)

62KB   M.S. Kachhawaha and A.K. Bhattacharya, Z. Anorg. Allg. Chem., 1962, 315, 104

62KP   R.S. Kolat and J.E. Powell, Inorg. Chem., 1962, 1, 293

62LJ   D.L. Leussing and J. Jayne, J. Phys. Chem., 1962, 66, 426

62LN   J.J. Lindberg, C.G. Nordstrom, and R. Lauren, Suomen Kem., 1962, B35, 182

62LZ   L.B. Levashova and V.L. Zolotavin, Russ. J. Inorg. Chem., 1962, 7, 418 (811)

62M   Y. Murakami, Bull. Chem. Soc. Japan, 1962, 35, 52

62Ma   Y. Murakami, J. Inorg. Nucl. Chem., 1962, 24, 679

62Mb   O. Makitie, Suomen Kem., 1962, B35, 1

62MD   D.L. McMasters, J.C. DiRaimondo, L.H. Jones, R.P. Lindley, and E.W. Zeltmann,
          J. Phys. Chem., 1962, 66, 249

62MM   P.G. Manning and C.B. Monk, Trans. Faraday Soc., 1962, 58, 938

62MR   I.N. Marov and D.I. Ryabchikov, Russ. J. Inorg. Chem., 1962, 7, 533 (1036)

62N   V.S.K. Nair, Talanta, 1962, 9, 27

62NF   V.A. Nazarenko and G.V. Flyantikova, Russ. J. Inorg. Chem., 1962, 7, 1210

62NM   R. Nasanen and P. Merilainen, Suomen Kem., 1962, B35, 79

62OU   Y. Oka, M. Umehara, and T. Nozoe, Nippon Kagaku Zasshi, 1962, 83, 703

62OUa   Y. Oka, M. Umehara, and T. Nozoe, Nippon Kagaku Zasshi, 1962, 83, 1197

62PA   V.M. Peshkova and P. Ang, Russ. J. Inorg. Chem., 1962, 7, 765 (1484)

62PB   N.P. Prokhorova and N.E. Brezhneva, Russ. J. Inorg. Chem., 1962, 7, 953 (1846)

62PL   K. Pan, Z.F. Lin, and P.J. Sun, J. Chinese Chem. Soc. (Taiwan), 1962, 9, 109

62PLB   M.G. Panova, V.I. Levin, and N.Ye. Brezhneva, Soviet Radiochem., 1962, 3, 14 (52)

62RA   G. Ross, D.A. Aikens, and C.N. Reilley, Anal. Chem., 1962, 34, 1766

62RM   D.I. Ryabchikov, I.N. Marov, and Y. Ko-min, Russ. J. Inorg. Chem., 1962, 7, 1415
          (2716)

62SA   D.J. Sawyer and R.T. Ambrose, Inorg. Chem., 1962, 1, 296

62SB   B. Salvesen and J. Bjerrum, Acta Chem. Scand., 1962, 16, 735

62SBa   D.D. Sharma and A.K. Bhattacharya, J. Indian Chem. Soc., 1962, 39, 299

62SBb   J. Stary and V. Balek, Coll. Czech. Chem. Comm., 1962, 27, 809

62SD   R.L. Seth and A.K. Dey, J. Indian Chem. Soc., 1962, 39, 724

62SDa   R.L. Seth and A.K. Dey, J. Indian Chem. Soc., 1962, 39, 773

62SK   Z.A. Sheka and E.E. Kriss, Russ. J. Inorg. Chem., 1962, 7, 333 (658)

62SKa   D.T. Sawyer and R.J. Kula, Inorg. Chem., 1962, 1, 303

62SM   A. Sobkowska and J. Minczewsky, Rocz. Chim., 1962, 36, 17

62SN   K. Schwabe and D. Nebel, Z. Phys. Chem. (Leipzig), 1962, 220, 339

62SS   G. Schwarzenbach and I. Szilard, Helv. Chim. Acta, 1962, 45, 1222

62SSB   K.N. Sahu, M.C. Saxena, and A.K. Bhattacharya, J. Indian Chem. Soc., 1962, 39, 731

62SSS   S.C. Saraiya, V.S. Srinivasan, and A.K. Sundaram, Curr. Sci. (India), 1962, 31,
          187

62SY   K. Suzuki and K. Yamasaki, J. Inorg. Nucl. Chem., 1962, 24, 1093

62T   H. Tsubota, Bull. Chem. Soc. Japan, 1962, 35, 640

62Ta   L.I. Tikhonova, Russ. J. Inorg. Chem., 1962, 7, 424 (822)

62TF   O.N. Temkin, R.M. Flid, and A.I. Malakhov, Kinetics and Catalysis, 1962, 3, 799
          (915)

62W   I. Wadso, Acta Chem. Scand., 1962, 16, 479

62WB   G. Wright and J. Bjerrum, Acta Chem. Scand., 1962, 16, 1262

62YB   K.B. Yatsimirskii and L.I. Budarin, Russ. J. Inorg. Chem., 1962, 7, 942 (1824)

62YP   K.B. Yatsimirskii and K.E. Prik, Russ. J. Inorg. Chem., 1962, 7, 821 (1589)

62YZ   K.B. Yatsimirskii and Yu.A. Zhukov, Russ. J. Inorg. Chem., 1962, 7, 818 (1583),
          1463 (2807)

62ZR   E.W. Zahnow and R.J. Robinson, J. Electroanal. Chem., 1962, 3, 263

63A   N.I. Ampelogova, Soviet Radiochem., 1963, 5, 520 (562) (see 73A)

63AL   G. Anderegg, F. L'Eplattenier, and G. Schwarzenbach, Helv. Chim. Acta, 1963, 46,
          1400

63ALa   G. Anderegg, F. L'Eplattenier, and G. Schwarzenbach, Helv. Chim. Acta, 1963, 46,
          1409

63AN   A.M. Andrianov and V.A. Nazarenko, Russ. J. Inorg. Chem., 1963, 8, 1192 (2276),
          1194 (2281)

63AT   P.J. Antikainen and K. Tevanen, Suomen Kem., 1963, B36, 199

63B   B. Budesinsky, Coll. Czech. Chem. Comm., 1963, 28, 2902

63BA   C.V. Banks and S. Anderson, Inorg. Chem., 1963, 2, 112

63BC   J.L. Bear, G.R. Choppin, and J.V. Quagliano, J. Inorg. Nucl. Chem., 1963, 25, 513

63BD   K. Burger and D. Dyrssen, Acta Chem. Scand., 1963, 17, 1489

63BG   S.K. Banerji and M. Garg, Z. Anorg. Allg. Chem., 1963, 325, 315

63BM   R.P. Bell and W.B.T. Miller, Trans. Faraday Soc., 1963, 59, 147

63BS   A.K. Babko and M.I. Shtokalo, Russ. J. Inorg. Chem., 1963, 8, 564 (1088)

63C   E.R. Clark, J. Inorg. Nucl. Chem., 1963, 25, 353

63Ca   E. Campi, Ann. Chim. (Rome), 1963, 53, 96

63Cb   E.G. Chikrgzova, Russ. J. Inorg. Chem., 1963, 8, 41 (83)

63CE   E. Coates, J.R. Evans, and B. Rigg, Trans. Faraday Soc., 1963, 59, 2369

63CR   A.N. Christensen and S.E. Rasmussen, Acta Chem. Scand., 1963, 17, 1315

63CRM   E. Coates, B. Rigg, R. Murton, and D.L. Smith, J. Soc. Dyers Col., 1963, 79, 465

63DG   A.F. Donda and A.M. Giuliani, Ric. Sci., 1963, 33 (II-A), 819

63DGa   S.P. Datta and A.K. Grzybowski, J. Chem. Soc., 1963, 6004

63DN   R.C. Das, R.K. Nanda, and S. Aditya, J. Indian Chem. Soc., 1963, 40, 739

63DP   M.M. Davis and M. Paabo, J. Res. Nat. Bur. Stand., 1963, 67A, 241

63DS   I. Danielsson and T. Suomineu, Acta Chem. Scand., 1963, 17, 979

63ED   L.E. Erikson and J.A. Dembo, J. Phys. Chem., 1963, 67, 707

63EM   Z.L. Ernst and J. Menashi, Trans. Faraday Soc., 1963, 59, 230

63EMa   Z.L. Ernst and J. Menashi, Trans. Faraday Soc., 1963, 59, 1794

63EMb   Z.L. Ernst and J. Menashi, Trans. Faraday Soc., 1963, 59, 1803

63EMc   Z.L. Ernst and J. Menashi, Trans. Faraday Soc., 1963, 59, 2838

63EW   L. Eberson and I. Wadso, Acta Chem. Scand., 1963, 17, 1552

63FK   I. Feldman and L. Koval, Inorg. Chem., 1963, 2, 145

63FL   V. Frei and J. Loub, Z. Phys. Chem., 1963, 222, 249

63FP   N.J. Friedman and R.A. Plane, Inorg. Chem., 1963, 2, 11

63FR   H.N. Farrer and F.J.C. Rossotti, Acta Chem. Scand., 1963, 17, 1824

63FV   Ya.D. Fridman, R.A. Veresova, N.V. Dolgashova, and R.I. Sorochan, Russ. J. Inorg.
           Chem., 1963, 8, 344 (676)

63G   S. Gobom, Acta Chem. Scand., 1963, 17, 2181

63Ga   I. Grenthe, Acta Chem. Scand., 1963, 17, 1814

63Gb   I. Grenthe. Acta Chem. Scand., 1963, 17, 2487

63GA   R.W. Green and P.W. Alexander, J. Phys. Chem., 1963, 67, 905

63GAa   Yu. P. Galaktionov and K.V. Astakhov, Russ. J. Inorg. Chem., 1963, 8, 1309 (2498)

63GJ   R.O. Gould and R.F. Jameson, J. Chem. Soc., 1963, 15

63GJa   R.O. Gould and R.F. Jameson, J. Chem. Soc., 1963, 5211

63GM   B.N. Ghosh, S.P. Moulik, K.K. Sengupta, and P.K. Pal, J. Indian Chem. Soc., 1963,
           40, 509

63GT   I. Grenthe and I. Tobiasson, Acta Chem. Scand., 1963, 17, 2101

63H   J. Halmekoski, Suomen Kem., 1963, B36, 29

63Ha   J. Halmekoski, Suomen Kem., 1963, B36, 46

63HO   J. Horak and A. Okac, Coll. Czech. Chem. Comm., 1963, 28, 2563

63HP   C.J. Hawkins and D.D. Perrin, Inorg. Chem., 1963, 2, 843

63HS   P. Hemmerich and Ch. Sigwart, *Experimentia*, 1963, 19, 488

63HZ   T.H. Handley, R.H. Zucal, and J.A. Dean, *Anal. Chem.*, 1963, 35, 1163

63I   Y.J. Israeli, *Canad. J. Chem.*, 1963, 41, 2710

63K   N.A. Kostromina, *Russ. J. Inorg. Chem.*, 1963, 8, 988 (1900)

63KE   H. Kroll, B. Elkind, and R. Davis, TID-19989, 1963

63KN   W.L. Koltun, L. Ng, and F.R.N. Gurd, *J. Biol. Chem.*, 1963, 238, 1367

63KP   A.S. Kereichuk and V.I. Paramonova, *Soviet Radiochem.*, 1963, 5, 427 (464)

63LC   A. Liberti, P. Curro, and G. Calabro, *Ric. Sci.*, 1963, 33 (II-A), 36

63LG   A.V. Lapitskii and I. Geletseanu, *Soviet Radiochem.*, 1963, 5, 298 (330)

63LGM   A.V. Lapitskii, I. Geletseanu, and Ya. Mink, *Soviet Radiochem.*, 1963, 5, 220

63LJ   J.G. Lanese and B. Jaselskis, *Anal. Chem.*, 1963, 35, 1878

63LK   E.J. Langmyhr and K.S. Klausen, *Anal. Chim. Acta*, 1963, 29, 149

63LM   S.H. Laurie and C.B. Monk, *J. Chem. Soc.*, 1963, 3343

63M   H. Miyata, *Bull. Chem. Soc. Japan*, 1963, 36, 382

63Ma   H. Miyata, *Bull. Chem. Soc. Japan*, 1963, 36, 386

63Mb   Y. Matsushima, *Chem. Pharm. Bull.*, 1963, 11, 566

63MA   H. Matsuda, Y. Ayabe, and K. Adachi, *Z. Elektrochem.*, 1963, 67, 593

63MN   Y. Murakami and K. Nakamura, *Bull. Chem. Soc. Japan*, 1963, 36, 1408

63MNa   A. McAuley and G.H. Nancollas, *J. Chem. Soc.*, 1963, 989

63MNS   V.K. Mathur, H.L. Nigam, and S.C. Srivastova, *Bull. Chem. Soc. Japan*, 1963, 36, 1658

63MNT   Y. Murakami, K. Nakamura, and M. Tokunada, *Bull. Chem. Soc. Japan*, 1963, 36, 669

63MP   N.V. Melchakova and V.M. Peshkova, *Russ. J. Inorg. Chem.*, 1963, 8, 663 (1280)

63MPa   R.-P. Martin and R.A. Paris, *Bull. Soc. Chim. France*, 1963, 1600

63MS   M. Muzaffaruddin, Salahuddin, and W.U. Malik, *J. Indian Chem. Soc.*, 1963, 40, 467

63MT   P.K. Migal and V.A. Tsiplyakova, *Russ. J. Inorg. Chem.*, 1963, 8, 319 (629)

63NA   R.K. Nanda and S. Aditya, *J. Indian Chem. Soc.*, 1963, 40, 660, 755

63NF   V.A. Nazarenko and G.V. Flyantikova, *Russ. J. Inorg. Chem.*, 1963, 8, 712 (1370); 1189 (2271)

63NK   K. Nagano, H. Kinoshita, and Z. Tamura, *Chem. Pharm. Bull.*, 1963, 11, 999

63NS   E. Nebel and K. Schwabe, *Z. Phys. Chem.* (Leipzig), 1963, 224, 29

63NT   K. Nagano, H. Tsukahara, H. Kinoshita, and Z. Tamura, *Chem. Pharm. Bull.*, 1963, 11, 797

630H   W.F. O'Hara, T. Hu, and L.G. Hepler, <u>J. Phys. Chem.</u>, 1963, <u>67</u>, 1933

630U   Y. Oka and M. Umehara, <u>Nippon Kagaku Zasshi</u>, 1963, <u>84</u>, 928

630Y   Y. Oka, K. Yamamoto, and M. Katagiri, <u>Nippon Kagaku Zasshi</u>, 1963, <u>84</u>, 259

63PB   M. Petek and M. Branica, <u>J. Polarog. Soc.</u>, 1963, <u>9</u>, 1

63PBa   V.I. Paramonova and S.A. Bartenev, <u>Russ. J. Inorg. Chem.</u>, 1963, <u>8</u>, 157 (311)

63R   A. Ringbon, <u>Complexation in Anal. Chem.</u>, Interscience, New York, 1963, 362

63RC   H. Roth, L. Czegledi and C.G. Macarovici, <u>Studii Cercet. Chim.</u> (Cluj), 1963, <u>14</u>, 69

63RM   S.M.F. Rahman and A.U. Malik, <u>Indian J. Chem.</u>, 1963, <u>1</u>, 424

63RP   R.A. Robinson and A. Peiperl, <u>J. Phys. Chem.</u>, 1963, <u>67</u>, 1723

63RS   M.L.N. Reddy and U.V. Seshaiah, <u>Indian J. Chem.</u>, 1963, <u>1</u>, 536

63S   L. Sommer, <u>Coll. Czech. Chem. Comm.</u>, 1963, <u>28</u>, 449

63Sa   L. Sommer, <u>Coll. Czech. Chem. Comm.</u>, 1963, <u>28</u>, 2102

63Sb   L. Sommer, <u>Coll. Czech. Chem. Comm.</u>, 1963, <u>28</u>, 2393

63Sc   L. Sommer, <u>Coll. Czech. Chem. Comm.</u>, 1963, <u>28</u>, 2716

63Sd   L. Sommer, <u>Coll. Czech. Chem. Comm.</u>, 1963, <u>28</u>, 3057

63Se   J. Stary, <u>Anal. Chim. Acta</u>, 1963, <u>28</u>, 132

63SA   G. Schwarzenbach and G. Anderegg, <u>Pharm. Acta Helv.</u>, 1963, <u>38</u>, 547

63SAP   V.M. Savostina, E.K. Astakhova, and V.M. Peshkova, <u>Vestn. Mosk. Univ. Ser. II</u>, <u>Khim.</u>, 1963, <u>18</u>, 43

63SC   J.T. Spence and H.H.Y. Chang, <u>Inorg. Chem.</u>, 1963, <u>25</u>, 319

63SD   S.C. Srivastava and A.K. Dey, <u>J. Inorg. Nucl. Chem.</u>, 1963, <u>25</u>, 217

63SDa   R.L. Seth and A.K. Dey, <u>Z. Anorg. Allg. Chem.</u>, 1963, <u>321</u>, 278

63SDb   S.P. Sangal and A.K. Dey, <u>J. Indian Chem. Soc.</u>, 1963, <u>40</u>, 464

63SDc   R.L. Seth and A.K. Dey, <u>J. Indian Chem. Soc.</u>, 1963, <u>40</u>, 794

63SDd   S.C. Srivastava and A.K. Dey, <u>Indian J. Chem.</u>, 1963, <u>1</u>, 200, 242

63SDe   S.P. Sangal and A.K. Dey, <u>Indian J. Chem.</u>, 1963, <u>1</u>, 270

63SDf   S.P. Sangal and A.K. Dey, <u>J. Prakt. Chem.</u>, 1963, <u>20</u>, 219

63SG   S.Ya. Shnaiderman and E.V. Galinker, <u>Russ. J. Inorg. Chem.</u>, 1963, <u>8</u>, 142 (279)

63SM   M. Sakaguchi, A. Mizote, H. Miyata, and K. Toel, <u>Bull. Chem. Soc. Japan</u>, 1963, <u>36</u>, 885

63SS   G. Schwarzenbach and K. Schwarzenbach, <u>Helv. Chim. Acta</u>, 1963, <u>46</u>, 1390

63SSa   S.C. Srivastava, S.N. Sinha, and A.K. Dey, <u>Bull. Chem. Soc. Japan</u>, 1963, <u>36</u>, 268

63SSb   S.P. Sangal, S.C. Srivastava, and A.K. Dey, J. Indian Chem. Soc., 1963, 40, 275

63SSc   S.C. Srivastava, S.N. Sinha, and A.K. Dey, J. Prakt. Chem., 1963, 20, 70

63SW   A. Swinarski and J. Wojtczakova, J. Phys. Chem. (Leipzig), 1963, 223, 345

63TM   S.K. Tobia and N.E. Milad, J. Chem. Soc., 1963, 734

63TS   N. Tanaka, Y. Saito, and H. Ogino, Bull. Chem. Soc. Japan, 1963, 36, 794

63TY   R.S. Tobias and M. Yasuda, Inorg. Chem., 1963, 2, 1307

63US   V.S. Ulyanov and R.A. Sviridova, Soviet Radiochem., 1963, 5, 386 (419)

63YT   M. Yasuda and R.S. Tobias, Inorg. Chem., 1963, 2, 207

63YZ   K.B. Yatsimirskii and Yu.A. Zhukov, Russ. J. Inorg. Chem., 1963, 8, 149 (295)

63ZA   Yu.A. Zolotov and I.P. Alimarin, J. Inorg. Nucl. Chem., 1963, 25, 691

63ZD   R.H. Zucal, J.A. Dean, and T.H. Handley, Anal. Chem., 1963, 35, 988

63ZG   V.K. Zolotukhin and Z.G. Galanets, Visn. L'vivsk. Derzh. Univ., Ser. Khim., 1963,
          91; Chem. Abs., 1965, 62, 4909a

64AM   D.W. Archer and C.B. Monk, J. Chem. Soc., 1964, 3117

64AS   E.K. Astakhova, V.M. Savostina, and V.M. Peshkova, Russ. J. Inorg. Chem., 1964,
          9, 452 (817)

64ASa   E.K. Astakhova, V.M. Savostina, and V.M. Peshkova, Russ. J. Phys. Chem., 1964,
          38, 1244 (2299)

64AT   P.J. Antikainen and K. Tevanen, Suomen Kem., 1964, B37, 6

64ATa   P.J. Antikainen and K. Tevanen, Suomen Kem., 1964, B37, 213

64B   B. Budesinsky, Z. Anal. Chem., 1964, 202, 96

64Ba   B. Budesinsky, Z. Anal. Chem., 1964, 206, 401

64BA   G.M. Brauer, H. Argentar, and G. Durany, J. Res. Nat. Bur. Stand., 1964, 68A, 619

64BB   A. Beauchamp and R.L. Benoit, Canad. J. Chem., 1964, 42, 2161

64BR   K.S. Bhatki, A.T. Rane, and M.B. Kabadi, J. Chem. Eng. Data, 1964, 9, 175

64BS   D. Banerjea and I.P. Singh, Z. Anorg. Allg. Chem., 1964, 331, 225

64BSa   B.M.L. Bansal and H.D. Sharma, J. Inorg. Nucl. Chem., 1964, 26, 799

64BSb   J.R. Brannan and D.T. Sawyer, J. Electroanal. Chem., 1964, 8, 286

64CC   G. Calabro and P. Curro, Atti Soc. Peloritana Sci. Fis. Met. Nat., 1964, 10, 357;
          Chem. Abs., 1966, 64, 15361c

64CK   R. Caletka, M. Kyrs, and J. Rais, J. Inorg. Nucl. Chem., 1964, 26, 1443

64COM   E. Campi, G. Ostacoli, M. Meirone, and G. Saini, J. Inorg. Nucl. Chem., 1964, 26,
          553

64COV   E. Campi, G. Ostacoli, A. Vanni, and E. Casorati, Ric. Sci., 1964, 34 (II-A, 6),
          341

64D   D. Dyrssen, Trans. Roy. Inst. Tech. Stockholm, 1964, No. 220

64Da   N.K. Davidenko, <u>Russ. J. Inorg. Chem.</u>, 1964, <u>9</u>, 859

64DA   R.C. Das and S. Aditya, <u>J. Indian Chem. Soc.</u>, 1964, <u>41</u>, 765

64DC   G. D'Amore, P. Curro, and G. Calabro, <u>Atti Soc. Peloritana Sci. Fis. Mat. Nat.</u>,
          1964, <u>10</u>, 229; <u>Chem. Abs.</u>, 1966, <u>64</u>, 15361b

64DM   S.N. Dubey and R.C. Mehrotra, <u>J. Inorg. Nucl. Chem.</u>, 1964 <u>26</u>, 1543

64DMa   S.N. Dubey and R.C. Mehrotra, <u>J. Less-Common Metals</u>, 1964, <u>7</u>, 169

64DV   H. Deelstra and F. Verbeek, <u>Anal. Chim. Acta</u>, 1964, <u>31</u>, 251

64E   V.I. Ermolenko, <u>Russ. J. Inorg. Chem.</u>, 1964, <u>9</u>, 25 (48)

64EH   Z.L. Ernst and F.G. Herring, <u>Trans. Faraday Soc.</u>, 1964, <u>60</u>, 1053

64EI   Z.L. Ernst, R.J. Irving, and J. Menashi, <u>Trans. Faraday Soc.</u>, 1964, <u>60</u>, 56

64EV   L. Eeckhaut, F. Verbeek, H. Deelstra, and J. Hoste, <u>Anal. Chim. Acta</u>, 1964, <u>30</u>,
          369

64G   I. Grenthe, <u>Acta Chem. Scand.</u>, 1964, <u>18</u>, 283

64H   J. Hala, <u>Coll. Czech. Chem. Comm.</u>, 1964, <u>29</u>, 905

64IN   R.J. Irving, L. Nelander, and I. Wadso, <u>Acta Chem. Scand.</u>, 1964, <u>18</u>, 769

64JG   D.S. Jain and J.N. Gaur, <u>Indian J. Chem.</u>, 1964, <u>2</u>, 503

64JJ   D.P. Joshi and D.V. Jain, <u>J. Indian Chem. Soc.</u>, 1964, <u>41</u>, 33

64KK   D.H. Klein and G.J. Kersels, <u>J. Inorg. Nucl. Chem.</u>, 1964, <u>26</u>, 1325

64KO   H.C.Ko, W.F. O'Hara, T. Hu, and L.G. Hepler, <u>J. Amer. Chem. Soc.</u>, 1964, <u>86</u>, 1003

64KS   G.I. Kurnevich and G.A. Shagisultanova, <u>Russ. J. Inorg. Chem.</u>, 1964, <u>9</u>,
          1383 (2559)

64L   I. Lundqvist, <u>Acta Chem. Scand.</u>, 1964, <u>18</u>, 858

64LS   D.L. Leussing and D.C. Shultz, <u>J. Amer. Chem. Soc.</u>, 1964, <u>86</u>, 4846

64LSA   W.F. Lee, N.K. Shastri, and E.S. Amis, <u>Talanta</u>, 1964, <u>11</u>, 685

64M   O. Makitie, <u>Suomen Kem.</u>, 1964, <u>B37</u>, 17

64Ma   W.A.E. McBryde, <u>Canad. J. Chem.</u>, 1964, <u>42</u>, 1917

64MA   F.J. Millero, J.C. Ahluwalis, and L.G. Hepler, <u>J. Chem. Eng. Data</u>, 1964, <u>9</u>, 192

64MD   K.N. Munshi and A.K. Dey, <u>J. Inorg. Nucl. Chem.</u>, 1964, <u>26</u>, 1603

64MDa   K.N. Munshi and A.K. Dey, <u>J. Indian Chem. Soc.</u>, 1964, <u>41</u>, 340

64MG   E.W. Malcolm, J.W. Green, and H.A. Swenson, <u>J. Chem. Soc.</u>, 1964, 4669

64MM   Y. Murakami and A.E. Martell, <u>J. Amer. Chem. Soc.</u>, 1964, <u>86</u>, 2119

64MP   R.-P. Martin and R.A. Paris, <u>Bull. Soc. Chim. France</u>, 1964, 80

64MS   O.B. Mathre and E.B. Sandell, <u>Talanta</u>, 1964, <u>11</u>, 295

64MSD   K.N. Munshi, S.P. Sangal, and A.K. Dey, J. Indian Chem. Soc., 1964, 41, 701

64MT   Y. Murakami and M. Takagi, Bull. Chem. Soc. Japan, 1964, 37, 268

64MTa   Y. Murakami and M. Tokunaga, Bull. Chem. Soc. Japan, 1964, 37, 1562

64MTb   P.K. Migal and V.A. Tsiplyakova, Russ. J. Inorg. Chem., 1964, 9, 333 (601)

64MZ   A.I. Moskvin, V.P. Zaitseva, and A.D. Gelman, Soviet Radiochem., 1964, 6, 206 (214)

64N   L. Nelander, Acta Chem. Scand., 1964, 18, 973

64NM   R.I. Novoselov, Z.A. Muzykantova, and B.V. Ptitsyn, Russ. J. Inorg. Chem., 1964, 9, 1399 (2590)

64OY   Y. Oka and K. Yamamoto, Nippon Kagaku Zasshi, 1964, 85, 779

64OYa   Y. Oka and M. Yanal, Bunseki Kagaku Tokyo (Japan Analyst), 1964, 13, 207

64OYA   Y. Oka, K. Yamamoto, and T. Aoki, Nippon Kagaku Zasshi, 1964, 85, 430

64OYS   Y. Oka, M. Yanai, and C. Suzuki, Nippon Kagaku Zasshi, 1964, 85, 873

64P   J. Prasilova, J. Inorg. Nucl. Chem., 1964, 26, 661

64PI   L.D. Pettit and H.M.N.H. Irving, J. Chem. Soc., 1964, 5336

64PK   R. Palmaeus and P. Kierkegaard, Acta Chem. Scand., 1964, 18, 2226

64PKK   J.E. Powell, R.H. Karraker, R.S. Kolat, and J.L. Farrell, Rare Earth Research II, ed. K.S. Vorres; Gordon and Breach, New York, 1964, 509

64PKP   J.E. Powell, R.S. Kolat, and C.S. Paul, Inorg. Chem., 1964, 13, 518

64PM   V.I. Paramonova, A.N. Mosevich, and M. Tzu-Kuang, Soviet Radiochem., 1964, 6, 663 (682)

64PP   J.E. Powell, G.S. Paul, B.D. Fleicher, W.R. Stagg, and Y. Suzuki, USAEC IS-900, Sept. 1964

64PPB   V.I. Paramonova, N.B. Platunova, and E.D. Baklanovskii, Soviet Radiochem., 1964, 6, 495 (513)

64PPD   V.I. Paramonova, N.B. Platunova, and V.S. Dubrovin, Soviet Radiochem., 1964, 6, 487 (505)

64PS   J.E. Powell and Y. Suzuki, Inorg. Chem., 1964, 3, 690

64R   R.A. Robinson, J. Res. Nat. Bur. Stand., 1964, 68A, 159

64RM   K.S. Rajan and A.E. Martell, J. Inorg. Nucl. Chem., 1964, 26, 789

64RMa   K.S. Rajan and A.E. Martell, J. Inorg. Nucl. Chem., 1964, 26, 1927

64RME   D.I. Ryabchikov, I.N. Marov, A.N. Ermakov, and V.K. Belgaeva, J. Inorg. Nucl. Chem., 1964, 26, 965

64RSa   M.L.N. Rdeey and U.V. Seshaiah, J. Indian Chem. Soc., 1964, 41, 289

64RZ   G.A. Rechnitz and S.B. Zamochnick, Talanta, 1964, 11, 1061

64S   S.P. Sangal, Chim. Anal., 1964, 46, 492

64Sa   S.P. Sangal, Vijnana Parishad Anusandhan Patrik, 1964, 7, 109

64Sb   M. Suryanarayana, Curr. Sci. (India), 1964, 33, 520

64SA   V.M. Savostina, E.K. Astakhova, and V.M. Peshkova, Russ. J. Inorg. Chem., 1964,
          9, 42 (80)

64SAa   I.E. Starik, N.I. Ampelogova, and B.S. Kuznetsov, Soviet Radiochem., 1964, 6,
          501 (519) (see 73Ab)

64SB   K.N. Sahu and A.K. Bhattacharya, J. Indian Chem. Soc., 1964, 41, 787

64SBE   P.W. Schneider, H. Brintzinger, and H. Erlenmeyer, Helv. Chim. Acta, 1964, 47,
          992

64SH   L. Sommer and J. Havel, Coll. Czech. Chem. Comm., 1964, 29, 690

64SM   Unpublished values quoted in L.G. Sillen and A.E. Martell, Stability Constants
          of Metal-Ion Complexes, Special Publication No. 17, The Chemical Soc., Lon-
          don, 1964

64SP   W.R. Stagg and J.E. Powell, Inorg. Chem., 1964, 3, 242

64SS   S.C. Srivastava, S.N. Sinha, and A.K. Dey, Mikrochim. Acta, 1964, 605

64T   C.F. Timberlake, J. Chem. Soc., 1964, 1229

64Ta   C.F. Timberlake, J. Chem. Soc., 1964, 5078

64TG   S.S. Tate, A.K. Grzybowski, and S.P. Datta, J. Chem. Soc., 1964, 1372

64TGa   S.S. Tate, A.K. Grzybowski, and S.P. Datta, J. Chem. Soc., 1964, 1381

64TM   S.K. Tobia and N.E. Milad, J. Chem. Soc., 1964, 1915

64TT   E.R. Tucci, F. Takahashi, V.A. Tucci, and N.C. Li, J. Inorg. Nucl. Chem., 1964,
          26, 1263

64WD   T.H. Wirth and N. Davidson, J. Amer. Chem. Soc., 1964, 86, 4322

64WDa   T.H. Wirth and N. Davidson, J. Amer. Chem. Soc., 1964, 86, 4325

64WI   D.P. Wrathall, R.M. Izatt, and J.J. Christensen, J. Amer. Chem. Soc., 1964, 86,
          4779

64WM   A.V. Willi and P. Mori, Helv. Chim. Acta, 1964, 47, 155

64YC   H. Yoneda, G.R. Choppin, J.L. Bear, and J.V. Quagliano, Inorg. Chem., 1964, 3,
          1642

64YK   K.B. Yatsimirskii and V.D. Korableva, Russ. J. Inorg. Chem., 1964, 9, 195 (337)

64ZT   O.E. Zuyogintsen and V.P. Tikhonov, Russ. J. Inorg. Chem., 1964, 9, 865 (1597)

65AE   D.W. Archer, D.A. East, and C.B. Monk, J. Chem. Soc., 1965, 720

65AK   I.P. Alimarin, S.A. Khamid, and I.V. Puzdrenkova, Russ. J. Inorg. Chem., 1965,
          10, 209 (389)

65AM   R.P. Agarwal and R.C. Mehrotra, J. Indian Chem. Soc., 1965, 42, 61

65B   H. Brintzinger, Helv. Chim. Acta, 1965, 48, 47

65Ba   B. Budesinsky, Z. Anal. Chem., 1965, 207, 105

65Bb   M. Bartusek, Coll. Czech. Chem. Comm., 1965, 30, 2746

65BB   B.A. Bezdekova and B. Budesinsky, Coll. Czech. Chem. Comm., 1965, 30, 811

65BC   E. Bottari and L. Ciavatta, Gazz. Chim. Ital., 1965, 95, 908

65BG   F. Baroncelli and G. Grosse, J. Inorg. Nucl. Chem., 1965, 27, 1085

65BH   B. Budesinsky and K. Haas, Z. Anal. Chem., 1965, 210, 263

65BK   T.A. Belyavskaya and I.F. Kolosova, Russ. J. Inorg. Chem., 1965, 10, 236 (441)

65BL   A.K. Babko, V.V. Lukachina, and B.I. Nabivanets, Russ. J. Inorg. Chem., 1965,
          10, 467 (865)

65BQ   S.K. Banerji and S.-Z. Qureshi, Bull. Chem. Soc. Japan, 1965, 38, 720

65BR   F. Brezina and J. Rosicky, Monat. Chem., 1965, 96, 1025

65BRP   F. Brezina, J. Rosicky, and R. Pastorek, Monat. Chem., 1965, 96, 553

65BS   M. Bartusok and L. Sommer, J. Inorg. Nucl. Chem., 1965, 27, 2397

65BSa   M. Bartusek and O. Stankova, Coll. Czech. Chem. Comm., 1965, 30, 3415

65BSb   B. Budesinsky and B. Savvin, Z. Anal. Chem., 1965, 214, 189

65BSc   R.F. Bauer and W.M. Smith, Canad. J. Chem., 1965, 43, 2755

65BV   B. Budesinsky and D. Vrzalova, Z. Anal. Chem., 1965, 210, 161

65BW   A. Basinski and Z. Warnke, Trans. Faraday Soc., 1965, 61, 129

65BWa   A. Basinski and Z. Warnke, Rocz. Chem., 1965, 39, 1776

65C   F.P. Cavasino, Ric. Sci., 1965, 35 (II-A), 1120

65CD   B. Carlqvist and D. Dyrssen, Acta Chem. Scand., 1965, 19, 1293

65CG   G.R. Choppin and A.J. Graffeo, Inorg. Chem., 1965, 4, 1254

65CM   J.P. Calmon and P. Maroni, Bull. Soc. Chim. France, 1965, 2525

65CO   E. Campi, G. Ostacoli, and A. Vanni, Gazz. Chim. Ital., 1965, 95, 796

65CV   L. Ciavatta and M. Villafiorita, Gazz. Chim. Ital., 1965, 95, 1247

65D   D.A. Doornbos, Thesis, Univ. of Rijks, Groninger, Neth., 1965

65Da   R.C. Das, Indian J. Chem., 1965, 3, 179

65DA   R.C. Das and S. Aditya, J. Indian Chem. Soc., 1965, 42, 15

65DD   M.N. Desai and B.M. Desai, J. Indian Chem. Soc., 1965, 42, 643

65DM   S.N. Dubey and R.C. Mehrotra, J. Indian Chem. Soc., 1965, 42, 685

65DMa   S.N. Dubey and R.C. Mehrotra, J. Less-Common Metals, 1965, 9, 123

65DN   R.C. Das, R.K. Nanda, and S. Aditya, J. Indian Chem. Soc., 1965, 42, 307

65DS   Y. Dutt and R.P. Singh, J. Indian Chem. Soc., 1965, 42, 767

65DSa   V.P. Devendran and M. Santappa, Curr. Sci. (India), 1965, 34, 145

65E   V.I. Ermolenko, Russ. J. Inorg. Chem., 1965, 10, 1423 (2617)

65EH   Z.L. Ernst and F.G. Herring, Trans. Faraday Soc., 1965, 61, 454

65F   V. Frei, Coll. Czech. Chem. Comm., 1965, 30, 1402

65FCW   M. Friedman, J.F. Cavins, and J.S. Wall, J. Amer. Chem. Soc., 1965, 87, 3672

65FK   A.B. Fasman, G.G. Kutyukov, and D. V. Sokolskii, Russ. J. Inorg. Chem., 1965,
         10, 727 (1338)

65FO   T. Fueno, T. Okuyama, T. Deguchi, and J. Furukawa, J. Amer. Chem. Soc., 1965,
         87, 170

65FP   D. Fleischer and J.E. Powell, USAEC IS-1121, Feb. 1965

65FS   V. Frei and A. Solcova, Coll. Czech. Chem. Comm., 1965, 30, 961

65G   G. Geier, Ber. Bunsenges. Phys. Chem. (Z. Elektrochem.), 1965, 69, 617

65GA   R.W. Green and P.W. Alexander, Aust. J. Chem., 1965, 18, 329

65GJ   J.N. Gaur and D.S. Jain, Aust. J. Chem., 1965, 18, 1687

65GJS   S.L. Gupta, J.N. Jaitly, and R.N. Soni, J. Indian Chem. Soc., 1965, 42, 384

65HF   H.S. Henderson and J.G. Fullington, Biochem., 1965, 4, 1599

65HS   H.E. Hellwege and G.K. Schweitzer, J. Inorg. Nucl. Chem., 1965, 27, 99

65IA   H.M.N.H. Irving and N.S. Al-Niami, J. Inorg. Nucl. Chem., 1965, 27, 419

65IM   D.J.G. Ives and P.D. Marsden, J. Chem. Soc., 1965, 649

65JK   M.C. Jain, A.A. Khan, and W.U. Malik, J. Indian Chem. Soc., 1965, 42, 597

65JL   K.M. Jones and E. Larsen, Acta Chem. Scand., 1965, 19, 1205

65JN   R.F. Jameson and W.F.S. Neillie, J. Chem. Soc., 1965, 2391

65JNa   R.F. Jameson and W.F.S. Neillie, J. Inorg. Nucl. Chem., 1965, 27, 2623

65JP   L.G. Joyce and W.F. Pickering, Aust. J. Chem., 1965, 18, 783

65KB   I.F. Kolosova and T.A. Belyavskaya, Russ. J. Inorg. Chem., 1965, 10, 411 (764)

65KC   S.S. Katiyar and V.B.S. Chauhan, J. Prakt. Chem., 1965, 30, 149

65KJ   J. Kratsmar-Smogrovic and V. Jokl, Chem. Zvesti, 1965, 19, 881

65KY   F.Ya. Kulba, Yu.B. Yakovlev, and V.E. Mironov, Russ. J. Inorg. Chem., 1965, 10,
         886 (1624)

65LA   S.Lormeau and M. Ahond, Bull. Soc. Chim. France, 1965, 505

65LF   R. Larsson and B. Folkeson, Acta Chem. Scand., 1965, 19, 53

65LJ   W. Lund and E. Jacobson, Acta Chem. Scand., 1965, 19, 1783

65LM   G.R. Lenz and A.E. Martell, Inorg. Chem., 1965, 4, 378

65LS   F.J. Langmyhr and T. Stumpe, Anal. Chim. Acta, 1965, 32, 535

65M   R.W. McGilvery, Biochem., 1965, 4, 1924

65Ma   C.B. Monk, J. Chem. Soc., 1965, 2456

65ME   M.H. Mills, E.M. Eyring, W.W. Epstein, R.E. Ostlund, J. Phys. Chem., 1965, 69,
          467

65MH   D.J. Macero, H.B. Herman, and A.J. Dukat, Anal. Chem., 1965, 37, 675

65MM   A.I. Moskvin and M.P. Mefodeva, Soviet Radiochem., 1965, 7, 411 (410)

65MN   M.K. Misra and R.K. Nanda, J. Indian Chem. Soc., 1965, 42, 267

65MS   G. Marcu and K. Samochocka, Studia Univ. Babes-Bolyai, 1965, 10, 71

65MSG   D.J. Miller, S.C. Srivastava, and M.L. Good, Anal. Chem., 1965, 37, 739

65N   V.S.K. Nair, J. Chem. Soc., 1965, 1450

65NC   M.H.T. Nyberg and M. Cefola, Arch. Biochem. Biophys., 1965, 111, 321

65ND   R.N. Nanda, R.C. Das, and R.K. Nanda, Indian J. Chem., 1965, 3, 278

65NK   O. Navratil and J. Kotas, Coll. Czech. Chem. Comm., 1965, 30, 2736

65NKK   H.L. Nigam, R.C. Kapoor, U. Kapoor, and S.C. Srivastava, Indian J. Chem., 1965,
          3, 443, 527

65NU   K. Nagata, A. Umayahara, and R. Tsuchiya, Bull. Chem. Soc. Japan, 1965, 38, 1059

65ON   Y. Oka, N. Nakazawa, and H. Harada, Nippon Kagaku Zasshi, 1965, 86, 1158

65ONa   Y. Oka, N. Nakazawa, ans H. Harada, Nippon Kagaku Zasshi, 1965, 86, 1162

65OS   H.G. Offner and D.A. Skoog, Anal. Chem., 1965, 37, 1018

65OY   Y. Oka and M. Yanai, Nippon Kagaku Zasshi, 1965, 86, 929

65PB   A. Polaczek and E. Baranowska, J. Inorg. Nucl. Chem., 1965, 27, 1649

65PP   G.S. Paul and J.E. Powell, USAEC IS-1122, 1965

65PPa   R.K. Patnaik and S. Pani, J. Indian Chem. Soc., 1965, 42, 527, 793

65PS   V.M. Peshkova, V.M. Savostina, E.K. Astakhova, and N.A. Minaeva, Tr. Komis. Anal.
          Khim. Akad. Nauk SSSR, Inst. Geokhim. Anal. Khim., 1965, 15, 104

65PV   B.V. Ptitsyn, L.I. Vinogradova, and E.A. Maksimyuk, Russ. J. Inorg. Chem., 1965,
          10, 1050 (1929)

65RM   K.S. Rajan and A.E. Martell, Inorg. Chem., 1965, 4, 462

65S   W.P. Schaefer, Inorg. Chem., 1965, 4, 642

65Sa   S.P. Sangal, Chim. Anal., 1965, 47, 288, 662

65Sb   T.D. Smith, J. Chem. Soc., 1965, 2145

65Sc   T. Sekine, Acta Chem. Scand., 1965, 19, 1476

65Sd   S.P. Sangal, J. Prakt. Chem., 1965, 30 (5-6), 314

65SB   K.N. Sahu and A.K. Bhattacharya, J. Indian Chem. Soc., 1965, 42, 247

65SF   J.S. Savic and I. Filipovic, Croat. Chem. Acta, 1965, 37, 91

65SK   V.M. Shulman and T.V. Kramareva, Russ. J. Inorg. Chem., 1965, 10, 890 (1632)

65SKP   N.A. Skorik, V.N. Kumok, E.I. Parov, K.P. Augustan, and V.V. Serebrennikov,
          Russ. J. Inorg. Chem., 1965, 10, 351 (653)

65SM   G.K. Schweitzer and S.W. McCarty, J. Inorg. Nucl. Chem., 1965, 27, 191

65SMa   R.G. Seys and C.B. Monk, J. Chem. Soc., 1965, 2452

65SMb   W.P. Schaefer and M.E. Mathisen, Inorg. Chem., 1965, 4, 431

65SMc   A.V. Stepanov and T.P. Makarova, Soviet Radiochem., 1965, 7, 669 (670)

65SN   S.C. Sinha and H.L. Nigam, Vijnana Parishad Anusandhan Patrika, 1965, 8, 83

65SS   G. Schwarzenbach and M. Schellenberg. Helv. Chim. Acta, 1965, 48, 28

65SSa   Z.A. Sheka and E.I. Sinyavskaya, Russ. J. Inorg. Chem., 1965, 10, 212 (394)

65SSb   L. Sommer, T. Sepel, and L. Kurilova, Coll. Czech. Chem. Comm., 1965, 30, 3426

65SSc   L. Sommer, T. Sepel, and L. Kurilova, Coll. Czech. Chem. Comm., 1965, 30, 3834

65SSR   A.V. Stepanov, V.P. Shvedov, and A.P. Rozhnov, Russ. J. Inorg. Chem., 1965, 10,
          750 (1379)

65TF   Yu.A. Treger, R.M. Flid, L.V. Antonova, and S.S. Spektor, Russ. J. Phys. Chem.,
          1965, 39, 1515 (2831)

65TG   S.S. Tate, A.K. Grzybowski, and S.P. Datta, J. Chem. Soc., 1965, 3905

65TMM   K. Toei, H. Miyata, and T. Mitsumata, Bull. Chem. Soc. Japan, 1965, 38, 1050

65TMS   K. Toei, H. Miyata, T. Shibata, and S. Miyamura, Bull. Chem. Soc. Japan, 1965,
          38, 334

65TP   N. Tripathy and R.K. Patnaik, J. Indian Chem. Soc., 1965, 42, 712

65TS   V.M. Tarayan and A.A. Sarkisyan, Russ. J. Inorg. Chem., 1965, 10, 1457 (2684)

65TSO   N. Tanaka, Y. Saito, and H. Ogino, Bull. Chem. Soc. Japan, 1965, 38, 984

65TV   H. Thun, F. Verbeek, and W. Vanderleen, J. Inorg. Nucl. Chem., 1965, 27, 1813

65VS   D.S. Veselinovic and M.V. Susic, Bull. Chem. Soc. Belgrade (TT65-50400), 1965,
          30, No. 2-3, 5 (63), 19 (79)

65VSa   D.G. Vartak and R.S. Shetiya, Indian J. Chem., 1965, 3, 533

65VT   F. Verbeek and H. Thun, Anal. Chim. Acta, 1965, 33, 378

66A   L. Avedikian, Bull. Soc. Chim. France, 1966, 2470

66AA   S. Aditya, S. Aditya, and S.K. Mukherjee, J. Electrochem. Soc. Japan, 1966, 34,
          203

66AM   D.W. Archer and C.B. Monk, J. Chem. Soc. (A), 1966, 1374

66AMa   D.W. Archer and C.B. Monk, Trans. Faraday Soc., 1966, 62, 1583

66AP   V.T. Athavale, L.H Prabhu, and D.G. Vartak, J. Inorg. Nucl. Chem., 1966, 28, 1237

66AT   P.J. Antikainen and K. Tevanen, Suomen Kem., 1966, B39, 2

66BB   A. Beauchamp and R.L. Benoit, Canad. J. Chem., 1966, 44, 1607, 1615

66BBa   A. Bezdekova and B.Budesinsky, Coll. Czech. Chem. Comm., 1966, 31, 199

66BBB   A. Barocas, F. Baroncelli, G.B. Biondi, and G. Grossi, J. Inorg. Nucl. Chem., 1966, 28, 2961

66BP   M. Bonnet and R.A. Paris, Bull. Soc. Chim. France, 1966, 747

66BR   M. Bartusek and J. Ruzickova, Coll. Czech. Chem. Comm., 1966, 31, 207

66BS   A.A. Biryukov, V.I. Shlenskaya, I.P. Alimarin, and T.I. Tikvinskaya, Russ. J. Inorg. Chem., 1966, 11, 897 (1679) (see also 72E)

66BZ   O. Budevsky and O. Zakharieva, Bulg. Akad. Wiss., Inst. Allg. Anorg. Chem., 1966, 4, 15

66CB   B. Cosovic and M. Branica, J. Polarog. Soc., 1966, 12, 5

66CF   G.R. Choppin and H.G Friedman, Inorg. Chem., 1966, 5, 1599

66CS   N.V. Chernaya and S.Ya. Shnaiderman, J. Gen. Chem. USSR, 1966, 36, 1179 (1165)

66D   M.N. Desai, Indian J. Chem., 1966, 4, 218

66DD   N.K. Davidenko and V.F. Deribon, Russ. J. Inorg. Chem., 1966, 11, 53 (99)

66DM   S.N. Dubey and R.C. Mehrotra, J. Indian Chem. Soc., 1966, 43, 73

66DMD   C.D. Dwivedi, K.N. Munshi, and A.K. Dey, J. Indian Chem. Soc., 1966, 43, 111

66DT   E. Doody, E.R. Tucci, R. Scruggs, and N.C. Li, J. Inorg. Nucl. Chem., 1966, 28, 833

66G   P. Gerding, Acta Chem. Scand., 1966, 20, 2624

66Ga   I. Galateanu, Canad. J. Chem., 1966, 44, 647

66GA   Yu.P. Galaktionov and K.V. Astakhov, Russ. J. Inorg. Chem., 1966, 11, 969 (1813)

66GC   A. Gregorowicz and J. Ciba, Rocz. Chem., 1966, 40, 1377

66GD   D.W. Gruenwedel and N. Davidson, J. Mol. Biol., 1966, 21, 129

66GDS   B.P. Gupta, Y. Dutt, and R.P. Singh, J. Indian Chem. Soc., 1966, 43, 610

66GG   M.A. Gouveia and R. Guedes DeCarvalho, J. Inorg. Nucl. Chem., 1966, 28, 1683

66GGJ   J.N. Gaur, N.K. Goswami, and D.S. Jain, Anal. Chem., 1966, 38, 626

66GS   S.L. Gupta, R.N. Soni, and J.N. Jaitly, J. Indian Chem. Soc., 1966, 43, 331

66HP   L.D. Hansen, J.A. Partridge, R.M. Izatt, and J.J. Christensen, Inorg. Chem., 1966, 5, 569

66HS   Y. Hasegawa and T. Sekine, Bull. Chem. Soc. Japan, 1966, 39, 2776

66Ia   A.A. Ivakin, J. Appl. Chem. USSR, 1966, 39, 2262 (2406)

66IR   R.M. Izatt, J.H. Rytting, L.D. Hansen, and J.J. Christensen, J. Amer. Chem. Soc.,
            1966, 88, 2641

66JG   D.S. Jain and J.N. Gaur, J. Polarog. Soc., 1966, 12, 59

66JGa   D.S. Jain and J.N. Gaur, J. Indian Chem. Soc., 1966, 43, 425

66JGb   D.S. Jain and J.N. Gaur, J. Electroanal. Chem., 1966, 11, 310

66JM   K.E. Jabalpurwala and R.M. Milburn, J. Amer. Chem. Soc., 1966, 88, 3224

66JN   R.F. Jameson and W.F.S. Neillie, J. Inorg. Nucl. Chem., 1966, 28, 2667

66KE   N.B. Kalinichenko, A.N. Ermakov, D.I. Ryabchikov, and I.N. Marov, Russ. J. Inorg.
            Chem., 1966, 11, 425 (781)

66KF   H. Koch and W. Falkenberg, Solvent Extr. Chem., Proc. Int. Conf., Gothenburg,
            Interscience Publ. (1967), 1966, 26

66KFa   Th. Kaden and S. Fallab, Chimia (Switz.), 1966, 20, 51

66KP   L.E. Kovar and J.E. Powell, USAEC IS-1450, Oct. 1966

66KPa   M. Krishnamurthy and N.S.K. Prasad, Indian J. Chem., 1966, 4, 316

66KSD   A.F. Krivis, G.R. Supp, and R.L. Doerr, Anal. Chem., 1966, 38, 936

66KSG   I.M. Korenman, F.R. Sheyanova, and Z.M. Gureva, Russ. J. Inorg. Chem., 1966, 11,
            1485 (2761)

66KZ   I.M. Korenman and N.V. Zaglyadimova, Russ. J. Inorg. Chem., 1966, 11, 1491 (2774)

66LH   D.L. Leussing and E.M. Hanna, J. Amer. Chem. Soc., 1966, 88, 693

66LHa   D.L. Leussing and E.M. Hanna, J. Amer. Chem. Soc., 1966, 88, 696

66LM   G.A. L'Heureux and A.E. Martell, J. Inorg. Nucl. Chem., 1966, 28, 481

66LN   S.J. Lyte and S.J. Naqui, J. Inorg. Nucl. Chem., 1966, 28, 2993

66M   O. Makitie, Suomen Kem., 1966, B39, 23

66Ma   O. Makitie, Suomen Kem., 1966, B39, 26

66Mb   O. Makitie, Suomen Kem., 1966, B39, 171

66Mc   O. Makitie, Suomen Kem., 1966, B39, 175

66Md   O. Makitie, Suomen Kem., 1966, B39, 218

66Me   P.G. Manning, Canad. J. Chem., 1966, 44, 3057

66Mf   S.E. Manahan, Inorg. Chem., 1966, 5, 482

66MC   A.K. Majumdar and A.B. Chatterjee, Talanta, 1966, 13, 821

66MK   P.B. Mikhelson and V.I. Kozachek, J. Anal. Chem. USSR, 1966, 21, 1112 (1255)

66MM   G.E. Mont and A.E. Martell, J. Amer. Chem. Soc., 1966, 88, 1387

66MMa   Y. Murakami and A.E. Martell, Bull. Chem. Soc. Japan, 1966, 39, 1077

66MN   A. McAuley, G.H. Nancollas, and K. Torrance, J. Inorg. Nucl. Chem., 1966, 28, 917

66MR   F. Maggio, V. Romano, and R. Cefalu, J. Inorg. Nucl. Chem., 1966, 28, 1979

66MS   E.G. Moorhead and N. Sutin, Inorg. Chem., 1966, 5, 1866

66MSD   K.N. Munshi, S.P. Sangal, and A.K. Dey, J. Indian Chem. Soc., 1966, 43, 115

66MSP   S.P. Mushran, P. Sanyal, and J.D. Pandey, J. Indian Chem. Soc., 1966, 43, 273

66MT   Y. Murakami and M. Takagi, Bull. Chem. Soc. Japan, 1966, 39, 122

66MTa   H. Muro and R. Tsuchiya, Bull. Chem. Soc. Japan, 1966, 39, 1589

66N   D. Nebel, J. Phys. Chem. (Leipzig), 1966, 232, 161, 368

66NU   D. Nebel and G. Urban, J. Phys. Chem. (Leipzig), 1966, 233, 73

66NV   M.S. Novakovskii and V.V. Voinova, Russ. J. Inorg. Chem., 1966, 11, 1409 (2624)

66OC   G. Ostacoli, E. Campi, A. Vanni, and E. Roletto, Ric. Sci., 1966, 36, 427

66OCa   G. Ostacoli, E. Campi, A. Vanni, and E. Roletto, Atti Accad. Sci. Torino, 1966,
            100, 723

66OT   K. Ogawa and N. Tobe, Bull. Chem. Soc. Japan, 1966, 39, 223

66OTa   K. Ogawa and N. Tobe, Bull. Chem. Soc. Japan, 1966, 39, 227

66P   M.V. Park, J. Chem. Soc. (A), 1966, 816; (Nature, 1963, 197, 283)

66PB   R. Pastorek and F. Brezina, Monat. Chem., 1966, 97, 1095

66PL   S. Petri and T. Lipiec, Rocz. Chem., 1966, 40, 1795

66PM   C. Postmus, Jr., L.B. Magnusson, and C.A. Craig, Inorg. Chem., 1966, 5, 1154

66PR   J.E. Prue and A.J. Read, Trans. Faraday Soc., 1966, 62, 1271

66PRa   A.T. Pilipenko and O.P. Ryabushko, Soviet Progr. Chem. (Ukr. Khim. Zh.), 1966,
            32, 477 (622)

66PRb   J.E. Powell and D.L.G. Rowlands, Inorg. Chem., 1966, 5, 819

66PRK   J.E. Powell, D.G.L. Rowlands, L.E. Kovar, J.L. Farrell, L.J. Wilson, and
            V.R. Schoeb, USAEC IS-1500, July 1966

66PRS   A.D. Pakhomova, Ya.L. Rudyakova, and I.A. Sheka, Russ. J. Inorg. Chem., 1966,
            11, 621 (1161)

66PS   V. Peshkova, V. Savostina, and E. Astakhova-Ivanova, Solvent Extr. Chem., Proc.
            Int. Conf., Gothenburg, Interscience Publ. (1967), 1966, 66

66RM   D.L.G. Rowlands and C.B. Monk, Trans. Faraday Soc., 1966, 62, 945

66RS   J.M. Rao and U.V. Seshaiah, Bull. Chem. Soc. Japan, 1966, 39, 2668

66S   A. Solladie-Cavallo, Compt. Rend. Acad. Sci. Paris, Sec. C, 1966, 263, 93

66Sa   S.P. Sangal, J. Prakt. Chem., 1966, 31, 68

66SB   V.I. Shlenskaya, A.A. Biryukov, and E.M. Moskovkina, <u>Russ. J. Inorg. Chem.</u>, 1966,
          <u>11</u>, 325 (600)

66SC   S.Ya. Shnaiderman and N.V. Chernaya, <u>Russ. J. Inorg. Chem.</u>, 1966, <u>11</u>, 72 (134)

66SD   R.P. Singh and Y. Dutt, <u>Indian J. Chem.</u>, 1966, <u>4</u>, 214

66SK   L. Sommer, L. Kurilova-Navratilova, and T. Sepel, <u>Coll. Czech. Chem. Comm.</u>, 1966,
          <u>31</u>, 1288

66SKa   T. Sekine, A. Koizumi, and M. Sakairi, <u>Bull. Chem. Soc. Japan</u>, 1966, <u>39</u>, 2681

66SL   A. Semb and F.J. Langmyhr, <u>Anal. Chim. Acta</u>, 1966, <u>35</u>, 286

66SLK   V.M. Shulman, S.V. Larionov, T.V. Kramareva, E.I. Arykova, and V.V. Yudina,
          <u>Russ. J. Inorg. Chem.</u>, 1966, <u>11</u>, 580 (1076)

66SM   P. Sanyal and S.P. Mushran, <u>Anal. Chim. Acta</u>, 1966, <u>35</u>, 400

66SN   S.C. Sinha, H.L. Nigam, and S.P. Sangal, <u>Chim. Anal.</u>, 1966, <u>48</u>, 515

66SP   I.H. Suffet and W.C. Purdy, <u>J. Electroanal. Chem.</u>, 1966, <u>11</u>, 302

66SPa   V.R. Schoeb and J.E. Powell, USAEC IS-1311, Jan. 1966

66SS   N.A. Skorik and V.V. Serebrennikov, <u>Russ. J. Inorg. Chem.</u>, 1966, <u>11</u>, 416 (764)

66SSa   A.M. Sorochan and M.M. Senyavin, <u>Russ. J. Inorg. Chem.</u>, 1966, <u>11</u>, 753 (1410)

66SSH   T. Sekine, M. Sakairi, and Y. Hasegawa, <u>Bull. Chem. Soc. Japan</u>, 1966, <u>39</u>, 2141

66TB   O.S. Tomar and P.K. Bhattacharya, <u>J. Indian Chem. Soc.</u>, 1966, <u>43</u>, 250

66TF   A.G. Tarasenko and V.M. Fedoseev, <u>Moscow Univ. Chem. Bull.</u>, 1966, <u>21</u>, 214 (75)

66TJ   N.V. Thakur, S.M. Jogdeo, and C.R. Kanekor, <u>J. Inorg. Nucl. Chem.</u>, 1966, <u>28</u>, 2997

66TK   K.V. Tserkasevich, N.P. Kfryushina, and N.S. Poluektov, <u>Russ. J. Inorg. Chem.</u>,
          1966, <u>11</u>, 49 (93)

66TP   K.K. Tripathy and R.K. Patnaik, <u>J. Indian Chem. Soc.</u>, 1966, <u>43</u>, 772

66TS   V.F. Toropova, M.K. Saikina, and R.Sh. Aleshov, <u>Russ. J. Inorg. Chem.</u>, 1966, <u>11</u>,
          605 (1130)

66TV   H. Thur, F. Verbeek, and W. Vanderleen, <u>J. Inorg. Nucl. Chem.</u>, 1966, <u>28</u>, 1949

66VS   D.S. Veselinovic and M.V. Susic, <u>Bull. Chem. Soc. Belgrade</u> (TT66-59000), 1966,
          <u>31</u>, No. 3, 5 (129); No. 4-5-6, 47 (229); No. 9-10, 37 (425)

66WB   Z. Warnke and A. Basinski, <u>Rocz. Chem.</u>, 1966, <u>40</u>, 1141

66ZO   F.G. Zharovskii and M.S. Ostrovskaya, <u>Soviet Progr. Chem.</u> (Ukr. Khim. Zh.), 1966,
          <u>32</u>, 674 (893)

67A   S. Aditya, <u>J. Inorg. Nucl. Chem.</u>, 1967, <u>29</u>, 1901

67Aa   L. Avedikian, <u>Bull. Soc. Chim. France</u>, 1967, 254

67AD   L. Avedikian and N. Dollet, <u>Bull. Soc. Chim. France</u>, 1967, 4551

67AM   V.T. Athavale, N. Mahadevan, P.K. Mathur, and R.M. Sathe, <u>J. Inorg. Nucl. Chem.</u>,
          1967, <u>29</u>, 1947

67AS   R.C. Aggarwal and A.K. Srivastava, <u>Indian J. Chem.</u>, 1967, <u>5</u>, 114

67B   M. Bartusek, Coll. Czech. Chem. Comm., 1967, 32, 116

67Ba   M. Bartusek, Coll. Czech. Chem. Comm., 1967, 32, 757

67BB   A. Beauchamp and R. Benoit, Bull. Soc. Chim. France, 1967, 672

67BE   M.S. Borisov, A.A. Elesin, I.A. Lebedev, E.M. Piskunov, V.T. Filimonov, and
           G.N. Yakovlev, Soviet Radiochem., 1967, 9, 164 (166)

67BH   M. Beran and S. Havelka, Coll. Czech. Chem. Comm., 1967, 32, 2944

67BHa   M. Bartusek and L. Havelkova, Coll. Czech. Chem. Comm., 1967, 32, 3853

67BHB   B. Budesinsky, K. Haas, and A. Bezdekova, Coll. Czech. Chem. Comm., 1967, 32,
           1528

67BK   M.K. Boreiko, E.I. Kazantsev, and I.I. Kalinichenko, Russ. J. Inorg. Chem., 1967,
           12, 137 (269)

67BP   O. Budevsky and E. Platikanova, Talanta, 1967, 14, 901

67BZ   M. Bartusek and J. Zelinka, Coll. Czech. Chem. Comm., 1967, 32, 992

67CA   D.K. Cabbiness and E.S. Amis, Bull. Chem. Soc. Japan, 1967, 40, 435

67CB   G. Choux and R.L. Benoit, Bull. Soc. Chim. France, 1967, 2920

67CBa   J.M. Conner and V.C. Bulgrin, J. Inorg. Nucl. Chem., 1967, 29, 1953

67CI   R.L. Carroll and R.R. Irani, Inorg. Chem., 1967, 6, 1994

67CIH   J.J. Christensen, R.M. Izatt, and L.D. Hansen, J. Amer. Chem. Soc., 1968, 89, 213

67CO   E.G. Chikryzova, B.A. Orgiyan, and L.Y. Kiriyak, Russ. J. Inorg. Chem., 1967,
           12, 1448 (2747)

67DM   S.N. Dubey and R.C. Mehrotra, Indian J. Chem., 1967, 5, 327

67EKG   N.P. Eromolaev, N.N. Krot, and A.D. Gelman, Soviet Radiochem., 1967, 9, 169 (171)

67EKM   A.N. Ermakov, N.B. Kalinichenko, and I.N. Marov, Russ. J. Inorg. Chem., 1967,
           12, 812 (1545)

67ES   V.A. Ermakov and I. Stary, Soviet Radiochem., 1967, 9, 195 (197)

67F   V. Frei, Coll. Czech. Chem. Comm., 1967, 32, 1815

67GA   R. Frostell and P.J. Antikainen, Suomen Kem., 1967, B40, 86

67G   P. Gerding, Acta Chem. Scand., 1967, 21, 2007

67Ga   P. Gerding, Acta Chem. Scand., 1967, 21, 2015

67GD   B.P. Gupta, Y. Dutt, and R.P. Singh, Indian J. Chem., 1967, 5, 322

67GDa   B.P. Gupta, Y. Dutt, and R.P. Singh, J. Inorg. Nucl. Chem., 1967, 29, 1806

67GDb   M.H. Gandhi and M.N. Desai, Anal. Chem., 1967, 39, 1643

67GP   E.C. Gruen and R.A. Plane, Inorg. Chem., 1967, 6, 1123

67GQ   R.J. Grabenstetter, O.T. Quimby, and T.J. Flautt, J. Phys. Chem., 1967, 71, 4194

67H   J. Hala, <u>Coll. Czech. Chem. Comm.</u>, 1967, <u>32</u>, 2565

67HH   K. Higashi, K. Hori, and R. Tsuchiya, <u>Bull. Chem. Soc. Japan</u>, 1967, <u>40</u>, 2569

67HM   A. Hulanicki and M. Minczewska, <u>Talanta</u>, 1967, <u>14</u>, 677

67HMS   Y. Hasegawa, K. Maki, and T. Sekine, <u>Bull. Chem. Soc. Japan</u>, 1967, <u>40</u>, 1845

67HV   F.R. Hartley and L.M. Venanzi, <u>J. Chem. Soc. (A)</u>, 1967, 333

67IH   R. Ishida and N. Hasegawa, <u>Bull. Chem. Soc. Japan</u>, 1967, <u>40</u>, 1153

67IW   D.T. Ireland and H.F. Walton, <u>J. Phys. Chem.</u>, 1967, <u>71</u>, 751

67K   M. Kodama, <u>Bull. Chem. Soc. Japan</u>, 1967, <u>40</u>, 2575

67Ka   C. Konecuy, <u>J. Phys. Chem.</u> (Leipzig), 1967, <u>235</u>, 39

67KE   M. Kodama and H. Ebine, <u>Bull. Chem. Soc. Japan</u>, 1967, <u>40</u>, 1857

67KH   H. Kelm and G.M. Harris, <u>Inorg. Chem.</u>, 1967, <u>6</u>, 706, 1743

67KHa   H. Kodama and K. Hayashi, <u>J. Electroanal. Chem</u>, 1967, <u>14</u>, 209

67KL   M.I. Kabachnik, R.P. Lastovskii, T.Ya. Medved, V.V. Medyntsev, I.D. Kolpakova, and N.M. Dyatlova, <u>Proc. Acad. Sci. USSR</u>, 1967, <u>177</u>, 1060 (582)

67KW   Y. Kanemura and J.I. Watters, <u>J. Inorg. Nucl. Chem.</u>, 1967, <u>29</u>, 1701

67LA   K. Lal and R.P. Agarwal, <u>J. Less-Common Metals</u>, 1967, <u>12</u>, 269

67LAa   K. Lal and R.P. Agarwal, <u>Bull. Chem. Soc. Japan</u>, 1967, <u>40</u>, 1148

67LC   T.-T. Lai, S.-N. Chen, and E. Lin, <u>Talanta</u>, 1967, <u>14</u>, 251

67LL   C.L. Liotta, K.H. Leavell, and D.F. Smith, Jr., <u>J. Phys. Chem.</u>, 1967, <u>71</u>, 3091

67LM   N.F. Lisenko and I.S. Mustafin, <u>J. Anal. Chem. USSR</u>, 1967, <u>22</u>, 20 (25)

67LN   S.J. Lyle and S.J. Naqvi, <u>J. Inorg. Nucl. Chem.</u>, 1967, <u>29</u>, 2441

67LR   J.M. Los, R.F. Rekker, and C.H.T. Tonsbeek, <u>Rec. Trav. Chim.</u>, 1967, <u>86</u>, 609

67M   O. Makitie, <u>Suomen Kem.</u>, 1967, <u>B40</u>, 27

67Ma   O. Makitie, <u>Suomen Kem.</u>, 1967, <u>B40</u>, 128, 267

67Mb   K.R. Manolov, <u>Russ. J. Inorg. Chem.</u>, 1967, <u>12</u>, 1431 (2715)

67Mc   R.M. Milburn, <u>J. Amer. Chem. Soc.</u>, 1967, <u>89</u>, 54

67Md   H. Miyata, <u>Bull. Chem. Soc. Japan</u>, 1967, <u>40</u>, 1875

67Me   P.G. Manning, <u>Canad. J. Chem.</u>, 1967, <u>45</u>, 1643

67Mf   M. Meier, Diss., ETH, Zurich, 1967 (see G. Anderegg, <u>Helv. Chim. Acta</u>, 1968, <u>51</u>, 1856

67MC   P.K. Migal and N.G. Chebotar, <u>Russ. J. Inorg. Chem.</u>, 1967, <u>12</u>, 630 (1190)

67ME   A.I. Moskvin and L.N. Essen, <u>Russ. J. Inorg. Chem.</u>, 1967, <u>12</u>, 359 (688)

67MI   S.E. Manahan and R.T. Iwamoto, J. Electroanal. Chem., 1967, 14, 213

67MM   Gh. Marcu and Gh. Murgu, Rev. Roum. Chim., 1967, 12, 957

67MN   C. Miyake and H.W. Nurnberg, J. Inorg. Nucl. Chem., 1967, 29, 2411

67MNa   E.A. Mazurenko and B.I. Nabivanets, Soviet Progr. Chem. (Ukr. Khim. Zh.), 1967, 33, No. 1, 86 (98)

67MNT   A. McAuley, G.H. Nancollas, and K. Torrance, Inorg. Chem., 1967, 6, 136

67MO   N.V. Melchakova, G.P. Ozerova, and V.M. Peshkova, Russ. J. Inorg. Chem., 1967, 12, 577 (1096)

67MP   S.P. Mushran, O. Prakash, and J.N. Awasthi, Anal. Chem., 1967, 39, 1307

67MS   M.B. Mishra, S.C. Sinha, and H.L. Nigam, Indian J. Chem., 1967, 5, 659

67MY   M. Murakami, T. Yoshino, and S. Harasawa, Talanta, 1967, 14, 1293

67NA   G. Niac and Z. Andrei, Rev. Roum. Chim., 1967, 12, 801

67NE   V.A. Nazarenko and L.D. Ermak, Russ. J. Inorg. Chem., 1967, 12, 335 (653), 1079 (2051)

67NEa   V.A. Nazarenko and L.D. Ermak, Russ. J. Inorg. Chem., 1967, 12, 1304 (2472)

67NG   B.I. Nabivanets, V.V. Grigoreva, and G.V. Molodid, Soviet Progr. Chem. (Ukr. Khim. Zh.), 1967, 33, No. 5, 78 (502)

67NMH   T. Nozaki, T. Mise, and K. Higaki, Nippon Kagaku Zasshi, 1967, 88, 1168

67NMT   S. Nakashima, H. Miyata, and K. Toei, Bull. Chem. Soc. Japan, 1967, 40, 870

67NN   B.M. Nikolova and G.St. Nikolov, J. Inorg. Nucl. Chem., 1967, 29, 1013

67NP   I.V. Nazarova and V.I. Prizbilevskaya, Russ. J. Inorg. Chem., 1967, 12, 1614 (3051)

67NS   A.N. Nevzorov and O.A. Songina, Russ. J. Inorg. Chem., 1967, 12, 1259 (2388)

67OH   Y. Oka and H. Harada, Nippon Kagaku Zasshi, 1967, 88, 441

67OT   W. Ooghe, H. Thun, and F. Verbeek, Anal. Chim. Acta, 1967, 39, 397

67OW   G. Ojelund and I. Wadso, Acta Chem. Scand., 1967, 21, 1408

67PB   P. Pichet and R.L. Benoit, Inorg. Chem., 1967, 6, 1505

67PM   O. Prakash and S.P. Mushran, Chim. Anal., 1967, 49, 473

67PN   J.E. Powell and W.F.S. Neillie, J. Inorg. Nucl. Chem., 1967, 29, 2371

67PNF   J.E. Powell, W.F.S. Neillie, J.L. Farell, and D.K. Johnson, USAEC IS-1600, C24, March 1, 1967

67PR   J.E. Powell and D.L.G. Rowlands, J. Inorg. Nucl. Chem., 1967, 29, 1729

67PRa   A.T. Pilipenko and O.P. Ryabushko, Soviet Progr. Chem. (Ukr. Khim. Zh.), 1967, 33, No. 4, 52 (403)

67PS   D.D. Perrin and I.G. Sayce, J. Chem. Soc. (A), 1967, 82

67PSS   D.D. Perrin, I.G. Sayce, and V.S. Sharma, J. Chem. Soc. (A), 1967, 1755

67RB   R.R. Rao and P.K. Bhattacharya, Curr. Sci. (India), 1967, 36, 71

67RBa   K.K. Rohatgi and P.K. Bhattacharya, Indian J. Chem., 1967, 5, 195

67RK   E.D. Romanenko and N.A. Kostromino, Russ. J. Inorg. Chem., 1967, 12, 266 (516)

67RM   K.S. Rajan and A.E. Martell, J. Inorg. Nucl. Chem., 1967, 29, 463

67RMa   K.S. Rajan and A.E. Martell, J. Inorg. Nucl. Chem., 1967, 29, 523

67S   S.P. Sangal, J. Prakt. Chem., 1967, 36, 126

67Sa   H. Schreck, KFK-672, 1967

67SB   M.B. Shchigol and N.B. Burchinskaya, Russ. J. Inorg. Chem., 1967, 12, 626 (1183)

67SBa   K.N. Sahu and A.K. Bhattacharya, Curr. Sci. (India), 1967, 36, 70

67SK   L. Sommer and V. Kuban, Coll. Czech. Chem. Comm., 1967, 32, 4355

67SKS   N.A. Skorik, V.N. Kumok, and V.V. Serebrennikov, Russ. J. Inorg. Chem., 1967,
          12, 1429 (2711)

67SM   A.V. Stepanov and T.P. Makarova, Russ. J. Inorg. Chem., 1967, 12, 1262 (2395)

67SMa   P. Sanyal and S.P. Mushran, Chim. Anal., 1967, 49, 231

67SP   A.J. Singh and N.S.K. Prasad, Indian J. Chem., 1967, 5, 573

67SPa   P. Spacu and S. Plostinaru, Rev. Roum. Chim., 1967, 12, 383

67SS   T. Sekine and M. Sakairi, Bull. Chem. Soc. Japan, 1967, 40, 261

67SSD   S.N. Sinha, S.P. Sangal, and A.K. Dey, J. Indian Chem. Soc., 1967, 44, 203

67ST   V.I. Shlenskaya, T.I. Tikvinskaya, A.A. Biryukov, and I.P. Alimarin, Bull. Acad.
          Sci. USSR, Ser. Chem., 1967, 2063 (2141)

67STV   H. Schurmans, H. Thun, F. Verbeek, J. Inorg. Nucl. Chem., 1967, 29, 1759

67TG   H. Thun, W. Guns, and F. Verbeck, Anal. Chim. Acta, 1967, 37, 332

67TK   T.V. Ternovaya, N.A. Kostromina, and E.D. Romanenko, Soviet Progr. Chem. (Ukr.
          Khim. Zh.), 1967, 33, No. 7, 1 (651)

67TKL   E.R. Tucci, C.H. Ke, and N.C. Li, J. Inorg. Nucl. Chem., 1967, 29, 1657

67TM   M.M. Taqui Khan and A.E. Martell, J. Amer. Chem. Soc., 1967, 89, 4176

67TMH   K. Toei, H. Miyata, and T. Harada, Bull. Chem. Soc. Japan, 1967, 40, 1141

67TMK   K. Toei, H. Miyata, and H. Kimura, Bull. Chem. Soc. Japan, 1967, 40, 2085

67TMN   K. Toei, H. Miyata, S. Nakashima, and S. Kiguchi, Bull. Chem. Soc. Japan, 1967,
          40, 1145

67TP   K.K. Tripathy and R.K. Patnaik, Indian J. Chem., 1967, 5, 511

67TT   N. Tripathy, K.K. Tripathy, and R.K. Patnaik, J. Indian Chem. Soc., 1967, 44,
          329

67VA   S.K. Verma and R.P. Agarwal, J. Less-Common Metals, 1967, 12, 221

67VD   W.E. VanderLinder and G. DenBoef, Anal. Chim. Acta, 1967, 37, 179

67VG   V.P. Vasilev and N.K. Grechina, Russ. J. Inorg. Chem., 1967, 12, 823 (1865)

67VK   V.P. Vasilev and L.A. Kochergina, Russ. J. Phys. Chem., 1967, 41, 1149 (2133);
          681 (1282); 1966, 40, 1622 (3024)

67VL   M. Vicedomini and A. Liberti, Gazz. Chim. Ital., 1967, 97, 1627

67VS   D.G. Vartak and R.S. Shetiya, J. Inorg. Nucl. Chem., 1967, 29, 1261

67VV   L.P. Varga and F.C. Veatch, Anal. Chem., 1967, 39, 1101

67W   H. Waki, "Selected Topics on Stability Constants of Metal Complexes," Proceed-
          ings of the Informal Meeting, Tenth International Conference on Coordination
          Chemistry, S. Misumi, Editor, Sept. 1967 (Inquires to Dr. N. Tanaka, Dept.
          of Chem., Faculty of Science, Tohoku Univ., Sendai, Japan), 15

67Wa   U. Weser, Z. Naturforsch., 1967, 22B, 457

67WG   J.M. Wilson, N.E. Gore, J.E. Sawbridge, and F. Cardenas-Cruz, J. Chem. Soc. (B),
          1967, 852

67WU   O. Wahlberg and P. Ulmgren, Acta Chem. Scand., 1967, 21, 2759

67YT   K. Yamamoto and K. Takamizawa, Nippon Kagaku Zasshi, 1967, 88, 345

67Z   A.A. Zavitsas, J. Chem. Eng. Data, 1967, 12, 94

67ZF   H.E. Zittel and T.M. Florence, Anal. Chem., 1967, 39, 320

67ZS   F.G. Zharovskii and R.I. Sukhomlin, Soviet Progr. Chem. (Ukr. Khim. Zh.), 1967,
          33, No. 5, 85 (509)

67ZSO   F.G. Zharovskii, R.I. Sukhomlin, and M.S. Ostrovskaya, Russ. J. Inorg. Chem.,
          1967, 12, 1306 (2476)

68AB   O.P. Afanasev, A.N. Bantysh, and D.A. Knyazev, Russ. J. Inorg. Chem., 1968, 13,
          182 (352)

68AL   A. Aziz, S.J. Lyle, and S.J. Naqvi, J. Inorg. Nucl. Chem., 1968, 30, 1013

68AO   P.J. Antikainen and H. Oksanen, Acta Chem. Scand., 1968, 22, 2867

68AP   P.J. Antikainen and I.P. Pitkanen, Suomen Kem., 1968, B41, 65

68APa   P.J. Antikainen and I.P. Pitkanen, Suomen Kem., 1968, B41, 108

68AS   V.T. Athavale, R.M. Sathe, and N. Mahadevan, J. Inorg. Nucl. Chem., 1968, 30,
          3107

68AT   I.P. Alimarin, T.I. Tikhvinskaya, V.I. Shlenskaya, and A.S. Biryukov, Bull. Acad.
          Sci. USSR, Ser. Chem., 1968, 2543 (2675)

68AV   P.J. Antikainen and M. Viro, Suomen Kem., 1968, B41, 206

68AVa   L.P. Adamovich and V.N. Vu, J. Anal. Chem. USSR, 1968, 23, 868 (994)

68B   E. Bottari, Monat. Chem., 1968, 99, 176

68BB   A.S. Berlyand, A.I. Byrke, and L.I. Martynenko, Russ. J. Inorg. Chem., 1968, 13,
          1089 (2106)

68BC   E. Bottari and L. Ciavatta, Gazz. Chim. Ital., 1968, 98, 1004

68BD   A. Banerjee and A.K. Dey, Anal. Chim. Acta, 1968, 42, 473

68BDa   A. Banerjee and A.K. Dey, J. Inorg. Nucl. Chem., 1968, 30, 995

68BDb   A. Banerjee and A.K. Dey, J. Inorg. Nucl. Chem., 1968, 30, 3134

68BL   E. Bottari and A. Liberti, Gazz. Chim. Ital., 1968, 98, 991

68BLD   A. Braibanti, E. Leporati, F. Dallavalle, and M.A. Pellinghelli, Inorg. Chim.
           Acta, 1968, 2, 449

68BN   E.A. Biryuk and V.A. Nazarenko, J. Anal. Chem. USSR, 1968, 23, 887 (1018)

68BR   E. Bottari and A. Rufolo, Monat. Chem., 1968, 99, 2383

68BRN   E.A. Buryuk, R.V. Ravitskaya, and V.A. Nazarenko, Soviet Progr. Chem. (Ukr. Khim.
           Zh.), 1968, 34, No. 10, 75 (1062)

68BS   W.G. Baldwin and D.R. Stranks, Aust. J. Chem., 1968, 21, 603

68BV   A. Bonniol and P. Vieles, J. Chim. Phys., 1968, 65, 414

68CC   J.P. Calmon, Y. Cazaux-Maraval, and P. Maroni, Bull. Soc. Chim. France, 1968,
           3779

68CH   E. Chiacchierini, J. Havel, and L. Sommer, Coll. Czech. Chem. Comm., 1968, 33,
           4215

68CI   R.L. Carroll and R.R. Irani, J. Inorg. Nucl. Chem., 1968, 30, 2971

68CM   G.R. Choppin and L.A. Martinez-Perez, Inorg. Chem., 1968, 7, 2657

68CO   J.J. Christensen, J.L. Oscarson, and R.M. Izatt, J. Amer. Chem. Soc., 1968, 90,
           5949

68CR   E. Coates, B. Rigg, and D.L. Smith, Trans. Faraday Soc., 1968, 64, 3255

68DD   S.S. Dube and S.S. Dhindsa, Curr. Sci. (India), 1968, 37, 642

68DK   R.A. Dyachkova, V.P. Khlebnikov, and V.I. Spitsyn, Russ. J. Inorg. Chem., 1968,
           13, 439 (836)

68DKa   R.A. Dyachkova, V.P. Khlebnikov, and V.I. Spitsyn, Soviet Radiochem., 1968, 10,
           17 (21)

68DM   M. Deneux, R. Meilleur, and R.L. Benoit, Canad. J. Chem., 1968, 46, 1383

68DY   T.M. Devyatova and M.Z. Yampolskii, J. Anal. Chem. USSR, 1968, 23, 1291 (1468)

68E   B. Egneus, Anal. Chim. Acta, 1968, 43, 53

68EN   Z.L. Ernst and P.J. Newman, Trans. Faraday Soc., 1968, 64, 1052

68ES   S.H. Eberle and J.B. Schaefer, Inorg. Nucl. Chem. Letters, 1968, 4, 283

68ESB   S.H. Eberle, J.B. Schaefer, and E. Brandau, Radiochim. Acta, 1968, 10, 91

68FI   V.I. Fadeeva, V.M. Ivanov, N.V. Eremeeva, and I.P. Alimarin, Russ. J. Inorg.
           Chem., 1968, 13, 1524 (2960)

68FK   T. Fueno, O. Kajimoto, and J. Furukawa, Bull. Chem. Soc. Japan, 1968, 41, 782

68FL   B. Folkesson and R. Larsson, Acta Chem. Scand., 1968, 22, 1953

68FP   I. Filipovic, I. Piljac, A. Medved, J. Savic, A. Bujak, B. Bach-Dragutinovic,
           and B. Mayer, Croat. Chem. Acta, 1968, 40, 131

68FV   Ya.D. Fridman and R.A. Veresova, Russ. J. Inorg. Chem., 1968, 13, 399 (762)

68G   P. Gerding, Acta Chem. Scand., 1968, 22, 1283

68Ga   R. Guillaumont, Bull. Soc. Chim. France, 1968, 1956

68GB   R. George and J. Bjerrum, Acta Chem. Scand., 1968, 22, 497

68GD   P.S. Gentile and A. Dadgar, J. Chem. Eng. Data, 1968, 13, 236

68GF   G. Gutnikov and H. Freiser, Anal. Chem., 1968, 40, 39

68GG   M.A. Gouveia and R. Guedes deCarvahlo, J. Inorg. Nucl. Chem., 1968, 30, 2219

68GGa   J.K. Gupta and C.M. Gupta, Monat. Chem., 1968, 99, 2526

68GGb   J.K. Gupta and C.M. Gupta, Indian J. Chem., 1968, 6, 50

68GH   A.M. Goeminne and M.A. Herman, Bull. Soc. Chim. Belges, 1968, 77, 227

68GHa   A.M. Goeminne and M.A. Herman, Anal. Chim. Acta, 1968, 41, 400

68GHE   A.M. Goeminne, M.A. Herman, and Z. Eeckhaut, Bull. Soc. Chim. Belges, 1968, 77,
           357

68GJ   P. Gerding and I. Jonsson, Acta Chem. Scand., 1968, 22, 2255

68GJP   J.N. Gaur, D.S. Jain, and M.M. Palrecha, J. Chem. Soc. (A), 1968, 2201

68GK   R.K. Gridasova, I.V. Kolosov, and B.N. Ivanov-Emin, Soviet Radiochem., 1968, 10,
           32 (37)

68GP   J.N. Gaur and M.M. Palrecha, Talanta, 1968, 15, 583

68GPa   K.I. Grigalashvili and I.V Pyatnitskii, Soviet Progr. Chem. (Ukr. Zh. Khim.),
           1968, 34, No. 4, 77 (402)

68GT   S.S. Goyal and J.P. Tandon, Aust. J. Chem., 1968, 21, 2433

68HB   L. Havelkova and M. Bartusek, Coll. Czech. Chem. Comm., 1968, 33, 385

68HBa   L. Havelkova and M. Bartusek, Coll. Czech. Chem. Comm., 1968, 33, 4188

68HL   P. Herman and K. Lemke, Z. Physiol. Chem., 1968, 349, 390

68JG   Z. Jablonski, J. Gornicki, and A. Lodzinska, Rocz. Chem., 1968, 42, 1809

68JH   E.S. Jayadevappa and P.B. Hukkeri, J. Inorg. Nucl. Chem., 1968, 30, 157

68JK   D.S. Jain, A. Kumar, and J.N. Gaur, J. Electroanal. Chem., 1968, 17, 201

68KH   Z. Kolarik, J. Hejna, and H. Pankova, J. Inorg. Nucl. Chem., 1968, 30, 2795

68KK   C.H. Ke, P.C. Kong, H.S. Cheng, and N.C. Li, J. Inorg. Nucl. Chem., 1968, 30, 961

68KS   M. Kodama and C. Sasaki, Bull. Chem. Soc. Japan, 1968, 41, 127

68KT   C.R. Kanekar, N.V. Thakar, and S.M. Jogdeo, Bull. Chem. Soc. Japan, 1968, 41,
           759

68LB   D.L. Leussing and K.S. Bai, Anal. Chem., 1968, 40, 575

68LC   T.T. Lai and T.Y. Chen, Anal. Chim. Acta, 1968, 43, 63

68LO   J.E. Land and C.V. Osborne, J. Less-Common Metals, 1968, 14, 349

68M   O. Makitie, Suomen Kem., 1968, B41, 31

68Ma   O. Makitie, Acta Chem. Scand., 1968, 22, 2703

68Mb   D.J. MacDonald, J. Org. Chem., 1968, 33, 4559

68Mc   W.A.E. McBryde, Canad. J. Chem., 1968, 46, 2385

68Md   J. Mach, Monat. Chem., 1968, 99, 1003

68Me   C. Moreau, Bull. Soc. Chim. France, 1968, 31

68Mf   K.R. Manolov, Monat. Chem., 1968, 99, 1774

68MD   S. Mandel and A.K. Dey, J. Inorg. Nucl. Chem., 1968, 30, 1221

68MDK   O.S. Musailov, B.A. Dunai, and N.P. Komar, J. Anal. Chem. USSR, 1968, 23, 123
            (157)

68MN   O.V. Mandzhgaladze and V.A. Nazarenko, Russ. J. Phys. Chem., 1968, 42, 1572
            (2957)

68MPC   L. Magon, R. Portanova, and A. Cassol, Inorg. Chim. Acta, 1968, 2, 237

68MPS   S.P. Mushran, O. Prakash, and P. Sanyal, J. Prakt. Chem., 1968, 38, 125

68MS   O. Makitie, K. Soininen, and H. Saarinen, Suomen Kem., 1968, B41, 246

68NA   V.A. Nazarenko and V.P. Antonovich, J. Anal. Chem. USSR, 1968, 23, 575 (668)

68NL   O. Navratil and J. Liska, Coll. Czech. Chem. Comm., 1968, 33, 987

68NM   S. Nakashima, H. Miyata, and K. Toei, Bull. Chem. Soc. Japan, 1968, 41, 2632

68OC   G. Ostacoli, E. Campi, and M.C. Gennaro, Gazz. Chim. Ital., 1968, 98, 301

68OF   Kh.K. Ospanov, S.N. Fedosov, and Z.B. Rozhdestvenskaya, J. Anal. Chem. USSR,
            1968, 23, 138 (175)

68OH   Y. Oka and Y. Hirai, Nippon Kagaku Zasshi, 1968, 89, 589

68OO   M. Oh-Eidhin and S. OCinneide, J. Inorg. Nucl. Chem., 1968, 30, 3209

68OOH   Y. Oka, M. Okamura, and H. Harada, Nippon Kagaku Zasshi, 1968, 89, 171

68OR   Kh.K. Ospanov, Z.B. Rozhdestuenskaya, S.N. Fedosov, and T.V. Vasileva, J. Anal.
            Chem. USSR, 1968, 23, 673 (779)

68OV   G. Ostacoli, A. Vanni, and E. Roletto, Ric. Sci., 1968, 38, 318

68OW   Y. Oka, I. Watanabe, and M. Hirai, Nippon Kagaku Zasshi, 1968, 89, 1220

68OY   E.R. Oskotskaya and M.Z. Yampolskii, J. Anal. Chem. USSR, 1968, 23, 1153 (1307)

68PF   J.E. Powell and J.L. Farrell, J. Inorg. Nucl. Chem., 1968, 30, 2135

68PFN   J.E. Powell, J.L. Farrell, W.F.S. Neillie, and R. Russell, J. Inorg. Nucl. Chem.,
            1968, 30, 2223

68PK   I.V. Pyatnitskii and L.F. Kravtsova, Soviet Progr. Chem. (Ukr. Khim. Zh), 1968,
            34, No. 1, 77 (86)

68PKa   I.V. Pyatnitskii and L.F. Kravtsova, Soviet Progr. Chem. (Ukr. Khim. Zh.), 1968,
         34, No. 3, 1 (231)

68PKb   I.V. Pyatnitskii and L.F. Kravtsova, Soviet Progr. Chem. (Ukr. Khim. Zh.), 1968,
         34, No. 7, 60 (706)

68PKc   I.V. Pyatnitskii and T.I. Kravchenko, Soviet Progr. Chem. (Ukr. Khim. Zh.), 1968,
         34, No. 12, 9 (1215)

68PM   A.T. Pilipenko, N.N. Maslei, and E.A. Karetnikova, J. Anal. Chem. USSR, 1968, 23,
         62 (80)

68PRM   A.T. Pilipenko, O.P. Ryabushko, and T.K. Makarenko, Soviet Progr. Chem. (Ukr.
         Khim. Zh.), 1968, No. 8, 59 (823)

68PRS   L.D. Pettit, A. Royston, C. Sherrington, and R.J. Whewell, J. Chem. Soc. (B),
         1968, 588

68PRW   L.D. Pettit, A. Royston, and R.J. Whewell, J. Chem. Soc. (A), 1968, 2009

68PS   L.D. Pettit and C. Sherrington, J. Chem. Soc. (A), 1968, 3078

68PSa   I.V. Pyatnitskii and E.S. Sereda, Soviet Progr. Chem. (Ukr. Khim. Zh.), 1968,
         34, No. 11, 62 (1162)

68PSW   L.D. Pettit, C. Sherrington, and R.J. Whewell, J. Chem. Soc. (A), 1968, 2204

68RS   S. Ramamoorthy and M. Santappa, J. Inorg. Nucl. Chem., 1968, 30, 1855

68RSa   S. Ramamoorthy and M. Santappa, J. Inorg. Nucl. Chem., 1968, 30, 2393

68RSb   S. Ramamoorthy and M. Santappa, Bull. Chem. Soc. Japan, 1968, 41, 1330

68RSc   N.P. Rudenko, A.I. Sevastyanov, and N.G. Lanskaya, Russ. J. Inorg. Chem., 1968,
         13, 821 (1566)

68S   A. Salahuddin, J. Prakt. Chem., 1968, 37, 290

68Sa   H. Schnecko, Anal. Chem., 1968, 40, 1391

68Sb   G.R. Supp, Anal. Chem., 1968, 40, 981

68SA   G.K. Schweitzer and M.M. Anderson, J. Inorg. Nucl. Chem., 1968, 30, 1051

68SD   K.K. Saxena and A.K. Dey, Anal. Chem., 1968, 40, 1280

68SDY   V.D. Salikhov, Yu.M. Dedkov, and M.Z. Yampolskii, J. Anal. Chem. USSR, 1968,
         23, 449 (529)

68SG   R.S. Saxena, K.C. Gupta, and M.L. Mittal, Aust. J. Chem., 1968, 21, 641

68SGa   R.S. Saxena, K.C. Gupta, and M.L. Mittal, Canad. J. Chem., 1968, 46, 311

68SGb   R.S. Saxena, K.C. Gupta, and M.L. Mittal, J. Inorg. Nucl. Chem., 1968, 30, 189

68SGc   R.S. Saxena and K.C. Gupta, J. Inorg. Nucl. Chem., 1968, 30, 3368

68SGd   R.S. Saxena, K.C. Gupta, and L. Mittal, Monat. Chem., 1968, 99, 1779

68SGe   A.I. Stetsenko, M.I. Gelfman, N.D. Mitkinova, and S.G. Strelin, Russ. J. Inorg.
         Chem., 1968, 13, 575 (1101)

68SH   A. Stefanovic, J. Havel, and L. Sommer, Coll. Czech. Chem. Comm., 1968, 33, 4198

68SK   F.D. Shevchenko and L.A. Kuzina, Soviet Progr. Chem. (Ukr. Khim. Zh.), 1968, 34,
         No. 5, 75 (499)

68SKM   K. Suzuki, C. Karaki, S. Mori, and K. Yamasaki, J. Inorg. Nucl. Chem., 1968, 30, 167

68SM   V.I. Spitsyn, L.I. Martynenko, Z.M. Rivina, K.N. Rivin, N.A. Dobrynina, and A.V. Nemukhin, Doklady Chem., 1968, 179, 364 (1348)

68SP   T.R. Sweet and H.W. Parlett, Anal. Chem., 1968, 40, 1885

68SY   V.D. Salikhov and M.Z. Yampolskii, J. Anal. Chem. USSR, 1968, 23, 150 (189)

68TF   B. Topuzovski and I. Filipovic, Croat. Chem. Acta, 1968, 40, 257

68TK   I.A. Tserkovnitskaya and N.A. Kustova, J. Anal. Chem. USSR, 1968, 23, 54 (72)

68TKT   T. Tomita, E. Kyuno, and R. Tsuchiya, Bull. Chem. Soc. Japan, 1968, 41, 1130

68TL   Yu.K. Tselinskii and E.V. Lapitskaya, Soviet Progr. Chem. (Ukr. Khim. Zh.), 1968, 34, No. 2, 65 (189)

68TM   C.A. Tyson and A.E. Martell, J. Amer. Chem. Soc., 1968, 90, 3379

68V   M. Vicedomini, Gazz. Chim. Ital., 1968, 98, 556

68Va   M. Vicedomini, Gazz. Chim. Ital., 1968, 98, 1152

68Vb   M. Vicedomini, Gazz. Chim. Ital., 1968, 98, 1161

68VA   S.K. Verma and R.P. Agarwal, J. Prakt. Chem., 1968, 38, 280

68VG   V.P. Vasilev and N.K. Grechina, Izvest. Vepsh. Ucheb. Zaved., Khim., 1968, 11, 1319

68WF   B. Wiberg and P.C. Ford, Inorg. Chem., 1968, 7, 369

68WM   R. Wake, H. Miyata, and K. Toei, Bull. Chem. Soc. Japan, 1968, 41, 1452

68WZ   K. Winkler and K.B. Zaborenko, Z. Phys. Chem. (Leipzig), 1968, 238, 348

68YN   K. Yamamoto and T. Nishio, Nippon Kagaku Zasshi, 1968, 89, 1214

68Z   S. Zommer, Rocz. Chem., 1968, 42, 1803

68ZK   E.A. Zekharova and V.N. Kumok, J. Gen. Chem. USSR, 1968, 38, 1868 (1922)

68ZS   Yu.A. Zolotov, N.V. Shakhova, and I.P. Alimarin, J. Anal. Chem. USSR, 1968, 23, 1164 (1321)

69AA   C. Andersson, S.O. Andersson, J.O. Liljenzin, H. Reinhardt, and J. Rydberg, Acta Chem. Scand., 1969, 23, 2781

69AI   A.N. Alybina, E.K. Ivanova, and V.M. Peshkova, Moscow Univ. Chem. Bull., 1969, 24, No. 1, 77 (93)

69AK   G. Ackermann and S. Koch, Talanta, 1969, 16, 284

69AN   V.P. Antonovich and V.A. Nazarenko, J. Anal. Chem. USSR, 1969, 24, 526 (676)

69AV   P.J. Antikainen, M. Viro, and L.R. Sahlstrom, Suomen Kem., 1969, B42, 178

69B   B. Budesinsky, Talanta, 1969, 16, 1277

69Ba   J.M. Blair, Eur. J. Biochem., 1969, 8, 287

69BB   J.C. Barnes and P.A. Bristow, J. Less-Common Metals, 1969, 18, 381

69BBa   G.V. Bakore and M.S. Bararia, Z. Phys. Chem. (Leipzig), 1969, 242, 102

69BBH   M. Bartusek, L. Brchan, and L. Havelkova, Spisy Prir. Fak. Univ. Purk. Brno,
           1969, No. 499, 19

69BD   A. Braibanti, F. Dallavalle, and E. Leporati, Inorg. Chim. Acta, 1969, 3, 459

69BF   A.I. Busev and V.Z. Filip, Moscow Univ. Chem. Bull., 1969, 24, No. 4, 58 (92)

69BFa   A.B. Blank and I.I. Fedorova, J. Anal. Chem. USSR, 1969, 24, 782 (978)

69BH   G.B. Briscoe and S. Humphries, Talanta, 1969, 16, 1403

69BL   G. Berthon and C. Luca, Bull. Soc. Chim. France, 1969, 432

69BLP   D. Barnes, P.G. Laye, and L.D. Pettit, J. Chem. Soc. (A), 1969, 2073

69BLR   E. Bottari, A. Liberti, and A. Rufolo, Inorg. Chim. Acta, 1969, 3, 201

69BM   A. Banerjee, S. Mandal, T. Singh, and A.K. Dey, J. Indian Chem. Soc., 1969, 46,
           824

69BR   A.I. Busev and V.V. Rozanova, Russ. J. Inorg. Chem., 1969, 14, 1770 (3358)

69BY   N.N. Basargin, P.Ya. Yakovlev, and I.A. Zanina, J. Anal. Chem. USSR, 1969, 24,
           649 (813)

69CB   E. Chiacchierini and M. Bartusek, Coll. Czech. Chem. Comm., 1969, 34, 530

69CM   D.L. Campbell and T. Moeller, J. Inorg. Nucl. Chem., 1969, 31, 1077

69CMa   G.F. Condike and A.E. Martell, J. Inorg. Nucl. Chem., 1969, 31, 2455

69CMT   A. Cassol, L. Magon, G. Tomat, and R. Portanova, Inorg. Chim. Acta, 1969, 3, 639

69CN   L. Ciavatta, G. Nunziata, and L.G. Sillen, Acta Chem. Scand., 1969, 23, 1637

69CP   A. Cassol, R. Portanova, and L. Magon, Inorg. Nucl. Chem. Lett., 1969, 5, 341

69CS   N.V. Chernaya and S.Ya. Shnaiderman, Russ. J. Inorg. Chem., 1969, 14, 384 (735)

69CV   B. Cosovic, M. Verzi, and M. Branica, Electroanal. Chem., 1969, 22, 325

69DD   S.S. Dube and S.S. Dhindsa, Indian J. Chem., 1969, 7, 823

69DDa   S.S. Dube and S.S. Dhindsa, J. Indian Chem. Soc., 1969, 46, 838

69DDb   S.S. Dube and S.S. Dhindsa, Z. Naturforschung, 1969, 24b, 967

69DDc   S.S. Dube and S.S. Dhindsa, Z. Naturforschung, 1969, 24b, 1234

69DM   A.G. Desai and R.M. Milburn, J. Amer. Chem. Soc., 1969, 91, 1958

69DS   N.K. Dutt and T. Seshadri, J. Inorg. Nucl. Chem., 1969, 31, 2153

69E   B. Egneus, Anal. Chim. Acta, 1969, 48, 291

69ES   S.H. Eberle and J.B. Schaefer, J. Inorg. Nucl. Chem., 1969, 31, 1523

69EV   I.P. Efimov, L.S. Voronets, L.G. Makarova, and V.M. Peshkova, Moscow Univ. Chem.
           Bull., 1969, 24, No. 4, 83 (121)

69F   V. Frei, <u>Coll. Czech. Chem. Comm.</u>, 1969, <u>34</u>, 1309

69FB   T.M. Florence and W.L. Belew, <u>Electroanal. Chem.</u>, 1969, <u>21</u>, 157

69FD   Ya.D. Fridman and T.V. Danilova, <u>Russ. J. Inorg. Chem.</u>, 1969, <u>14</u>, 370

69FT   S. Funahashi and M. Tanaka, <u>Inorg. Chem.</u>, 1969, <u>8</u>, 2159

69GG   J.K. Gupta and C.M. Gupta, <u>J. Indian Chem. Soc.</u>, 1969, <u>45</u>, 491

69GGa   J.K. Gupta and C.M. Gupta, <u>J. Prakt. Chem.</u>, 1969, <u>311</u>, 635

69GGR   I. Grenthe, G. Gardhammar, and E. Rundcrantz, <u>Acta Chem. Scand.</u>, 1969, <u>23</u>, 93

69GH   I. Grenthe and E. Hansson, <u>Acta Chem. Scand.</u>, 1969, <u>23</u>, 611

69GL   D.R. Goddard, B.D. Lodam, S.O. Ajayi, and M.J. Campbell, <u>J. Chem. Soc. (A)</u>, 1969,
           506

69GS   S.L. Gupta and R.N. Soni, <u>J. Indian Chem. Soc.</u>, 1969, <u>46</u>, 561, 763

69GSa   S.L. Gupta and R.N. Soni, <u>J. Indian Chem. Soc.</u>, 1969, <u>46</u>, 761

69GW   J. Guilleme and B. Wojtkowiak, <u>Bull. Soc. Chim. France</u>, 1969, 3007, 3013

69H   G.M. Habashy, <u>Electroanal. Chem.</u>, 1969, <u>21</u>, 357

69Ha   H. Harada, <u>Nippon Kagaku Zasshi</u>, 1969, <u>90</u>, 267

69Hb   J. Havel, <u>Coll. Czech. Chem. Comm.</u>, 1969, <u>34</u>, 3248

69HB   L. Havelkova and M. Bartusek, <u>Coll. Czech. Chem. Comm.</u>, 1969, <u>34</u>, 2919

69HBa   L. Havelkova and M. Bartusek, <u>Coll. Czech. Chem. Comm.</u>, 1969, <u>34</u>, 3722

69HBb   G.M. Husain and P.K. Bhattacharya, <u>J. Indian Chem. Soc.</u>, 1969, <u>46</u>, 875

69HE   G.M. Habashy and Z. El-Shafie Ahmed, <u>Electroanal. Chem.</u>, 1969, <u>20</u>, 129

69HF   H.G. Hamilton and H. Freiser, <u>Anal. Chem.</u>, 1969, <u>41</u>, 1310

69HG   P.R. Huber, R. Griesser, B. Prijs, and H. Sigel, <u>Eur. J. Biochem.</u>, 1969, <u>10</u>, 238

69HH   J. Havel, L. Havelkova, and M. Bartusek, <u>Chem. Zvesti.</u>, 1969, <u>23</u>, 582

69HI   J.O. Hill and R.J. Irving, <u>J. Chem. Soc. (A)</u>, 1969, 2759

69HK   J. Horak and M. Kratka, <u>Coll. Czech. Chem. Comm.</u>, 1969, <u>34</u>, 395

69HO   H. Harada and Y. Oka, <u>Nippon Kagaku Zasshi</u>, 1969, <u>90</u>, 898

69HS   J. Havel and L. Sommer, <u>Coll. Czech. Chem. Comm.</u>, 1969, <u>34</u>, 2674

69HSB   J. Haladjian, R. Sabbah, and P. Bianco, <u>J. Chim. Phys.</u>, 1969, <u>66</u>, 1824, 1833

69IV   A.A. Ivakin and E.M. Voronova, <u>Trudy Inst. Khim. Uralsk. Fil. Akad. Nauk SSSR</u>,
           1969, No. 17, 104

69IY   H. Imai, Y. Yaehashi, and Y. Mizuno, <u>Nippon Kagaku Zasshi</u>, 1969, <u>90</u>, 1048

69JA   S.J. Joris, K.I. Aspila, and C.L. Chakrabarti, <u>Anal. Chem.</u>, 1969, <u>41</u>, 1441

69JC   A.D. Jones and G.R. Choppin, J. Inorg. Nucl. Chem., 1969, 23, 3523

69JJ   D.P. Joshi and D.V. Jain, J. Indian Chem. Soc., 1969, 46, 91

69JR   K.C. Jain, N.G. Rathi, and S.N. Banerji, Indian J. Chem., 1969, 7, 806

69KA   S.H. Khan and A. Azizkhan, Electroanal. Chem., 1969, 22, 424

69KAa   I.V. Kolosov and Z.F. Andreeva, Russ. J. Inorg. Chem., 1969, 14, 346 (664)

69KAM   S.H. Khan, A.Azizkhan, and W.U. Malik, J. Electroanal. Chem., 1969, 23, 327

69KM   K.F. Karlysheva, L.A. Malinko, I.A. Sheka, and R.P. Bogushevskaya, Russ. J.
            Inorg. Chem., 1969, 14, 85 (164)

69KP   Ts.B. Konunova, M.S. Popov, and K. Chan, Russ. J. Inorg. Chem., 1969, 14, 1084
            (2066)

69KS   C. Keller and H. Schreck, J. Inorg. Nucl. Chem., 1969, 31, 1121

69KT   J. Knoeck and J.K. Taylor, Anal. Chem., 1969, 41, 1730

69L   J.O. Liljenzin, Acta Chem. Scand., 1969, 23, 3592

69LA   K. Lal and R.P. Agarwal, J. Indian Chem. Soc., 1969, 46, 49

69LM   G.H.Y. Lin and D.J. MacDonald, J. Inorg. Nucl. Chem., 1969, 31, 3233

69LP   D. Litchinsky, N. Purdie, M.B. Tomson, and W.D. White, Anal. Chem., 1969, 41,
            1726

69LR   G. Limb and R.J. Robinson, Anal. Chim. Acta, 1969, 47, 451

69LS   F.I. Lobanov, V.M. Savostina, and V.M. Peshkova, Moscow Univ. Chem. Bull., 1969,
            24, No. 1, 80 (95)

69LSa   F.I. Lobanov, V.M. Savostina, and V.M. Peshkova, Moscow Univ. Chem. Bull., 1969,
            24, No. 2, 86 (121)

69LW   B. Lenarcik and Z. Warnke, Rocz. Chem., 1969, 43, 1329

69M   V.A. Mikhailov, Russ. J. Inorg. Chem., 1969, 14, 1119 (2133)

69Ma   J. Mach, Monat. Chem., 1969, 100, 806

69MB   R. Meilleur and R.L. Benoit, Canad. J. Chem., 1969, 47, 2569

69MBa   V. Mihailova and M. Bonnet, Bull. Soc. Chim. France, 1969, 4258

69MBJ   G.S. Manku, A.N. Bhat, and B.D. Jain, J. Inorg. Nucl. Chem., 1969, 31, 2533

69MD   K.N. Munshi and A.K. Dey, J. Indian Chem. Soc., 1969, 46, 695

69MF   E. Mantovani and C. Furlani, Z. Anorg. Allg. Chem., 1969, 364, 322

69MK   O. Makitie and V. Konttinen, Acta Chem. Scand., 1969, 23, 1459

69MKS   M.P. Mefodeva, N.N. Krot, T.V. Smirnova, and A.D. Gelman, Soviet Radiochem.,
            1969, 11, 187 (193)

69MN   M.G. Mushkina and M.S. Novakovskii, Soviet Progr. Chem. (Ukr. Khim. Zh.), 1969,
            35, No. 5, 61 (517)

69MS   O. Makitie and H. Saarinen, Suomen Kem., 1969, B42, 86

69MSa   O. Makitie and H. Saarinen, Suomen Kem., 1969, B42, 394

69MSb   O. Makitie and H. Saarinen, Anal. Chim. Acta, 1969, 46, 314

69MSc   O. Makitie and H. Saarinen, Ann. Acad. Sci. Fenn., 1969, A2, No. 149

69MV   F. Manok, C. Varhelyi, and S. Kiss-Rajhoua, Rev. Roum. Chim., 1969, 14, 1251

69NA   N.V. Nikolaeva and L.P. Adamovich, J. Gen. Chem. USSR, 1969, 39, 929 (959)

69NB   V.A. Nazarenko and E.A. Biryuk, J. Anal. Chem. USSR, 1969, 24, 35 (44)

69NY   M. Nanjo and T. Yamasaki, Bull. Chem. Soc. Japan, 1969, 42, 968, 972

69OC   G. Ostacoli, E. Campi, and M.C. Gennaro, Atti Accad. Sci. Torino, 1969, 103, 839

69OF   T. Okuyama and T. Fueno, Bull. Chem. Soc. Japan, 1969, 42, 3106

69OO   M. O.h-Eidhin and S. O.Cinneide, J. Inorg. Nucl. Chem., 1969, 31, 2845

69P   T.N. Pliev, Doklady Chem., 1969, 184, 123 (1113)

69Pa   C.E. Plock, Anal. Chim. Acta, 1969, 47, 27

69Pb   M. Petit-Ramel, Theses, Univ. Lyon, 1969

69PC   J.E. Powell, A.R. Chughtai, and J.W. Ingemanson, Inorg. Chem., 1969, 8, 2216

69PD   A.V. Pavlinova and L.S. Demyanchuk, Russ. J. Inorg. Chem., 1969, 14, 500 (959)

69PG   M.M. Palrecha and J.N. Gaur, Indian J. Chem., 1969, 7, 1035

69PJ   J.E. Powell and D.K. Johnson, J. Chromatog., 1969, 44, 212

69PJa   L. Pajdowski and E. John, Rocz. Chem., 1969, 43, 1125

69PK   D.V. Pakhomova, V.N. Kumok, and V.V. Serebrennikov, Russ. J. Inorg. Chem., 1969,
         14, 752 (1434)

69PL   S.B. Pirkes and A.V. Lapitskaya, Russ. J. Inorg. Chem., 1969, 14, 84 (161)

69PM   G. Pilloni and F. Milani, Inorg. Chim. Acta, 1969, 3, 689

69PP   L.J. Porter and D.D. Perrin, Aust. J. Chem., 1969, 22, 267

69PPa   R.K. Patnik and S. Pani, J. Indian Chem. Soc., 1969, 46, 953

69PR   L.D. Pettit and A. Royston, J. Chem. Soc. (A), 1969, 1570

69PS   I.V. Pyatnitskii and E.S. Sereda, Soviet Progr. Chem. (Ukr. Khim. Zh.), 1969,
         35, No. 11, 68 (1192)

69PV   A.V. Pavlinova and T.D. Vysotskaya, Soviet Progr. Chem. (Ukr. Khim. Zh.), 1969,
         35, No. 1, 36 (37)

69RB   M.V. Reddy and P.K. Bhattacharya, Indian J. Chem., 1969, 7, 282

69RBa   M.V. Reddy and P.K. Bhattacharya, J. Indian Chem. Soc., 1969, 46, 1058

69RF   R. Roulet, J. Feuz, and T. VuDuc, Helv. Chim. Acta, 1969, 52, 2154

69RRS   S. Ramamoorthy, A. Raghavan, and M. Santappa, J. Inorg. Nucl. Chem., 1969, 31,
         1765

69RRV   S. Ramamoorthy, A. Raghavan, V.R. Vijayaraghavan, and M. Santappa, J. Inorg.
          Nucl. Chem., 1969, 31, 1851

69RS   S. Ramamoorthy and M. Santappa, Bull. Chem. Soc. Japan, 1969, 42, 411

69S   A. Sandell, Acta Chem. Scand., 1969, 23, 478

69SC   G.K. Schweitzer and F.C. Clifford, Anal. Chim. Acta, 1969, 45, 57

69SCF   P.R. Subbaraman, M. Cordes, and H. Freiser, Anal. Chem., 1969, 41, 1878

69SD   D. Suznjevic, J. Dolezal, and M. Kopanica, J. Electroanal. Chem., 1969, 20, 279

69SF   T. Sakai and Y. Funaki, Bull. Chem. Soc. Japan, 1969, 42, 2272

69SG   R.S. Saxena and K.C. Gupta, J. Indian Chem. Soc., 1969, 46, 190

69SGa   R.S. Saxena and K.C. Gupta, J. Indian Chem. Soc., 1969, 46, 258

69SGb   R.S. Saxena and K.C. Gupta, J. Indian Chem. Soc., 1969, 46, 303

69SGc   R.S. Saxena and K.C. Gupta, J. Indian Chem. Soc., 1969, 46, 1045

69SGd   R.S. Saxena and K.C. Gupta, Z. Phy. Chem. (Leipzig), 1969, 241, 169

69SGe   R.S. Saxena and K.C. Gupta, Indian J. Chem., 1969, 7, 371

69SGM   R.S. Saxena, K.C. Gupta, and M.L. Mittal, Indian J. Chem., 1969, 7, 374

69SH   H.B. Singh, J. Havel, and L. Sommer, Coll. Czech. Chem. Comm., 1969, 34, 3277

69SK   N.A. Skorik and V.N. Kumok, Russ. J. Inorg. Chem., 1969, 14, 52 (98)

69SKa   V.M. Savostina and S.O. Kobyakova, Moscow Univ. Chem. Bull., 1969, 24, No. 1,
          36 (46)

69SKK   S.Ya. Shnaiderman, E.P. Klimenko, E.N. Knyazeva, and N.V. Chernaya, Russ. J.
          Inorg. Chem., 1969, 14, 648 (1238)

69SL   V.S. Smelov and V.P. Lanin, Soviet Radiochem., 1969, 11, 433 (445)

69SMK   Sh.A. Sherif, N.E. Milad, and A.A. Khedr, J. Inorg. Nucl. Chem., 1969, 31, 3225

69SMM   H. Saarinen, M. Melnik, and O. Makitie, J. Inorg. Nucl. Chem., 1969, 23, 2542

69SS   T. Sekine and M. Sakairi, Bull. Chem. Soc. Japan, 1969, 42, 2712

69SSa   B. Sethuram and E.V. Sundaram, Indian J. Chem., 1969, 7, 415

69SV   Vl. Simeon, K. Voloder, and O.A. Weber, Anal. Chim. Acta, 1969, 44, 309

69T   B. Topuzovski, God. Zb., Prir.-Mat. Fak. Univ., Skopje, Mat., Fiz. Hem., 1969,
          19, 71; Chem. Abs., 1971, 75, 113264f

69TA   J.A. Thomson and G.F. Atkinson, Anal. Chim. Acta, 1969, 47, 380

69TM   M.M. TaquiKhan and A.E. Martell, J. Amer. Chem. Soc., 1969, 91, 4668

69TW   P.H. Tedesco and H.F. Walton, Inorg. Chem., 1969, 8, 932

69U   K. Uesugi, Bull. Chem. Soc. Japan, 1969, 42, 2051

69Ua   K. Uesugi, Bull. Chem. Soc. Japan, 1969, 42, 2398

69Ub   K. Uesugi, Bull. Chem. Soc. Japan, 1969, 42, 2998

69US   V.S. Ulyanov, R.A. Sviridova, and A.I. Zarubin, Soviet Radiochem., 1969, 11, 11
        (13)

69VD   S.P. Vorobev, I.P. Davydov, and I.V. Shilin, Russ. J. Inorg. Chem., 1969, 14, 536

69VI   E.M. Voronova and A.A. Ivakin, Russ. J. Inorg. Chem., 1969, 14, 818 (1564)

69VOP   V.G. Voden, M.E. Obukhova, and M.F. Pushlenkov, Soviet Radiochem., 1969, 11,
        633 (644)

69VOR   A. Vanni, G. Ostacoli, and E. Roletto, Ann. Chim. (Rome), 1969, 59, 847

69VP   E. Verdier and J. Piro, Ann. Chim., 1969, 4, 213

69W   Z. Warnke, Rocz. Chem., 1969, 43, 1939

69ZS   F.G. Zharovskii, R.I. Sukhomlin, and M.S. Ostrovskaya, Russ. J. Inorg. Chem.,
        1969, 14, 129 (249)

70A   P.J. Antikainen, Suomen Kem., 1970, B43, 31

70AB   J. Ascanio and F. Brito, An. Quim., 1970, 66, 617

70AD   B.K. Avinashi, C.D. Dwivedi, and S.K. Banerji, J. Inorg. Nucl. Chem., 1970, 32,
        2641

70AJ   K.I. Aspila, S.J. Joris, and C.L. Chakrabarti, J. Phys. Chem., 1970, 74, 3625

70AK   A.M. Andrianov and V.P. Koryukova, Russ. J. Inorg. Chem., 1970, 15, 230 (445)

70AKa   L.P. Adamovich and M.S. Kravchenko, J. Anal. Chem. USSR, 1970, 25, 1434 (1668)

70AKN   A. Adin, P. Klotz, and L. Newman, Inorg. Chem., 1970, 9, 2499

70AL   H.F. Aly and R.M. Latimer, Radiochim. Acta, 1970, 14, 27

70AM   G. Anderegg and S. Malik, Helv. Chim. Acta, 1970, 53, 577

70AT   P.J. Antikainen and K. Tevanen, J. Inorg. Nucl. Chem., 1970, 32, 1915

70BA   N.N. Basargin, M.K. Akhmedli, and A.A. Kafarova, J. Anal. Chem. USSR, 1970, 25,
        1292 (1497)

70BB   J.C. Barnes and P.A. Bristow, J. Less-Common Metals, 1970, 22, 463

70BC   G. Besse, J.L. Chabard, G. Voissiere, J. Pettit, and J.A. Berger, Bull. Soc.
        Chim. France, 1970, 4166

70BG   M. Bartusek, B. Grebenova, and L. Sommer, Spisy Prir. Fak. Univ. Purk. Brne,
        1970, No. 517, 389

70BJ   S.P. Banerjee and P.C. Jain, J. Indian Chem. Soc., 1970, 47, 986

70BL   A. Braibanti, E. Leporati, and F. Dallavalle, Inorg. Chim. Acta, 1970, 4, 529

70BR   E.A. Biryuk and R.V. Ravitskaya, J. Anal. Chem. USSR, 1970, 25, 494 (576)

70BRa   E.A. Biryuk and R.V. Ravitskaya, J. Anal. Chem. USSR, 1970, 25, 1417 (1643)

70BRS   N.M. Ballash, E.B. Robertson, and M.D. Sokolowski, Trans. Faraday Soc., 1970,
        66, 2622

70BS   J. Borak, Z. Slovak, and J. Fisher, Talanta, 1970, 17, 215

70BT   J.W. Bunting and K.M. Thong, Canad. J. Chem., 1970, 48, 1654

70BTL   S.A. Bajue, G.A. Taylor, and G.C. Lalor, Rev. Latinoamer. Quim., 1970, 1, 102

70C   J.M. Conner, J. Inorg. Nucl. Chem., 1970, 32, 3545

70CB   M.E. Clark and J.L. Bear, J. Inorg. Nucl. Chem., 1970, 32, 3569

70CF   Y. Couturier and J. Faucherre, Bull. Soc. Chim. France, 1970, 1323

70CG   L. Ciavatta, M. Grimaldi, and G. Paoletta, Gazz. Chim. Ital., 1970, 100, 100

70CL   A.S. Carson, P.G. Laye, and W.V. Steele, J. Chem. Thermodynamics, 1970, 2, 757

70CM   D.L. Campbell and T. Moeller, J. Inorg. Nucl. Chem., 1970, 32, 945

70CRa   J.J. Christensen, J.H. Rytting, and R.M. Izatt, J. Chem. Soc. (B), 1970, 1646

70CS   G.R. Choppin and J.K. Schneider, J. Inorg. Nucl. Chem., 1970, 32, 3283

70CSa   N.V. Chernaya and S.Ya. Shnaiderman, Soviet Progr. Chem. (Ukr. Khim. Zh.), 1970,
          36, No. 4, 7 (321)

70CSb   E. Chiacchierini, T. Sepel, and L. Sommer, Coll. Czech. Chem. Comm., 1970, 35,
          794

70CSS   J.J. Christensen, M.D. Slade, D.E. Smith, R.M. Izatt, and J. Tsang, J. Amer.
          Chem. Soc., 1970, 92, 4164

70CV   E.G. Chikryzova and I.I. Vataman, Russ. J. Inorg. Chem., 1970, 15, 219 (424)

70CVa   E. Campi and A. Vanni, Atti Accad. Sci. Torino, 1970, 104, 99

70DD   S.S. Dube and S.S. Dhindsa, J. Inorg. Nucl. Chem., 1970, 32, 543

70DDa   S.S. Dube and S.S. Dhindsa, J. Inorg. Nucl. Chem., 1970, 32, 1041

70DDb   S.S. Dube and S.S. Dhindsa, Canad. J. Chem., 1970, 48, 1007

70DS   N.K. Dutt and U.V.M. Sharma, J. Inorg. Nucl. Chem., 1970, 32, 1035

70DY   S.N. Drozdova and M.Z. Yampolskii, Russ. J. Inorg. Chem., 1970, 15, 307 (595)

70E   V.N. Epimakhov, J. Gen. Chem. USSR, 1970, 40, 2130 (2144)

70EM   C.E. Evans and E.B. Monk, Trans. Faraday Soc., 1970, 66, 1491

70FB   I. Filipovic, A. Bujak, and V. Vukicevic, Croat. Chem. Acta, 1970, 42, 493

70FK   Y. Fukuda, E. Kyuno, and R. Tsuchiya, Bull. Chem. Soc. Japan, 1970, 43, 745

70FM   I. Filipovic, T. Matusinovic, B. Mayer, I. Piljac, B. Bach-Dragutinovic, and
          A. Bujak, Croat. Chem. Acta, 1970, 42, 541

70FR   K.F. Fouche, H.J. leRoux, and F. Phillips, J. Inorg. Nucl. Chem., 1970, 32, 1949

70G   J. Gross, KFK-1339, 1970

70GB   V.I. Grebenshchikova, R.V. Bryzgalova, and Yu.M. Rogozin, Soviet Radiochem.,
          1970, 12, 252 (279)

70GF   B. Grabaric and I. Filipovic, Croat. Chem. Acta, 1970, 42, 479

70GI   P. Gabor-Klatsmanyi, J. Inczedy, and L. Erdey, Acta Chim. Acad. Sci. Hung., 1970,
          57, 5

70GM   W.J. Geary and D.E. Malcolm, J. Chem. Soc. (A), 1970, 797

70GMa   A.P. Gerbeleu and P.C. Migal, Proc. 13th I.C.C.C., Poland, 1970, vol. I, 57 (69)

70GMb   B.D. Gupta and W.U. Malik, J. Indian Chem. Soc., 1970, 47, 145, 770

70GN   R. Ghosh and V.S.K. Nair, J. Inorg. Nucl. Chem., 1970, 32, 3025

70GNa   R. Ghosh and V.S.K. Nair, J. Inorg. Nucl. Chem., 1970, 32, 3033

70GO   T. Goina and M. Olaru, Rev. Roumaine Chim., 1970, 15, 1546

70GP   J.N. Gaur and M.M. Palrecha, J. Inorg. Nucl. Chem., 1970, 32, 1375

70GS   R.O. Gould and H.M. Sutton, J. Chem. Soc. (A), 1970, 1184

70GSa   R.O. Gould and H.M. Sutton, J. Chem. Soc. (A), 1970, 1439

70GSb   R. Griesser and H. Sigel, Inorg. Chem., 1970, 9, 1238

70GT   A.K. Grzybowski, S.S. Tate, and S.P. Datta, J. Chem. Soc. (A), 1970, 241

70H   H. Harada, Nippon Kagaku Zasshi, 1970, 91, 1064

70Ha   M. Hara, Bull. Chem. Soc. Japan, 1970, 43, 89

70HL   T.M. Hseu, L.S. Lin, and H.H. Sun, J. Chinese Chem. Soc. (Taiwan), 1970, 17, 214

70HO   M. Hirai and Y. Oka, Bull. Chem. Soc. Japan, 1970, 43, 778

70HS   J. Havel and L. Sommer, Coll. Czech. Chem. Comm., 1970, 35, 45

70IM   D.J.G. Ives and P.G.N. Moseley, J. Chem. Soc. (B), 1970, 1655

70IP   D.J.G. Ives and D. Prasad, J. Chem. Soc. (B), 1970, 1649

70IPa   D.J.G. Ives and D. Prasad, J. Chem. Soc. (B), 1970, 1652

70IS   H.M.N.H. Irving and S.P. Sinha, Anal. Chim. Acta, 1970, 49, 449

70IV   A.A. Ivakin and E.M. Voronova, Trudy Inst. Khim. Uralsk. Fil. Akad. Nauk SSSR,
          1970, 17, 134

70IVa   A.A. Ivakin and E.M. Voronova, Trudy Inst. Khim. Uralsh, Fil. Akad. Nauk SSSR,
          1970, 17, 139

70IVC   A.A. Ivakin, E.M. Voronova, and L.V. Chashchina, Russ. J. Inorg. Chem., 1970,
          15, 945 (1839)

70JJ   D.P. Joshi and D.V. Jain, J. Indian Chem. Soc., 1970, 47, 1109

70K   L.N. Klatt, Anal. Chem., 1970, 42, 1837

70Ka   S. Kilmartin, M.S. Thesis, Nat. Univ. Ireland, 1970 (see 740a and 740b)

70KA   P.V. Khadikar, R.L. Ameria, and W.V. Bhagwat, Sci. Culture, 1970, 36, 482

70KAa   P.V. Khadikar, R.L. Ameria, and W.V. Bhagwat, Sci. Culture, 1970, 36, 675

70KAb   P.V. Khakikar and R.L. Ameria, J. Indian Chem. Soc., 1970, 47, 1201

70KC   G.C. Kugler and G.H. Carey, Talanta, 1970, 10, 907

70KCa   S.S. Katiyar and V.B.S. Chauhan, J. Indian Chem. Soc., 1970, 47, 796

70KK   Ts.B. Konunova and L.S. Kachkar, Russ. J. Inorg. Chem., 1970, 15, 1543 (2964)

70KKM   K.F. Karlysheva, A.V. Koshel, L.A. Malinko, and I.A. Sheka, Russ. J. Inorg. Chem., 1970, 15, 937 (1825)

70KL   I.M. Kaganskii and N.P. Lopatina, Russ. J. Inorg. Chem., 1970, 15, 1208 (2333)

70KLa   N. Konopik and W. Luf, Monat. Chem., 1970, 101, 1591

70KM   P.V. Khadikar, S.N. Mehrotra, and M.G. Kanungo, Sci. Culture, 1970, 36, 570

70KN   I. Kawai, R. Nakajima, and T. Hara, Bull. Chem. Soc. Japan, 1970, 43, 749

70KP   K. Kustin and R. Pizer, Inorg. Chem., 1970, 9, 1536

70KS   V.N. Kumok and N.A. Skorik, Russ. J. Inorg. Chem., 1970, 15, 153 (291)

70L   J.O. Liljenzin, Acta Chem. Scand., 1970, 24, 1655

70La   T. Lengyel, Acta Chim. Acad. Sci. Hung., 1970, 58, 313

70Lb   T. Lengyel, Acta Chim. Acad. Sci. Hung., 1970, 60, 225

70Lc   T. Lengyel, Acta Chim. Acad. Sci. Hung., 1970, 60, 373

70LG   C.S. Leung and E. Grunwald, J. Phys. Chem., 1970, 74, 687

70LS   J.O. Liljenzin and J. Stary, J. Inorg. Nucl. Chem., 1970, 32, 1357

70LSF   F.I. Lobanov, V.M. Savostina, V.M. Feskova, O.A. Shpigun, and V.M. Peshkova, Russ. J. Inorg. Chem., 1970, 15, 81 (161)

70MC   P.K. Migal and N.G. Chebotar, Russ. J. Inorg. Chem., 1970, 15, 625 (1218)

70MKP   N.V. Melchakova, N.A. Krasnyanskaya, and V.M. Peshkova, J. Anal. Chem. USSR, 1970, 25, 1756 (2046)

70MKS   S.A. Merkusheva, V.N. Kumok, N.A. Skorik, and V.V. Serebrennikov, Soviet Radio-chem., 1970, 12, 155 (175)

70MM   O. Makitie and H. Mattinen, Suomen Kem., 1970, B43, 504

70MMT   O. Makitie, H. Mattinen, and R. Tikkanen, Suomen Kem., 1970, B43, 235

70MN   M.B. Mishra and H.L. Nigam, Acta Chim. Acad. Sci. Hung., 1970, 57, 1

70MR   W.A.E. McBryde, J.L. Rohr, J.S. Penciner, and J.A. Page, Canad. J. Chem., 1970, 48, 2574

70MS   O. Makitie and H. Saarinen, J. Inorg. Nucl. Chem., 1970, 32, 2800

70MSa   A. Meenakshi and M. Santappa, Indian J. Chem., 1970, 8, 467

70MSL   O. Makitie, H. Saarinen, L. Lindroos, and K. Seppovaara, Acta Chem. Scand., 1970, 24, 740

70MSM   O. Makitie, H. Saarinen, H. Mattinen, and K. Seppovaara, Suomen Kem., 1970, B43, 340

70MW   R.G. McGregor and D.R. Wiles, J. Chem. Soc. (A), 1970, 323

70NK   T. Nozaki, M. Kadowake, D. Sagawa, and K. Orita, Nippon Kagaku Zasshi, 1970,
          91, 64

70NL   A. Napoli and A. Liberti, Gazz. Chim. Ital., 1970, 100, 906

70NLa   V.A. Nazarenko, N.V. Lebedeva, and L.I. Vinarova, Russ. J. Inorg. Chem., 1970,
          15, 330 (643)

70NLb   V.A. Nazarenko, N.V. Lebedeva, and L.I. Vinarova, Russ. J. Inorg. Chem., 1970,
          15, 1557 (2990)

70NM   M.S. Novakovskii and M.G. Mushkina, J. Anal. Chem. USSR, 1970, 25, 946 (1092)

70NP   V.S.K. Nair and S. Parthasarathy, J. Inorg. Nucl. Chem., 1970, 32, 3289

70NPa   V.S.K. Nair and S. Parthasarathy, J. Inorg. Nucl. Chem., 1970, 32, 3293

70NPb   V.S.K. Nair and S. Parthasarathy, J. Inorg. Nucl. Chem., 1970, 32, 3297

70NPC   B.P. Nikolskii, V.V. Palchevskii, A.D. Chegodaeva, and N.K. Barkova, Dokl. Chem.,
          1970, 194, 744 (1100)

70NT   R. Nasanen, P. Tilus, and P. Helander, Suomen Kem., 1970, B43, 331

70OM   M.D. Olczak and J. Maslowska, Rocz. Chem., 1970, 44, 17

70OT   A.L. Oleinikova, O.N. Temkin, M.I. Bogdanov, and R.M. Flid, Russ. J. Phys. Chem.,
          1970, 44, 1373

70OV   G. Ostacoli, A. Vanni, and E. Roletto, Gazz. Chim. Ital., 1970, 100, 350

70PB   D.C. Patel and P.K. Bhattacharya, Indian J. Chem., 1970, 8, 835

70PC   R. Portanova, A. Cassol, L. Magon, and G. Tomat, J. Inorg. Nucl. Chem., 1970, 32,
          221

70PK   D.V. Pakhomova, V.N. Kumok, and V.V. Serebrennikov, Russ. J. Inorg. Chem., 1970,
          15, 622

70PL   S.B. Pirkes and A.V. Lapitskaya, Russ. J. Inorg. Chem., 1970, 15, 1683 (3228)

70PLS   N.S. Poluektov, R.S. Lauer, and M.A. Sandu, J. Anal. Chem. USSR, 1970, 25, 1821
          (2118)

70PP   J. Podlaha and J. Podlahova, Inorg. Chim. Acta, 1970, 4, 521

70PPa   R.K. Patnaik and S. Pani, J. Indian Chem. Soc., 1970, 47, 613

70PS   R.G. Pearson and D.A. Sweigart, Inorg. Chem., 1970, 9, 1167

70PSa   A.D. Parulekar and P.R. Subbaraman, Indian J. Chem., 1970, 8, 266

70PSb   K. Pan and H.Y. Su, J. Chinese Chem. Soc. (Taiwan), 1970, 17, 57

70PTM   R. Portanova, G. Tomat, L. Magon, and A. Cassol, J. Inorg. Nucl. Chem., 1970,
          32, 2343

70PTB   A.N. Petrov, O.N. Temkin, and M.I. Bogdanov, Russ. J. Phys. Chem., 1970, 44, 1574

70RB   M.V. Reddy and P.K. Bhattacharya, J. Prakt. Chem., 1970, 312, 69

70RF   J.H. LeRoux and K.F. Fouche, J. Inorg. Nucl. Chem., 1970, 32, 3059

70RFV   R. Roulet, J. Feuz, and T. VuDuc, Helv. Chim. Acta, 1970, 53, 1876

70RJ   N.G. Rathi, K.C. Jain, and S.N. Banerji, J. Indian Chem. Soc., 1970, 47, 788

70RS   A. Raghavan and M. Santappa, Curr. Sci. (India), 1970, 39, 302

70RSa   S. Ramamoorthy and M. Santappa, J. Inorg. Nucl. Chem., 1970, 32, 1623

70RV   R. Roulet and T. VuDuc, Helv. Chim. Acta, 1970, 53, 1873

70S   A. Sandell, Acta Chem. Scand., 1970, 24, 1561

70Sa   A. Sandell, Acta Chem. Scand., 1970, 24, 1718

70Sb   A. Sandell, Acta Chem. Scand., 1970, 24, 3391

70Sc   T. Seshadri, Indian J. Chem., 1970, 8, 282

70SA   G.M. Sergeev, L.G. Astrashkova, and N.N. Yagodinskaya, Soviet Radiochem., 1970,
          12, 356 (392)

70SB   T.R. Sweet and D. Brengartner, Anal. Chim. Acta, 1970, 52, 173

70SF   H. Scheidegger, W. Felty, and D.L. Leussing, J. Amer. Chem. Soc., 1970, 92, 808

70SG   O.N. Shrivastava and C.M. Gupta, Indian J. Chem., 1970, 8, 1007

70SH   L.M. Schwartz and L.O. Howard, J. Phys. Chem., 1970, 74, 4374

70SK   N.V. Sistkova, Z. Kolarik, and V. Chotivka, J. Inorg. Nucl. Chem., 1970, 32, 637

70SM   H. Saarinen and O. Makitie, Acta Chem. Scand., 1970, 24, 2877

70SMa   G. Saini and E. Mentasti, Inorg. Chem. Acta, 1970, 4, 210

70SMb   G. Saini and E. Mentasti, Inorg. Chim. Acta, 1970, 4, 585

70SP   A.V. Stepanov and E.M. Pazukhin, Russ. J. Inorg. Chem., 1970, 15, 761 (1483)

70SR   I.S. Sklyarenko and N.S. Radionova, Russ. J. Inorg. Chem., 1970, 15, 53 (103)

70SS   R.S. Saxena and P. Singh, Indian J. Chem., 1970, 8, 76

70SSa   R.S. Saxena and P. Singh, J. Indian Chem. Soc., 1970, 47, 1076

70SSK   L.D. Shtenke, N.A. Skorik, and V.N. Kumok, Russ. J. Inorg. Chem., 1970, 15, 623,
          (1214)

70ST   H. Schurmans, H. Thun, and F. Verbeek, Electroanal. Chem., 1970, 26, 299

70TB   T.I. Tikvinskaya, A.A. Biryukov, V.I. Shlenskaya, and N.K. Gordynskaya, Russ. J.
          Inorg. Chem., 1970, 15, 65 (128)

70TG   P.H. Tedesco and J.A. Gonzalez Quintana, J. Inorg. Nucl. Chem., 1970, 32, 2689

70TK   V.Sh. Telyakova and O.I. Khotsyanovskii, J. Gen. Chem. USSR, 1970, 40, 1419
          (1433)

70TM   K. Toei, H. Miyata, and T. Ozaki, Nippon Kagaku Zasshi, 1970, 91, 1148

70TP   K.K. Tripathy and R.K. Patnaik, J. Indian Chem. Soc., 1970, 47, 331

70VE   L.S. Voronets, I.P. Efimov, and V.M. Peshkova, Russ. J. Inorg. Chem., 1970, 15,
          451 (886)

70VH   G.L. VandeCappelle and M.A. Herman, <u>Bull. Soc. Chim. Belges</u>, 1970, <u>79</u>, 585

70VHE   G.L. VandeCappelle, M.A. Herman, and Z. Eeckhaut, <u>Bull. Soc. Chim. Belges</u>, 1970,
            <u>79</u>, 421

70VS   E. Vrachnou-Astra, P. Sakellaridis, and D. Katakis, <u>J. Amer. Chem. Soc.</u>, 1970,
            <u>92</u>, 811

70WB   J.M. Wilson, A.G. Briggs, J.E. Sawbridge, and P. Tickle, <u>J. Chem. Soc. (A)</u>, 1970,
            1024

70WW   E.M. Wooley, R.W. Wilton, and L.G. Hepler, <u>Canad. J. Chem.</u>, 1970, <u>48</u>, 3249

70Z   A.V. Zholnin, <u>Russ. J. Inorg. Chem.</u>, 1970, <u>15</u>, 655 (1277)

70ZP   E.M. Zhurenkov and D.N. Pobezhimovskaya, <u>Soviet Radiochem.</u>, 1970, <u>12</u>, 89 (105)

71A   J. Alexa, <u>Coll. Czech. Chem. Comm.</u>, 1971, <u>36</u>, 3370

71AA   I.P. Alimarin, N.V. Arslanova, and F.P. Sudakov, <u>J. Anal. Chem. USSR</u>, 1971, <u>26</u>,
            2136 (2383)

71AD   A.N. Alybina, A.E. Dorfman, E.K. Ivanova, and V.M. Peshkova, <u>Russ. J. Inorg.
            Chem.</u>, 1971, <u>16</u>, 233 (446)

71AG   S.O. Ajayi and D.R. Goddard, <u>J. Chem. Soc. (A)</u>, 1971, 2673

71AK   P.J. Antikainen and R. Katila, <u>Suomen Kem.</u>, 1971, <u>B44</u>, 256

71AKa   S. Ahrland and L. Kullberg, <u>Acta Chem. Scand.</u>, 1971, <u>25</u>, 3677

71AKB   A.I. Astakhov, E.N. Knyazeva, Ya.I. Bleikherands, and S.Ya. Shnaiderman, <u>Russ. J.
            Inorg. Chem.</u>, 1971, <u>16</u>, 521 (980)

71AL   A. Aziz and S.J. Lyle, <u>J. Inorg. Nucl. Chem.</u>, 1971, <u>33</u>, 3407

71ALN   A. Aziz, S.J. Lyle, and J.E. Newbery, <u>J. Inorg. Nucl. Chem.</u>, 1971, <u>33</u>, 1757

71AO   S. Akalin and U.Y. Ozer, <u>J. Inorg. Nucl. Chem.</u>, 1971, <u>33</u>, 4171

71AW   P.J. Antikainen and U. Witikainen, <u>Suomen Kem.</u>, 1971, <u>B44</u>, 173

71B   M. Beran, <u>J. Inorg. Nucl. Chem.</u>, 1971, <u>33</u>, 3885

71BA   P. Bianco, M. Asso, and J. Haladjian, <u>Bull. Soc. Chim. France</u>, 1971, 3943

71BC   J.L. Bear and M.E. Clark, <u>J. Inorg. Nucl. Chem.</u>, 1971, <u>33</u>, 3805

71BCP   E. Boitard, G. Carpeni, R. Pilard, and C. Rousset, <u>J. Chim. Phys.</u>, 1971, <u>68</u>, 41

71BFA   O.D. Bonner, H.B. Flora, and H.W. Aitken, <u>J. Phys. Chem.</u>, 1971, <u>75</u>, 2492

71BFP   D.S. Barnes, G.J. Ford, L.D. Pettit, and C. Sherrington, <u>J. Chem. Soc. (A)</u>, 1971,
            2883

71BJ   J. Becka and J. Jokl, <u>Coll. Czech. Chem. Comm.</u>, 1971, <u>36</u>, 2467, 3263

71BL   G. Berthon and C. Luca, <u>Chim. Anal.</u> (Paris), 1971, <u>53</u>, 40

71BV   E. Bottari and M. Vicedomini, <u>Gazz. Chim. Ital.</u>, 1971, <u>101</u>, 671

71BVa   E. Bottari and M. Vicedomini, <u>Gazz. Chim. Ital.</u>, 1971, <u>101</u>, 860

71BVb   E. Bottari and M. Vicedomini, <u>Gazz. Chim. Ital.</u>, 1971, <u>101</u>, 871

71BVc   E. Bottari and M. Vicedomini, J. Inorg. Nucl. Chem., 1971, 33, 1463

71CA   T.A. Chernova and K.V. Astakhov, Russ. J. Phys. Chem., 1971, 45, 624 (1114)

71CD   F.P. Cavasino and E. DiDio, J. Chem. Soc. (A), 1971, 3176

71CN   J. Canonne, G. Nowogrocki, and G. Tridot, Bull. Soc. Chim. France, 1971, 1121

71CS   M.C. Chattopadhyaya and R.S. Singh, Indian J. Chem., 1971, 9, 490

71DC   A. Dadgar and G.R. Choppin, J. Coord. Chem., 1971, 1, 179

71DD   S.S. Dube and S.S. Dhindsa, Monat. Chem., 1971, 102, 598

71DG   I. Dellien and I. Grenthe, Acta Chem. Scand., 1971, 25, 1387

71DM   H.S. Dunsmore and D. Midgley, J. Chem. Soc. (A), 1971, 3238

71DR   W.C. Duer and R.A. Robinson, J. Chem. Soc. (B), 1971, 2375

71DS   S.C. Dhupar, K.C. Srivastava, and S.K. Banerji, J. Indian Chem. Soc., 1971, 48,
          921

71DV   H.F. DeBrabander, L.C. VanPoucke, and Z. Eeckhaut, Inorg. Chim. Acta, 1971, 5,
          473

71EI   H. Einaga and H. Ishii, Anal. Chim. Acta, 1971, 54, 113

71EP   N. Elenkova, C. Palasev, and L. Ilceva, Talanta, 1971, 18, 355

71F   H.K. Frensdorff, J. Amer. Chem. Soc., 1971, 93, 600

71FN   D.P. Fay, A.R. Nichols, Jr., and N. Sutin, Inorg. Chem., 1971, 10, 2096

71FP   G.J. Ford, L.D. Pettit, and C. Sherrington, J. Inorg. Nucl. Chem., 1971, 33, 4119

71G   R.I. Gelb, Anal. Chem., 1971, 43, 1110

71Ga   G. Gardhammar, Acta Chem. Scand., 1971, 25, 158

71GB   R. Guillaumont and L. Bourderie, Bull. Soc. Chim. France, 1971, 2806

71GC   R.J. Grabenstetter and W.A. Cilley, J. Phys. Chem., 1971, 75, 676

71GCa   K.K.S. Gupta and A.K. Chatterjee, Z. Anorg. Allg. Chem., 1971, 384, 280

71GH   J. Gillet, G. Huyge-Tiprez, J. Nicole, and G. Tridot, Compt. Rend. Acad. Sci.
          Paris, Ser. C, 1971, 273, 1743

71GK   V.I. Gordienko, L.P. Khudyakova, and V.I. Sidorenko, J. Anal. Chem. USSR, 1971,
          26, 2051 (2284)

71GM   V.I. Gordienko, Yu.I. Mikhailyuk, and V.I. Sidorenko, J. Gen. Chem. USSR, 1971,
          41, 501 (507)

71GS   R. Griesser and H. Sigel, Inorg. Chem., 1971, 10, 2229

71GSB   S. Goldman, P. Sagner, and R.G. Bates, J. Phys. Chem., 1971, 75, 826

71H   G.M. Habashy, Electroanal. Chem., 1971, 30, 315

71Ha   B. Hurnik, Rocz. Chem., 1971, 45, 147

71Hb   H. Harada, Bull. Chem. Soc. Japan, 1971, 44, 3459

71HK   T.M. Hseu and S.Y. Ko, J. Chinese Chem. Soc. (Taiwan), 1971, 18, 203

71HL   A. Hilton and D.L. Leussing, J. Amer. Chem. Soc., 1971, 93, 6831

71HM   K. Horizawa and S. Masuyama, Bull. Chem. Soc. Japan, 1971, 44, 2697

71IN   R.M. Izatt, D.P. Nelson, J.H. Rytting, B.L. Haymore, and J.J. Christensen,
        J. Amer. Chem. Soc., 1971, 93, 1619

71JB   G. Jung, E. Breitmaier, and W. Voelter, Eur. J. Biochem., 1971, 24, 438

71JF   M.R. Jaffe, D.P. Fay, M. Cefola, and N. Sutin, J. Amer. Chem. Soc., 1971, 93,
        2878

71JM   A.P. Joshi and K.N. Munshi, J. Indian Chem. Soc., 1971, 48, 1

71JS   W.P. Jencks and K. Salvesen, J. Amer. Chem. Soc., 1971, 93, 4433

71K   A.S. Kereichuk, Russ. J. Inorg. Chem., 1971, 16, 1346 (2523)

71Ka   H. Krakauer, Biopolymers, 1971, 10, 2459

71KC   B. Kuznik and D.M. Czakis-Sulikowska, Rocz. Chem., 1971, 45, 959

71KG   P.V. Khadikar and S.K. Gupta, Sci. Culture, 1971, 37, 213

71KK   N.P. Komar and Yu.M. Khoroshevskii, J. Anal. Chem. USSR, 1971, 26, 1986 (2222)

71KKK   J. Kotek, H. Klierova, M. Kopanica, and J. Dolezal, J. Electroanal. Chem., 1971,
        31, 451

71KL   N. Konopik and W. Luf, Monat. Chem., 1971, 102, 896

71KLa   J. Krajewski and T. Lipiec, Rocz. Chem., 1971, 46, 1613

71KLM   V.P. Kerentseva, M.D. Lipanova, and I.S. Mustafin, J. Anal. Chem. USSR, 1971,
        26, 1025 (1144)

71KM   K. Kina, H. Miyata, and K. Toei, Bull. Chem. Soc. Japan, 1971, 44, 1855

71KMa   K. Kina, H. Miyata, and K. Toei, Bull. Chem. Soc. Japan, 1971, 44, 2710

71KMb   T. Katayama, H. Miyata, and K. Toei, Bull. Chem. Soc. Japan, 1971, 44, 2712

71KO   H. Koshimura and T. Okubo, Anal. Chim. Acta, 1971, 55, 163

71KT   M. Kodama and S. Takahashi, Bull. Chem. Soc. Japan, 1971, 44, 697

71KTa   K. Kina and K. Toei, Bull. Chem. Soc. Japan, 1971, 44, 2416

71LD   T.S. Lobanova, K.M. Dunaeva, and E.A. Ippolitova, Moscow Univ. Chem. Bull., 1971,
        26, No. 2, 71 (229)

71LDa   T.S. Lobanova, K.M. Dunaeva, and E.A. Ippolitova, Moscow Univ. Chem. Bull., 1971,
        26, No. 3, 40 (321)

71LDb   J.C. Londesborough and K. Dalziel, Eur. J. Biochem., 1971, 23, 194

71LG   N.M. Lukovskaya and M.I. Gerasimenko, J. Anal. Chem. USSR, 1971, 26, 1928 (2159)

71LM   L.F. Loginova, V.V. Medyntsev, and B.I. Khomutov, J. Gen. Chem. USSR, 1971, 42,
        732 (739)

71LS   P. Lingaiah and E.V. Sundaram, Indian J. Chem., 1971, 9, 852

71M   L.B. Magnusson, J. Inorg. Nucl. Chem., 1971, 33, 3602

71MC   P.K. Migal, N.G. Chebotar, and A.M. Sorochinskaya, Russ. J. Inorg. Chem., 1971,
           16, 53 (102)

71MG   A.I. Mikhailichenko, N.N. Guseva, E.V. Sklenskaya, and M.Kh. Karapetyants, Russ.
           J. Inorg. Chem., 1971, 16, 1645 (3101)

71MM   O. Makitie and S. Mirttinen, Suomen Kem., 1971, B44, 155

71MMa   O. Makitie and S. Mirttinen, Acta Chem. Scand., 1971, 25, 1146

71MMb   Y. Masuda and S. Misumi, Nippon Kagaku Zasshi, 1971, 92, 710

71MMB   O. Menis, B.E. McClellan, and D.S. Bright, Anal. Chem., 1971, 43, 431

71MMW   J.P. Manners, K.G. Morallee, and R.J.P. Williams, J. Inorg. Nucl. Chem., 1971,
           33, 2085

71MP   M. Morin, M.R. Paris, and J.P. Scharff, Anal. Chim. Acta, 1971, 57, 123

71MS   O. Makitie and H. Saarinen, Suomen Kem., 1971, B44, 180

71MSa   O. Makitie and H. Saarinen, Suomen Kem., 1971, B44, 209

71MSb   P.K. Migal and A.M. Sorochinskaya, Russ. J. Inorg. Chem., 1971, 16, 1717 (3243)

71MSP   O. Makitie, H. Saarinen, R. Pelkonen, and J. Maki, Suomen Kem., 1971, B44, 410

71MST   A.I. Mogilyanskii, T.G. Sukhova, O.N. Temkin, and R.M. Flid, Russ. J. Inorg.
           Chem., 1971, 16, 1074 (2017)

71N   G. Norheim, Acta Chem. Scand., 1971, 25, 987

71NM   S. Nagamori and H. Miyata, Bull. Chem. Soc. Japan, 1971, 44, 3476

71NP   V.S.K. Nair and S. Parthasarathy, J. Inorg. Nucl. Chem., 1971, 33, 3019

71NPP   B.P. Nikolskii, V.V. Palchevskii, A.A. Pendin, E.Kh. Tkachuk, S.N. Isaeva, and
           Kh.M. Yakubov, Doklady Chem., 1971, 196, 101 (609)

71NS   O. Navratil and J. Smola, Coll. Czech. Chem. Comm., 1971, 36, 1649

71NV   L.V. Nazarova and Zh.Yu. Vaisbein, Russ. J. Inorg. Chem., 1971, 16, 1704 (3216)

71OB   U.Y. Ozer and R.F. Bogucki, J. Inorg. Nucl. Chem., 1971, 33, 4143

71OO   E. Ohyoshi and A. Ohyoshi, J. Inorg. Nucl. Chem., 1971, 33, 4265

71P   H. Persson, Acta Chem. Scand., 1971, 25, 1774

71PB   D.C. Patel and P.K. Bhattacharya, J. Inorg. Nucl. Chem., 1971, 33, 529

71PJ   J.E. Powell and D.K. Johnson, J. Inorg. Nucl. Chem., 1971, 33, 3586

71PK   I.V. Pyatnitskii and T.I. Kravchenko, Soviet Progr. Chem. (Ukr. Khim. Zh.), 1971,
           37, No. 10, 79 (1054)

71PKa   I.V. Pyatnitskii and T.I. Kravchenko, Soviet Progr. Chem. (Ukr. Khim. Zh.), 1971,
           37, No. 12, 66 (1273)

71PL   K. Pan, C.C. Lai, and T.S. Huang, Bull. Chem. Soc. Japan, 1971, 44, 93

71PP   J. Podlaha and J. Podlahova, Inorg. Chim. Acta, 1971, 5, 413

71PPa   J. Podlaha and J. Podlahova, Inorg. Chim. Acta, 1971, 5, 420

71PPb   R.P. Patel and R.D. Patell, J. Indian Chem. Soc., 1971, 48, 521

71PR   J.E. Prue, A.J. Read, and G. Romeo, Trans. Faraday Soc., 1971, 67, 420

71PS   A.N. Pant, R.N. Soni, and S.L. Gupta, Indian J. Chem., 1971, 9, 270

71PSa   A.N. Pant, R.N. Soni, and S.L. Gupta, J. Inorg. Nucl. Chem., 1971, 33, 3202

71PSb   E.M. Pazukhin and G.V. Sterlyadkina, Russ. J. Inorg. Chem., 1971, 16, 612 (1155)

71PT   A.N. Petrov, O.N. Temkin, and M.I. Bogdanov, Russ. J. Phys. Chem., 1971, 45, 19
         (37)

71R   N.A. Rumbaut, Bull. Soc. Chim. Belges, 1971, 80, 63

71RB   S. Rani and S.K. Banerji, J. Indian Chem. Soc., 1971, 48, 1039

71RBS   S. Ramamoorthy, M.S. Balakrishnan, and M. Santappa, J. Inorg. Nucl. Chem., 1971,
         33, 2713

71RC   R. Roulet, R. Chenaux, and T. VuDuc, Helv. Chim. Acta, 1971, 54, 916

71RG   G.H. Rizvi, B.P. Gupta, and R.P. Singh, Indian J. Chem., 1971, 9, 372

71RR   R.R. Reeder and P.H. Rieger, Inorg. Chem., 1971, 10, 1258

71RS   J. Rais, P. Selucky, and S. Drazanova, J. Inorg. Nucl. Chem., 1971, 33, 3087

71S   H. Saarinen, Suomen Kem., 1971, B44, 100

71Sa   H. Saarinen, Suomen Kem., 1971, B44, 264

71Sb   R.N. Soni, Coll. Czech. Chem. Comm., 1971, 36, 1650

71Sc   A. Sandell, Acta Chem. Scand., 1971, 25, 1795

71Sd   A. Sandell, Acta Chem. Scand., 1971, 25, 2609

71Se   A. Sandell, Acta Chem. Scand., 1971, 25, 3172

71Sf   A.V. Stepanov, Russ. J. Inorg. Chem., 1971, 16, 1583 (2981)

71SB   R.N. Soni and M. Bartusek, J. Inorg. Nucl. Chem., 1971, 33, 2557

71SC   R. Stampfli and G.R. Choppin, J. Coord. Chem., 1971, 1, 173

71SE   I.A. Shikhova, M.I. Ermakova, and N.I. Latosh, J. Gen. Chem. USSR, 1971, 41,
         1337 (1329)

71SG   R.N. Soni and K.L. Gupta, Coll. Czech. Chem. Comm., 1971, 36, 2371

71SH   L.M. Schwartz and L.O. Howard, J. Phys. Chem., 1971, 75, 1798

71SI   T. Sekine and N. Ihara, Bull. Chem. Soc. Japan, 1971, 44, 2942

71SIa   A.S. Solovkin and A.I. Ivantsov, Russ. J. Inorg. Chem., 1971, 16, 1199 (2247)

71SK   N.A. Skorik and V.N. Kumok, Russ. J. Inorg. Chem., 1971, 16, 1643 (3098)

71SKI   E.M. Shvarts, L.I. Korchenenkova, and A.F. Ievinsk, Russ. J. Inorg. Chem., 1971,
         16, 486 (913)

71SS   A. Swinarski and W. Szczepaniak, Rocz. Chem., 1971, 45, 323

71SSa   R.S. Saxena and R. Singh, J. Indian Chem. Soc., 1971, 48, 787

71SSb   R.S. Saxena and R. Singh, Monat. Chem., 1971, 102, 956

71SSP   V.M. Savostin, O.A. Shpigun, and V.M. Peshkova, J. Anal. Chem. USSR, 1971, 26,
          1828 (2044)

71ST   J.P. Shukla and S.G. Tandon, Aust. J. Chem., 1971, 24, 2701

71TD   P.H. Tedesco and V.B. DeRumi, J. Inorg. Nucl. Chem., 1971, 33, 969

71TDa   P.H. Tedesco and V.B. DeRumi, J. Inorg. Nucl. Chem., 1971, 33, 3833

71TDG   P.H. Tedesco, V.B. DeRumi, and J.A. Gonzalez Quintana, J. Inorg. Nucl. Chem.,
          1971, 33, 3839

71TM   V.A. Tsiplyakova, P.K. Migal, and N.V. Suen, Russ. J. Inorg. Chem., 1971, 16,
          1248 (2341)

71TS   R. Tunaboylu and G. Schwarzenbach, Helv. Chim. Acta, 1971, 54, 2166

71TZ   L.G. Tebelev, B.N. Zaitsev, V.M. Levedev, R.F. Melkaya, and V.M. Nikolaev,
          Russ. J. Inorg. Chem., 1971, 16, 519 (976)

71UK   L.N. Usherenko, N.V. Kulikova, N.A. Skorik, and V.N. Kumok, Russ. J. Inorg.
          Chem., 1971, 16, 1711 (3230)

71VK   V.P. Vasilev, L.A. Kochergina, and V.I. Eremenko, Russ. J. Phys. Chem., 1971,
          45, 1196 (2102)

71VS   A. Varadarajulu and U.V. Seshaiah, J. Indian Chem. Soc., 1971, 48, 1065

71VSa   Z. Voznakova and F. Strafelda, Coll. Czech. Chem. Comm., 1971, 36, 2993

71W   O. Wahlberg, Acta Chem. Scand., 1971, 25, 1045, 1064, 1079

71WC   R. Wojtas and D.M. Czakis-Sulikowska, Rocz. Chem., 1971, 45, 737

71WF   H. Wada and Q. Fernando, Anal. Chem., 1971, 43, 751

71WH   E.M. Woolley and L.G. Hepler, Canad. J. Chem., 1971, 49, 3054

71YT   M. Yamane, T. Iwachido, and K. Toei, Bull. Chem. Soc. Japan, 1971, 44, 745

71ZB   J. Zelinka and M. Bartusek, Coll. Czech. Chem. Comm., 1971, 36, 2615

71ZG   V.K. Zolotukhin, O.M. Gnatyshin, and M.I. Koshik, Russ. J. Inorg. Chem., 1971,
          16, 1359 (2550)

71ZP   A.V. Zholnin and V.N. Podchainova, Russ. J. Inorg. Chem., 1971, 16, 616 (1162)

72AB   B.K. Avinashi and S.K. Banerji, Indian J. Chem., 1972, 10, 213

72AD   G.M. Armitage and H.S. Dunsmore, J. Inorg. Nucl. Chem., 1972, 34, 2811

72ADa   H.G. Asawa and L. Dhoot, Z. Phys. Chem. (Leipzig), 1972, 250, 180

72AN   M. Aplincourt, D. Noizet, and R. Hugel, Bull. Soc. Chim. France, 1972, 26

72AV   D. Alexandersson and N.G. Vannerberg, Acta Chem. Scand., 1972, 26, 1909

72B   M. Beran, J. Inorg. Nucl. Chem., 1972, 34, 1043

72Ba   B.W. Budesinsky, Z. Anal. Chem., 1972, 260, 351

72BD   C. Bifano Rizzuti and R. Diaz Cadavieco, An. Quim., 1972, 68, 147

72BF   P.D. Bolton, K.A. Fleming, and F.M. Hall, J. Amer. Chem. Soc., 1972, 94, 1033

72BK   T.V. Beloedova, L.V. Kazakova, and N.A. Skorik, Russ. J. Inorg. Chem., 1972, 17, 816 (1580)

72BM   A. Braibandi, G. Mori, F. Dallavalle, and E. Leporati, Inorg. Chim. Acta, 1972, 6, 106

72BS   B.W. Budesinsky and J. Svec, Talanta, 1972, 19, 87

72BT   F. Bertin and G. Thomas, Bull. Soc. Chim. France, 1972, 1665

72BTL   S.A. Bajue, G.A. Taylor, and G.C. Lalor, J. Inorg. Nucl. Chem., 1972, 34, 1353

72BV   E. Bottari and M. Vicedomini, Gazz. Chim. Ital., 1972, 102, 902

72BVa   E. Bottari and M. Vicedomini, J. Inorg. Nucl. Chem., 1972, 34, 921, 1897

72CA   E. Casassas and J.J. Arias, J. Chim. Phys., 1972, 69, 1261

72CAa   E. Casassas and J.J. Arias, J. Chim. Phys., 1972, 69, 1262

72CB   J.C. Chang and J. Bjerrum, Acta Chem. Scand., 1972, 26, 815

72CD   G.R. Choppin and G. Degischer, J. Inorg. Nucl. Chem., 1972, 34, 3473

72CDP   A. Cassol, P. DiBernardo, R. Portanova, and L. Magon, Gazz. Chim. Ital., 1972, 102, 1118

72CH   G.R. Cayley and D.N. Hague, J. Chem. Soc. Faraday I, 1972, 68, 2259

72CS   M.C. Chattopadhyaya and R.S. Singh, Indian J. Chem., 1972, 10, 850

72CSK   L.G. Cilindro, E. Stadlbauer, and C. Keller, J. Inorg. Nucl. Chem., 1972, 34, 2577

72CSS   J.J. Christensen, D.E. Smith, M.D. Slade, and R.M. Izatt, Thermochim. Acta, 1972, 4, 17

72DC   A. Dadgar and G.R. Choppin, J. Inorg. Nucl. Chem., 1972, 34, 1297

72DCa   G. Degischer and G.R. Choppin, J. Inorg. Nucl. Chem., 1972, 34, 2823

72DM   H.S. Dunsmore and D. Midgley, J. Chem. Soc. Dalton, 1972, 64

72DMa   H.S. Dunsmore and D. Midgley, J. Chem. Soc. Dalton, 1972, 1138

72DN   A.R. Das and V.S.K. Nair, J. Inorg. Nucl. Chem., 1972, 34, 1271

72DSB   S.C. Dhupar, K.C. Srivastava, and S.K. Banerji, J. Indian Chem. Soc., 1972, 49, 935

72DSS   N.K. Dutt, S. Sanyal, and U.U.M. Sharma, J. Inorg. Nucl. Chem., 1972, 34, 2261

72DT   G. Duc and G. Thomas, Bull. Soc. Chim. France, 1972, 4439

72DV   H.F. DeBrabander, L.C. VanPoucke, and Z. Eeckhaut, Inorg. Chim. Acta, 1972, 6, 459

72E   B. Egneus, Talanta, 1972, 19, 1387

72Ea   L.G. Egorova, J. Gen. Chem. USSR, 1972, 42, 2237 (2240)

72EM   S.H. Eberle and F. Moattar, Inorg. Nucl. Chem. Letters, 1972, 8, 265

72ES   G. Eisenman, G. Szabo, S.G.A. McLaughlin, and S.M. Ciane, in "Symposium on Mole-
        cular Mechanisms of Antibiotic Action on Protein Biosynthesis and Membranes,"
        D. Vasquez, ed., Springer Verlag, New York, 1972

72ET   N.G. Elenkova, R.A. Tsoneva, and D. Dzhimbizova, Russ. J. Inorg. Chem., 1972, 17,
        356 (681)

72EZ   A.A. Elesin and A.A. Zaitsev, Soviet Radiochem., 1972, 14, 381 (370)

72EZa   A.A. Elesin, A.A. Zaitsev, V.A. Karaseva, I.I. Nazarova, and I.V. Petukhova,
        Soviet Radiochem., 1972, 14, 385 (374)

72EZb   A.A. Elesin, A.A. Zaitsev, S.S. Kazakova, and G.N. Yakovlev, Soviet Radiochem.,
        1972, 14, 558 (541)

72EZc   A.A. Elesin, A.A. Zaitsev, N.A. Ivanovich, V.A. Karaseva, and G.N. Yakovlev,
        Soviet Radiochem., 1972, 14, 563 (546)

72FD   Ya.D. Fridman and T.V. Danilova, Russ. J. Inorg. Chem., 1972, 17, 492 (946)

72FG   G.J. Ford, P. Gans, L.D. Pettit, and C. Sherrington, J. Chem. Soc. Dalton, 1972,
        1763

72FP   A.A. Fedorov and A.V. Pavlinova, J. Anal. Chem. USSR, 1972, 27, 619 (708)

72GB   P.K. Govil and S.K. Banerji, Indian J. Chem., 1972, 10, 538

72GBa   P.K. Govil and S.K. Banerji, J. Chinese Chem. Soc. (Taiwan), 1972, 19, 83

72GD   P.K. Govil, C.D. Dwivedi, and S.K. Banerji, Indian J. Chem., 1972, 10, 211

72GDa   P.K. Govil, C.D. Dwivedi, and S.K. Banerji, Israel J. Chem., 1972, 10, 685

72GK   J. Gross and C. Keller, J. Inorg. Nucl. Chem., 1972, 34, 725

72GKa   R.P. Guseva and V.N. Kumok, Russ. J. Inorg. Chem., 1972, 17, 1680 (3195)

72GKK   Z. Gregorowicz, G. Kwapulinska, Z. Klima, and E. Ziaja, Coll. Czech. Chem. Comm.,
        1972, 37, 119

72GM   V.I. Gordienko and Yu.I. Mikhailyuk, J. Anal. Chem. USSR, 1972, 27, 959 (1069)

72GMa   G.P. Gupta and K.N. Munshi, J. Indian Chem. Soc., 1972, 49, 217

72GO   I. Grenthe and H. Ots, Acta Chem. Scand., 1972, 27, 1217, 1229

72GP   K. Garbett, G.W. Partridge, and R.J.P. Williams, Bioinorg. Chem., 1972, 1, 309

72GS   N.N. Guseva, E.V. Sklenskaya, M.Kh. Karapetyants, and A.I. Mikhailichenko,
        Soviet Radiochem., 1972, 14, 132 (132)

72HK   E.J. Hakoila, J.J. Kankare, and T. Skarp, Anal. Chem., 1972, 44, 1857

72HS   J. Hala and J. Smola, J. Inorg. Nucl. Chem., 1972, 34, 1039

72I   T. Iwachido, Bull. Chem. Soc. Japan, 1972, 45, 432

72IM   V.A. Ivanov and P.K. Migal, Russ. J. Inorg. Chem., 1972, 17, 553 (1067)

72JK   D.V. Jahagirdar and D.D. Khanolkar, J. Indian Chem. Soc., 1972, 49, 1105

72JM   C.R. Jejurkar, I.P. Mavani, and P.K. Bhattacharya, Indian J. Chem., 1972, 10, 1190

72JW   R.F. Jameson and M.F. Wilson, J. Chem. Soc. Dalton, 1972, 2610

72JWa   R.F. Jameson and M.F. Wilson, J. Chem. Soc. Dalton, 1972, 2614

72JWb   R.F. Jameson and M.F. Wilson, J. Chem. Soc. Dalton, 1972, 2617

72K   J.J. Kankare, Anal. Chem., 1972, 44, 2376

72Ka   W. Kaminski, Rocz. Chem., 1972, 46, 339

72KA   P.V. Khadikar and R.L. Ameria, J. Indian Chem. Soc., 1972, 49, 717

72KC   A.S. Kereichuk and I.M. Churikova, Russ. J. Inorg. Chem., 1972, 17, 1300 (2486)

72KE   T.S. Kas and A.M. Egorov, Russ. J. Phys. Chem., 1972, 46, 1232

72KEa   I.E. Kalinichenko and N.A. Emtsova, J. Gen. Chem. USSR, 1972, 42, 2227 (2231)

72KEM   N.A. Krasnyanskaya, I.I. Eventova, N.V. Melchakova, and V.M. Peshkova, J. Anal. Chem. USSR, 1972, 27, 1672 (1842)

72KG   S.C. Khurana and C.M. Gupta, Talanta, 1972, 19, 1235

72KL   N. Konopik and W. Luf, Monat. Chem., 1972, 103, 355

72KLa   N. Konopik and W. Luf, Monat. Chem., 1972, 103, 1091

72KLM   V.P. Kerentseva, M.D. Lipanova, and L.I. Masko, J. Anal. Chem. USSR, 1972, 27, 628 (719)

72KM   K. Kurzelina-Cedzynska and J. Maslowska, Rocz. Chem., 1972, 46, 1215

72KMK   N.P. Komar, V.V. Melnik, and A.G. Kozachenko, Russ. J. Phys. Chem., 1972, 46, 928 (1615)

72KMP   N.A. Krasnyanskaya, N.V. Melchakova, and V.M. Peshkova, J. Anal. Chem. USSR, 1972, 27, 1470 (1620)

72KN   N.P. Komar and M.T. Nguyen, J. Anal. Chem. USSR, 1972, 27, 1517 (1669)

72KS   J. Krishnamacharyulu and U.V. Seshaiah, Indian J. Chem., 1972, 10, 1192

72L   N.M. Lukovskaya, Soviet Progr. Chem. (Ukr. Khim. Zh.), 1972, 38, No. 5, 73 (485)

72LN   R. Larsson and G. Nunziata, Acta Chem. Scand., 1972, 26, 1971

72LP   P. Lumme and K. Ponkala, Suomen Kem., 1972, B45, 52

72LPN   P. Lumme, K. Ponkala, and K. Nieminen, Suomen Kem., 1972, B45, 170

72LPS   P. Lingaiah, G. Punnaiah, and E.V. Sundaram, Indian J. Chem., 1972, 10, 521

72LS   P. Lingaiah and E.V. Sundaram, Indian J. Chem., 1972, 10, 670

72LSH   C.L. Liotta, D.F. Smith, Jr., H.P. Hopkins, Jr., and K.A. Rhodes, J. Phys. Chem., 1972, 76, 1909

72M   M.Y. Maftoon, Coll. Czech. Chem. Comm., 1972, 37, 688

72MA   O. Makitie and A. Aholainen, Suomen Kem., 1972, B45, 281

72MB   L. Magon, A. Bismondo, G. Tomat, and A. Cassol, Radiochim. Acta, 1972, 17, 164

72MC   G.S. Manku and R.C. Chadha, J. Inorg. Nucl. Chem., 1972, 34, 357

72MD   A. Mahan and A.K. Dey, J. Indian Chem. Soc., 1972, 49, 939

72MF   G. Molle and J.C. Fenyo, Compt. Rend. Acad. Sci. Paris, Ser. C, 1972, 274, 11

72MG   H. Metivier and R. Guillaumont, Radiochem. Radioanal. Letters, 1972, 10, 239

72MGa   H. Metivier and R. Guillaumont, Radiochem. Radioanal. Letters, 1972, 11, 165

72MGb   Yu.I. Mikhailyuk and V.I. Gordienko, Russ. J. Inorg. Chem., 1972, 17, 668 (1287)

72MGS   R.C. Mehrotra, V.D. Gupta, and C.K. Sharma, Indian J. Chem., 1972, 10, 433

72MJ   I.P. Mavani, C.R. Jejurkar, and P.K. Bhattacharya, J. Indian Chem. Soc., 1972,
          49, 469

72MK   G. Markovits, P. Klotz, and L. Newman, Inorg. Chem., 1972, 11, 2405

72ML   O. Makitie and A. Lehto, Acta Chem. Scand., 1972, 26, 2141

72MP   O. Makitie and R. Pelkonen, Suomen Kem., 1972, B45, 243

72MPA   O. Makitie, R. Petrola, and U.M. Aarnisalo, Suomen Kem., 1972, B45, 31

72MPM   O. Makitie, R. Petrola, P. Maenpaa, and U. M. Aarnisalo, Ann. Acad. Sci. Fenn.,
          Ser. A, 1972, No. 162

72MPZ   L. Magon, R. Portanova, B. Zarli, and A. Bismondo, J. Inorg. Nucl. Chem., 1972,
          34, 1971

72MS   O. Makitie and H. Saarinen, Suomen Kem., 1972, B45, 308

72MSa   M. Morin and J.P. Scharff, Anal. Chim. Acta, 1972, 60, 101

72MSb   E. Mentasti and G. Saini, Atti Accad. Sci. Torino, 1972, 106, 561

72MSD   S. Mandal, T. Singh, and A.K. Dey, J. Indian Chem. Soc., 1972, 49, 333

72MSS   T.P. Makarova, G.S. Sinitsyna, I.A. Shestakova, A.V. Stepanov, and
          B.I. Shestakov, Soviet Radiochem., 1972, 14, 852 (822)

72N   A. Napoli, J. Inorg. Nucl. Chem., 1972, 34, 987

72Na   A. Napoli, J. Inorg. Nucl. Chem., 1972, 34, 1225

72Nb   A. Napoli, J. Inorg. Nucl. Chem., 1972, 34, 1347

72Nc   A. Napoli, Gazz. Chim. Ital., 1972, 102, 273

72NB   R. Nath, K.K. Banerji, and P. Nath, J. Indian Chem. Soc., 1972, 49, 643

72NK   B.I. Nabivanets and L.V. Kalabina, J. Anal. Chem. USSR, 1972, 27, 1015 (1134)

72NP   H.L. Nigam and K.B. Pandeya, Curr. Sci. (India), 1972, 41, 485

72NS   M.S. Novakovskii, G.V. Samoilenko, and E.I. Vail, J. Gen. Chem. USSR, 1972, 42,
          40 (44)

720   M.M. Osman, Helv. Chim. Acta, 1972, 55, 239

72P   L. Pettersson, Acta Chem. Scand., 1972, 26, 4067

72PA   S. Parthasarathy and S. Ambujavalli, Electrochim. Acta, 1972, 17, 1219

72PAL   L. Pettersson, I. Andersson, L. Lyhamn, and N. Ingri, Trans. Roy. Inst. Tech.
          Stockholm, 1972, No. 256

72PB   B.R. Panchal and P.K. Bhattacharya, Indian J. Chem., 1972, 10, 857

72PBa   M.M. Petit-Ramel and C.M. Blanc, J. Inorg. Nucl. Chem., 1972, 34, 1233, 1241, 1253

72PG   R. Panossian, J. Galea, and M. Asso, Ann. Chim. (France), 1972, 7, 223

72PM   S.P. Pande and K.N. Munshi, Curr. Sci. (India), 1972, 41, 330

72PMa   S.P. Pande and K.N. Munshi, J. Indian Chem. Soc., 1972, 49, 533

72PP   P. Petras and J. Podlaha, Inorg. Chim. Acta, 1972, 6, 253

72PR   B.R. Panchal, M.V. Reddy, and P.K. Bhattacharya, Indian J. Chem., 1972, 10, 218

72PS   A.N. Pant, R.N. Soni, and S.L. Gupta, Indian J. Chem., 1972, 10, 90

72PSa   A.D. Parulekar and P.R. Subbaraman, Indian J. Chem., 1972, 10, 205

72PSb   A.N. Pant, R.N. Soni, and S.L. Gupta, Indian J. Chem., 1972, 10, 632

72PSc   A.N. Pant, R.N. Soni, and S.L. Gupta, Indian J. Chem., 1972, 10, 724

72PSd   A.N. Pant, R.N. Soni, and G.L. Gupta, Indian J. Chem., 1972, 10, 859

72PSL   S.B. Pirkes, M.T. Shestakova, and A.V. Lapitskaya, Russ. J. Inorg. Chem., 1972,
          17, 206 (395)

72PSS   Pushparaja, M. Sudersanan, and A.K. Sundaram, Curr. Sci. (India), 1972, 41, 633

72PT   N. Purdie, M.B. Tomson, and G.K. Cook, Anal. Chem., 1972, 44, 1525

72PTa   R. Portanova, G. Tomat, A. Cassol, and L. Magon, J. Inorg. Nucl. Chem., 1972, 34,
          1685

72PTB   A.N. Petrov, O.N. Temkin, and M.I. Bogdanov, Russ. J. Phys. Chem., 1972, 46, 1043

72PTM   R. Portanova, G. Tomat, L. Magon, and A. Cassol, J. Inorg. Nucl. Chem., 1972, 34,
          1768

72PTR   N. Purdie, M.B. Tomson, and N. Riemann, J. Solution Chem., 1972, 1, 465

72PZ   V.I. Paramonova, V.Ya. Zamanskii, and V.B. Kolychev, Russ. J. Inorg. Chem., 1972,
          17, 542 (1042)

72R   D.L. Rabenstein, Canad. J. Chem., 1972, 50, 1036

72RG   P.C. Rawat and C.M. Gupta, J. Inorg. Nucl. Chem., 1972, 34, 951

72RGa   P.C. Rawat and C.M. Gupta, J. Inorg. Nucl. Chem., 1972, 34, 1621

72RGb   P.C. Rawat and C.M. Gupta, Talanta, 1972, 19, 706

72RL   R. Rauhamaki, K. Lehtonen, and O. Makitie, Suomen Kem., 1972, B45, 185

72RM   S. Ramamoorthy and P.G. Manning, J. Inorg. Nucl. Chem., 1972, 34, 1977, 1989,
          1997

72RMG   S. Ramamoorthy, P.G. Manning, and C. Guarnaschelli, J. Inorg. Nucl. Chem., 1972,
          34, 3443

72RV   E. Roletto, A. Vanni, and G. Ostacoli, J. Inorg. Nucl. Chem., 1972, 34, 2817

72S   L.L. Soni, J. Indian Chem. Soc., 1972, 49, 341

72SC   R.S. Saxena and U.S. Chaturvedi, J. Inorg. Nucl. Chem., 1972, 34, 913

72SCa   R.S. Saxena and U.S. Chaturvedi, J. Inorg. Nucl. Chem., 1972, 34, 2964

72SCb   R.S. Saxena and U.S. Chaturvedi, J. Inorg. Nucl. Chem., 1972, 34, 3272

72SCc   R.S. Saxena and U.S. Chaturvedi, J. Indian Chem. Soc., 1972, 49, 321

72SI   V. Simeon, N. Ivicic, and M. Tkalcec, Z. Phys. Chem. (Frankfort), 1972, 78, 1

72SL   P.J. Sun and C.Y. Liu, J. Chinese Chem. Soc. (Taiwan), 1972, 19, 35

72SM   R. Sarin and K.N. Munshi, Aust. J. Chem., 1972, 25, 929

72SMa   R. Sarin and K.N. Munshi, J. Inorg. Nucl. Chem., 1972, 34, 581

72SMD   T. Singh, A. Mahan, and A.K. Dey, J. Inorg. Nucl. Chem., 1972, 34, 2551

72SMS   J.P. Shukla, V.K. Manchanda, and M.S. Subramanian, J. Electroanal. Chem., 1972,
            40, 431

72SR   D.Z. Suznjevic and L.M. Rajkovic, J. Electroanal. Chem., 1972, 38, 203

72SS   M. Sudersanan and A.K. Sundaram, Proc. Indian Acad. Sci., 1972, 75A, 151

72SSa   G.N. Shabanova and N.A. Skorik, J. Gen. Chem. USSR, 1972, 42, 198 (204)

72SSb   P.V. Selvaraj and M. Santappa, Curr. Sci. (India), 1972, 41, 872

72SSF   J. Savic, M. Savic, and I. Filipovic, Croat. Chem. Acta, 1972, 44, 305

72SSP   P.K. Spitsyn, V.S. Shvarev, and T.P. Popyvanova, Russ. J. Inorg. Chem., 1972,
            17, 502 (966)

72ST   J.P. Shukla and S.G. Tandon, J. Indian Chem. Soc., 1972, 49, 83

72STa   H. Schurmans, H. Thun, F. Verbeek, and W. Vanderleen, J. Electroanal. Chem.,
            1972, 34, 109

72STb   H. Schurmans, H. Thun, F. Verbeek, and W. Vanderleen, J. Electroanal. Chem.,
            1972, 38, 209

72TA   P.H. Tedesco and M.C. Anon, J. Inorg. Nucl. Chem., 1972, 34, 2271

72TM   G. Tomat, L. Magon, R. Portanova, and A. Cassol, Z. Anorg. Allg. Chem., 1972,
            393, 184

72TP   S.C. Tripathi and S. Paul, Indian J. Chem., 1972, 10, 113

72TPP   K.K. Tripathy, R.K. Patnaik, and S. Pani, J. Indian Chem. Soc., 1972, 49, 345

72TS   K. Tunaboylu and G. Schwarzenbach, Helv. Chim. Acta, 1972, 55, 2065

72TSa   D. Tuhtar, J. Savic, and M. Savic, Glas. Hem. Tehnol. Bosne Hercegovine,
            1971-1972, 19-20, 11; Chem. Abs., 1975, 82, 35470y

72TSb   D. Tuhtar, M. Savic, and J. Savic, Glas. Hem. Tehnol. Bosne Hercegovine,
            1971-1972, 19-20, 19; Chem. Abs., 1975, 82, 35471z

72UC   M. Urdaneta and M.P. Collados deDiaz, An. Quim., 1972, 68, 235

72WF H. Wada and Q. Fernando, *Anal. Chem.*, 1972, **44**, 1640

72WH E.M. Woolley and L.G. Hepler, *Anal. Chem.*, 1972, **44**, 1521

72WV W.D. Wakley and L.P. Varga, *Anal. Chem.*, 1972, **44**, 169

72YV R.Yu. Yuryavichus and Ch.A. Valyukyavichyus, *J. Anal. Chem. USSR*, 1972, **27**, 1006 (1125)

72ZG P.M. Zaitsev, V.N. Grechishkina, L.S. Chasovskikh, and Z.V. Zaitseva, *J. Gen. Chem. USSR*, 1972, **42**, 1075 (1084)

73A R. Aruga, *Atti Accad. Sci. Torino*, 1973, **107**, 207

73Ab N.I. Ampelogova, *Soviet Radiochem.*, 1973, **15**, 823 (813)

73AD G.M. Armitage and H.S. Dunsmore, *J. Inorg. Nucl. Chem.*, 1973, **35**, 817

73AG S.O. Ajayi and D.R. Goddard, *J. Chem. Soc. Dalton*, 1973, 1751

73AH P.J. Antikainen and E. Huttunen, *Suomen Kem.*, 1973, **B46**, 184

73AV D. Alexandersson and N.G. Vannerberg, *Acta Chem. Scand.*, 1973, **27**, 3499

73AW P.J. Antikainen and U. Witikainen, *Acta Chem. Scand.*, 1973, **27**, 2075

73B M. Bartusek, *Coll. Czech. Chem. Comm.*, 1973, **38**, 2255

73BB M. Baranowska-Zralko and J. Biernat, *Electrochem. Acta*, 1973, **17**, 1877

73BD A. Bellomo, D. DeMarco, and A. DeRobertis, *Talanta*, 1973, **20**, 1225

73BL E. Bottari, A. Liberti, and M. Vicedomini, *Gazz. Chim. Ital.*, 1973, **103**, 859

73BPC M.A.A. Butt, R. Parkash, R.M. Chandhary, and M.L. Mittal, *Coll. Czech. Chem. Comm.*, 1973, **38**, 2425

73BPS E.M. Belousova, A.F. Pozharitskii, I.I. Seifullina, and M.M. Bobrovskaya, *Russ. J. Inorg. Chem.*, 1973, **18**, 1470 (2766)

73BR M. Barres, J.P. Redoute, R. Romanetti, H. Tachoire, and C. Zahra, *Compt. Rend. Acad. Sci. Paris*, Ser. C, 1973, **276**, 363

73BS C.E. Bamberger and A. Suner, *Inorg. Nucl. Chem. Letters*, 1973, **9**, 1005

73BSa B.W. Budesinsky and M. Sagat, *Talanta*, 1973, **20**, 228

73BSb J.F.C. Boodts and W. Saffioti, *Compt. Rend. Acad. Sci. Paris*, Ser. C, 1973, **276**, 755

73BV E. Bottari and M. Vicedomini, *J. Inorg. Nucl. Chem.*, 1973, **35**, 1269, 2447

73BVa E. Bottari and M. Vicedomini, *J. Inorg. Nucl. Chem.*, 1973, **35**, 1657

73CA E. Casassas, J.J. Arias, and A. Mederos, *An. Quim.*, 1973, **69**, 1121

73CB A. Cassol, P. DiBernardo, R. Portanova, and L. Magon, *Inorg. Chim. Acta*, 1973, **7**, 353

73CDB E. Chiacchierini, G. DeAngelis, and F. Balestrieri, *Gazz. Chim. Ital.*, 1973, **103**, 387

73CDC E. Chiacchierini, G. DeAngelis, and P. Coccanari, *Gazz. Chim. Ital.*, 1973, **103**, 413

73CDS G.R. Choppin, A. Dadgar, and R. Stampfli, *J. Inorg. Nucl. Chem.*, 1973, **35**, 875

73CG   E. Chiacchierini, R. Gocchieri, and L. Sommer, <u>Coll. Czech. Chem. Comm.</u>, 1973,
<u>38</u>, 1478

73CS   E.I. Chubakova and N.A. Skorik, <u>Russ. J. Inorg. Chem.</u>, 1973, <u>18</u>, 1446 (2723)

73CSM   V.A. Cortinez, J.P. Santagata, and C.B. Marone, <u>An. Quim.</u>, 1973, <u>69</u>, 343

73D   I. Dellien, <u>Acta Chem. Scand.</u>, 1973, <u>27</u>, 733

73DF   P. DeMaria, A. Fini, and F.M. Hall, <u>J. Chem. Soc. Perkin II</u>, 1973, 1969

73DG   I. Dellien, I. Grenthe, and G. Hessler, <u>Acta Chem. Scand.</u>, 1973, <u>27</u>, 2431

73DM   I. Dellien and L. Malmsten, <u>Acta Chem. Scand.</u>, 1973, <u>27</u>, 2877

73DMa   J. Duda and J. Maslowska, <u>Rocz. Chem.</u>, 1973, <u>47</u>, 1337

73DP   E.A. Didenko and S.B. Pirkes, <u>Russ. J. Inorg. Chem.</u>, 1973, <u>18</u>, 36

73DS   S.C. Dhupar, K.C. Srivastava, and S.K. Banerji, <u>J. Indian Chem. Soc.</u>, 1973, <u>50</u>,
19

73EB   O. Enea and G. Berthon, <u>Thermochim. Acta</u>, 1973, <u>6</u>, 47

73EM   K.J. Ellis and A. McAuley, <u>J. Chem. Soc. Dalton</u>, 1973, 1533

73ER   L.G. Egorova and V.L. Nirenburg, <u>J. Gen. Chem. USSR</u>, 1973, <u>43</u>, 1384 (1396)

73ERa   L.G. Egorova and V.L. Nirenburg, <u>J. Gen. Chem. USSR</u>, 1973, <u>43</u>, 1533 (1548)

73ES   H.M. ElFatatry, A.W. vonSmolinski, C.E. Gracias, and D.A. Coviello, <u>Talanta</u>,
1973, <u>20</u>, 923

73FB   J.C. Fenyo, J. Beaumais, E. Selegny, M. Petit-Ramel, and R.P. Martin, <u>J. Chim.
Phys.</u>, 1973, <u>70</u>, 299

73FJ   O. Forsberg, K. Johansson, P. Ulmgren, and O. Wahlberg, <u>Chem. Scripta</u>, 1973, <u>3</u>,
153

73FP   I. Filipovic, I. Piljac, B. Bach-Dragutinovic, I. Kruhak, and B. Grabaric,
<u>Croat. Chem. Acta</u>, 1973, <u>45</u>, 447

73FS   B. Finlayson and A. Smith, <u>J. Chem. Eng. Data</u>, 1973, <u>18</u>, 368

73GB   P.K. Govil and S.K. Banerji, <u>J. Inorg. Nucl. Chem.</u>, 1973, <u>35</u>, 3932

73GK   V.I. Gordienko and L.P. Khudyakova, <u>J. Gen. Chem. USSR</u>, 1973, <u>43</u>, 323 (326)

73GP   B. Grabaric, I. Piljac, and I. Filipovic, <u>Anal. Chem.</u>, 1973, <u>45</u>, 1932

73H   B. Hurnik, <u>Rocz. Chem.</u>, 1973, <u>47</u>, 1065

73HH   M.H. Hutchinson and W.C.E. Higginson, <u>J. Chem. Soc. Dalton</u>, 1973, 1247

73HHG   S. Hubert, M. Hussonnois, and R. Guillaumont, <u>J. Inorg. Nucl. Chem.</u>, 1973, <u>35</u>,
2923

73I   T. Iwachido, <u>Bull. Chem. Soc. Japan</u>, 1973, <u>46</u>, 2761

73IT   S. Ito, H. Tomiyasu, and H. Ohtaki, <u>Bull. Chem. Soc. Japan</u>, 1973, <u>46</u>, 2238

73IV   A.A. Ivakin and E.M. Voronova, <u>Russ. J. Inorg. Chem.</u>, 1973, <u>18</u>, 817 (1552)

73JA   A.K. Jain, V.P. Agrawal, P. Chand, and S.P. Garg, <u>J. Indian Chem. Soc.</u>, 1973,
<u>50</u>, 517

73JK   D.V. Jahagirdar and D.D. Khanolkar, J. Inorg. Nucl. Chem., 1973, 35, 921

73K   J.J. Kankare, Anal. Chem., 1973, 45, 2050

73KA   P.V. Khadikar and R.L. Ameria, J. Indian Chem. Soc., 1973, 50, 458

73KC   S.C. Khurana, D.N. Chaturvedi, and C.M. Gupta, J. Inorg. Nucl. Chem., 1973, 35,
          1645

73KD   M.K. Koul and K.P. Dubey, J. Inorg. Nucl. Chem., 1973, 35, 2571

73KE   T.S. Kas and A.M. Egorov, Russ. J. Phys. Chem., 1973, 47, 157 (273)

73KG   S.C. Khurana and C.M. Gupta, J. Inorg. Nucl. Chem., 1973, 35, 209

73KGa   S.C. Khurana and C.M. Gupta, Talanta, 1973, 20, 789

73KGG   S.C. Khurana, J.K. Gupta, and C.M. Gupta, Electrochim. Acta, 1973, 18, 59

73KJ   V.D. Khanolkar, D.V. Jahagirdar, and D.D. Khanolkar, J. Inorg. Nucl. Chem., 1973,
          35, 931

73KK   Ts.B. Konunova and L.S. Kachkar, Russ. J. Inorg. Chem., 1973, 18, 805 (1527)

73KL   K. Kustin and S.T. Liu, Inorg. Chem., 1973, 12, 2362

73KLa   K. Kustin and S.T. Liu, J. Amer. Chem. Soc., 1973, 95, 2487

73KM   M. Komatsu, Y. Masuda, and S. Misumi, Nippon Kagaku Kaishi, 1973, 1461

73KP   S.S. Kalinina, Z.N. Prozorovskaya, L.N. Komissarova, L.I. Yuranova, and V.I.
          Spitsyn, Russ. J. Inorg. Chem., 1973, 18, 776

73KS   A.K. Kalra, H.B. Singh, and R.P. Singh, J. Inorg. Nucl. Chem., 1973, 35, 187

73KT   K. Kustin and D.L. Toppen, Inorg. Chem., 1973, 12, 1404

73L   Y.H. Lee, Acta Chem. Scand., 1973, 27, 1807

73La   S. Lasztity, Radiochem. Radioanal. Letters, 1973, 14, 277

73LP   A.V. Lapitskaya and S.B. Pirkes, Russ. J. Inorg. Chem., 1973, 18, 635 (1204)

73LR   S. Libich and D.L. Rabenstein, Anal. Chem., 1973, 45, 118

73LS   B.M. Lowe and D.G. Smith, J. Chem. Soc. Faraday I, 1973, 69, 1934

73LSa   L.P. Lisovaya and N.A. Skorik, Russ. J. Inorg. Chem., 1973, 18, 599 (1134)

73LU   L.P. Lisovaya, L.N. Usherenko, N.A. Skorik, and V.N. Kumok, Russ. J. Inorg.
          Chem., 1973, 18, 505 (961)

73MB   C.B. Marone and G. Bianchi, An. Quim., 1973, 69, 205

73MBM   L. Magon, A. Bismondo, L. Maresca, G. Tomat, and R. Portanova, J. Inorg. Nucl.
          Chem., 1973, 35, 4237

73MBP   L. Magon, A. Bismondo, G. Bandoli, and R. Portanova, J. Inorg. Nucl. Chem., 1973,
          35, 1995

73MD   A. Mahan and A.K. Dey, Anal. Chim. Acta, 1973, 63, 85

73MDa   A. Mahan and A.K. Dey, J. Inorg. Nucl. Chem., 1973, 35, 3263

73MH   T. Matsui and L.G. Hepler, Canad. J. Chem., 1973, 51, 1941, 3789

73MI   P.K. Migal and V.A. Ivanov, Russ. J. Inorg. Chem., 1973, 18, 536 (1019)

73MP   S.P. Mushran, O. Prakash, and R. Murti, J. Inorg. Nucl. Chem., 1973, 35, 2119

73MS   O. Makitie, H. Saarinen, T. Jantunen, P. Jyske, and M. Suonpaa, Suomen Kem.,
           1973, B46, 336

73N   A. Napoli, J. Inorg. Nucl. Chem., 1973, 35, 3360

73NK   B.P. Nikolskii, L.I. Krylov, B.B. Zakhvataev, and R.I. Lyubtsev, Soviet Radio-
           chem., 1973, 15, 814 (804)

73NKa   B.P. Nikolskii, L.I. Krylov, and B.B. Zakhvataev, Soviet Radiochem., 1973, 15,
           820 (810)

73NP   S. Nushi, I. Piljac, B. Grabaric, and I. Filipovic, Croat. Chem. Acta, 1973, 45,
           453

73O   E. Ohyoshi, Bull. Chem. Soc. Japan, 1973, 46, 2758

73OO   C. O.Nuallain and S. O.Cinneide, J. Inorg. Nucl. Chem., 1973, 35, 2871

73PA   L. Pettersson and I. Andersson, Acta Chem. Scand., 1973, 27, 977

73PAa   L. Pettersson and I. Andersson, Acta Chem. Scand., 1973, 27, 1019

73PAb   S. Parthasarathy and S. Ambujavalli, Electrochim Acta, 1973, 18, 68

73PAc   S. Parthasarathy and S. Ambujavalli, Thermochim. Acta, 1973, 7, 225

73PAT   N.S. Poluektov, L.A. Alakaeva, and M.A. Tishchenko, Russ. J. Inorg. Chem., 1973,
           18, 40 (81)

73PD   R.G. Pearson and D.G. DeWit, J. Coord. Chem., 1973, 2, 175

73PDB   R. Portanova, G. DePaoli, A. Bismondo, and L. Magon, Gazz. Chim. Ital., 1973,
           103, 691

73PG   I. Piljac, B. Grabaric, and I. Filipovic, J. Electroanal. Chem., 1973, 42, 433

73PL   S.K. Pal and S.C. Lahiri, Z. Phys. Chem. (Leipzig), 1973, 252, 177

73PM   R. Petrola and O. Makitie, Suomen Kem., 1973, B46, 10

73PP   O. Prochazkova, J. Podlahova, and J. Podlaha, Coll. Czech. Chem. Comm., 1973,
           38, 1120

73PPa   J. Podlaha and J. Podlahova, Coll. Czech. Chem. Comm., 1973, 38, 1730

73PPb   P. Petras, J. Podlahova, and J. Podlaha, Coll. Czech. Chem. Comm., 1973, 38, 3221

73PS   A.N. Pant, R.N. Soni, and S.L. Gupta, J. Inorg. Nucl. Chem., 1973, 35, 1390

73PT   N. Purdie and M.B. Tomson, J. Amer. Chem. Soc., 1973, 95, 48

73RD   M.N. Rusina and N.M. Dyatlova, Russ. J. Inorg. Chem., 1973, 18, 1419 (2672)

73RM   S. Ramamoorthy and P.G. Manning, J. Inorg. Nucl. Chem., 1973, 35, 1571

73RS   A. Rokosz and R. Stepak, Rocz. Chem., 1973, 47, 1767

73RSM   A. Rokosz, R. Stepak, and M. Meus, Rocz. Chem., 1973, 47, 1073

73RT   R. Robbrecht, H. Thun, and F. Verbeek, Bull. Soc. Chim. Belg., 1973, 82, 397

73S   H. Saarinen, Suomen Kem., 1973, B46, 250

73Sa   H. Saarinen, Suomen Kem., 1973, B46, 333

73Sb   H. Saarinen, Ann. Acad. Sci. Fenn., 1973, A2, No. 170

73SA   W.R. Stagg and B.L. Andrews, Proc. Tenth Rare Earth Res. Conf., Carefree, Arizona, 1973, 342

73SB   V. Stejskal and M. Bartusek, Coll. Czech. Chem. Comm., 1973, 38, 3103

73SBa   V.G. Selyanina and V.F. Barkovskii, J. Anal. Chem. USSR, 1973, 28, 420 (473)

73SBb   V.G. Selyanina and V.F. Barkovskii, J. Anal. Chem. USSR, 1973, 28, 847 (956)

73SBL   R. Sabbah, F. Bugueret, and M. Laffitte, Z. Phys. Chem. (Frankfurt), 1973, 85, 5

73SHG   H. Sigel, P.R. Huber, R. Griesser, and B. Prijs, Inorg. Chem., 1973, 12, 1198

73SHI   T. Sekine, Y. Hasegawa, and N. Ihara, J. Inorg. Nucl. Chem., 1973, 35, 3968

73SJ   E.G. Sase and D.V. Jahagirdar, J. Indian Chem. Soc., 1973, 50, 378

73SM   R. Sarin and K.N. Munshi, J. Inorg. Nucl. Chem., 1973, 35, 201

73SN   B.D. Struck and H.W. Nurnberg, J. Electroanal. Chem., 1973, 48, 175

73SNH   B.D. Struck, H.W. Nurnberg, and H.N. Hechner, J. Electroanal. Chem., 1973, 46, 171

73SO   Z. Soylemez and U.Y. Ozer, J. Inorg. Nucl. Chem., 1973, 35, 545

73SS   R. Sundaresan and A.K. Sundaram, Proc. Indian Acad. Sci. (A), 1973, 78, 218

73SSC   R.R. Scharfe, V.S. Sastri, and C.L. Chakrabarti, Anal. Chem., 1973, 45, 413

73TD   P.H. Tedesco, V.B. DeRumi, and J.A. Gonzalez Quintana, J. Inorg. Nucl. Chem., 1973, 35, 285

73TDa   P.H. Tedesco, V.B. DeRuni, and J.A. Gonzalez Quintana, J. Inorg. Nucl. Chem., 1973, 35, 287

73TP   S.C. Tripathi and S. Paul, J. Inorg. Nucl. Chem., 1973, 35, 2465

73TPa   K.K. Tripathy and R.K. Patnaik, J. Inorg. Nucl. Chem., 1973, 35, 1050

73TPb   K.K. Tripathy and R.K. Patnaik, Acta Chim. Acad. Sci. Hung., 1973, 79, 279

73UW   P. Ulmgren and O. Wahlberg, Chem. Scripta, 1973, 3, 159, 193

73VB   R.S. Vaidya and S.N. Banerji, J. Indian Chem. Soc., 1973, 50, 747

73VG   K. Vadasdi and I. Gaal, J. Inorg. Nucl. Chem., 1973, 35, 658

73VN   Sh.E. Vassershtein and N.V. Nam, Russ. J. Inorg. Chem., 1973, 18, 541 (1028)

73VP   Z.A. Vladimirova, Z.N. Prozorovskaya, and L.N. Komissarova, Russ. J. Inorg. Chem., 1973, 18, 368 (704)

73VS   V.P. Vasilev, L.D. Shekhanova, and L.A. Kochergina, J. Gen. Chem. USSR, 1973, 43, 967 (971)

73VV   V.N. Vasileva, V.P. Vasilev, T.K. Korbut, and L.I. Bukoyazova, Russ. J. Inorg. Chem., 1973, 18, 974 (1843)

73WK   Z. Warnke and E. Kwiatkowski, Rocz. Chem., 1973, 47, 467

73YB   O. Yamauchi, H. Benno, and A. Nakahara, Bull. Chem. Soc. Japan, 1973, 46, 3458

73ZG   V.K. Zolotukhin and O.M. Gnatyshin, Russ. J. Inorg. Chem., 1973, 18, 1467 (2761)

74AB   M. Asso and D. Benlian, Comp. Rend. Acad. Sci. Paris, Ser. C, 1974, 278, 1373

74AH   M. Aplincourt and R. Hugel, J. Inorg. Nucl. Chem., 1974, 35, 345

74AL   L.I. Antropov, V.M. Ledovskikh, V.A. Molodtsova, and O.I. Khotsyanovskii, J. Gen. Chem. USSR, 1974, 44, 929 (967)

74AN   M. Atchayya, O.G.B. Nambiar, and P.R. Subbaraman, Indian J. Chem., 1974, 12, 633

74AR   A. Arevalo, J.C. Rodriguez Placeres, A. Cabrera Gonzalez, and J. Segura, An. Quim., 1974, 70, 824

74AS   Y.K. Agrawal and J.P. Shukla, Z. Phys. Chem. (Leipzig), 1974, 255, 889

74AV   D. Alexandersson and N.G. Vannerberg, Acta Chem. Scand., 1974, A28, 423

74BL   N.N. Basargin and G.E. Lunina, Russ. J. Inorg. Chem., 1974, 19, 1119 (2042)

74BLa   N.N. Basargin and G.E. Lunina, Russ. J. Inorg. Chem., 1974, 19, 1333 (2441)

74BS   T.N. Briggs and J.E. Stuehr, Anal. Chem., 1974, 46, 1517

74BSa   M.S. Balakrishnan and M. Santappa, J. Inorg. Nucl. Chem., 1974, 36, 3813

74BV   E. Bottari and M. Vicedomini, Gazz. Chim. Ital., 1974, 104, 523

74CC   J.N. Cape, D.H. Cook, and D.R. Williams, J. Chem. Soc. Dalton, 1974, 1849

74CH   A.K. Covington, M.L. Hassall, and D.E. Irish, J. Solution Chem., 1974, 3, 629

74CJ   K.K. Choudhary, D.S. Jain, and J.N. Gaur, Indian J. Chem., 1974, 12, 655

74CP   G. Carpeni, S. Poize, N. Sabiani, and G. Perinet, J. Chim. Phys., 1974, 71, 311

74CS   M.C. Chattopadhyaya and R.S. Singh, Anal. Chim. Acta, 1974, 70, 49

74CSa   P.B. Chakrawarti and H.N. Sharma, Sci. Culture, 1974, 40, 407

74CT   A.K. Covington and R. Thompson, J. Solution Chem., 1974, 3, 603

74DF   P. DeMaria, A. Fini, and F.M. Hall, J. Chem. Soc. Perkin II, 1974, 1443

74DS   A.A. Dundorina and A.N. Sergeeva, Russ. J. Inorg. Chem., 1974, 19, 181 (334)

74DT   J.F. DeBrabander, J.J. Tombeux, and L.C. VanPoucke, J. Coord. Chem., 1974, 4, 87

74DV   H.F. DeBrabander, L.C. VanPoucke, and Z. Eeckhaut, Anal. Chim. Acta, 1974, 70, 401

74ET   N.G. Elenkova and R.A. Tsoneva, J. Anal. Chem. USSR, 1974, 29, 1161 (1344)

74FF   V.A. Fedorov, A.V. Fedorova, G.G. Nifanteva, and L.I. Gruber, Russ. J. Inorg. Chem., 1974, 19, 538 (990)

74FM   T.B. Field, J.L. McCourt, and W.A.E. McBryde, Canad. J. Chem., 1974, 52, 3119

74FP   S. Friedman, B. Pace, and R. Pizer, J. Amer. Chem. Soc., 1974, 96, 5381

74FR   P. Fiordiponti, F. Rallo, and F. Rodante, Gazz. Chim. Ital., 1974, 104, 649

74FS   B. Finlayson and A. Smith, J. Chem. Eng. Data, 1974, 19, 94

74FSD   Ya.D. Fridman, O.P. Svanidze, N.V. Dolgashova, and P.V. Gogorshvili, Russ. J.
           Inorg. Chem., 1974, 19, 1809 (3304)

74G   S. Gobom, Acta Chem. Scand., 1974, A28, 1180

74GE   V.P. Gruzdev and V.L. Ermolaev, Russ. J. Inorg. Chem., 1974, 19, 1446 (2648)

74GF   E.A. Gyunner and A.M. Fedorenko, Russ. J. Inorg. Chem., 1974, 19, 979 (1797)

74GG   I. Grenthe and G. Gardhammer, Acta Chem. Scand., 1974, A28, 125

74GK   V.I. Gordienko, L.P. Khudyakova, and V.I. Sidorenko, J. Anal. Chem. USSR, 1974,
           29, 1274 (1467)

74GM   V.I. Gordienko and Yu.I. Mikhailyuk, J. Anal. Chem. USSR, 1974, 29, 179 (210)

74GMP   B. Grabaric, B. Mayer, I. Piljac, and I. Filipovic, J. Inorg. Nucl. Chem., 1974,
           36, 3809

74HB   J. Hala and L. Bednarova, Coll. Czech. Chem. Comm., 1974, 39, 1843

74HBP   J. Haladjian, P. Bianco, and R. Pilard, J. Chim. Phys., 1974, 71, 1251

74HH   S. Hubert, M. Hussonnois, L. Brillard, G. Goby, and R. Guillaumont, J. Inorg.
           Nucl. Chem., 1974, 36, 2361

74HO   S. Harada, Y. Okuue, H. Kan, and T. Yasunaga, Bull. Chem. Soc. Japan, 1974, 47,
           769

74I   A. Ivaska, Talanta, 1974, 21, 387

74Ia   A. Ivaska, Talanta, 1974, 21, 1175

74Ib   P.B. Issopoulos, Ann. Chim. (France), 1974, 9, 157

74J   D.V. Jahagirdar, J. Inorg. Nucl. Chem., 1974, 36, 2388

74Ja   E. John, Rocz. Chem., 1974, 48, 1809

74JB   M.M. Jones, A.J. Banks, and C.H. Brown, J. Inorg. Nucl. Chem., 1974, 36, 1833

74JP   M.M. Jones, T.H. Pratt, and C.H. Brown, J. Inorg. Nucl. Chem., 1974, 36, 1213

74K   S. Krzewska, Rocz. Chem., 1974, 48, 555

74KA   V.P. Koryukova, A.M. Andrianov, and L.K. Oleinik, Russ. J. Inorg. Chem., 1974,
           19, 194 (358)

74KBM   A.P. Kreshkov, L.N. Balyatinskaya, Yu.F. Milyaev, and A.V. Filopova, Russ. J.
           Inorg. Chem., 1974, 19, 43 (77)

74KBZ   F.Ya. Kulba, N.A. Babkina, and A.P. Zharkov, Russ. J. Inorg. Chem., 1974, 19,
           365 (674)

74KC   A.S. Kereichuk and I.M. Churikova, Russ. J. Inorg. Chem., 1974, 19, 1180 (2154)

74KG   A. Kumar and J.N. Gaur, J. Electroanal. Chem., 1974, 49, 317

74KK   N.P. Komar and Yu.M. Khoroshevskii, J. Anal. Chem. USSR, 1974, 29, 314 (376)

74KKa   N.P. Komar and Yu.M. Khoroshevskii, Russ. J. Phys. Chem., 1974, 48, 456 (784)

74KKb   E.E. Kriss and G.P. Kurbatova, Russ. J. Inorg. Chem., 1974, 19, 692 (1273)

74KKc   Ts.B. Konunova and L.S. Kachkar, Russ. J. Inorg. Chem., 1974, 19, 1246 (2279)

74KL   O.I. Khotsyanovskii, V.M. Ledovskikh, and V.A. Molodtsova, J. Gen. Chem. USSR,
         1974, 44, 698 (727)

74KM   N.P. Komar, L.S. Manzhelii, and L.I. Gorodilova, J. Anal. Chem. USSR, 1974, 29,
         2055 (2391)

74KPD   A.S. Kereichuk, R.S. Pobegai, and L.I. Dodova, Russ. J. Inorg. Chem., 1974, 19,
         300 (553)

74KS   F. Kai and Y. Sadakane, J. Inorg. Nucl. Chem., 1974, 36, 1404

74KT   N.P. Komar and N.M. Tkhuet, Russ. J. Phys. Chem., 1974, 48, 1122 (1887)

74L   R. Lundqvist, Acta Chem. Scand., 1974, A28, 243, 358

74La   S. Lasztity, Radiochem. Radioanal. Letters, 1974, 17, 189

74Lb   S. Lasztity, Radiochem. Radioanal. Letters, 1974, 20, 7

74LD   M.C. Langlois and M. Devaud, J. Chim. Phys., 1974, 71, 605

74LP   C.L. Liotta, E.M. Perdue, and H.P. Hopkins, Jr., J. Amer. Chem. Soc., 1974, 96,
         7981

74LS   P. Lingaiah and E.V. Sundaram, Indian J. Chem., 1974, 12, 539

74MH   T. Matsui, L.G. Hepler, and E.M. Woolley, Canad. J. Chem., 1974, 52, 1910

74MK   T. Matsui, H.C. Ko, and L.G. Hepler, Canad. J. Chem., 1974, 52, 2906

74MKa   T. Matsui, H.C. Ko, and L.G. Hepler, Canad. J. Chem., 1974, 52, 2912

74MM   M.H. Mihailov, V.Ts. Mihailova, and V.A. Khalkin, J. Inorg. Nucl. Chem., 1974,
         35, 121

74MMa   M.H. Mihailov, V.Ts. Mihailova, and V.A. Khalkin, J. Inorg. Nucl. Chem., 1974,
         35, 127

74MMb   M.H. Mihailov, V.Ts. Mihailova, and V.A. Khalkin, J. Inorg. Nucl. Chem., 1974,
         35, 133

74MMc   M.H. Mihailov, V.Ts. Mihailova, and V.A. Khalkin, J. Inorg. Nucl. Chem., 1974,
         35, 145

74MS   H. Moriya and T. Sekine, Bull. Chem. Soc. Japan, 1974, 47, 747

74MSM   M. Miyazaki, Y. Shimoishi, H. Miyata, and K. Toei, J. Inorg. Nucl. Chem., 1974,
         36, 2033

74MT   L. Magon, G. Tomat, A. Bismondo, R. Portanova, and U. Croatto, Gazz. Chim. Ital.,
         1974, 104, 967

74ND   R. Nayan and A.K. Dey, J. Inorg. Nucl. Chem., 1944, 36, 2545

74NM   S.A. Nikolaeva, N.V. Melchakova, and V.M. Peshkova, J. Anal. Chem. USSR, 1974,
         29, 1768 (2055)

74NV   V.A. Nazarenko, L.I. Vinarova, and N.V. Lebedeva, Russ. J. Inorg. Chem., 1974,
         19, 1295 (2371)

74O   C. O.Nuallain, J. Inorg. Nucl. Chem., 1974, 36, 339

740a   C. O.Nuallain, <u>J. Inorg. Nucl. Chem.</u>, 1974, <u>36</u>, 1420

740b   C. O.Nuallain, <u>J. Inorg. Nucl. Chem.</u>, 1974, <u>36</u>, 2325

740D   G. Ostacoli, P.G. Daniele, and A. Vanni, <u>Atti Accad. Sci. Torino</u>, 1974, <u>108</u>, 539

74PD   R. Portanova, P. DiBernardo, A. Cassol, E. Tondello, and L. Magon, <u>Inorg. Chim. Acta</u>, 1974, <u>8</u>, 233

74PJ   N.G. Palaskar, D.V. Jahagirdar, and D.D. Khanolkar, <u>Indian J. Chem.</u>, 1974, <u>12</u>, 197

74PK   J.E. Powell and S. Kulprathipanja, <u>Inorg. Chim. Acta</u>, 1974, <u>11</u>, 31

74PKa   M.M. Petit-Ramel and I. Khalil, <u>Bull. Soc. Chim. France</u>, 1974, 1255

74PKb   M.M. Petit-Ramel and I. Khalil, <u>Bull. Soc. Chim. France</u>, 1974, 1259

74PL   S.K. Pal and S.C. Lahiri, <u>Z. Phys. Chem.</u> (Leipzig), 1974, <u>255</u>, 910

74PS   Pushparaja and M. Sudersanan, <u>Proc. Indian Acad. Sci.</u>, <u>Sec. A</u>, 1974, <u>80</u>, 278

74PV   N.I. Pechurova, G.P. Vakhramova, and V.I. Spitsyn, <u>Russ. J. Inorg. Chem.</u>, 1974, <u>19</u>, 1136 (2074)

74R   E. Roletto, <u>J. Chim. Phys.</u>, 1974, <u>71</u>, 931

74RD   C.E. Rodriguez and C.D. Devine, <u>Talanta</u>, 1974, <u>21</u>, 1313

74RG   P.C. Rawat and C.M. Gupta, <u>Indian J. Chem.</u>, 1974, <u>12</u>, 174

74RM   S. Ramamoorthy and P.G. Manning, <u>Inorg. Nucl. Chem. Letters</u>, 1974, <u>10</u>, 109

74RR   F. Rodante, F. Rallo, and P. Fiordiponti, <u>Thermochim. Acta</u>, 1974, <u>9</u>, 269

74RS   C.B. Riolo, T.F. Soldi, G. Gallotti, and M. Pesavento, <u>Gazz. Chim. Ital.</u>, 1974, <u>104</u>, 193

74S   H. Saarinen, <u>Acta Chem. Scand.</u>, 1974, <u>A28</u>, 589

74SJ   H. Saarinen, T. Jantunen, O. Jarvinen, S. Raikas, and O. Makitie, <u>Finn. Chem. Lett.</u>, 1974, 146

74SM   K.M. Stelting and S.E. Manahan, <u>Anal. Chem.</u>, 1974, <u>46</u>, 2118

74SMN   T. Sekine, R. Murai, M. Niitsu, and N. Ihara, <u>J. Inorg. Nucl. Chem.</u>, 1974, <u>36</u>, 2569

74SN   V.P. Shilov, V.B. Nikolaevskii, and N.N. Krot, <u>Russ. J. Inorg. Chem.</u>, 1974, <u>19</u>, 254 (469)

74SR   H. Saarinen and T. Raikas, <u>Finn. Chem. Lett.</u>, 1974, 63

74SS   R.S. Saxena and S.S. Sheelwant, <u>J. Prakt. Chem.</u>, 1974, <u>316</u>, 517

74ST   S.P. Singh and J.P. Tandon, <u>Acta Chim. Acad. Sci. Hung.</u>, 1974, <u>80</u>, 425

74TG   P.H. Tedesco and J. Gonzales Quintana, <u>J. Inorg. Nucl. Chem.</u>, 1974, <u>36</u>, 2628

74UW   P. Ulmgren and O. Wahlberg, <u>Acta Chem. Scand.</u>, 1974, <u>A28</u>, 631

74UY   D.R. Underdown, S.S. Yun, and J.L. Bear, <u>J. Inorg. Nucl. Chem.</u>, 1974, <u>36</u>, 2043

74VP   G.P. Vakhramova, N.N. Pechurova, and V.I. Spitsyn, <u>Moscow Univ. Chem. Bull.</u>, 1974, <u>29</u>, No. 6, 30 (682)

74VPa   G.P. Vakhramova, N.I. Pechurova, and V.I. Spitsyn, <u>Bull. Acad. Sci. USSR</u>, <u>Chem.</u>, 1974, <u>23</u>, 2276 (2361)

74W   O.A. Weber, <u>J. Inorg. Nucl. Chem.</u>, 1974, <u>36</u>, 1341

74ZB   J. Zelinka, M. Bartusek, and A. Okac, <u>Coll. Czech. Chem. Comm.</u>, 1974, <u>39</u>, 83

74ZBD   E.V. Zakharova, V.D. Balukova, and V.V. Dubrovskaya, <u>Russ. J. Inorg. Chem.</u>, 1974, <u>19</u>, 675 (1241)

75AD   H.G. Asawa and L.N. Dhoot, <u>Z. Phys. Chem.</u> (Leipzig), 1975, <u>256</u>, 841

75AH   M. Aplincourt and R. Hugel, <u>Bull. Soc. Chim. France</u>, 1975, 138

75AHa   S.J. Angyal and R.J. Hickman, <u>Aust. J. Chem.</u>, 1975, <u>28</u>, 1279

75B   E. Bottari, <u>Monat. Chem.</u>, 1975, <u>106</u>, 451

75Ba   B.W. Budesinsky, <u>Anal. Chem.</u>, 1975, <u>47</u>, 560

75BC   S.C. Baghel, K.K. Choudhary, and J.N. Gaur, <u>J. Inorg. Nucl. Chem.</u>, 1975, <u>37</u>, 2513

75BP   E.M. Belousova, A.F. Pozharitskii, I.I. Seifullina, and I.N. Nazarova, <u>Russ. J. Inorg. Chem.</u>, 1975, <u>20</u>, 1542 (2787)

75BS   M.S. Balakrishnan and M. Santappa, <u>J. Inorg. Nucl. Chem.</u>, 1975, <u>37</u>, 1229

75BSa   T.N. Briggs and J.E. Stuehr, <u>Anal. Chem.</u>, 1965, <u>47</u>, 1916

75BSb   P.F. Brun and K.H. Schroder, <u>J. Electroanal. Chem.</u>, 1975, <u>66</u>, 9

75BSP   Yu.A. Barbalat, E.P. Shapovalenko, and V.M. Peshkova, <u>J. Anal. Chem. USSR</u>, 1975, <u>30</u>, 410 (483)

75BU   J. Biernat, I. Urbanska, and M. Zralko, <u>Rocz. Chem.</u>, 1975, <u>49</u>, 2095

75C   J.S. Chazhoor, <u>Proc. Indian Acad. Sci.</u>, 1975, <u>81A</u>, 93

75CG   R. Cali, S. Gurrieri, E. Rizzarelli, and S. Sammartano, <u>Thermochim. Acta</u>, 1975, <u>12</u>, 19

75CS   R.D. Cannon and J.S. Stillman, <u>Inorg. Chem.</u>, 1975, <u>14</u>, 2202

75CT   P. Chaudhuri and R.S. Taylor, <u>Anal. Chim. Acta</u>, 1975, <u>78</u>, 451

75DG   J.F. DeBrabander, A.M. Goeminne, and L.C. VanPoucke, <u>J. Inorg. Nucl. Chem.</u>, 1975, <u>37</u>, 799

75DI   A.W.L. Dudeney and R.J. Irving, <u>J. Chem. Soc. Faraday I</u>, 1975, <u>71</u>, 1215

75DJ   D.G. Dhuley, D.V. Jahagirdar, and D.D. Khanolkar, <u>J. Inorg. Nucl. Chem.</u>, 1975, <u>37</u>, 2135

75DN   A.R. Das and V.S.K. Nair, <u>J. Inorg. Nucl. Chem.</u>, 1975, <u>37</u>, 991

75DNa   A.R. Das and V.S.K. Nair, <u>J. Inorg. Nucl. Chem.</u>, 1975, <u>37</u>, 995

75DNb   A.R. Das and V.S.K. Nair, <u>J. Inorg. Nucl. Chem.</u>, 1975, <u>37</u>, 2121

75DNc   A.R. Das and V.S.K. Nair, <u>J. Inorg. Nucl. Chem.</u>, 1975, <u>37</u>, 2125

75DS   N.K. Dutt and U.U.M. Sarma, <u>J. Inorg. Nucl. Chem.</u>, 1975, <u>37</u>, 606

75F   A.A. Fedorov, <u>J. Gen. Chem. USSR</u>, 1975, <u>45</u>, 1058 (1072)

75FC   T.B. Field, J. Coburn, J.L. McCourt, and W.A.E. McBryde, Anal. Chim. Acta, 1975, 74, 101

75FF   V.A. Fedorov, A.V. Fedorova, and G.F. Nifanteva, Russ. J. Inorg. Chem., 1975, 20, 978 (1748)

75FP   S. Friedman and R. Pizer, J. Amer. Chem. Soc., 1975, 97, 6059

75FPG   I. Filipovic, I. Piljac, B. Grabaric, and B. Mayer, Anal. Chim. Acta, 1975, 76, 224

75GG   V.V. Grigoreva and I.V. Golubeva, Russ. J. Inorg. Chem., 1975, 20, 526 (941)

75GH   C. Gerard and R. Hugel, Bull. Soc. Chim. France, 1975, 2404

75GJ   J.N. Gaur, D.S. Jain, and A. Kumar, Indian J. Chem., 1975, 13, 165

75GM   B. Grabaric, B. Mayer, I. Piljac, and I. Filipovic, Electrochim. Acta, 1975, 20, 799

75GS   S. Grewal, B.S. Sekhon, and S.L. Chopra, Thermochim. Acta, 1965, 11, 315

75GT   B. Grabaric, M. Tkalcec, I. Piljac, and I. Filipovic, Anal. Chim. Acta, 1975, 74, 147

75HS   J. Hala and J. Smola, Coll. Czech. Chem. Comm., 1975, 40, 3329

75IM   J. Inczedy and J. Marothy, Acta Chim. Acad. Sci. Hung., 1975, 86, 1

75IP   M. Israeli and L.D. Pettit, J. Chem. Soc. Dalton, 1975, 414

75JB   M.M. Jones, A.J. Banks, and C.H. Brown, J. Inorg. Nucl. Chem., 1975, 37, 761

75JK   D.V. Jahagirdar and D.D. Khanolkar, Indian J. Chem., 1975, 13, 168

75KA   P.V. Khadikar and R.L. Ameria, Sci. Culture, 1975, 41, 343

75KAa   P.V. Khadikar and R.L. Ameria, Indian J. Chem., 1975, 13, 525

75KB   F.Ya. Kulba, N.A. Babkina, and A.P. Zharkov, Russ. J. Inorg. Chem., 1975, 20, 1461 (2640)

75KI   A.S. Kereichuk and L.M. Ilicheva, Russ. J. Inorg. Chem., 1975, 20, 1291 (2330)

75KIF   Y. Kawai, T. Imamura, and M. Fujimoto, Bull. Chem. Soc. Japan, 1975, 48, 3142

75KL   G.E. Kodina, V.I. Levin, and V.S. Novoselov, Russ. J. Inorg. Chem., 1975, 20, 1140 (2049)

75KM   B.I. Kim, C. Miyake, and S. Imoto, Bull. Chem. Soc. Japan, 1975, 48, 349

75KN   S.C. Khurana, I.J. Nigam, S.P. Saxena, and C.M. Gupta, Aust. J. Chem., 1975, 28, 1617

75KO   H.C. Ko and W.F. O'Hara, Thermochim. Acta, 1975, 11, 94

75KPT   I.I. Kuselman, M.S. Podolskii, and Yu.K. Tselinskii, Russ. J. Inorg. Chem., 1975, 20, 1414 (2554)

75KPV   Ts.B. Konunova, M.S. Popov, and A.S. Venichenko, Russ. J. Inorg. Chem., 1975, 20, 861 (1540)

75KS   F. Kai, Y. Sadakane, N. Tanaka, and T. Matsuda, J. Inorg. Nucl. Chem., 1975, 37, 1311

75L   L.H.J. Lajunen, Finn. Chem. Lett., 1975, 1

75La   L.H.J. Lajunen, Finn. Chem. Lett., 1975, 71

75Lb   G.C. Lalor, Inorg. Chim. Acta, 1975, 14, 179

75Lc   R. Lundqvist, Acta Chem. Scand., 1975, A29, 231

75LB   W.E. VanderLinden and C. Beers, Talanta, 1975, 22, 89

75LK   P. Lumme and E. Kari, Acta Chem. Scand., 1975, A29, 117

75LKa   P. Lumme and E. Kari, Acta Chem. Scand., 1975, A29, 125

75LP   D.K. Laing and L.D. Pettit, J. Chem. Soc. Dalton, 1975, 2297

75LS   B.M. Lowe and D.G. Smith, J. Chem. Soc. Faraday I, 1975, 71, 389

75M   V. Michaylova, J. Inorg. Nucl. Chem., 1975, 37, 2317

75MD   J. Maslowska and J. Duda, Rocz. Chem., 1975, 49, 1971

75MG   P.K. Migal, A.P. Gerbeleu, and G.G. Muntyanu, Russ. J. Inorg. Chem., 1975, 20,
          1320 (2383)

75MH   H. Matsushita and H. Hironaka, Nippon Kagaku Kaishi, 1975, 1252

75MJ   A.K. Maheshwari, D.S. Jain, and J.N. Gaur, J. Inorg. Nucl. Chem., 1975, 37, 805

75MJa   A.K. Maheshwari, D.S. Jain, and J.N. Gaur, J. Inorg. Nucl. Chem., 1975, 37, 2319

75ML   O.A. Makitie and K.V.O. Lajunen, Talanta, 1975, 22, 1053

75MR   C. Makni, B. Regaya, M. Aplincourt, and C. Kappenstein, Comp. Rend. Acad. Sci.
          Paris, Ser. C, 1975, 280, 117

75MT   M. Miyazaki and K. Toei, Talanta, 1975, 22, 929

75NA   B.P. Nikolskii, A.M. Antonova, V.V. Palchevskii, and V.N. Solntsev, Doklady Phys.
          Chem., 1975, 221, 308 (669)

75NF   H. Nakayama and M. Fujimoto, Bull. Chem. Soc. Japan, 1975, 48, 3399

75NM   N.A. Nepomnyashchaya, A.A. Menkov, and A.S. Lenskii, Russ. J. Inorg. Chem., 1975,
          20, 1010 (1810)

75NW   G. Nakagawa, H. Wada, and T. Hayakawa, Bull. Chem. Soc. Japan, 1975, 48, 424

75O   W.F. O'Hara, J. Solution Chem., 1975, 4, 793

75OM   M.A. Olatunji and A. McAuley, J. Chem. Soc. Dalton, 1975, 682

75OO   E. Ohyoshi, J. Oda, and A. Ohyoshi, Bull. Chem. Soc. Japan, 1975, 48, 227

75OS   A. Olin and P. Svanstrom, Acta Chem. Scand., 1975, A29, 849

75P   P.B. Issopoulos, Compt. Rend. Acad. Sci. Paris, Ser. C, 1975, 280, 1359

75PA   S. Parthasarathy and S. Ambujavalli, Electrochim. Acta, 1975, 20, 611

75PAa   S. Parthasarathy and S. Ambujavalli, Electrochim. Acta, 1975, 20, 887

75PC   K.N. Pearce and L.K. Creamer, Aust. J. Chem., 1975, 28, 2409

75PD   R. Portanova, P. DiBernardo, O. Traverso, G.A. Mazzocchin, and L. Magon, J.
          Inorg. Nucl. Chem., 1975, 37, 2177

75PF   J.E. Powell, J.L. Farrell, and S. Kulprathipanja, Inorg. Chem., 1975, 14, 786

75PJ   T.H. Pratt and M.M. Jones, J. Inorg. Nucl. Chem., 1975, 37, 2403

75PL   S.B. Pirkes, A.V. Lapitskaya, and T.V. Zakharova, Russ. J. Inorg. Chem., 1975,
          20, 1621 (2929)

75PM   M. Pajula and O. Makitie, Finn. Chem. Lett., 1975, 75

75PMa   S.P. Pande and K.N. Munshi, Indian J. Chem., 1975, 13, 90

75PS   A.F. Pozharitskii, I.I. Seifullina, and E.M. Belousova, J. Gen. Chem. USSR, 1975,
          45, 1026 (1038)

75PSa   A.F. Pozharitskii, I.I. Seifullina, and E.M. Belousova, J. Gen. Chem. USSR, 1975,
          45, 1285 (1311)

75PSb   A.A. Popel and V.A. Shchukin, Russ. J. Inorg. Chem., 1975, 20, 1068 (1917)

75PSL   S.B. Pirkes, M.T. Shestakova, and A.V. Lapitskaya, Russ. J. Inorg. Chem., 1975,
          20, 795 (1415)

75SD   J.T. Smith and V.M. Doctor, J. Inorg. Nucl. Chem., 1975, 37, 775

75SG   J.P. Scharff and R. Genin, Anal. Chim. Acta, 1975, 78, 201

75SGY   L.M. Schwartz, R.I. Gelb, and J.O. Yardley, J. Phys. Chem., 1975, 79, 2246

75SJ   E.G. Sase and D.V. Jahagirdar, J. Inorg. Nucl. Chem., 1975, 37, 985

75SN   E. Shchori, N. Nae, and J. Jagur-Grodzinski, J. Chem. Soc. Dalton, 1975, 2381

75SP   I.I. Seifullina, A.F. Pozharitskii, E.M. Belousova, and M.M. Bobrovskaya, Russ.
          J. Inorg. Chem., 1975, 20, 1798 (3256)

75SS   R.M. Sanyal, P.C. Srivastava, and B.K. Banerjee, J. Inorg. Nucl. Chem., 1975, 37,
          343

75ST   S.P. Singh and J.P. Tandon, Monat. Chem., 1975, 106, 271

75STa   S.P. Singh and J.P. Tandon, Monat. Chem., 1975, 106, 871

75SV   F. Secco and M. Venturini, Inorg. Chem., 1975, 14, 1978

75TB   S.C. Tripathi and S. Browmik, Indian J. Chem., 1975, 13, 846

75TD   P.H. Tedesco and V.B. DeRumi, J. Inorg. Nucl. Chem., 1975, 37, 1833

75TG   P.H. Tedesco and J.A. Gonzalez Quintana, J. Inorg. Nucl. Chem., 1975, 37, 1798

75TK   Yu.K. Tselinskii, L.Ya. Kvyatkovskaya, and A.M. Nechaevskii, Russ. J. Inorg.
          Chem., 1975, 20, 1480 (2676)

75TM   J.G. Travers, K.G. McCurdy, D. Dolman, and L.G. Hepler, J. Solution Chem., 1974,
          4, 267

75V   B. Viossat, Bull. Soc. Chim. France, 1975, 113

75VB   J. Votava and M. Bartusek, Coll. Czech. Chem. Comm., 1975, 40, 2050

75VG   A. Vanni, M.C. Gannaro, and G. Ostacoli, J. Inorg. Nucl. Chem., 1975, 37, 1443

75VH   J. Votava, J. Havel, and M. Bartusek, Scripta Fac. Sci. Nat. Univ. Purk. Brun.,
          Chem. 1, 1975, 5, 71

75VS   V. Vajgand and T. Suranyi-Mihajlovic, Talanta, 1975, 22, 803

75VSK   F. Vlacil, B.M. Sayeh, and J. Koncky, Coll. Czech. Chem. Comm., 1975, 40, 1345

75W   R. Wojtas, Rocz. Chem., 1975, 49, 1231

75Y   S.A. Yajnik, Proc. Indian Acad. Sci., 1975, 81A, 143

75YB   S.S. Yun and J.L. Bear, J. Inorg. Nucl. Chem., 1975, 37, 1757

75ZB   J. Zelinka, M. Bartusek, and A. Okac, Coll. Czech. Chem. Comm., 1975, 40, 390

76HM   W.R. Harris and A.E. Martell, Inorg. Chem., 1976, 15, 713

76MM   R.J. Motekaitis and A.E. Martell, Report to Monsanto Chemical Company, 1976

LIGAND FORMULA INDEX

Order of elements: C,H,O,N, others in alphabetical order.

LIGAND NAME INDEX